W0043559

Michael Torrens   John F. B. Morrison (Eds.)

The Physiology of the Lower Urinary Tract

With 226 Figures

Springer-Verlag
London Berlin Heidelberg New York
Paris Tokyo

Michael Torrens, MB, BS, MPhil, ChM, FRCS
Consultant Neurosurgeon, Department of Neurosurgery, Frenchay Hospital,
Bristol BS16 1LE, UK.

John F. B. Morrison, MB, ChB, BSc, PhD
Reader in Physiology, The Department of Physiology, The Worsley Medical and
Dental Building, The University, Leeds LS2 9NQ, UK.

ISBN-13:978-1-4471-1451-2   e-ISBN-13:978-1-4471-1449-9
DOI: 10.1007/978-1-4471-1449-9

British Library Cataloguing in Publication Data
The Physiology of the lower urinary tract.
  1. Urinary organs
  I. Torrens, Michael   II. Morrison, John F.B.
  612′.467 QP248
ISBN-13:978-1-4471-1451-2

Library of Congress Cataloging-in-Publication Data
The Physiology of the lower urinary tract.
Includes bibliographies and index.
1. Urinary organs. 2. Urinary organs—Innervation. I. Torrens, Michael, 1942–   . II. Morrison,
J. F. B. (John Finlay Benzie) [DNLM: 1. Urinary Tract—physiology. WJ 102 P578] QP248.P47   1987
612′. 46 87–12924
ISBN-13:978-1-4471-1451-2 (U.S.)

This work is subject to copyright. All rights are reserved, whether the whole or part of the material is
concerned, specifically the rights of translation, reprinting, reuse of illustrations, recitation, broadcasting,
reproduction on microfilms or in other ways, and storage in data banks. Duplication of this publication
or parts thereof is only permitted under the provisions of the German Copyright Law of September 9,
1965, in its version of June 24, 1985, and a copyright fee must always be paid. Violations fall under the
prosecution act of the German Copyright Law.

© Springer-Verlag Berlin Heidelberg 1987
Softcover reprint of the hardcover 1st edition 1987

The use of names, trademarks, etc. in this publication does not imply, even in the absence of a specific
statement, that such names are exempt from the relevant protective laws and regulations and therefore
free for general use.

Product Liability: The publisher can give no guarantee for information about drug dosage and application
thereof contained in this book. In every individual case the respective user must check its accuracy by
consulting other pharmaceutical literature.

Filmset and printed by Butler and Tanner Ltd, Frome, Somerset.

2128/3916–543210

Preface

In this book we attempt a synthesis of knowledge from two investigative extremes. On the one hand, neurophysiology and neuropharmacology are progressing via the single neuron to a subcellular level; on the other, clinicians are studying the function of the human urinary system in vivo as a whole. A special effort must be made over the next decade to bridge this gap. We hope that the information summarized here will catalyse the process.

In 1968, de Groat and Ryall published a group of papers in the *Journal of Physiology* in which modern quantitative electrophysiological techniques were applied to the study of the reflexes that regulate bladder function. These papers represent a landmark in the history of bladder neurophysiology, forming a dividing line between old and new. The earlier techniques of lesioning and stimulation of nervous structures yielded mainly qualitative information which was open to criticism because of lack of precise control over what was actually being destroyed or stimulated. Much of this earlier work was reviewed in an authoritative volume by Bors and Comarr in 1971, entitled *Neurological Urology*. The subsequent 16 years have seen great advances in our understanding of the control of lower urinary tract function.

The particular benefit from modern recording techniques is the ability to measure the responses of individual neurons and to examine how their behaviour changes during normal activity and as a consequence of a variety of experimental interventions. Naturally there has been a greater emphasis on the properties of individual neural elements rather than of gross pathways, which were often found to be quite heterogeneous in their composition. This cellular approach has resulted in the functional dissection of many of the classical pathways and the discovery of new connections. There have been major changes in the understanding of function in the sensory pathways from the bladder, in the relationship between supraspinal and spinal reflex circuits, in the interactions between sympathetic, parasympathetic and somatic nerves and in the processes of neurotransmission in the spinal cord, peripheral autonomic ganglia and neuromuscular junctions.

Electrophysiological studies have also improved our understanding of smooth muscle function: electrical recordings of transmembrane potentials show that smooth muscle cells form a functional syncytium, and their membranes exhibit properties which vary from those of nerve and skeletal muscle. Various types of smooth muscles are unable to generate action potentials; they are electrically inexcitable, yet they can respond to neurotransmitters or drugs. These differences can now be explained in terms of intracellular chemistry and membrane permeability to different ions.

Recordings from excitable tissues are only one part of a multidisciplinary approach which pervaded experimentation in the neurosciences throughout the 1970s. New anatomical techniques for the identification of nerve pathways and cellular morphology were complemented by histochemical studies which allowed the indirect visualization of neurotransmitters. In more recent years, beautiful immunocytochemical studies have revealed the distribution of several new putative neurotransmitters in the bladder and in parts of the central or peripheral nervous system concerned with its regulation. Radioimmunoassay and peptide iontophoresis, together with the study of ligand distribution, are providing alternative chemical and pharmacological methods for dissection of the nervous system, which can be combined with the electrophysiological approach.

Meanwhile, improvements in instrumentation have allowed physiological measurement to become widespread in clinical medicine. Such investigation of the lower urinary tract has been dignified with the title 'Urodynamics'. Numerous observations of bladder and urethral function have been made in disease states, but with relatively few studies of normal persons. These investigations have concentrated on the hydrodynamics of urine transport and storage.

Rigorous application of the scientific method is often difficult in the clinical situation. Urodynamics has been criticized for its inadequate controls and a lack of scientific authority. Cooperation between clinicians and experimental physiologists will be one way to redress the balance. Until now this has been less frequent than is desirable and we hope that the book will stimulate such a dialogue.

Bristol and Leeds   Michael Torrens
1987   John F. B. Morrison

Acknowledgements

Many people have provided inspiration and practical help during the preparation of this book. I wish to thank the contributors for their enthusiasm and the publishers, especially Roger Dobbing and Judy Watt, for their efficiency. Particular acknowledgement is also due to Derek Griffiths, Larry Kavanagh, Tony Desmond and the staff of the Clinical Investigation Unit at Ham Green Hospital, Bristol. My patient and tireless secretary Sally Normanton deserves special thanks.

Michael Torrens

I would like to acknowledge the collaboration of colleagues and students in the Department of Physiology at Leeds who have contributed to the work described in the neurophysiological chapters in this volume. In particular, the contributions of Keith Floyd, Verity Hick, Juthika Koley, Geralyn Collins, Steve McMahon, Kathy Spillane, Sam Cadden, Bridget Lumb, Andy Wright, Nigel Womack and Alan Nimmo at different phases in the work of my laboratory are gratefully acknowledged. The technical assistance of Graham Tate, Nick Clements, David Wilkinson, Jon Preston, Dave Johanson and the late George Thompson has been particularly appreciated. The collaboration of the following has also been much appreciated: Julia Polak, Stephen Bloom, Sally Gibson, Giorgio Terenghi, Marion Blank and Praveen Anand from the Departments of Histochemistry and Medicine at the Royal Postgraduate Medical School, London; Norman Williams (formerly of the Department of Surgery at Leeds General Infirmary and now of the Department of Surgery at the London Hospital); and the staff of the Spinal Injuries Unit at Pinderfields Hospital, Wakefield. Finally, the encouragement and patience of my wife, Lynn, has been a constant support throughout the work itself and in the writing of the neurophysiological chapters in this book: they are dedicated to her.

John F. B. Morrison

Contents

Contributors

Alison Brading   University Department of Pharmacology, South Parks Road, Oxford OX1 3QT, UK.

John Dixon   Department of Anatomy, University of Manchester, Stopford
John Gosling   Building, Oxford Road, Manchester M13 9PT, UK.

Ahmad Elbadawi   Department of Pathology, College of Medicine, Health Science Center, 750 East Adams Street, Syracuse, NY 13210, USA.

Clare J. Fowler   The Reta Lila Weston Institute of Neurological Studies, The
Chris Fowler   Middlesex Hospital Medical School, University of London, London W1N 8AA, UK.

John F. B. Morrison   Department of Physiology, The Worsley Medical and Dental Building, The University, Leeds LS2 9NQ, UK.

Michael Torrens   Department of Neurosurgery, Frenchay Hospital, Bristol BS16 1LE, UK.

NEUROMORPHOLOGY

Structure and Innervation in the Human
Comparative Neuromorphology in Animals
Neural Connections Between the Lower Urinary Tract and the Spinal Cord

Introduction and Terminology

A sound understanding of morphology is fundamental to the appraisal of functional studies. We have been fortunate in obtaining contributions to this book from the foremost authorities on the neuromorphology of the lower urinary tract. These have been arranged to give accounts of human and animal studies separately, while retaining appropriate observations derived from comparative morphology.

There is now a substantial amount of information available about the afferent and efferent fibres connecting the lower urinary tract and the spinal cord, the pathways that they take, and their histochemical characteristics. The afferents and efferents in the parasympathetic and sympathetic nerve trunks and in the pudendal nerve are considered individually, and the differences between species is outlined. The latter is of some importance to those undertaking experimental work on the role of the extrinsic nerves to the lower urinary tract. In addition, the animal work has provided a scientific base for new morphological and functional concepts that fit in with existing knowledge of the innervation in humans and give extra insight into the mechanisms and interactions that may occur within these pathways.

It is not surprising that there are areas of controversy—not all of which can be explained by species differences in the material examined. Some of these controversies have been identified and emphasized by editing together reciprocal comments by the authors concerned. Such inclusions, when not by the authors of the chapter concerned, are printed in small type.

A fundamental confusion may arise in relation to terminology. The nomenclature used by authors in their respective chapters is outlined in Fig. 1.0. In summary, while all authors agree that there are structural and histochemical differences between the detrusor and urethral smooth muscle, there is disagreement as to whether the deep trigone should be included with the detrusor (Gosling/Dixon) or urethra (Elbadawi).

Elbadawi considers that "all available evidence" indicates that the bladder base, periureteral sheaths and proximal urethral smooth muscle together form a coordinated muscle unit designated the lissosphincter, which controls both the internal urethral meatus and the ureterovesical junction to prevent incontinence and vesicoureteric reflex. This conclusion depends particularly on histochemical studies in animals which suggest a species variable gradation with increasing adrenergic innervation of the base detrusor.

Dixon and Gosling argue that in the human the junctional zone is clearly seen at the bladder neck where the morphology of the smooth muscle fibres changes to a smaller calibre in the urethra and where, in the male only, the adrenergic innervation appears in a significant amount. It is probably still too early in our understanding of the function of the bladder base and bladder neck to be making firm functional conclusions from morphology. The discussion that follows should point the way to further investigation of what remains one of the great unsolved problems in urology—how does the bladder neck work?

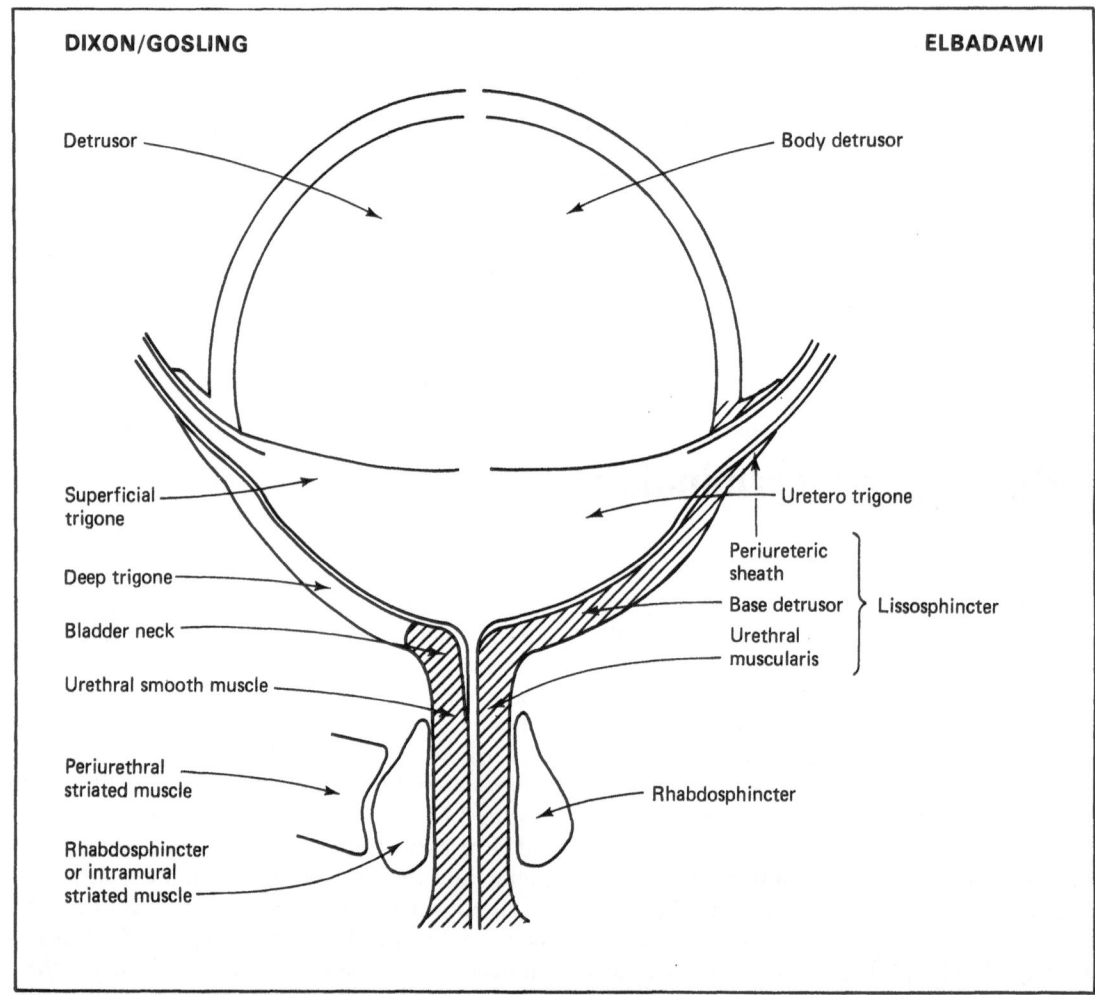

Fig. 1.0. Comparison of terminology used by the various authors in Section 1: Neuromorphology.

1 · Structure and Innervation in the Human

John Dixon and John Gosling

Until recently the smooth musculature of the bladder and urethra was considered to be controlled by the parasympathetic division of the autonomic nervous system with very little influence by the sympathetic component. This concept has subsequently been modified following the results of numerous recent investigations using a variety of neurohistochemical, electron microscopic and immunologic techniques. Studies of the anatomy, physiology and pharmacology of the vesicourethral musculature in experimental animals and humans have provided a wealth of new information and it is now evident that the processes of urine storage and micturition are considerably more complex than was once believed. This chapter provides a summary of current knowledge on the structure of the human urinary bladder and urethra, including their extrinsic and intrinsic innervation.

Extrinsic Innervation of the Urinary Bladder and Urethra

Efferent sympathetic and parasympathetic fibres are conveyed to the bladder and urethra via the hypogastric and pelvic nerves respectively. These nerves also convey afferent (sensory) fibres to the spinal cord. The sympathetic fibres are derived from the lower two thoracic and upper two lumbar segments of the spinal cord. The parasympathetic fibres arise from the second to the fourth sacral segments of the spinal cord (the nervi erigentes or pelvic splanchnic nerves). The hypogastric and pelvic nerves of either side meet and branch to form the pelvic plexus which lies lateral to the rectum, internal genital organs and the bladder. That part of each pelvic plexus specifically related to the urinary bladder is referred to as the vesical plexus of autonomic nerves and contains both sympathetic and parasympathetic ganglion cells together with occasional small intensely fluorescent (SIF) cells.

The parasympathetic neurons of the vesical plexus provide the cholinergic motor innervation to the musculature of the bladder wall and urethra. These neurons receive a preganglionic input not only from cholinergic (excitatory) axons, but also from noradrenergic (possibly inhibitory) nerve terminals.

Afferent impulses arising from sensory nerve endings in the wall of the bladder and urethra pass to the spinal cord via the pudendal, pelvic and hypogastric nerves. The pudendal nerve transmits sensation mainly from the urethral mucosa (in addition to that from the skin of the genital area and the anal canal) and also proprioceptive impulses from the striated muscle of the pelvic floor. The afferent pathway of the micturition reflex is carried in the pelvic nerves (Learmonth 1931), together with those afferents concerned with bladder mucosal pain and lower ureteric pain. The part played by the hypogastric nerves in relaying sensation from the lower urinary tract is not well defined. No alteration in bladder or urethral sensation occurs after presacral neurectomy (Learmonth 1931), although some sensation of fullness has occasionally been noted to be preserved after lower spinal injury (Head and Riddoch 1917).

Structure of the Urinary Bladder

The urinary bladder is a hollow muscular organ lined by a mucous membrane and covered on its outer aspect partly by peritoneal serosa and partly by fascia. The muscularis of the urinary bladder is formed of smooth muscle cells which comprise the detrusor muscle.

Detrusor Muscle

The muscle coat of the bladder is often described as consisting of three layers, the muscle fibres of the outer and inner layers tending to be orientated longitudinally, while those of the middle layer are circularly disposed. However, the constituent muscle bundles frequently branch and reunite with one another to form an interlacing meshwork so that discrete layers are not readily discernible (Fig. 1.1). Thus, from a functional viewpoint, the detrusor muscle comprises an integrated unit of interconnected muscle bundles which, on contraction, will cause a reduction in all dimensions of the bladder lumen (Gosling 1979; Gosling et al. 1983).

Posteriorly, some of the outer longitudinal muscle bundles extend over the bladder base and merge with the capsule of the prostate (in the male) or with the anterior vaginal wall (in the female). Other bundles extend onto the anterior aspect of the rectum to form the rectovesical muscle. Anteriorly, some outer longitudinal muscle bundles continue into the pubovesical ligaments and contribute to the muscular component of these structures.

The body and fundus of the bladder are supple, mobile and highly distensible, and are capable of expansion into the abdomen according to the volume of contained urine. In contrast, the bladder base, closely related to the genital tissues in both male and female, is relatively indistensible.

Elbadawi provides a more intricate description of the bladder base. The base detrusor is not separable from the body detrusor by any anatomical or histological boundary. The base region is sizable and well defined in the human bladder, and its junction with the urethra is easily identified by appreciable narrowing of the lumen. Continuing into the base region, the bundles of the body detrusor become organized as a ventral and a dorsal group of longitudinal bundles.

Two lateral, an outer anterior, and an inner anterior bundle constitute the ventral component of the base detrusor. Uniting across the ventral midline plane of the bladder neck, the right and left lateral bundles form the precervical arc. This is transversally orientated across the ventral aspect of the bladder neck, has a posterior concavity, and lies superficial to the detrusor loop. The outer anterior component is in part attached to the posterior surface of the public bone, but is largely inserted into the precervical arc. The inner anterior bundle is slender, suburothelial, and plexiform, and extends distally as the inner longitudinal fibres of the ventral urethral muscularis.

Dorsal longitudinal bundles of the base detrusor comprise a prominent posterior, and two less prominent posterolateral bundles. The posterior bundle is anchored in the posterior midline to the cranial aspect of the prostate gland, or the urethro-vaginal septum, is in part inserted into the circularly orientated trigonal base detrusor, and provides some fibres to the lateral part of the external "layer" of the urethral muscularis. The posterolateral longitudinal bundles extend along the right and left sides of the posterior bundle, each splitting at the internal urethral orifice into a medial and a lateral moiety: the pars interna et pars externa of Gil Vernet (1968). The medial parts of the right and left posterolateral longitudinal bundles join the trigonal base detrusor; their lateral counterparts arch ventrally and join to form the detrusor loop. This is a stout, transversally orientated, horseshoe-shaped, smooth muscle loop which is concave dorsocephalically and is completed on the dorsal aspect of the base by the similarly circularly orientated trigonal base detrusor. Detrusor loop bundles splay out as caudal and cranial loops that lie, respectively, caudal and cranial to the precervical arc.

The trigonal component of the base detrusor is represented by a triangular muscular structure that lies in the dorsal bladder base between its outer longitudinal bundles and the uretero-

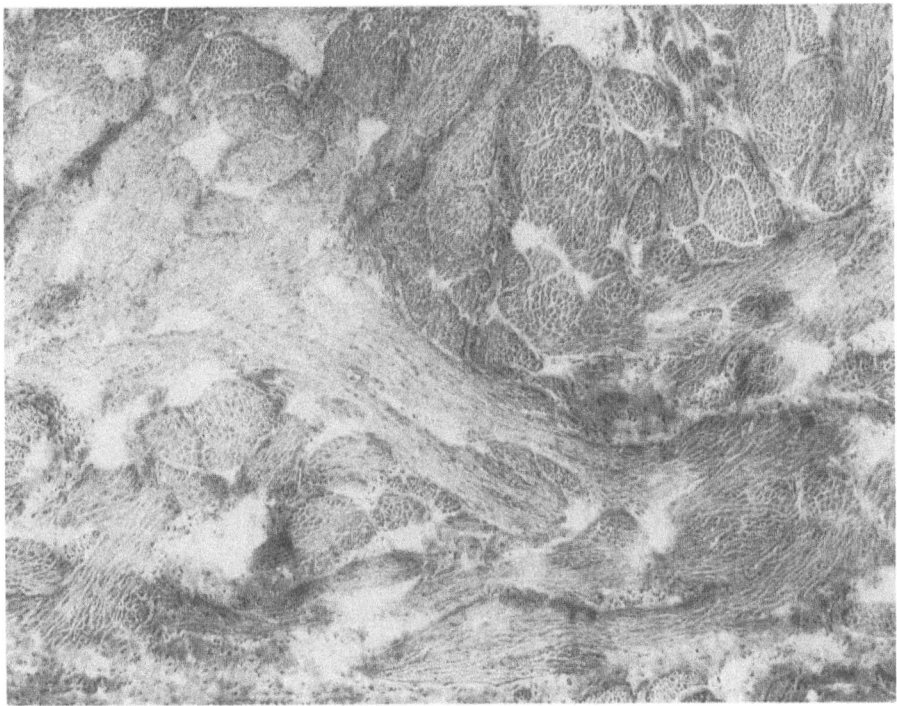

Fig. 1.1. The human detrusor consists of an interlacing meshwork of relatively large-diameter muscle bundles. (Masson's trichrome preparation, × 150)

trigone. It is attached on each side to the detrusor loop, completing the "middle circular layer" of the bladder base muscularis.

An editorial comment at this point is appropriate. Many authors have described complex arrangements of loops and slings around the bladder neck and the reader is referred to the work of Donker et al. (1976), Hutch (1972), Tanagho and Smith (1966) and Woodburne (1967). Little progress has been made in the correlation of structure and function in this area. Most interpretations are extremely speculative. A new approach to the problem is needed.

Fine Structure of Detrusor Smooth Muscle Cells

Under the electron microscope smooth muscle cells of the human urinary bladder are similar in appearance to those of other muscular organs. The cells are spindle-shaped, varying in length from 150 to 300 μm and from 6 to 8 μm in diameter. An elongated nucleus occupies the widest part of each cell. The sarcoplasm is packed with myofilaments among which numerous electron-dense bodies are scattered. Organelles such as mitochondria, Golgi membranes and granular cisternae are relatively sparse and tend to cluster around either end of the elongated cell nucleus. The sarcolemma is characterized by rows of flask-shaped caveolae interspersed with subsarcolemmal electron-dense material. An electron-dense basal lamina surrounds each smooth muscle cell except at certain "junctional" regions. The most frequently observed type of junction between detrusor smooth muscle cells is the region of close approach (Fig. 1.2) at which an intercellular separation of 10–20 nm occurs over distances sometimes in excess of 1 μm. Junctions of the "peg and socket" and "intermediate" types are also present, although they are far less numerous than regions of close approach. True gap junctions (nexuses) are apparently absent from the detrusor muscle of humans. This is discussed in more detail in Chapter 6 (p. 164).

Since electrotonic spread of excitation occurs in the smooth muscle of the bladder wall, the regions of close approach may represent the morphological feature which enables this physiological event to take place. Within each muscle bundle, the individual cells are closely packed together such that the basal lamina of one cell very often becomes confluent with that of its neighbours.

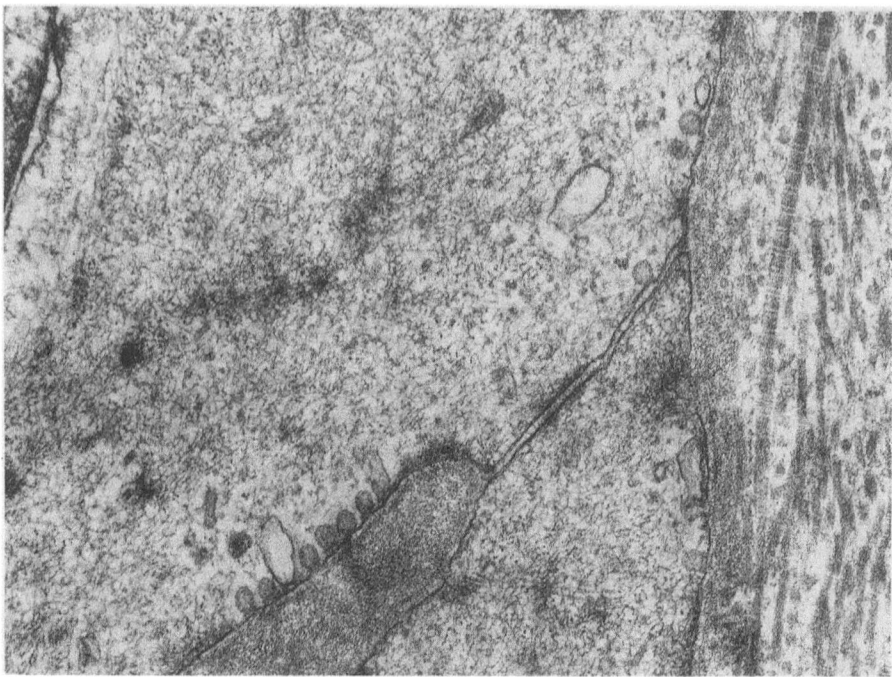

Fig. 1.2. Electron micrograph of a region of "close approach" between neighbouring smooth muscle cells from the human detrusor. (× 45 000)

Intrinsic Innervation of the Urinary Bladder

The urinary bladder is profusely supplied with autonomic nerve fibres which form a dense plexus among the detrusor smooth muscle cells. The majority of these nerves contain acetylcholinesterase (Fig. 1.3) (Mobley et al. 1966; Nyo 1969; Ek et al. 1977; Alm 1978; Gosling 1979; Klück 1980; McConnell et al. 1982), and while they occur in profusion throughout the muscle coat of the bladder, some muscle bundles appear to be more richly innervated than others. In the electron microscope the ratio of nerve axons to detrusor smooth muscle cells has been shown to be at least 1:1 (Daniel et al. 1983). The majority of axonal varicosities lying adjacent to detrusor smooth muscle cells possess features characteristic of cholinergic nerve terminals, i.e. they contain clusters of small (50 nm diameter) agranular vesicles together with occasional large (80–160 nm diameter) granulated vesicles and small mitochondria (Fig. 1.4). Such terminal regions approach to within 20 nm of the muscle cells' surface and are either partially surrounded by or more often totally denuded of neurilemmal cell cytoplasm. Thus, the majority of the autonomic nerves innervating the detrusor muscle are considered to be excitatory cholinergic in type.

Unlike several other species (notably the cat) the human detrusor possesses an exceedingly sparse supply of sympathetic noradrenergic nerves (Sundin et al. 1977). Although nerves of this type generally accompany the vascular supply to the bladder they rarely extend among the smooth muscle cells of the detrusor.

Recent research has led to the proposal that a third component occurs in the autonomic innervation of the urinary bladder (Ambache and Zar 1970). This nonadrenergic, noncholinergic (NANC) nerve-mediated effect has been linked to several active substances, including vasoactive intestinal polypeptide, prostaglandins and adenosine triphosphate, and it has been claimed that this NANC component represents a third type of effector nerve which provides a separate innervation to the bladder. Alternatively it may be that other neurotransmitters are released by the cholinergic nerves which innervate the detrusor muscle. Under the electron microscope, axons containing a predominance of large granulated vesicles (80–160 nm diameter) together with occasional small agranular vesicles have been observed, albeit very infrequently, lying adjacent to detrusor smooth muscle cells. Whether such axons actually represent

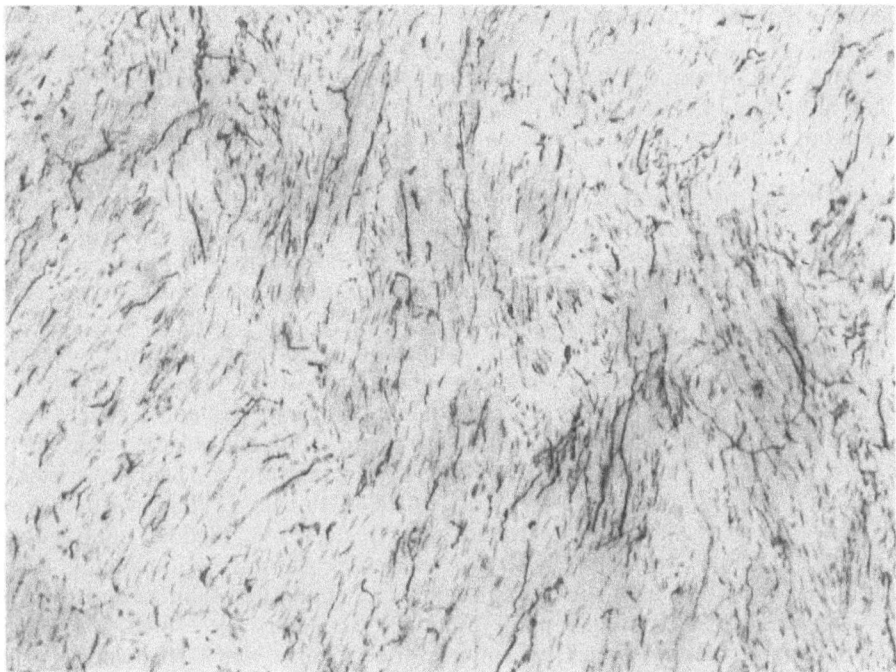

Fig. 1.3. An extensive plexus of acetylcholinesterase-positive nerve fibres occurs throughout the detrusor muscle. (Acetylcholinesterase preparation, × 200)

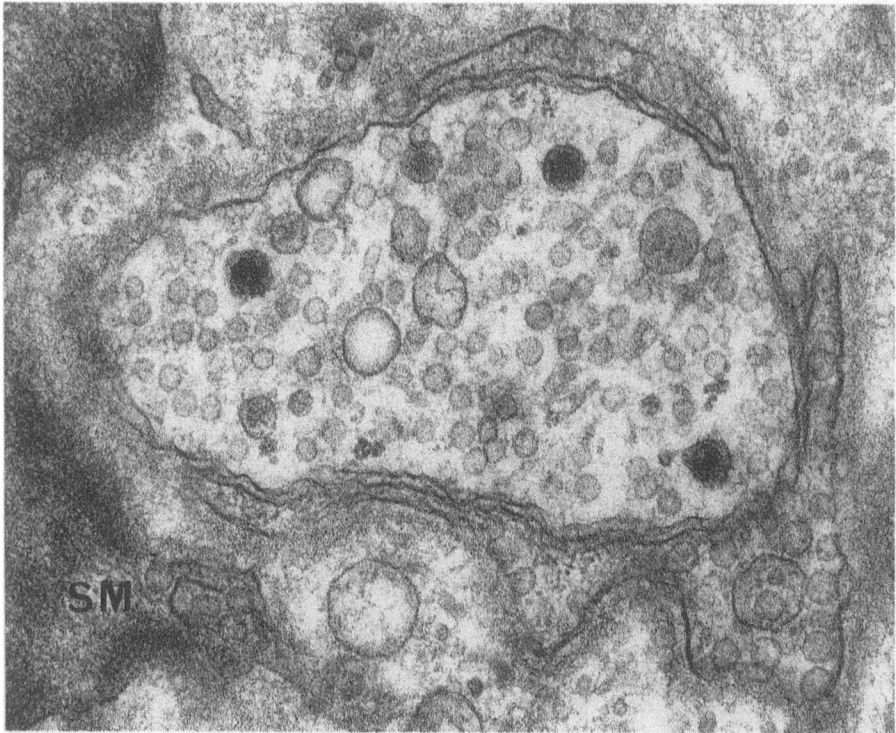

Fig. 1.4. Electron micrograph of a presumptive cholinergic nerve terminal lying among detrusor smooth muscle cells (*SM*). The axonal varicosity contains numerous small electron-lucent vesicles together with some larger granulated vesicles. (× 56 000)

the proposed "peptidergic" type of autonomic effector nerve (the proportions of the two types of vesicle are reversed in a "typical" cholinergic nerve terminal) or whether they are merely chance sections through cholinergic axons has yet to be determined.

In addition, profiles containing a predominance of small mitochondria may occasionally be found among the smooth muscle cells of the bladder wall. It has been suggested that similar mitochondria-filled profiles in other tissues represent "sensory" nerve terminals (Iwayama et al. 1970). However, such profiles are relatively rare in the detrusor muscle of the bladder and may simply be sections through mitochondria-rich areas of smooth muscle cells.

Intramural Ganglia

Small clusters of autonomic ganglion cells occur throughout all regions of the human bladder wall, being especially numerous in the adventitia (Fig. 1.5). These intramural ganglia contain from 1 to 20 or more acetylcholinesterase-positive, presumptive cholinergic neurons, similar to those occurring in the vesical plexus (Dixon et al. 1983). Such neurons

receive a preganglionic input from both cholinergic (excitatory) and noradrenergic (possibly inhibitory) nerve terminals (de Groat and Booth 1980).

The presence of two types of axosomatic terminal at the surface of the intramural ganglion cells has been confirmed by electron microscopy (Gilpin 1985). Typical cholinergic preganglionic terminals contain numerous small agranular vesicles while preganglionic noradrenergic terminals are distinguished by the presence of small, dense-cored vesicles.

Inhibition of detrusor muscle contraction may occur by inhibition of transmission to excitatory nerves within the intramural bladder ganglia, e.g. by the action of noradrenaline on adrenergic receptors on the cholinergic intramural neurons (Fletcher and Bradley 1978). In cats, it is known that sympathetic activity depresses bladder contractions both by an action on muscle (involving adrenoceptors) and an action to inhibit transmission in parasympathetic ganglia (also involving adrenoceptors) (de Groat and Theobald 1976). However, in humans there is little direct sympathetic innervation of detrusor muscle so that inhibition of ganglionic transmission appears to be the major mechanism of inhibition of detrusor smooth muscle (Daniel et al. 1983).

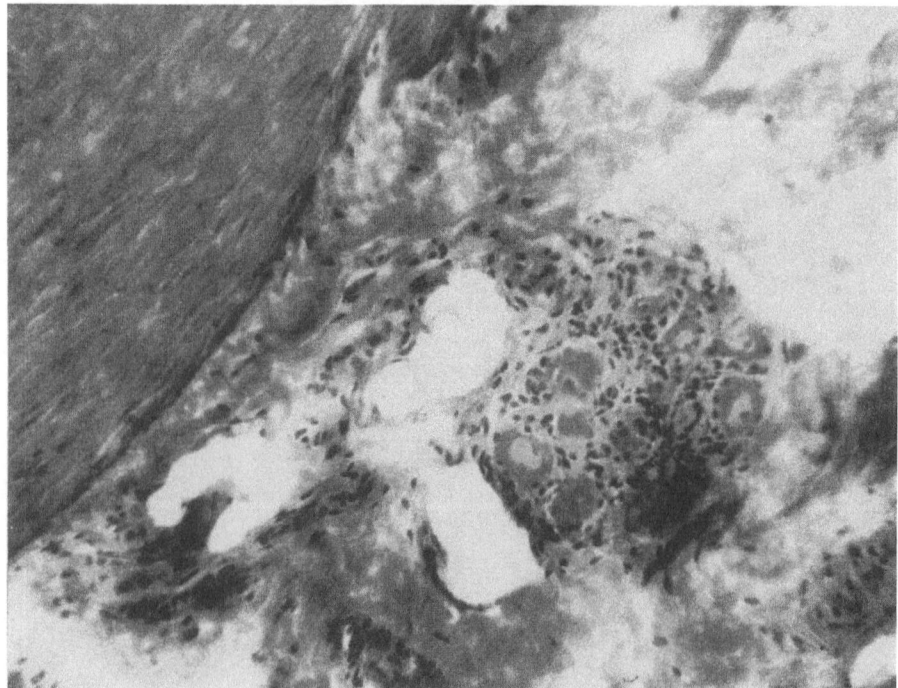

Fig. 1.5. Small autonomic ganglia occur frequently within the detrusor muscle. (Masson's trichrome preparation, ×250)

The ultrastructural features of the intramural ganglion cells are similar to those of other parasympathetic neurons that occur elsewhere. However, a number of unusual features have been described for the intramural neurons of the human bladder wall (Gilpin et al. 1983). First, the investing layer of satellite cell cytoplasm is often incomplete, leaving large areas of the nerve cell plasmalemma in direct contact with connective tissue. Second, groups of three or four nerve cells have been observed in close proximity to one another without an intervening layer of satellite cell cytoplasm (Fig. 1.6). Nerve cells in this configuration thus form ephaptic connections with their neighbours over quite extensive areas of their surface. However, the functional significance of this arrangement is not clear. Third, many of the blood capillaries that course through the intramural ganglia have been shown to be fenestrated so that adjacent nerve cells may be directly influenced by circulating hormones such as noradrenaline.

Electron microscopy has confirmed the presence of occasional small intensely fluorescent (SIF) cells in the intramural bladder ganglia. Cells of this type may be distinguished from ordinary ganglion cells by their content of numerous, large membrane-limited granules which are believed to represent amine storage sites.

The Submucosal Nerve Plexus

The urinary bladder possesses a nerve plexus which lies immediately beneath the epithelial lining, forming the so-called submucous or suburothelial nerve plexus (Gosling et al. 1983). The constituent nerves are acetylcholinesterase positive (Fig. 1.7), and, while some accompany submucosal blood vessels, others lie free in the connective tissue and occasionally penetrate between the basal urothelial cells. The suburothelial plexus is relatively sparse in the dome of the bladder, but becomes progressively denser as the bladder neck is approached and is particularly prominent in the trigone.

Under the electron microscope the submucosal nerves are seen to contain from one to eight axons, the majority of which are nonmyelinated. Many single axons occur immediately beneath the urothelium, some of which are completely or partially invested in neurilemmal cell cytoplasm while others appear completely naked (Fig. 1.8). The suburothelial axons possess varicose regions which are packed with axonal vesicles and thus appear very electron dense at low magnification. The vesicles are of two types, namely small (approximately 50 nm

diameter) agranular vesicles and large (approximately 100 nm diameter) vesicles, most of which possess a central dense granule (Dixon and Gilpin 1987). Occasional axons have been observed to penetrate the urothelial basal lamina and be in close proximity to the urothelial cells. When compared with nerve terminals among the detrusor muscle cells, the suburothelial axons have been shown to possess a greater "packing density" of axonal vesicles and to contain a significantly higher proportion of large granulated vesicles (over 10% of the vesicle population compared with less than 3% for the presumptive cholinergic terminals within the muscle coat). Thus, the suburothelial axonal varicosities differ structurally from typical cholinergic nerve endings and, in the absence of any obvious neuroeffector sites, are considered to be sensory in nature. The high proportion of large granulated vesicles in such nerves may reflect their known content of substance P, a neuropeptide that has been shown to occur in primary sensory neurons, especially those involved in nociception (Hockfelt et al. 1975).

Recent experimental evidence has provided further support for the hypothesis that the suburothelial nerves of the urinary bladder are sensory in nature. Treatment of rat urinary bladder with capsaicin not only reduces the level of substance P in the bladder wall (Holzer et al. 1982), and abolishes substance P-containing nerves in the bladder as demonstrated immunocytochemically (Sharkey et al. 1983), but also markedly alters the sensory threshold of micturition, causing urine retention (Santicioli et al. 1985).

In addition to their probable sensory function it has also been suggested that the suburothelial nerves may exert a trophic influence upon the transitional epithelial cells of the mucosa (Alm et al. 1978) although there is no experimental evidence to support this hypothesis.

Structure of the Ureterovesical Junctions and Trigone

The distal 1–2 cm of each ureter is surrounded by an incomplete collar of detrusor smooth muscle which forms a sheath (of Waldeyer) separated from the ureteric muscle coat by a connective tissue sleeve. The ureters pierce the posterior aspect of the bladder and run obliquely through its wall for a distance of 1.5–2.0 cm before terminating at the

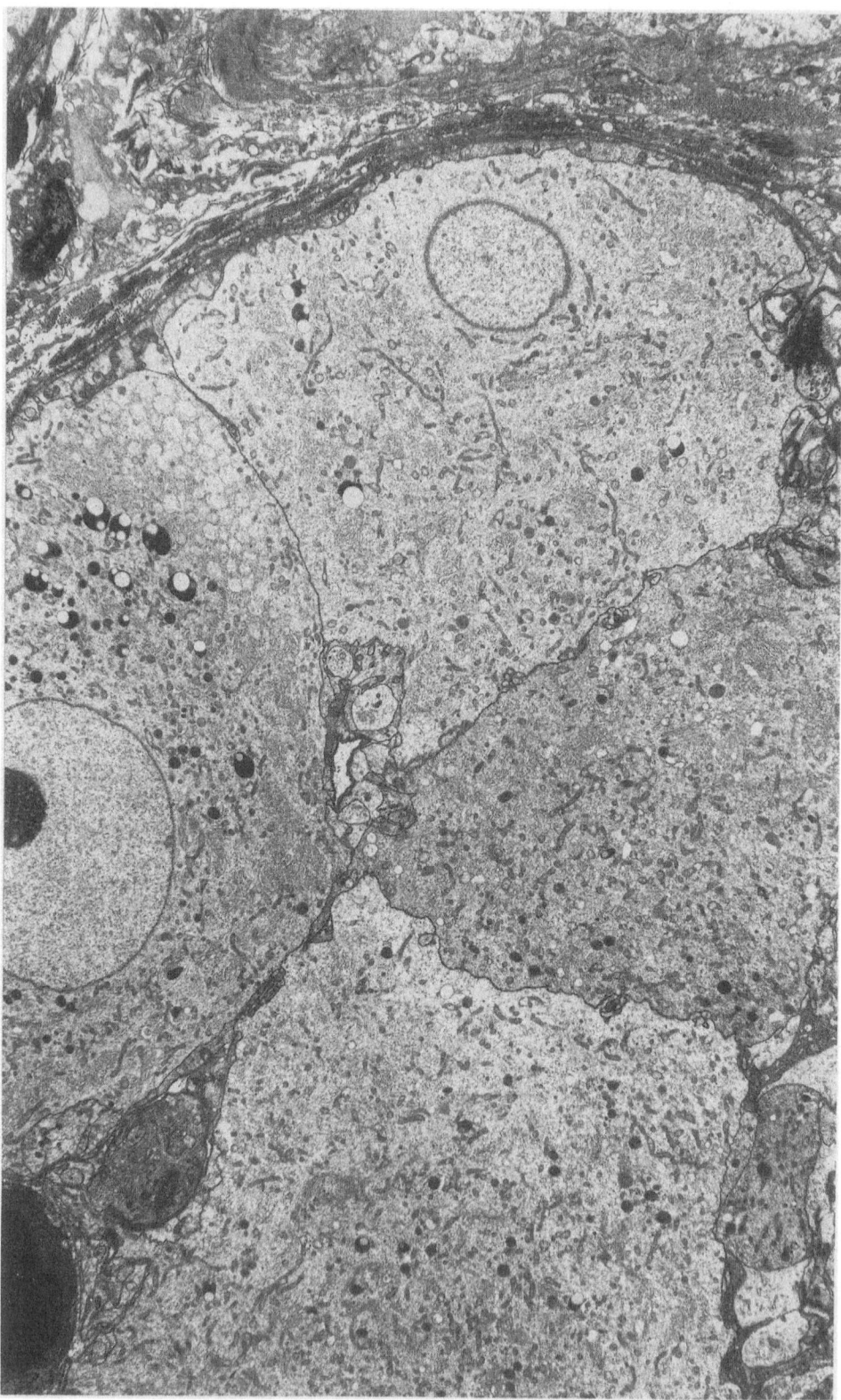

Fig. 1.6. Electron micrograph of a group of four intramural neurons which lie in close association with one another without intervening satellite cell cytoplasm. (× 4500)

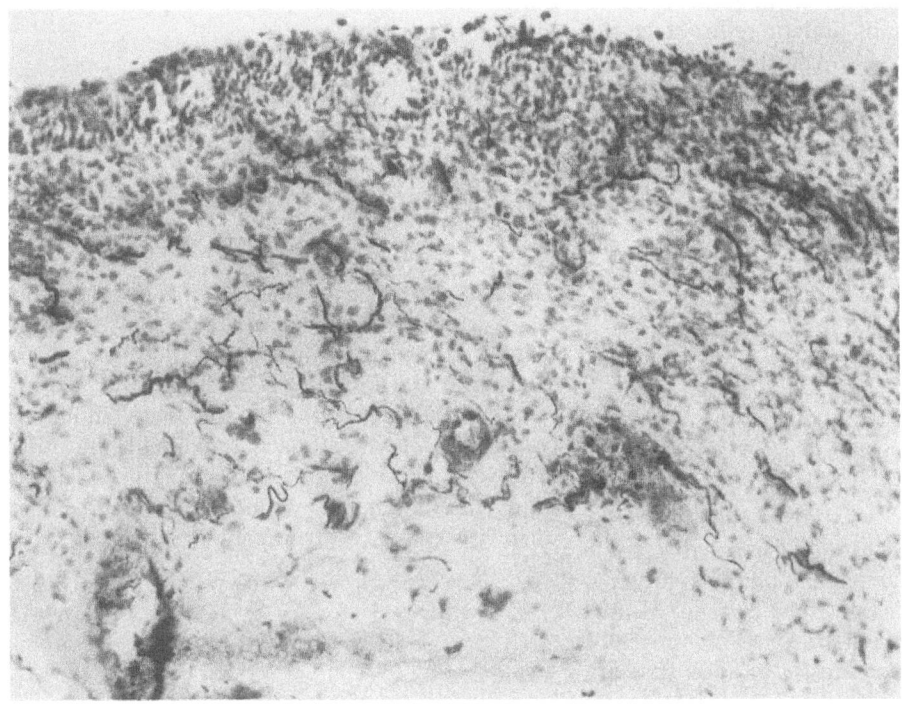

Fig. 1.7. A plexus of acetylcholinesterase-positive nerves occurs beneath the urothelial lining of the human urinary bladder. (Acetylcholinesterase preparation, ×200)

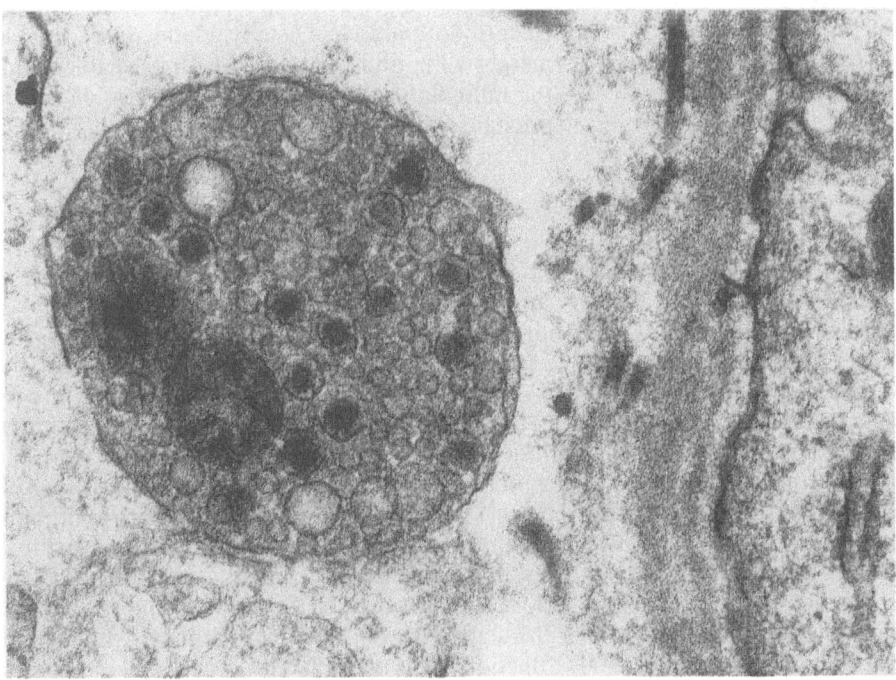

Fig. 1.8. Electron micrograph of a vesicle-packed suburothelial (presumptive sensory) axon from the human urinary bladder. (×56 000)

ureteric orifices. This arrangement is believed to assist in the prevention of ureteric reflux since the intramural ureters are thought to be occluded during increases in bladder pressure. The terminal portion of each ureter forms a visible ridge on the interior surface of the bladder known as the ureteric fold (see also Chap. 2, p. 24).

The trigone is defined as a triangular area of the posterior bladder wall which lies between the ureteric orifices and the interior urethral meatus. The smooth muscle of this region consists of two distinct layers, often termed the superficial and deep trigonal muscles. The latter is composed of muscle cells which are indistinguishable from those of the detrusor. Hence this deep trigonal muscle is merely the posteroinferior portion of the detrusor muscle proper and confusion might be avoided if the term "deep trigonal muscle" was abandoned in favour of the more accurate definition as "trigonal detrusor muscle". The superficial trigonal muscle represents a morphologically distinct component of the trigone which, unlike the detrusor, is composed of relatively small diameter muscle bundles that are continuous proximally with those of the intramural ureters. The smooth muscle layer comprising the superficial trigone is relatively thin, but becomes thickened along its superior border to form the interureteric crest (Mercier's bar). Similar thickenings occur along the lateral edges of the superficial trigone (Bell's muscles). In both sexes the superficial trigone muscle becomes continuous with the smooth muscle of the proximal urethra, extending in the male along the urethral crest as far as the openings of the ejaculatory ducts.

Innervation

The autonomic innervation of the deep trigonal muscle is identical to that of the remainder of the detrusor, being rich in cholinergic nerves and relatively sparse in noradrenergic nerves. In contrast the superficial trigonal muscle possesses relatively few cholinergic (parasympathetic) nerves while those of the noradrenergic (sympathetic) variety occur frequently. This contrasting pattern of innervation may reflect the different histochemical properties of the smooth muscle cells of the superficial trigone which are essentially ureteric in nature.

It should be emphasized that the superficial trigonal muscle forms a very minor part of the total muscle mass of the bladder neck and proximal urethra and is probably of little significance in the physiological mechanisms which control these regions. However, it may well be of importance

during micturition in helping to prevent vesico-ureteral reflux of urine.

Structure of the Bladder Neck

The smooth muscle of this region is histologically, histochemically and pharmacologically distinct from that which comprises the detrusor proper (Nergardh and Boréus 1972; Klück 1980; Gosling et al. 1983). Hence the bladder neck should be considered as a separate functional unit. The arrangement of smooth muscle in this region is quite different in males and females and consequently each sex will be described separately.

Male

At the male bladder neck, the smooth muscle cells form a complete circular collar which extends distally to surround the preprostatic portion of the urethra. Because of the location and orientation of its constituent fibres, the terms internal, proximal or preprostatic urethral sphincter are suitable alternatives for this particular component of urinary tract smooth muscle. Distally, bladder neck muscle merges with, and becomes indistinguishable from, the musculature in the stroma and capsule of the prostate gland.

Innervation

In the male, the bladder neck smooth muscle is supplied with cholinergic (parasympathetic) nerves and also possesses a rich noradrenergic (sympathetic) innervation (Fig. 1.9; Gosling et al. 1977). A similar distribution of autonomic nerves occurs in the smooth muscle of the prostate gland, seminal vesicles and ductus deferentes. On stimulation the sympathetic nerves cause contraction of smooth muscle in the wall of the genital tract, resulting in seminal emission. Concomitant sympathetic stimulation of bladder neck muscle causes sphincteric closure of the region, thereby preventing reflux of ejaculate into the bladder. Although this genital function of the male bladder neck is well established it is not known whether the smooth muscle of this region plays an active role in maintaining urinary continence.

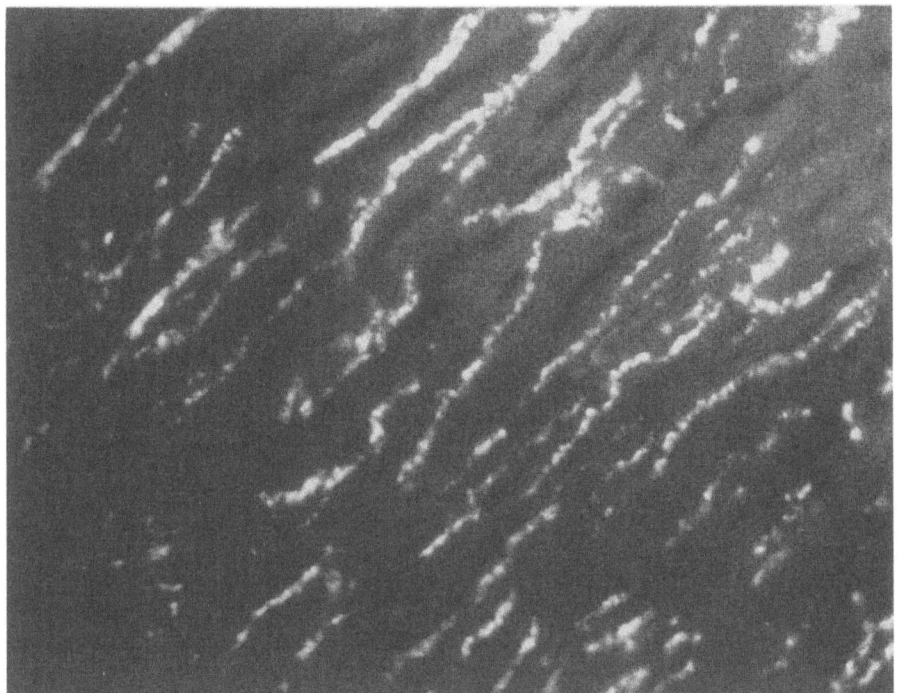

Fig. 1.9. At the male bladder neck numerous fluorescent noradrenergic nerves are present amongst the circularly arranged smooth muscle bundles which form the preprostatic sphincter. (Formaldehyde-induced fluorescence preparation, × 250)

Female

The female bladder neck also consists of morphologically distinct smooth muscle, since the large-diameter muscle bundles characteristic of the detrusor are replaced in the region of the bladder neck by those of small diameter (Fig. 1.10). However, unlike the circularly orientated preprostatic smooth muscle of the male, the majority of muscle bundles in the female bladder neck extend obliquely or longitudinally into the urethral wall. The female does not therefore possess a smooth muscle sphincter at the bladder neck, and it is unlikely that active "sphincteric" contraction of this region plays a significant part in the maintenance of female urinary continence.

Innervation

In contrast to the rich sympathetic innervation in the male, the smooth muscle of the female bladder neck possesses relatively few noradrenergic nerves, but is well supplied with presumptive cholinergic fibres. The sparse supply of sympathetic nerves presumably relates to the absence of a functional "genital" portion incorporated within the wall of the female urethra. The role played by presumed cholinergic fibres in the bladder neck in either sex remains uncertain.

Structure of the Male Urethra

The male urethra may be considered in four regional parts: preprostatic, prostatic, membranous and penile. Except during the passage of fluid, the urethral canal is a mere slit; in the prostatic part the slit is transversely arched; in the membranous portion, stellate; in the spongiose portion, transverse, while at the external orifice it is sagittal.

Elbadawi divides the smooth muscle of the human male urethra into vesicocervical and prostatourethral groups of longitudinal bundles. The vesicocervical group consists of inner suburothelial and outer mural bundles. The former peter out around the midprostatic urethra, and consist of anterior, lateral and posterior bundles that represent the caudal extensions of the inner anterior longitudinal bundles of the base detrusor, the inner lateral longitudinal bundles of the base detrusor, and the ureterotrigone, respectively. Bundles of the outer longitudinal mural group arise from the posterior longitudinal bundle of the base detrusor, extend as vesicourethral prostatic bundles along the

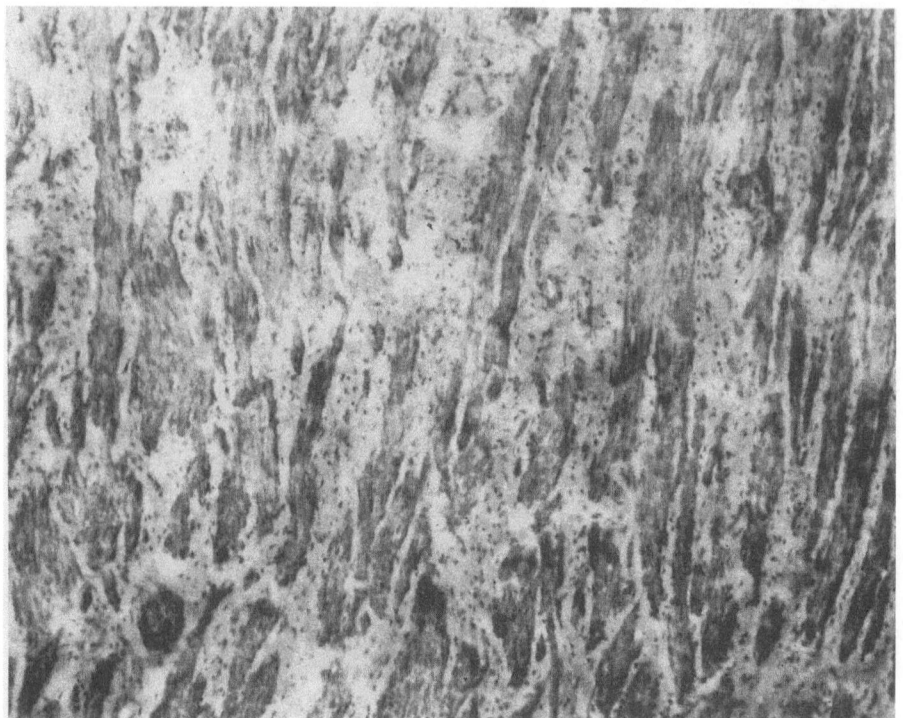

Fig. 1.10. The musculature of the female bladder neck consists of relatively small-diameter muscle bundles, most of which are orientated longitudinally or obliquely. (Masson's trichrome preparation, × 150)

lissosphincteric urethra, and have a component that lies external (i.e. superficial) to the urethral rhabdosphincter. The prostatourethral fibre group consists of one posterior and two anterolateral bundles that "arise" in the suburothelium of the prostatic urethra and end in that of the membranous urethra.

One group of the circular muscle fibres of the male urethra is represented by the "internal sphincter" of Gil Vernet (1968). This forms a ring that lies immediately distal to the bladder neck and is sandwiched between the vesicocervical longitudinal urethral fibres caudally, and the precervical arc of the base detrusor cranially. Hutch (1972) identified the structure commonly known as the capsule of the prostate as a part of the outer circular bundles of the muscularis of the prostatic urethra. Another part of the circular urethral muscularis forms an arc in the anterior wall of the membranous urethra, lies internal (i.e. deep) to the urethral rhabdosphincter, and embraces the prostatourethral longitudinal urethral bundles.

Preprostatic Urethra

The preprostatic urethra is approximately 1–1.5 cm in length, extending almost vertically from the bladder neck to the base of the prostate gland. The smooth muscle bundles of the preprostatic urethra are arranged circularly and are continuous proximally with bladder neck muscle and distally with the capsule of the prostate gland. The bundles which combine with those of the bladder neck to form this preprostatic sphincter are separated by connective tissue containing many elastic fibres. Unlike the detrusor, the smooth muscle surrounding the proximal urethra possesses relatively few parasympathetic cholinergic nerves, but is richly supplied with sympathetic noradrenergic nerves (see Fig. 1.9). Similar nerves also supply the smooth muscle of the prostate, ductus deferentes and seminal vesicles and are involved in causing smooth muscle contraction at the time of ejaculation (Learmonth 1931). As previously mentioned, contraction of the preprostatic sphincter serves to prevent the retrograde flow of ejaculate through the proximal urethra into the bladder.

Prostatic Urethra

The prostatic urethra is approximately 3–4 cm in length and extends through the substance of the prostate closer to the anterior than the posterior surface of the gland. It is continuous above with the preprostatic urethra and emerges from the prostate slightly anterior to its apex. Throughout most of its length the posterior wall possesses a midline ridge, the urethral crest, which projects into the lumen, causing it to appear crescentic in transverse section. Distally the prostatic urethra possesses a layer of circularly disposed striated muscle cells which are continuous with a prominent collar of striated muscle (the rhabdosphincter) within the wall of the membranous urethra.

Puboprostatic Ligaments

A pair of fibromuscular ligaments firmly anchor the anterior aspect of the prostate gland to the posteroinferior surface of the symphysis pubis, close to the midline. These are the so-called puboprostatic ligaments which are continuous superiorly with the pubovesical ligaments. In addition to many collagen fibres, the puboprostatic ligaments contain smooth muscle bundles which receive a rich cholinergic innervation, similar to that supplying the detrusor muscle.

Membranous Urethra

The membranous urethra is the shortest, least distensible, and, with the exception of the external orifice, normally the narrowest section of the urethra. It descends with a slight ventral concavity from the prostate to the bulb of the penis, passing through the perineal membrane about 2.5 cm postero-inferior to the pubic symphysis. The wall of the membranous urethra contains a muscle coat which is separated from the urethral epithelium by a narrow layer of fibroelastic connective tissue. This muscle coat consists of a relatively thin inner layer of smooth muscle bundles which are continuous proximally with those of the prostatic urethra, together with a closely apposed outer layer of circularly orientated striated muscle fibres which form the rhabdosphincter.

Elbadawi reports that the human male rhabdosphincter has caudal, intermediate prostatic and cranial components. The caudal component forms a complete ring of obliquely orientated fibres that surround the membranous urethra, and extend cranially as the intermediate prostatic component. The latter consists of internal and external leaflets. The short internal leaflet lies external to and interdigitates with the smooth muscularis of the urethra in the apical part of the prostate gland. The external leaflet extends cranially, with variable degrees of obliquity, mainly on the ventral surface of the prostate. The cranial component of the rhabdosphincter extends from the ventral surface of the prostate to each side of the bladder neck, and behind the seminal vesicles, to interdigitate with the posterior longitudinal bundle of the bladder base detrusor.

The striated muscle fibres which comprise this rhabdosphincter (Fig. 1.11) are unusually small in cross section with diameters of only 15–20 μm (Von Hayek 1969). The fibres are histochemically all of the slow twitch type (Gosling et al. 1981), unlike the pelvic floor musculature which is a heterogeneous mixture of slow and fast twitch fibres of larger diameter (Parks et al. 1977). Moreover, the rhabdosphincter is devoid of muscle spindles and is probably supplied by the pelvic splanchnic nerves, further distinguishing it from the periurethral levator ani muscle (Donker et al. 1976). The slow twitch fibres of the external sphincter are capable of sustained contraction over relatively long periods of time and actively contribute to the tone which closes the urethra and maintains urinary continence. The difference between slow and fast twitch fibres is explained further in Table 1.1.

Table 1.1. Classification of striated muscle fibres

Slow twitch	Fast twitch
Type 1—red muscle	Type 2—pale muscle
Fairly uniform	Less uniform
Twitch tension lasts ~100 ms	Twitch tension lasts ~30 ms
Low tetanic tension	Wide range of tetanic tension
Resistant to fatigue	2a—resistant to fatigue
	2b—highly fatiguable
Usually innervated by tonic α-motoneuron	Usually innervated by phasic α-motoneuron
Many mitochondria	Few mitochondria
Little sarcoplasmic reticulum	Abundant sarcoplasmic reticulum
Weak ATPase at pH 9.4	Strong ATPase at pH 9.4
Strong oxidative enzymes (NADH, SDH, LDH)	Weak oxidative enzymes

Innervation of Rhabdosphincter

The rhabdosphincter is supplied by motor nerves whose cell bodies lie in the intermediolateral columns of the second, third and fourth sacral segments of the spinal cord. There is still some debate as to whether the nerve fibres are conveyed by the pelvic splanchnic nerves (Gosling 1979) or the pudendal nerves as classically described. It is possible that there is variation from individual to individual and that both theories are correct with a proportion of the fibres travelling via either route. The clinical relevance of this arrangement is that pudendal blockade or neurectomy may not achieve total denervation since not all the motor innervation to the rhabdosphincter will be affected by such procedures.

Studies in numerous animal species have led to the proposal that the rhabdosphincter receives a

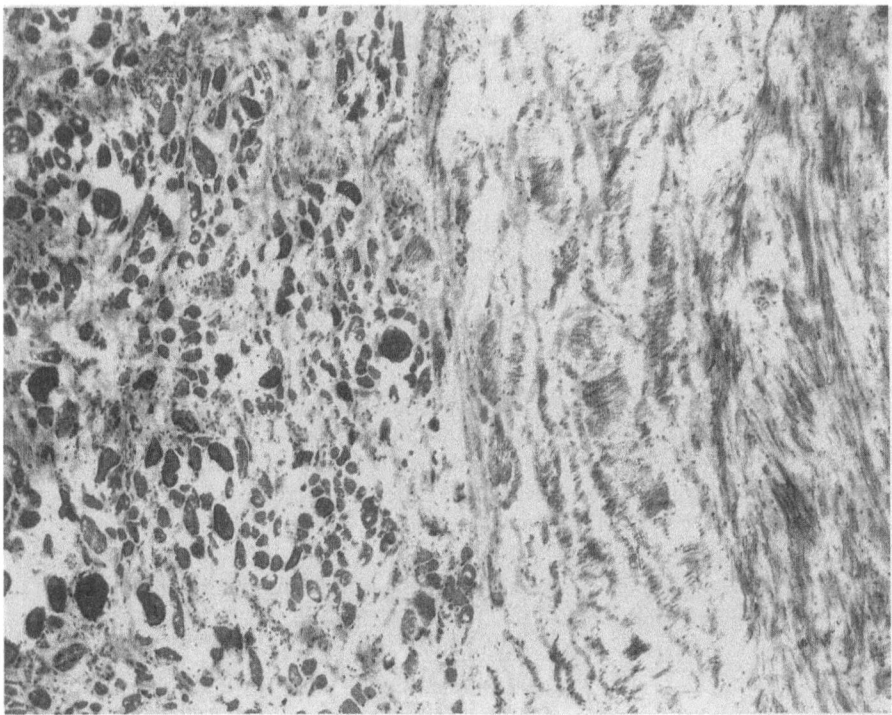

Fig. 1.11. The wall of the male membranous urethra contains an inner smooth muscle layer (on the right) and an outer layer of circularly disposed striated muscle, the rhabdosphincter (on the left). (Masson's trichrome preparation, × 150)

triple innervation composed of somatic motor fibres and also both cholinergic and noradrenergic autonomic nerves (Elbadawi 1982; see also Chap. 2, pp. 45–48). However, since a layer of smooth muscle cells occurs on the inner aspect of the striated muscle of the rhabdosphincter then any autonomic nerves observed within the latter may simply be en route to this smooth muscle tissue and may not have a direct influence on the striated muscle cells. In addition, there is considerable intermingling of smooth and striated cells in the junctional region between the rhabdosphincter and the inner collar of smooth muscle so that fluorescent noradrenergic nerves observed adjacent to striated muscle cells may in fact be innervating adjacent smooth muscle cells (Dixon and Gosling 1977). Certainly there is as yet no published morphological evidence to suggest that a triple innervation of the rhabdosphincter occurs in the human.

Pelvic Floor (Levator Ani)

The medial parts of the levator ani muscles (pubococcygeus) are related to, but anatomically separate from, the membranous urethra of the male. This periurethral striated muscle is a heterogeneous mixture of slow and fast twitch fibres (Parks et al. 1977), and may be distinguished from the much smaller diameter fibres of the rhabdosphincter. The pelvic floor musculature is capable of providing an additional occlusive force on the wall of the membranous urethra, its motor innervation being conveyed by the pudendal nerve.

Structure of the Female Urethra

The adult female urethra is approximately 4 cm long and 6 mm in diameter. It begins at the internal urethral orifice of the bladder and extends anteroinferiorly behind the symphysis pubis, embedded in the anterior wall of the vagina. It traverses the perineal membrane and terminates at the external urethral orifice. Except during the passage of urine the anterior and posterior walls of the urethra are in apposition, and the epithelium forms extensive longitudinal folds.

The wall of the female urethra comprises an outer muscle coat and an inner mucous membrane which

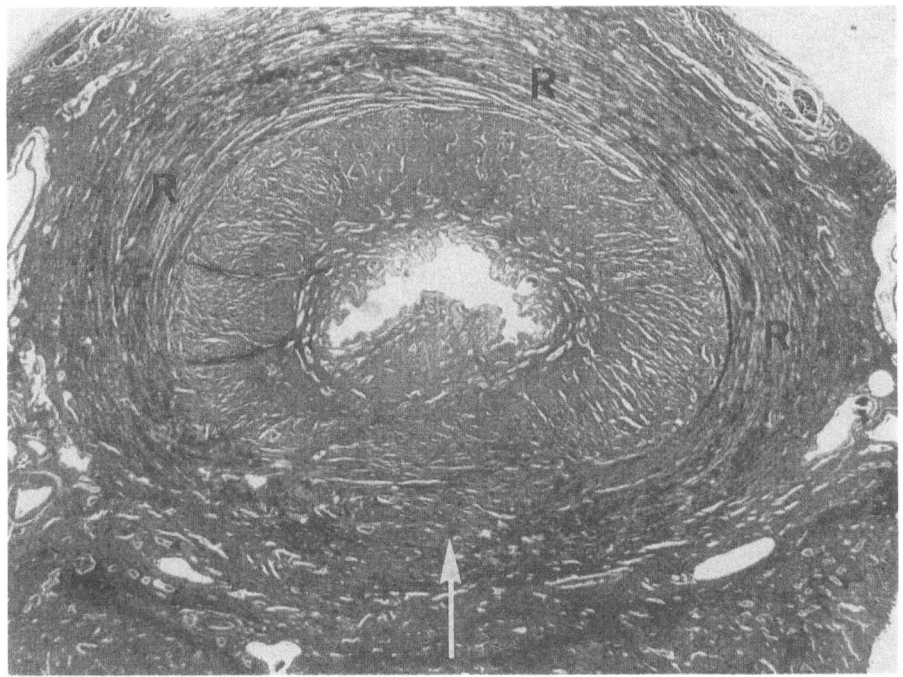

Fig. 1.12. A transverse section through the midportion of the female urethra. The circularly orientated rhabdosphincter (*R*) is incomplete posteriorly (*arrow*) and the urethral lumen appears as a transverse slit. (Masson's trichrome preparation, × 75)

lines the lumen and is continuous with that of the bladder. The muscle coat consists of an outer sleeve of striated muscle (the rhabdosphincter) together with an inner coat of smooth muscle fibres. The female rhabdosphincter is anatomically separate from the adjacent periurethral striated muscle of the anterior pelvic floor. The constituent fibres of this sphincter are circularly disposed and form a sleeve which is thickest in the middle third of the urethra. In this region striated muscle completely surrounds the urethra although the posterior portion lying between the urethra and vagina is relatively thin (Fig. 1.12). The striated muscle extends into the anterior wall of both the proximal and distal thirds of the urethra, but is deficient posteriorly in these regions. The muscle cells forming the rhabdosphincter are histochemically all of the slow twitch variety (Fig. 1.13). As in the male, cells of the rhabdosphincter are unusually small with diameters of about 15–20 μm. Although the thickness of the rhabdosphincter in the female is less than that of the male, its constituent fibres are able to exert tone upon the urethral lumen over prolonged periods, especially in relation to the middle third of its length. Periurethral striated muscle (pubococcygeus) aids urethral closure during events which require rapid, albeit short-lived, elevation of urethral resistance.

The smooth muscle coat extends throughout the length of the urethra and consists of slender muscle bundles, the majority of which are orientated obliquely or longitudinally. A few circularly arranged muscle fibres occur in the outer aspect of the smooth muscle layer and intermingle with the inner part of the rhabdosphincter. Proximally the urethral smooth muscle is continuous with that of the bladder neck. This region in the female is devoid of a well-defined circular smooth muscle component comparable with the preprostatic sphincter of the male. When traced distally urethral smooth muscle bundles terminate in the subcutaneous adipose tissue surrounding the external urethral meatus. The smooth muscle of the female urethra is associated with relatively few noradrenergic nerves, but receives an extensive presumptive cholinergic parasympathetic nerve supply identical in appearance to that which supplies the detrusor (Ek et al. 1977). From a functional viewpoint, in the absence of an anatomical sphincter, it seems unlikely that competence of the female bladder neck and proximal urethra is solely the result of smooth muscle activity. The innervation and longitudinal orien-

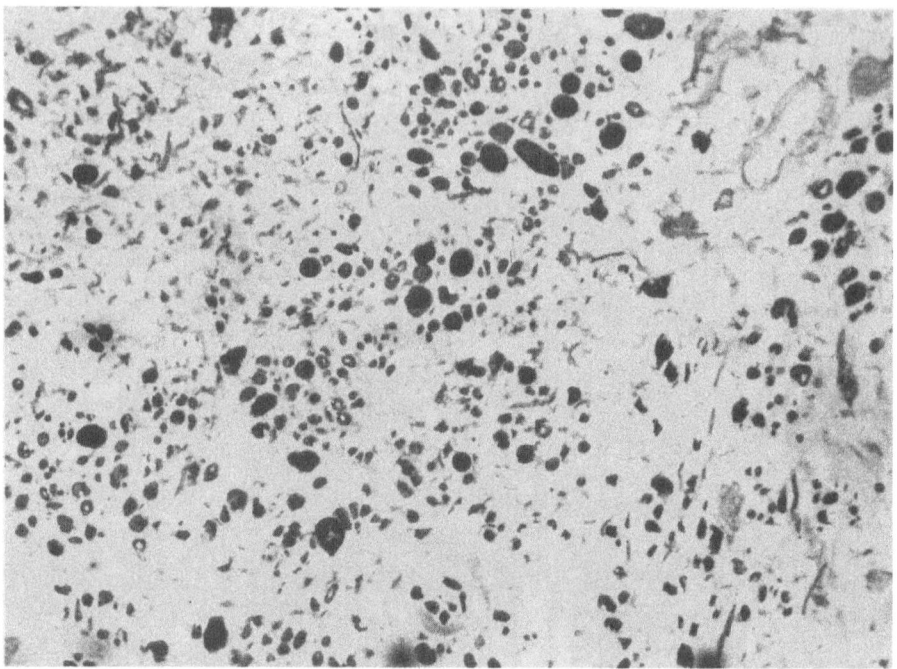

Fig. 1.13. A section through the female rhabdosphincter to demonstrate the rich acid-stable myosin ATPase content of the constituent striated muscle fibres. (Myosin ATPase preparation, × 150)

tation of the majority of the muscle fibres suggest that urethral smooth muscle in the female may be active during micturition, serving to shorten and widen the urethral lumen.

Pelvic Floor (Levator Ani)

The medial parts of the levator ani muscles (sphincter vaginae) are related to (but structurally separate from) the urethral wall. These periurethral fibres consist of an admixture of large-diameter fast and slow twitch fibres (Gosling et al. 1981), together with muscle spindles. Therefore, unlike the rhabdosphincter, periurethral muscle possesses morphological features that are similar to other "typical" voluntary muscles.

The levator ani plays an important part in urinary continence by providing an additional occlusive force on the urethral wall, particularly during events that are associated with an increase in intra-abdominal pressure, such as coughing and sneezing. This urethral occlusive force in the female is maximum at a level immediately distal to the maximum urethral pressure generated by the external urethral sphincter. Thus, in addition to providing support for the pelvic viscera, the periurethral parts of the levator ani also play an import-

ant active role in the urethral mechanisms that maintain continence of urine.

Innervation of Rhabdosphincter and Levator Ani

The motor cell bodies of the nerves supplying the rhabdosphincter lie in the intermediolateral columns of the second, third and fourth sacral segments of the spinal cord. The nerve fibres probably travel via the pelvic splanchnic nerves and not the pudendal nerves as often described. As in the male the innervation of the levator ani is provided by the pudendal nerve and consequently electromyographic recordings obtained from this muscle should not be assumed to represent the activity of the adjacent less accessible and differently innervated striated muscle of the rhabdosphincter. Accurate electromyographic evaluation of the rhabdosphincter can only be obtained by direct measurement of the activity of its constituent fibres.

Pubourethral Ligaments

Close to the midline a pair of fibromuscular ligaments firmly anchor the anterior aspect of the

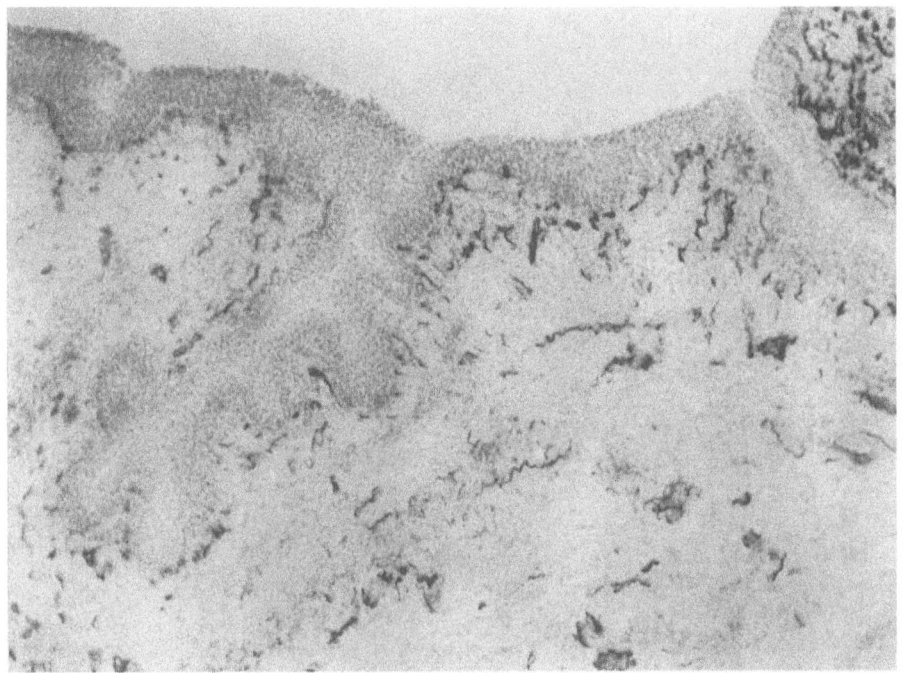

Fig. 1.14. Acetylcholinesterase-positive nerve fibres form a plexus beneath the epithelium of the female urethra. (Acetylcholinesterase preparation, × 250)

female urethra to the posteroinferior surface of the symphysis pubis. These so-called pubourethral ligaments are continuous superiorly with the pubovesical ligaments. As in the puboprostatic ligaments of the male, the pubourethral ligaments contain smooth muscle bundles (Zacharin 1963), which receive a rich presumptive cholinergic innervation (Wilson et al. 1983) similar to that of the detrusor muscle.

Functionally the pubovesical and pubourethral ligaments provide an important supportive role for the female bladder neck and anterior aspect of the urethral wall. In addition, the smooth muscle component of these ligaments may contract at the same time as the detrusor, thereby maintaining the position of the urethra relative to the pubis at the time of micturition. DeLancey (1986a,b) has shown that the "pubourethral" ligaments arise mainly from the lateral vaginal wall and insert into the pubococcygeus muscle, thus making their support more active.

Submucosal Nerves of the Urethra

The submucosa of both the male and female urethra contains a plexus of acetylcholinesterase-positive nerve fibres (Fig. 1.14) similar to that described the bladder (see p. 9). This nerve plexus is especially dense immediately beneath the epithelial lining of the urethra and is believed to subserve a sensory function, the urethral mucosa being sensitive to touch, thermal stimulation, distension and pain (Nathan and Smith 1951).

In the electron microscope single nerve axons may be observed in close proximity to the basal aspect of the urethral epithelium (Fig. 1.15). Many appear varicose and contain axonal vesicles which are of two morphologically distinct types. The majority have diameters of 40–60 nm and have a "clear" interior, being referred to as agranular vesicles. The second type of vesicle is larger, with diameters of 80–120 nm and is characterized by the presence of a central dense granule. Specialized sensory endings similar to those in the dermis of the skin have not been reported to occur in the wall of the urethra.

Functional Considerations

The urinary bladder performs a dual function: acting at times as a reservoir for urine accumulating

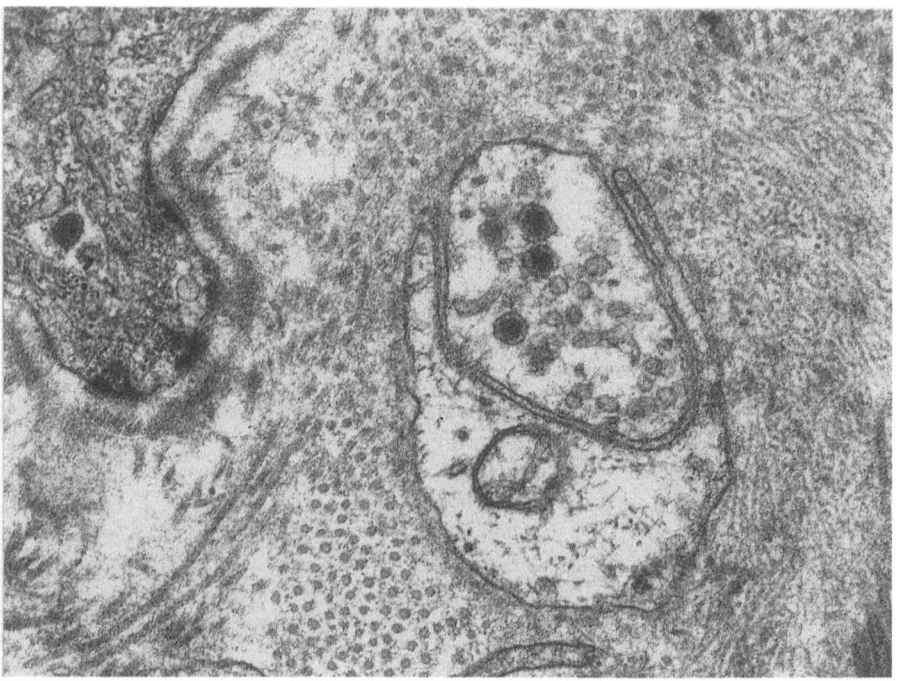

Fig. 1.15. Electron micrograph of a subepithelial axon from the female urethra. The presumptive sensory axon contains numerous small agranular vesicles and a few large granulated vesicles. ($\times 42\,000$)

within its lumen, and at others as a contractile organ actively expelling its contents into the urethra. In the following account the tissue components and, where appropriate, their neurological control are considered.

Continence of Urine

To achieve urinary continence, the bladder acts as a passive reservoir retaining fluid because the forces acting on the urethra produce an intra-urethral pressure greater than the bladder pressure. Several tissue components play a part in generating this urethral resistance, providing either an active or passive contribution. Since the smooth muscle of the bladder is replaced in the bladder neck region by a morphologically distinct type of smooth muscle, the detrusor muscle itself is not considered to play a part in closing the proximal urethra.

In the male, a distinct collar of circularly orientated smooth muscle occurs in the bladder neck and preprostatic urethra which is continuous distally with the muscular components of the genital tract. This smooth muscle sphincter is supplied by a rich plexus of sympathetic nerve fibres which, on stimulation, cause the sphincter to contract, thereby pre-

venting retrograde flow of semen into the urinary bladder at ejaculation. Despite this well-defined genital role, it is not known whether the smooth muscle of the bladder neck region and preprostatic urethra play an active part in the maintenance of continence. Intramural collagen and elastic fibres within the wall of the bladder neck, proximal urethra and prostate generate passive forces which help to close the urethral lumen. However, postoperative incontinence of urine does not usually follow radical surgical excision of the bladder neck, the preprostatic urethra and the prostate gland, suggesting that these structures provide only a partial contribution to urinary continence.

In the female, a smooth muscle sphincter cannot be anatomically recognized in the wall of the bladder neck and proximal urethra. Consequently it is even less likely that active smooth muscle contraction should be considered as an important factor in the continence of urine in women. However, the bladder neck and proximal urethra possess within their walls innumerable elastic fibres which are of particular importance in producing passive occlusion of the urethral lumen. Indeed, it has been suggested that the passive elastic resistance offered by the urethral wall is the most important single factor responsible for the closure of the bladder neck and proximal urethra in the continent woman.

In both sexes the urethra contains within its walls the rhabdosphincter, the location of which corresponds anatomically to the zone where maximal urethral closure pressure is normally recorded. This striated muscle sphincter is morphologically adapted to maintain tone over relatively long periods without fatigue and plays an important active role in producing urethral occlusion at rest. It remains to be determined, however, whether the force exerted by the sphincter is maximal at all times between two consecutive acts of micturition, or whether additional motor units are recruited during coughing, sneezing, etc. to enhance the occlusive force on the urethra during these events. The rhabdosphincter is innervated by nerve fibres, most of which travel via the pelvic splanchnic nerves.

Concerning the role of periurethral muscle in the maintenance of continence, the medial parts of the levator ani muscles in both sexes are related to (but structurally separate from) the urethral wall. These periurethral fibres are innervated by the pudendal nerve and consist of an admixture of large-diameter fast and slow twitch fibres. Therefore, unlike the rhabdosphincter, periurethral muscle possesses morphological features which are similar to other "typical" voluntary muscles. This pelvic floor musculature plays an important part (especially in the female) by providing an additional occlusive force on the urethral wall, particularly during events which are associated with an increase in intra-abdominal pressure. In addition, the muscles provide support for the pelvic viscera.

Micturition

To enable fluid to flow along the urethra it is necessary for the pressure in the urinary bladder to exceed that within the urethral lumen. Under normal circumstances, in order to initiate micturition, a fall in urethral pressure immediately precedes a rise in pressure within the lumen of the bladder. This pressure rise is usually produced by active contraction of detrusor smooth muscle at the onset of micturition. The detrusor muscle coat consists of innumerable interlacing muscle bundles forming a complex meshwork of smooth muscle which, on contraction, reduces all dimensions of the bladder. The muscle coat is collectively involved and it is unnecessary to attach special significance to the precise orientation of individual bundles within the wall of the viscus.

The preganglionic nerve supply travels in the pelvic splanchnic nerves before synapsing on neurons located within the vesical part of the pelvic plexuses and also within the wall of the bladder. These peripheral neurons supply nerve fibres which ramify throughout the thickness of the detrusor smooth muscle coat. The profuse distribution of these motor nerves emphasizes the importance of the autonomic nervous system in initiating and sustaining bladder contraction during micturition. For micturition to occur the pressure differential between the bladder and urethra must overcome the elastic resistance of the bladder neck. Immediately prior to the onset of micturition, the tonus of the rhabdosphincter is reduced by central inhibition of its motor neurons located in the second, third and fourth sacral spinal segments. Such inhibition is mediated by descending spinal pathways originating in higher centres of the central nervous system. Concomitantly, other descending pathways activate (either directly or via sacral interneurons) the preganglionic parasympathetic motor outflow to the urinary bladder. This central integration of the nervous control of the bladder and urethra is essential for normal micturition.

References

Alm P (1978) Cholinergic innervation of the human urethra and urinary bladder. Acta Pharmacol Toxicol (Kbh) 43 (Suppl 2): 56–62

Alm P, Alumets J, Brodin E, Hakanson R, Nilsson G, Sjoberg N-O, Sundler F (1978) Peptidergic (substance P) nerves in the genito-urinary tract. Neurosci 3: 419–425

Ambache N, Zar MA (1970) Non-cholinergic transmission by post-ganglionic motor neurones in the mammalian bladder. J Physiol (Lond) 210: 761–783

Daniel EEL, Cowan W, Daniel VP (1983) Structural bases for neural and myogenic control of human detrusor muscle. Can J Physiol Pharmacol 61: 1247–1273

de Groat W, Booth AM (1980) Inhibition and facilitation in parasympathetic ganglia of the urinary bladder. Fed Proc 39: 2990–2996

de Groat WC, Theobald RJ (1976) Reflex activation of sympathetic pathways to reseal smooth muscle and parasympathetic ganglia by electrical stimulation of vesical afferents (cat). J Physiol (Lond) 259: 223–239

DeLancey JOL (1986a) Anatomy of the extrinsic continence mechanism. Proceedings of the 16th Annual Meeting of the International Continence Society, Boston, pp 22–24

DeLancey JOL (1986b) A correlative study of paraurethral anatomy. Obstet Gynecol (in press)

Dixon JS, Gilpin CJ (1987) Presumptive sensory axons of the human urinary bladder. A fine structural study. J Anat (to be published)

Dixon JS, Gosling JA (1977) Light and electron microscopic observations on noradrenergic nerves and striated muscle cells of the guinea pig urethra. Am J Anat 149: 121–126

Dixon JS, Gilpin SA, Gilpin CJ, Gosling JA (1983) Intramural ganglia of the human urinary bladder. Br J Urol 55: 195–198

Donker PJ, Dröes JThPM, van Ulden BM (1976) Anatomy of the musculature and innervation of the bladder and the urethra. In: Williams DI, Chisholm GO (eds) Scientific foundations of urology, vol II. Heinemann, London, pp 32–39

Ek A, Alm P, Andersson K-E, Persson CGA (1977) Adrenergic and cholinergic nerves of the human urethra and urinary bladder. A histochemical study. Acta Physiol Scand 99: 345–352

Elbadawi A (1982) Neuromorphologic basis of vesicourethral function. I. Histochemistry, ultrastructure and function of intrinsic nerves of the bladder and urethra. Neurourol Urodyn 1: 3–50

Fletcher TF, Bradley WF (1978) Neuroanatomy of the bladder-urethra. J Urol 119: 153–160

Gilpin CJ (1985) Fine structure of the intrinsic autonomic innervation of the human urinary bladder. MSc thesis, Manchester University

Gilpin CJ, Dixon JS, Gilpin SA, Gosling JA (1983) The fine structure of autonomic neurons in the wall of the human urinary bladder. J Anat (Lond) 137: 705–713

Gil Vernet S (1968) Morphology and function of vesico-prostatourethral musculature. Canova, Treviso

Gosling JA (1979) The structure of the bladder and urethra in relation to function. Urol Clin North Am 6: 31–38

Gosling JA, Dixon JS, Critchley HOD, Thompson SA (1981) A comparative study of the human external sphincter and periurethral levator ani muscles. Br J Urol 53: 35–41

Gosling JA, Dixon JS, Lendon RG (1977) The autonomic innervation of the human male and female bladder neck and proximal urethra. J Urol 118: 302–305

Gosling JA, Dixon JS, Humpherson JA (1983) Functional anatomy of the urinary tract: an integrated text and colour atlas. Churchill Livingstone, Edinburgh

Head H, Riddoch G (1917) The automatic bladder, excessive sweating and some other reflex conditions in gross injuries of the spinal cord. Brain 40: 188–263

Hockfelt T, Kellerth J-O, Nilson G, Pernow B (1975) Experimental immunohistochemical studies on the localisation and distribution of substance P in cat primary sensory neurons. Brain Res 100: 235–252

Holzer P, Bucsics A, Lembeck F (1982) Distribution of capsaicin sensitive nerve fibres containing substance P in cutaneous and visceral tissue of the rat. Neurosci Lett 31: 253–257

Hutch JA (1972) Anatomy and physiology of the bladder trigone and urethra. Meredith, New York

Iwayama T, Furness JB, Burnstock G (1970) Dual adrenergic and cholinergic innervation of the cerebral arteries of the rat.

Circ Res 26: 635–646

Klück P (1980) The autonomic innervation of the human urinary bladder, bladder neck and urethra. A histochemical study. Anat Rec 198: 439–447

Learmonth JR (1931) A contribution to the neurophysiology of the urinary bladder in man. Brain 54: 147–176

McConnell J, Benson GS, Wood JG (1982) Autonomic innervation of the urogenital system. Adrenergic and cholinergic elements. Brain Res Bull 9: 679–694

Mobley TL, Elbadawi A, McDonald DF, Schenk EA (1966) Innervation of the human urinary bladder. Surg Forum 27: 505–506

Nathan PW, Smith MC (1951) The centripetal pathway from the bladder and urethra within the spinal cord. J Neurol Neurosurg Psychiatry 14: 262–280

Nergardh A, Boréus LO (1972) Autonomic receptor function in the lower urinary tract of man and cat. Scand J Urol Nephrol 6: 32–36

Nyo MM (1969) Innervation of the bladder and urethra. J Anat 105: 210

Parks AG, Swash M, Urich H (1977) Sphincter denervation in anorectal incontinence and rectal prolapse. Gut 18: 656–665

Santicioli P, Maggi CA, Meli A (1985) The effect of capsaicin pretreatment on the cystometrograms of urethane anesthetized rats. J Urol 133: 700–703

Sharkey KA, Williams RG, Schultzberg M, Dockray GJ (1983) Sensory substance P innervation of the urinary bladder. Possible site of action of capsaicin in causing urine retention in rats. Neurosci 10: 861–868

Sundin T, Dahlström A, Norlén L, Svedmyr N (1977) The sympathetic innervation and adrenoreceptor function of the human lower urinary tract in the normal state and after parasympathetic denervation. Invest Urol 14: 322–328

Tanagho EA, Smith DR (1966) The anatomy and function of the bladder neck. Br J Urol 38: 54–71

Von Hayek H (1969) Die Muskulatur des Beckenbodens. In: Alken CE, Dix VW, Goodwin WE, Wildbolz E (eds) Handbuch der Urologie, vol I. Springer, Berlin Heidelburg, pp 279–288

Wilson PD, Dixon JS, Brown ADG, Gosling JA (1983) Posterior pubourethral ligaments in normal and genuine stress incontinent women. J Urol 130: 802–805

Woodburne RT (1967) Anatomy of the bladder. In: Boyarsky S (ed) The neurogenic bladder. Williams and Wilkins, Philadelphia, pp 1–17

Zacharin RF (1963) The suspensory mechanism of the female urethra. J Anat 97: 423–427

2 · Comparative Neuromorphology in Animals

Ahmad Elbadawi

Structural Organization of the Muscular Apparatus of Micturition

Innervation of the Vesicourethral Smooth Muscularis

Anatomy of Peripheral Neural Pathways

Microstructure of Peripheral Neural Pathways

Innervation of Urethral Rhabdosphincter

Structural Organization of the Muscular Apparatus of Micturition

Anatomical, neuromorphological and physio-pharmacological studies on the musculature of the organs of micturition, especially in the past two decades, have suggested the following concepts of its structural organization. First, although the detrusor is anatomically a single muscle, it is neuromorphologically divisible into body and base units (Elbadawi and Schenk, 1966). Second, muscle bundles of the base detrusor are anatomically continuous with the urethral muscularis distally (Hutch 1972) and provide the superficial and, in part, the deep periureteric sheaths of the distal segment of each ureter (Elbadawi 1972; Elbadawi et al. 1973). Third, upon its termination in the bladder wall, the muscularis of the distal ureters fan out beyond the ureteric orifices, and conjointly form the trigonal muscle (Elbadawi and Schenk 1971a). Fourth, the striated urethral sphincter, currently known as the rhabdosphincter (Elbadawi 1982b), is an integral part of the urethral wall, and is a distinct entity, being different and separate from the outlying musculature of the pelvic floor.

The concept of functional division of the detrusor into body and base units was introduced on the basis of regional differences in its intrinsic adrenergic innervation (see p. 27) (Elbadawi and Schenk 1966, 1968a). As originally defined, the body detrusor lies above and the base detrusor below the level at which the ureters enter into the posterior wall of the bladder. Because of their anatomical continuity, and their similar intrinsic muscular innervation patterns, the base detrusor, urethral muscularis and periureteric sheaths have been proposed as the components of a distinct unit of the micturition muscular apparatus, named the lissosphincter (Elbadawi 1982b).

Comparative Anatomy

Except for the periureteric sheaths, the anatomy of the human vesicourethral muscularis has been studied in detail only in humans (Uhlenhuth et al. 1953; Woodburne 1964, 1967; Gil Vernet 1968; Hutch 1972; Young 1972). The account which follows is based largely on my observations in the dog, cat, rabbit and rat. The muscular apparatus of micturition in these species consists of the same four units already described. The concept of functional division of the detrusor into a body and a base region has been confirmed in all species (Elbadawi and Schenk 1966; Raezer et al. 1973; Sundin and Dalström 1973; Slack et al. 1982).

Body Detrusor

As in humans, muscle bundles of the body detrusor in the four animal species have a "plexiform" organization without discernible "layers". The ventral surface of the feline body detrusor has a stout vertical midline bundle that extends from its dome to near its junction with the base region; short, transversely orientated bundles derived from the outlying detrusor muscularis are inserted in each side of this midline bundle. A comparable, but much thinner, bundle exists in the rat.

Lissosphincter: Base Detrusor

The base detrusor is not separable from the body detrusor by any discernible anatomical or histological boundary. The base region is sizable and well defined in the human bladder, and its junction with the urethra is easily identified by appreciable narrowing of the lumen. The bladder base is also well defined and relatively sizable in the dog and rabbit, but is very short in both cat and rat, because of the very caudal vesical insertion of the ureters in both species. The junction of the bladder base and urethra in the rabbit is very gradual and almost imperceptible, but in the other three animal species, it is easily identifiable by rapid tapering of the base as it joins a urethra of a uniformly narrow external diameter.

Lissosphincter: Periureteric Sheaths

In humans, the distal end of each ureter is surrounded by craniolateral extensions of muscle bundles of the base detrustor. These extensions are arranged as a superficial and a deep periureteric sheath (Elbadawi 1972; Elbadawi et al. 1973). The superficial sheath includes the sheath of Waldeyer (1892, 1899) around the juxtavesical (prehiatal), bundles of the ureteric hiatus around the intravesical (hiatal), and the transureteric bundles crossing the "submucosal" (posthiatal) segments of the ureter (Fig. 2.1). Bundles of the two periureteric sheaths are orientated longitudinally and helically around the ureter. At multiple levels, slender muscle bundles connect the superficial to the deep sheath, and the deep sheath to the ureteric muscularis.

The base detrusor provides the periureteric sheaths of the distal ureters in the dog, cat, rabbit and rat (Elbadawi et al. 1973), but unlike the human, only the equivalent of the superficial sheath is easily discernible by gross dissection and histologically in the four animal species. The transureteric bundle component of the superficial sheath is prominent in the dog, but is rudimentary or absent in the other three species.

Lissosphincter: Urethral Muscularis

As in humans, the canine prostate surrounds the most cranial part of the male urethra, abutting against the vesicourethral junction. The rat prostate is huge relative to the size of the bladder, and conceals its very short base region. The rabbit prostate is represented by fragmented components related to the subvesical portion of the urethra. The feline prostate, on the other hand, is relatively small and is located very deeply in the pelvis, dorsal to the caudal part of the symphysis pubis, and is separated from the vesicourethral junction by a 3–4 cm long preprostatic urethra. As in humans, the urethral smooth muscularis in all four animal species is composed of longitudinally and circularly orientated components that tend to form quasidistinct "layers". In these species, urethral smooth muscle fascicles frequently change their planes of orientation, from longitudinal to circular and vice versa, with many "obliquely" orientated "layer-connecting" bundles.

Ureterotrigone

The predominantly longitudinal muscle bundles of each distal human ureter extend beyond the ureteric orifice into the trigonal region. The extensions from both sides fan out the blend to form a thin triangular sheet of compactly arranged slender bundles, that was originally designated as the superficial trigone (Tanagho and Pugh 1963), and, later, simply as the trigonal muscle (Elbadawi and Schenk 1971a; Elbadawi 1982b). The human trigonal muscle is difficult to separate from the overlying urothelium. Its base is generally in the form of a stout bundle that extends between the two ureteric orifices (Mercier's bar), and its apex lies at the posterior midline of the vesicourethral junction. Each lateral border of the trigonal muscle (Bell's muscle) receives slim bundles derived from the out-

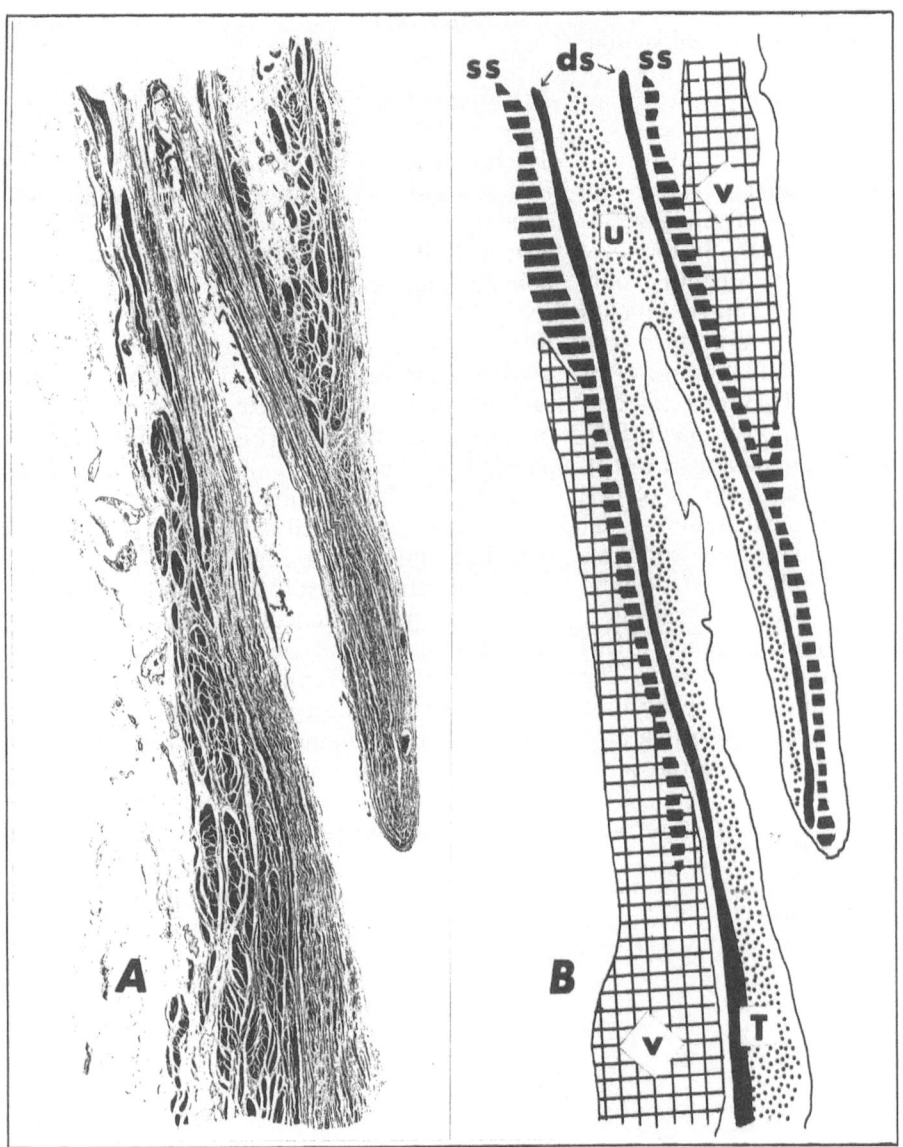

Fig. 2.1a, b. Dual periureteric sheath: longitudinal section of ureterovesical junction. **a** Photomicrograph of histological section. **b** Diagrammatic reconstruction by tracing of muscle bundles in section. *u*, ureteric muscularis; *ds*, deep sheath; *ss*, superficial sheath; *v*, muscularis of bladder base; *T*, trigonal region. (Elbadawi 1972; © The Williams & Wilkins Co., Baltimore, Maryland)

lying base detrusor. The muscle also receives bundles cranially from the body detrusor, and dorsally from the underlying trigonal part of the bladder base detrusor (deep trigone: Tanagho and Pugh 1963) (Elbadawi 1982b; A. Elbadawi, E.O. Amaku and I.N. Frank 1970, unpublished work). The trigonal muscle extends distally as far as the verumontanum in the form of a slender bundle that lies superficial to the muscularis of the posterior urethral wall.

Extension of the ureteric muscularis into the trigonal region is present in all four animal species. Unlike humans, however, the trigonal muscle in animals is poorly developed, is not in the form of a triangular sheet, and peters out close to the vesicourethral junction with very short or no caudal urethral extensions. In both dog and cat, the trigonal muscle is represented by two flimsy separate bundlets, each extending from one ureteric orifice to the vesicourethral junction. The trigonal muscle of both rabbit and rat is even less developed, and is practically indiscernible, even under magnification.

Urethral Rhabdosphincter

A striated muscle structure comparable to the human urethral rhabdosphincter is present in a simplified form in nonhuman mammals. In the male rat, the rhabdosphincter can be recognized only microscopically as a few striated muscle bundles at the junction of the prostatic and postprostatic segments of the urethra. The rhabdosphincter of the male cat is visible grossly and has three anatomically continuous components (A. Elbadawi, J.W. Downie and M.A. Atta 1984, unpublished work). The preprostatic component is composed of helically orientated fibres that surround the most distal (±1 cm) portion of the preprostatic urethra, especially ventrally and laterally. The prostatic component is located in the prostatic "capsule", and is composed of circularly and obliquely orientated fibres, also mainly ventrally and laterally. The postprostatic component is represented by the most cranial portion of the circularly orientated urethralis muscle surrounding the postprostatic urethra (Martin et al. 1974; A. Elbadawi, J.W. Downie and M.A. Atta 1984, unpublished work).

Ultrastructure

Lissosphincter of Cat and Rat

The muscularis of the bladder base and urethra in both cat and rat conforms with the known ultrastructure of mammalian visceral smooth muscle in general (Elbadawi 1982a). In both of these components of the lissosphincter, the muscle cell profiles are grouped as fascicles, separated by connective tissue that contains collagen, elastin, blood vessels, and Schwann cell-ensheathed axon bundles. The sarcoplasm of each muscle cell is packed with bundles of myofilaments (Fig. 2.2). Short electron-dense (cigar) bodies are uniformly distributed parallel to the myofilaments throughout the sarcoplasm. Cellular organelles, including sarcoplasmic reticulum, mitochondria, and the Golgi complex, are usually present next to each pole of the nucleus. Sacs of endoplasmic reticulum and occasional mitochondria are present beneath the sarcolemma.

The sarcolemma is ensheathed by a ±20 nm thick basal lamina, with an intervening electron-lucent narrow space. Contiguous muscle cells are separated by 0.7–1.0 μm wide intercellular spaces that contain particulate electron-dense material, but no collagen or elastin. The sarcolemma has 200–

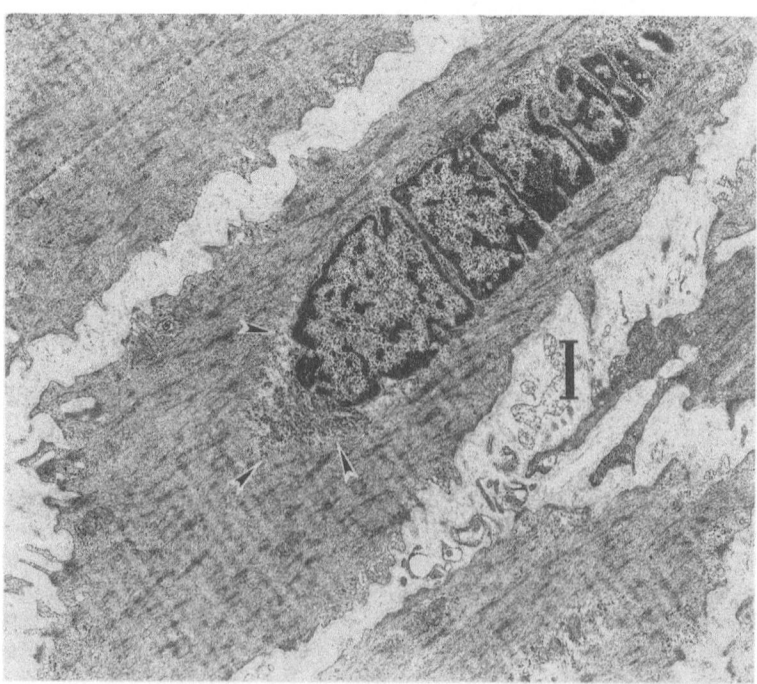

Fig. 2.2. Longitudinally sectioned smooth muscle cell in cat bladder base. Sarcoplasm has evenly distributed linear dense bodies; cell organelles next to pole of nucleus marked by *arrowheads*. Scale = 1 μm. (Elbadawi 1982a; © The Williams & Wilkins Co., Baltimore, Maryland)

400 nm long dense bands on the sarcoplasmic side, and alternating caveolae, which are flask-shaped indentations that have narrow necks and are usually arranged in rows.

Frequent adherens, less common intrusion and rare probable nexus-type (gap) muscle cell junctions are present in the vesicourethral lissosphincter of both cat and rat (Elbadawi 1982a). The most common form of the adherens junction resembles the zonula adherens (intermediate junction), which has a 5–25 nm wide separation gap, without a central linear density, and parallel sarcolemmal densities along the zone of close sarcolemmal apposition (Fig. 2.3a). A less common form resembles the desmosome (macula adherens) which has a 30–60 nm wide separation gap, with a central linear density, and parallel sarcolemmal densities—in which myofilaments are inserted—along the zone of close apposition (Fig. 2.3b). The intrusion junction is formed by a bulbous muscle cell process that characteristically has an "empty" ultrastructural profile, and protrudes in a "peg-in-socket" manner into an invagination in an adjacent muscle cell, with a 5–20 nm sarcolemmal separation gap. Areas of very close sarcolemmal contact, with 2–5 nm separation gaps, are not uncommon in the lissosphincter. Gap junctions probably exist along some of these very close contacts, although none with the characteristic penta- or heptalaminar structure has so far been demonstrated unequivocally.

Feline Rhabdosphincter

The rhabdosphincter of the male cat is composed of a mixture of fast and slow twitch myofibres that have the same ultrastructure as in ordinary striated muscle (Elbadawi and Atta 1985, 1986; Elbadawi 1987). The rhabdosphincter myofibre is a cylindrical cell that has multiple subsarcolemmal nuclei, and a ~20 nm thick basal lamina. The sarcoplasm is occupied by parallel myofibrils, each of which consists of a series of sarcomeres with the characteristic arrangement of Z-discs, A- and I-bands, H-zones and M-lines (Figs. 2.4, 2.5). Sacs of sarcoplasmic reticulum extend in the form of a reticular cylinder around the entire length of each myofibril.

Transverse (T)-tubules cross each myofibril, particularly at the junctions of the A- and I-bands; T-tubules and sarcoplasmic reticulum sacs form the triad system at the points of their intersection across each myofibril. Each unit of this system is represented by a central T-tubule sandwiched between two laterally located sacs of sarcoplasmic reticulum that contain osmiophilic material. The sarcoplasm between contiguous myofibrils has a variable content of mitochondria, ribosomes, glycogen particles and lipid droplets.

As described previously (Murata and Ogata 1969; Padykula and Gauthier 1970; Lentz 1971), the fast twitch myofibre has abundant sarcoplasmic reticulum and glycogen, a small content of mitochondria and tubular elements, thin straight Z-discs and well-defined straight M-lines (Fig. 2.4). In contrast, the slow twitch myofibre is generally smaller, contains relatively little endoplasmic reticulum, glycogen and tubular elements, has abundant and prominent mitochondria that tend to be aligned in rows between adjacent sarcomeres, and has thick, jagged or zig-zag Z-discs, but less defined M-lines (Fig. 2.5).

As in ordinary striated muscle, myosatellite cells abut against myofibres of the male feline rhabdosphincter. The myosatellite cell is sandwiched between the sarcolemma and its basal lamina, and has a single nucleus and scanty cytoplasm that contains ribosomes and mitochondria, but no myofilaments (Elbadawi and Atta 1986).

Innervation of the Vesicourethral Smooth Muscularis

Histochemistry of Intrinsic Efferent Innervation

The bladder of the cat, dog, rabbit and rat has a dual muscular innervation by cholinergic and adrenergic nerves. Regional differences in the density and distribution pattern of the latter are the basis for the concept of functional division of the bladder into a body and a base region (Elbadawi and Schenk 1966; 1968a).

Cholinergic muscular innervation is uniformly rich (Fig. 2.6) in the body detrusor, base detrusor and urethral muscularis of the cat, dog and rabbit (Elbadawi and Schenk 1966, 1974). In the rat, however, this innervation is more dense in the ventral than either the dorsal or lateral wall of the bladder body, and in the dorsal than either the ventral or lateral wall of the bladder base (Elbadawi 1982b).

Adrenergic muscular innervation varies regionally in the four species (Fig. 2.7). In general, it is virtually absent in the bladder dome, is more abundant in the urethra than the base detrusor, the base than the body detrusor, and the outer than the inner parts of the wall in any region of the body detrusor. In the cat, the base detrusor has commensurate adrenergic innervation in its ventral, lateral and dorsal walls. In both dog and pig, adre-

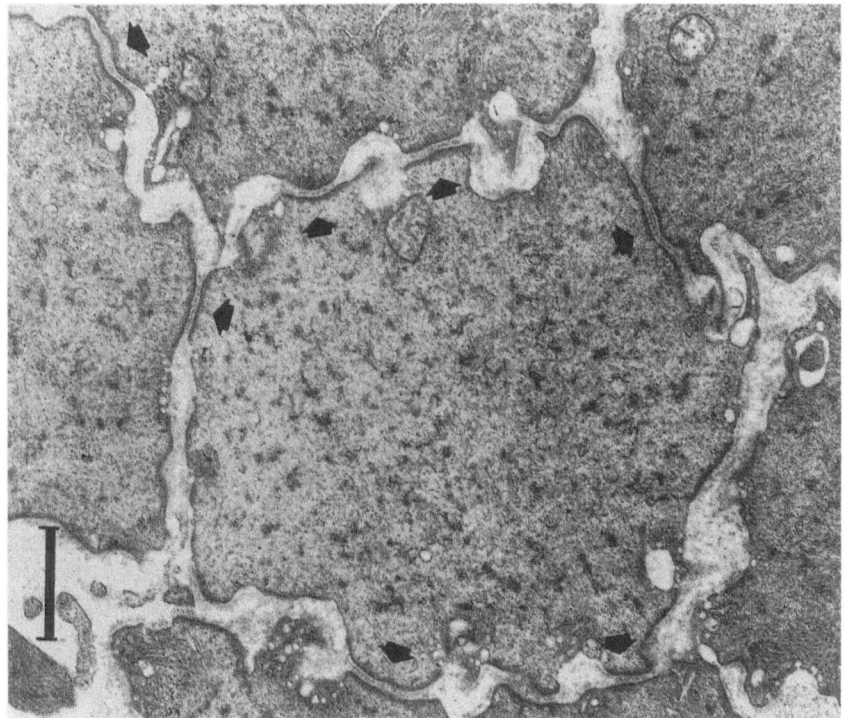

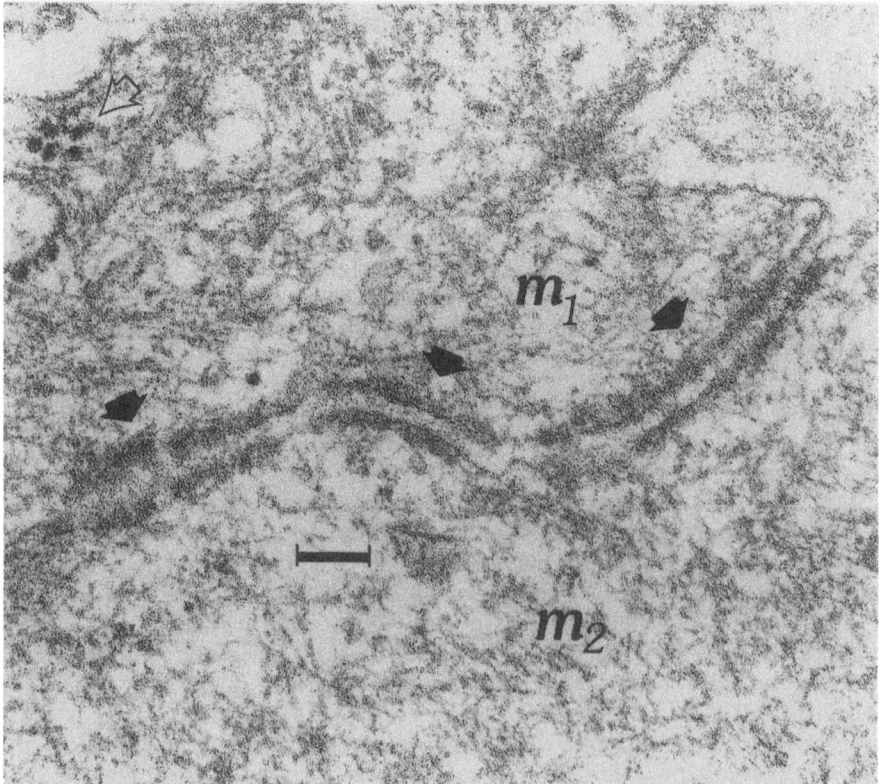

Fig. 2.3. a Transversely sectioned smooth muscle cells in cat bladder base; intermediate cell junctions are marked with *arrows*. **b** High magnification of intermediate junctions (*solid arrows*) between muscle cells m_1 and m_2; glycogen particles marked with *open arrow*. Scale = 100 nm. (Elbadawi 1982a; © The Williams & Wilkins Co., Baltimore, Maryland)

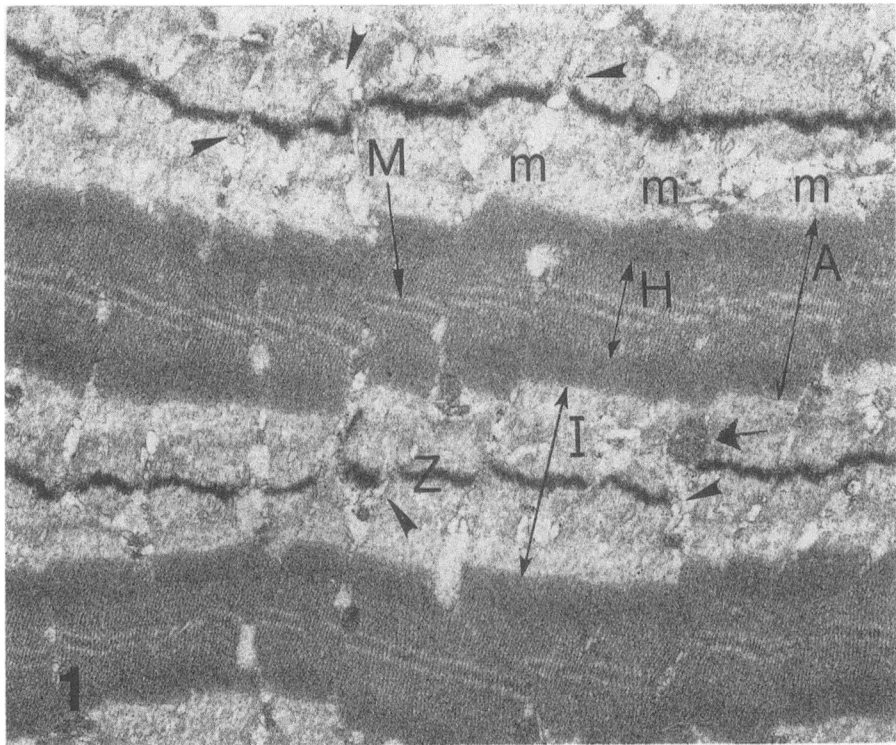

Fig. 2.4. Fast myofibre in cat rhabdosphincter. *m*, myofibrils; band pattern marked *A*, *H*, *I*, *Z* and *M* (see text for details). Sacs of endoplasmic reticulum marked with *arrowheads*, and mitochondrion with *arrow*. (Elbadawi and Atta 1986.) (×18 400)

nergic innervation of the base is similar, but less prominent (Elbadawi and Schenk 1966, 1971a). In both the rabbit and rat, the density of adrenergic innervation of the body detrusor increases progressively as it is followed at any level from its ventral to its dorsal midline (Elbadawi and Schenk 1966; Slack et al. 1982). The body detrusor of the rat has a considerably weaker innervation than other animal species, except close to its junction with the base detrusor (Elbadawi and Schenk 1966; Alm and Elmér 1975; Elmér 1975), and in its vertical, ventral midline bundle, in which adrenergic innervation is as dense as in the dorsal wall of the base detrusor (Elbadawi 1982b).

The periureteric sheath in the cat, dog, rabbit, rat and pig has the same density and distribution pattern of both cholinergic and adrenergic muscular innervation as in the base detrusor (Elbadawi and Schenk 1971a; Elbadawi 1982b). On the other hand, the muscularis of the pelvic ureters and ureterotrigone have a much less dense commensurate dual innervation that, in the cat, is still more dense than that of the abdominal ureter (Elbadawi and Schenk 1969, 1971a; Elbadawi 1982b).

Neurohistochemical studies on the vesicourethral muscularis in humans have confirmed its dual adrenergic and cholinergic innervation, but have yielded conflicting observations on the regional variations in density of both nerve types (Elbadawi 1982b). One group of authors described sparse cholinergic and rich adrenergic innervation of the "bladder neck" in the male, and the opposite pattern in the female (Gosling et al. 1977; Gosling 1979). No such sex differences in muscular innervation of the vesicourethral junction, however, have been observed in any animal so far studied (Elbadawi and Schenk 1966; Sundin and Dahlström 1973; Elmér 1975; Elbadawi 1982b), or in other studies on the human bladder (Ek et al. 1977; Alm 1978; Klück 1980; Nordling and Christensen 1978).

J. Dixon (personal communication) reports that the original findings of Gosling et al. (1977) in 4 cases have been confirmed in a further series of 12 cases. In addition to species differences it is likely that the efficiency of collection of fresh material and the vagaries of staining technique may explain the conflicting results. In relation to this contentious issue the interested reader is advised to refer to the original literature and come to his or her own conclusion.

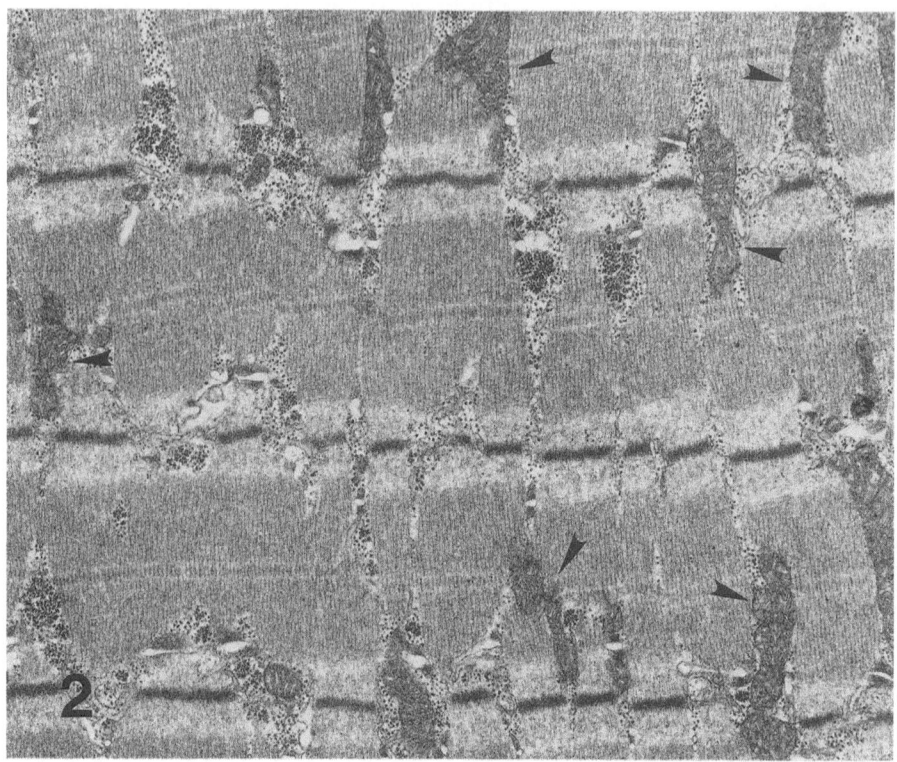

Fig. 2.5. Slow myofibre in cat rhabdosphincter. Abundant mitochondria stretched along myofibrils are marked with *arrowheads*. (Elbadawi and Atta 1986.) (× 18 500)

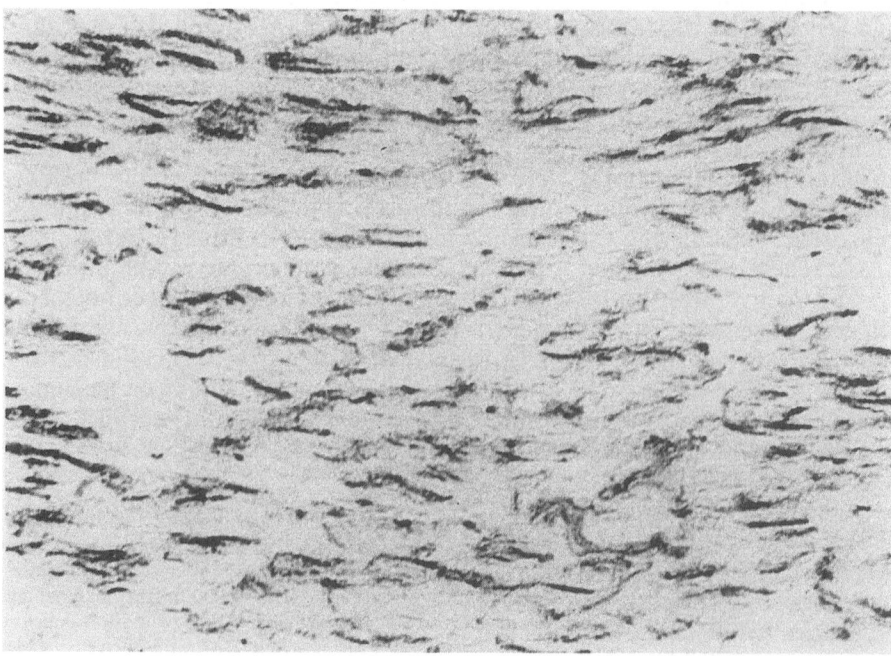

Fig. 2.6. Rich cholinergic nerves in cat bladder base muscularis. (Elbadawi–Schenk acetylcholinesterase stain)

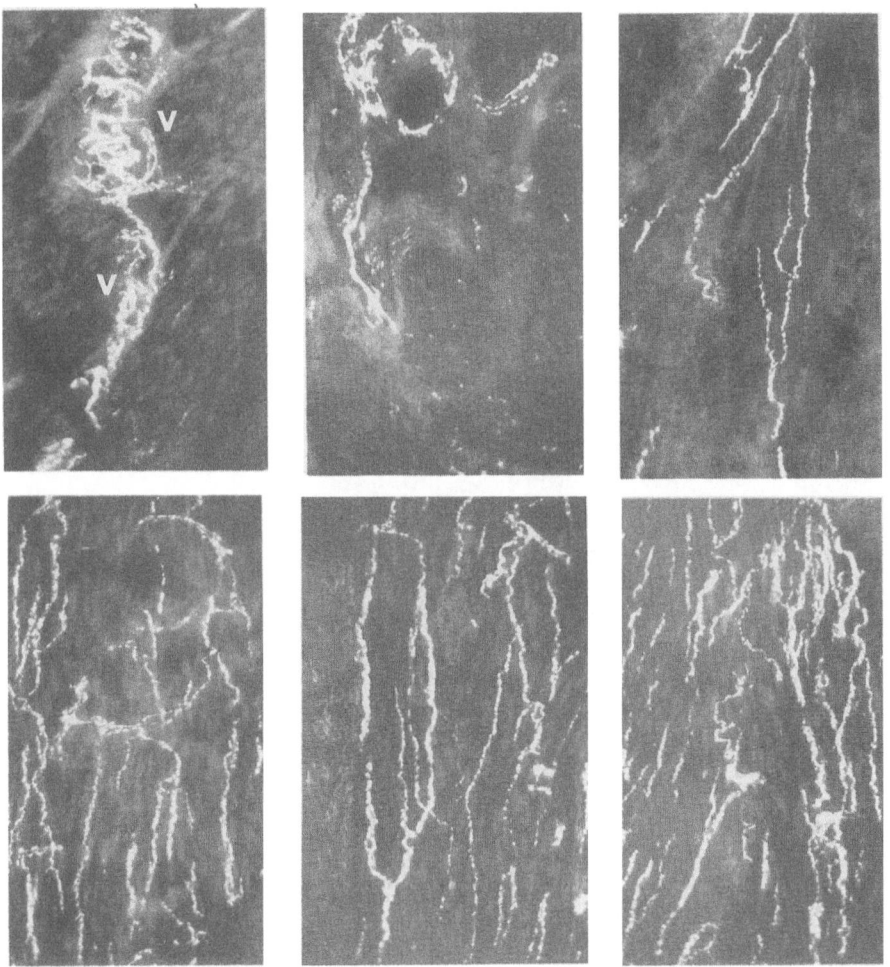

Fig. 2.7a–f. Adrenergic innervation of vesicourethral muscularis. **a** Bladder dome; **b** upper bladder body; **c** lower bladder body; **d** bladder base; **e** vesicourethral junction (bladder neck); **f** proximal urethra. Innervation in dome is restricted to blood vessels (*v*), and gradually increases in density as bladder is traced from dome towards urethra; urethral muscular innervation is slightly more dense than in bladder base. (All photographs same magnification; Elbadawi–Schenk formaldehyde-fluorescence method for norepinephrine)

Ultrastructure of Intrinsic Efferent Innervation

Nerve trunks supplying the bladder and urethra of both cat and rat run in their adventitia, mainly dorsolaterally and laterally. Composed of myelinated and Schwann cell-ensheathed nonmyelinated adrenergic and cholinergic axons, these trunks divide into smaller branches of similar nerve structure (Fig. 2.8) that penetrate the vesicourethral muscularis (Elbadawi 1982b). Within the muscularis, these branches repeatedly subdivide into progressively thinner nerve bundles, and lose their myelinated axonal component as they approach the muscle cells that they innervate (Elbadawi 1982a,b). The smallest axon bundles within muscle fascicles are composed of 4–10 adrenergic and/or cholinergic axons, and provide adrenergic and cholinergic terminals that innervate one or more muscle cells, separately or conjointly (Fig. 2.9). Throughout their course, the intrinsic nerves of the vesicourethral muscularis are closely associated with, and innervate the blood vessels of both the vesical and urethral walls (Elbadawi and Schenk 1966, 1974).

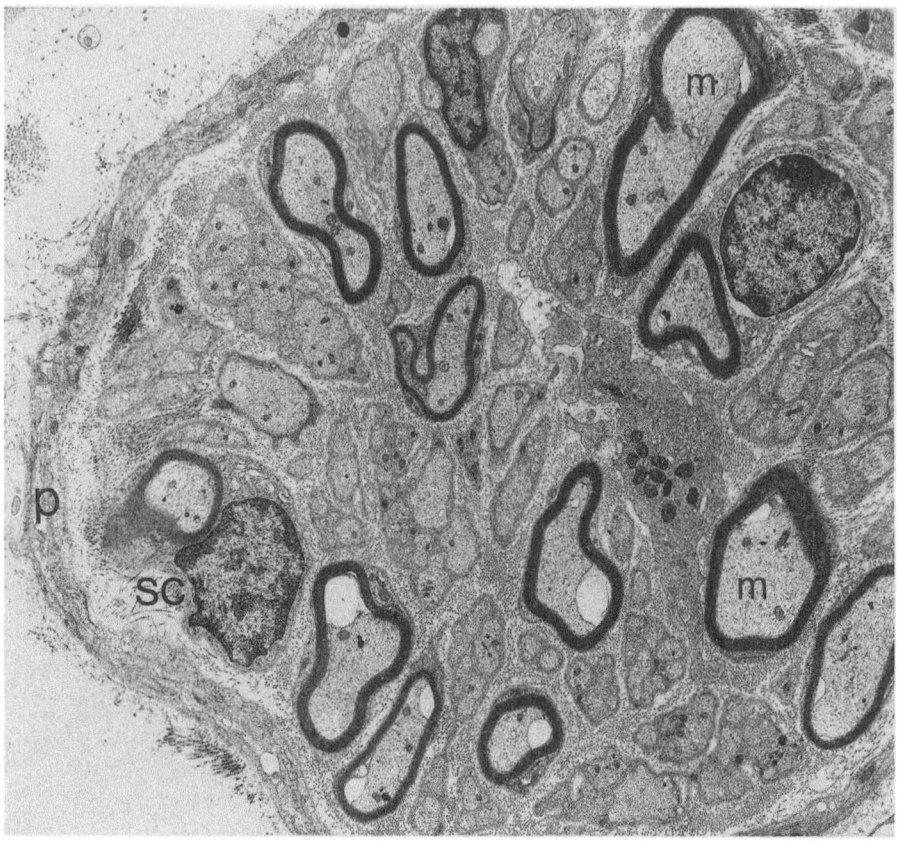

Fig. 2.8. Nerve trunk, adventitia of cat bladder base. Trunk is composed of myelinated (*m*) and nonmyelinated axons. *p*, perineurium; *sc*, Schwann cell. (Elbadawi 1982b.) (×4290)

Neuroeffector Junctions in the Lissosphincter of the Cat and Rat

Neuroeffector junctions with smooth muscle cells of the lissosphincter in both the cat and rat are established by terminals and/or varicosities of adrenergic and cholinergic axons (Figs. 2.9–2.11). Most of the junctions are formed by contact of an axon with the surface of a muscle cell, but some are invaginated, with an axon terminal occupying a deep groove in the cell. The neuroeffector junctions have 10–80 nm wide gaps, and thus may be close, almost close, or en passant, with gap widths, respectively, of 10–25, >25–40, and >40–80 nm. The junctions are rarely associated with symmetrical axolemmal and sarcolemmal densities similar to those of zonulae adherentes (Fig. 2.10).

Most neuroeffector junctions are unitary, comprising one axonal and one muscle cell profile (Fig. 2.12a). Some are multieffector, with one axon in contact with 2–4 spatially intimate muscle cells (Fig.

2.12b). Other junctions are multiaxonal, being formed by 2–4 adrenergic or cholinergic axon terminals in contact with one muscle cell (Fig. 2.12c). Some multiaxonal junctions are diautonomic, with a cholinergic and an adrenergic axon in contact with the same muscle cell (Fig. 2.12d).

Even the richly innervated lissosphincter of the cat does not appear to be innervated by either cholinergic or adrenergic axons on a 1:1 nerve:muscle cell basis (Elbadawi 1982b), as suggested in earlier histochemical studies (Elbadawi and Schenk 1966). In the lissosphincter of both cat and rat, some muscle cells have only cholinergic, some only adrenergic, some both cholinergic and adrenergic (diautonomic), and some no neuroeffector junctions of any type (Elbadawi 1982a, 1987). Excitation of the lissosphincter in these species, therefore, is almost certainly both by neuromuscular transmission and muscle cell coupling (Elbadawi 1982a,b, 1983, 1987), as in the mammalian vas deferens (Merrillees et al. 1963; Merrillees 1968).

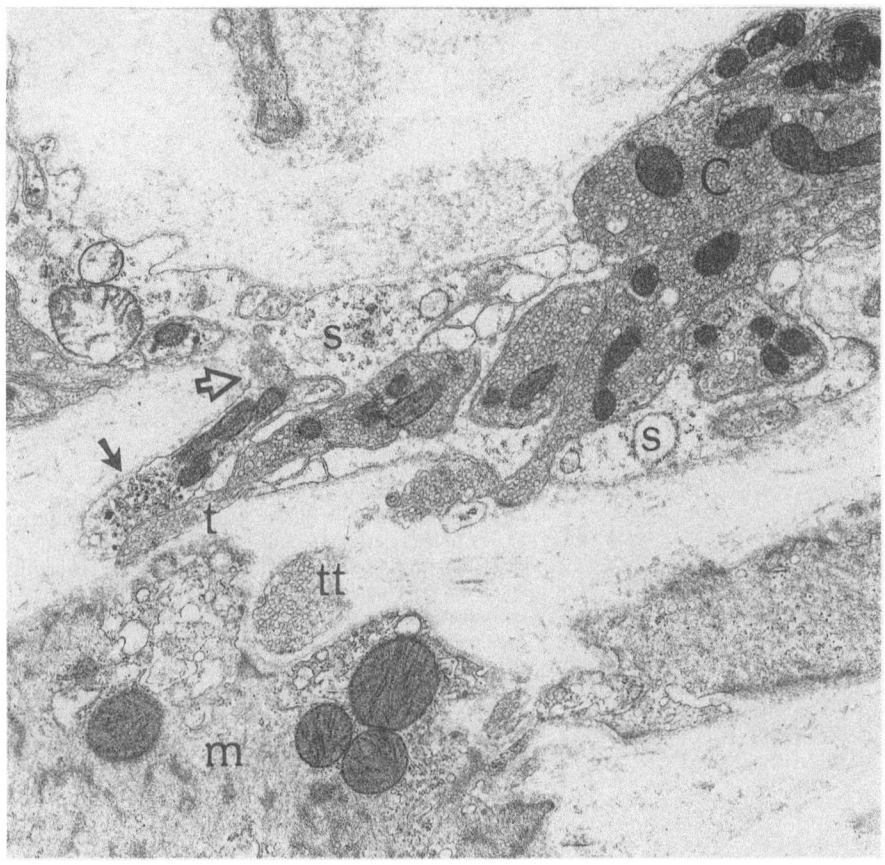

Fig. 2.9. Neuroeffector field, prediaphragmatic urethra of male rat. Preterminal axon bundle contains cholinergic axon (*c*) and four of its branches, plus branching adrenergic axon (*open arrow*), all ensheathed by Schwann cell (*s*). Terminals of two branch cholinergic axons (*t*; *tt*) form multiaxonal neuroeffector junction with muscle cell (*m*). Terminal t also forms axoaxonal synapse (15 nm cleft; 700 nm contact zone) with adjacent adrenergic terminal (*arrow*), which in serial sections forms adrenergic neureffector junction with same muscle cell. (Elbadawi 1984.) (× 1700)

Concept of the Axoaxonal Synapses

Intrinsic adrenergic and cholinergic axons innervating the lissosphincter of both the cat and rat run very closely together, but are generally separated by the cytoplasm and/or cytomembranes of their ensheathing Schwann cell somas or processes (see Fig. 2.9). Some spatially intimate axons within axon bundles and preterminal neuroeffector fields in the lissosphincter, however, have direct axolemmal contacts without any interposed Schwann cell elements (see Figs. 2.9, 2.14), (Elbadawi 1984, 1985). With rare exceptions, such contacts are not specialized as the classical chemical synaptic complex (Elbadawi 1984); this complex has axolemmal densities in the form of a presynaptic grid in one axon and an irregular postsynaptic density in the other, a separation gap of 20–30 nm that con-

tains finely particulate electron-dense material, and an active synaptic length of 400–500 nm (Fig. 2.13; Elbadawi 1984). By contrast, most axoaxonal synapses involve 10 nm–4 um stretches of two directly apposed axons that have no axolemmal densities, with 15–25 nm separation gaps that contain no electron-dense material (Fig. 2.14; see also Fig. 2.9). Most axoaxonal contacts involve a cholinergic and an adrenergic axon profile, but some are formed by two cholinergic, and a few by two adrenergic or a cholinergic and a probable SIF cell-derived profile (Elbadawi 1984, 1985). Based on available physiopharmacological observations on peripheral autonomic neurotransmission, it has been proposed that the adrenergic/cholinergic axoaxonal synapses mediate reciprocal prejunctional sympathetic/parasympathetic inhibition of transmission in postganglionic axons innervating the lissosphincter

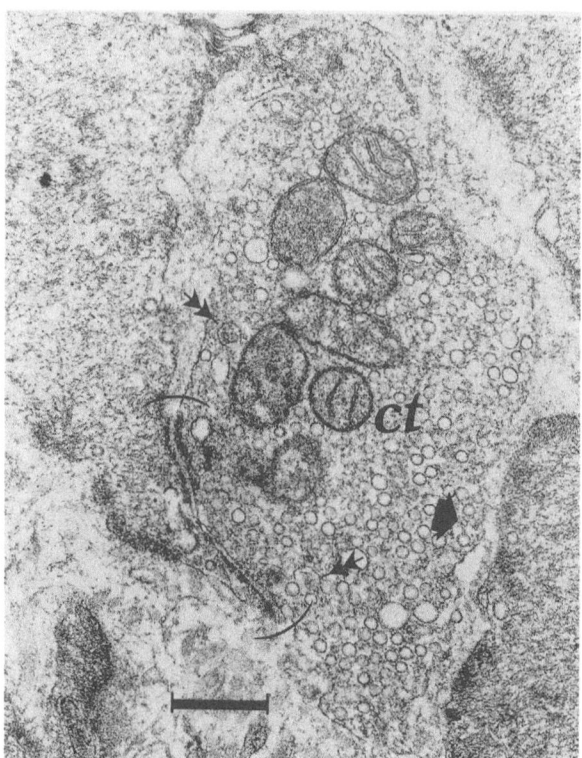

Fig. 2.10. Cholinergic neuroeffector junction, cat bladder base. Cholinergic axon terminal (*ct*), packed with small clear (agranular) vesicles, contains abundant mitochondria and a few large dense-core (granular) vesicles (*double arrows*) and forms close neuroeffector junction (15 nm cleft; 1 μm contact zone: *between parentheses*) with one muscle cell and en passant neuroeffector junction (44–70 nm cleft) with another. Apposed axolemma and sarcolemma at close junction are symmetrically thickened to form adherens junction. Scale = 400 nm. (Elbadawi 1982a; © The Williams & Wilkins Co., Baltimore, Maryland)

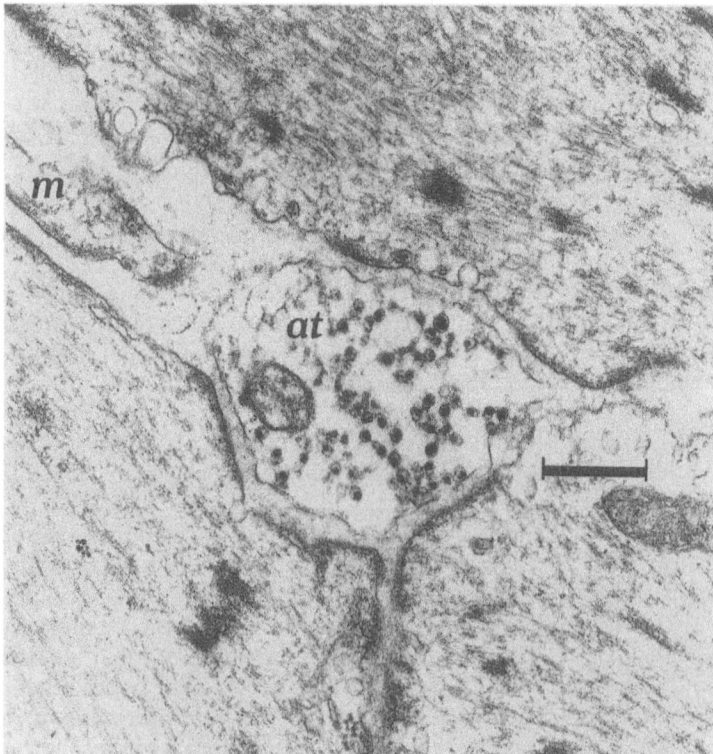

Fig. 2.11. Adrenergic neuroeffector junction, rat bladder base. Adrenergic axon terminal (*at*) contains abundant small dense-core (granular) vesicles and a mitochondrion, forms neuroeffector junctions with three smooth muscle cells (30–60 nm clefts; 0.8–1 μm contact zones) and—in serial sections—with a fourth cell (*m*). Scale = 400 nm. (Elbadawi 1982a; © The Williams & Wilkins Co., Baltimore, Maryland)

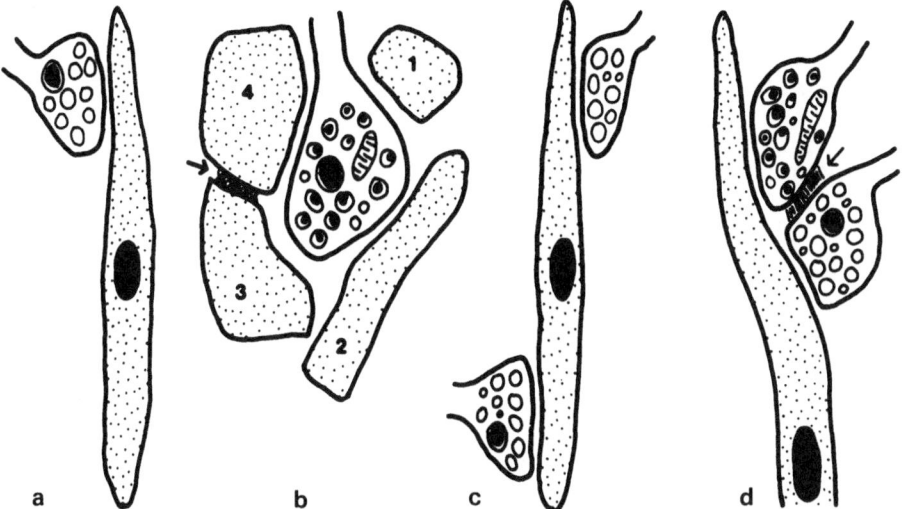

Fig. 2.12a–d. Types of neuroeffector junctions in vesicourethral muscularis. **a** Cholinergic unitary junction. **b** Adrenergic multi-effector junction—close with muscle cells *2* and *4*, almost close with cell *1*, and en passant with cell *4*; *arrow* marks cell junction between *3* and *4*. **c** Multiaxonal cholinergic junction. **d** Multiaxonal diautonomic junction, with axoaxonal synapse (*arrow*) between cholinergic and adrenergic axon terminals. (Elbadawi 1983)

of both the cat and rat (Elbadawi 1984; Mattiasson et al. 1987). Such inhibition has already been demonstrated pharmacologically in many smooth muscle systems (Westfall 1977), as well as the urethral muscularis of man and rabbit (Mattiasson et al. 1984).

Intrinsic Nonadrenergic Noncholinergic Nerves?

Evidence has been presented both in favour of and against physiological excitatory neurotransmission by the purine adenosine triphosphate (ATP) in the vesicourethral muscularis (for review see Elbadawi 1982b). It is doubtful that "purinergic nerves" exist as a distinct and separate entity in this muscularis (Elbadawi 1982b). It is possible, if not likely, that ATP has a cotransmitter role in adrenergic and/or cholinergic neuroeffector transmission in the muscular apparatus of micturition, since it occurs naturally in both autonomic axon types innervating various mammalian smooth muscle systems (Nathanson 1977; Richards and da Prada 1977; Iijma 1981). A cotransmitter role of ATP has already been demonstrated pharmacologically in the guinea pig vas deferens (Sneddon and Westfall 1984).

Neuropeptide-containing nerves have been identified immunohistochemically in the musculature of the lower urinary tract of humans and various animals. VIP-immunoreactive nerves are present in the bladder base, in the "trigone", around the ureteric orifices, and at the urethrovesical junction in both the cat and guinea pig (Alm et al. 1977; Alumets et al. 1980). "VIPergic" muscular nerves are present throughout the human bladder, and appear to be more dense in the trigonal part of the base detrusor than in the bladder dome (Gu et al. 1984; Islam et al. 1984). Substance P-immunoreactive nerves are present in the suburothelium and around muscle cells of the proximal urethra of the guinea pig, and in the human body detrusor (Alm et al. 1977; Gu et al. 1984). Neuropeptide Y-immunoreactive nerves have also been demonstrated in the posterior base detrusor, and in a greater density in the body detrusor of the human bladder (Gu et al. 1984; Islam et al. 1984).

Neuropeptides probably act as cotransmitters in or modulators of adrenergic or cholinergic neuroeffector transmission (Cuello 1982; Lundberg et al. 1982; Lundberg and Hökfelt 1983). They are probably localized in large dense-core axonal vesicles which occur in small numbers in both cholinergic and adrenergic axon terminals and varicosities (Larsson 1977; Chan-Palay et al. 1978; Atta et al. 1984).

Intrinsic Afferent Innervation

So far, there is no light or electron microscopic method that demonstrates preferentially, and with

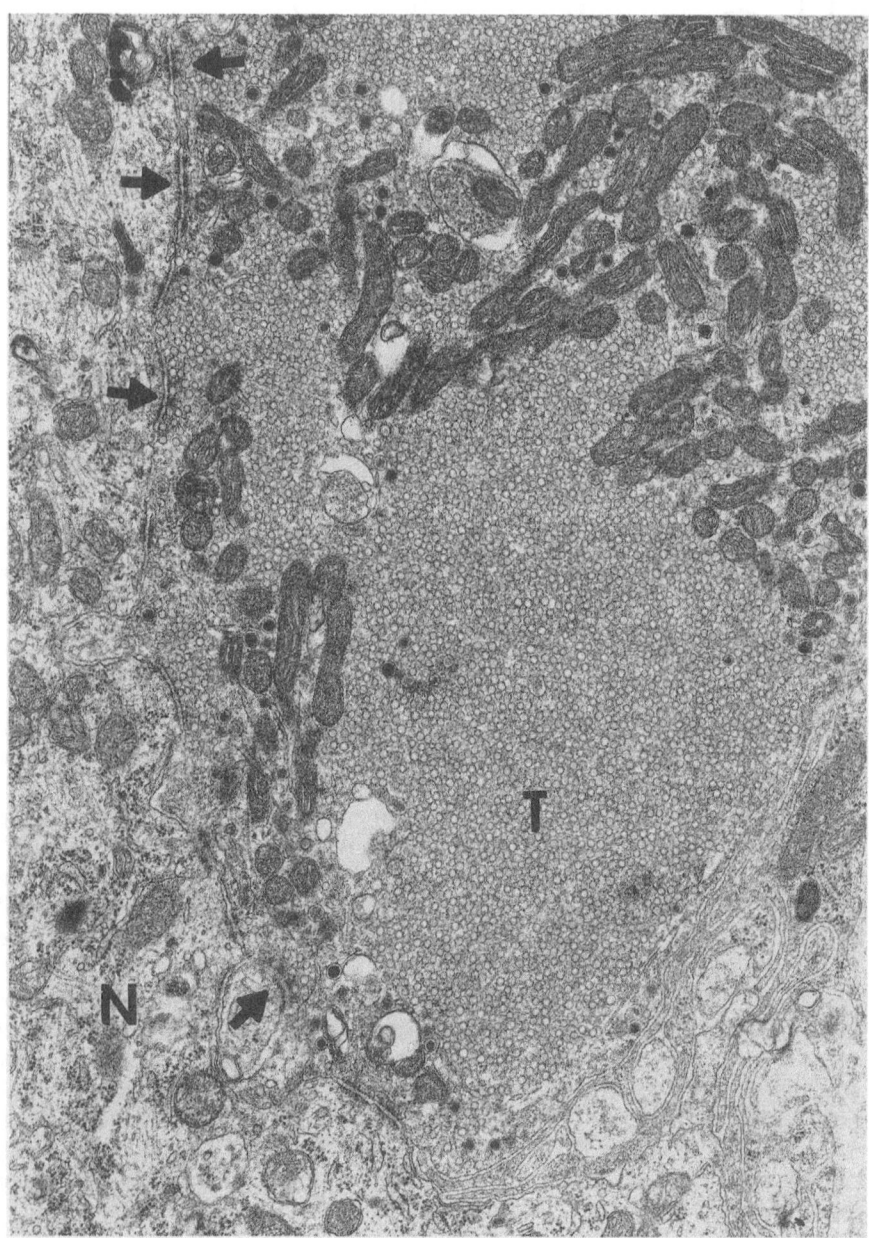

Fig. 2.13. Preganglionic synapse, pelvic plexus ganglion, cat. Preganglionic cholinergic terminal *T* is packed with small clear (agranular) vesicles, contains abundant mitochondria and some large dense-core vesicles (*arrowheads*), and establishes a series of chemical synapses with adjacent principal neuron *N* (*arrows*). These synapses have periodic pyramidal membrane densities on the axonal, and irregular flat membrane density on the neuronal cell body side. Additional short areas of symmetrical axolemmal and neuronolemmal thickening represent puncta adherentia. (× 23 400)

certainty, afferent autonomic nerves (Elbadawi 1982b, 1987). A variety of capsulated and non-capsulated free-ending sensory bodies have been described in the suburothelium and muscularis of the lower urinary tract in silver-stained preparations (Yoshida 1957; Seto 1963; Kuru 1965;

Bradley and Teague 1968; Stach et al. 1970; Fletcher and Bradley 1978). It is known, however, that silver staining techniques have no *specific* affinity for any particular nerve entity (Elbadawi 1982b).

The sensory body that can be identified unequivocally by both light and electron microscopy in

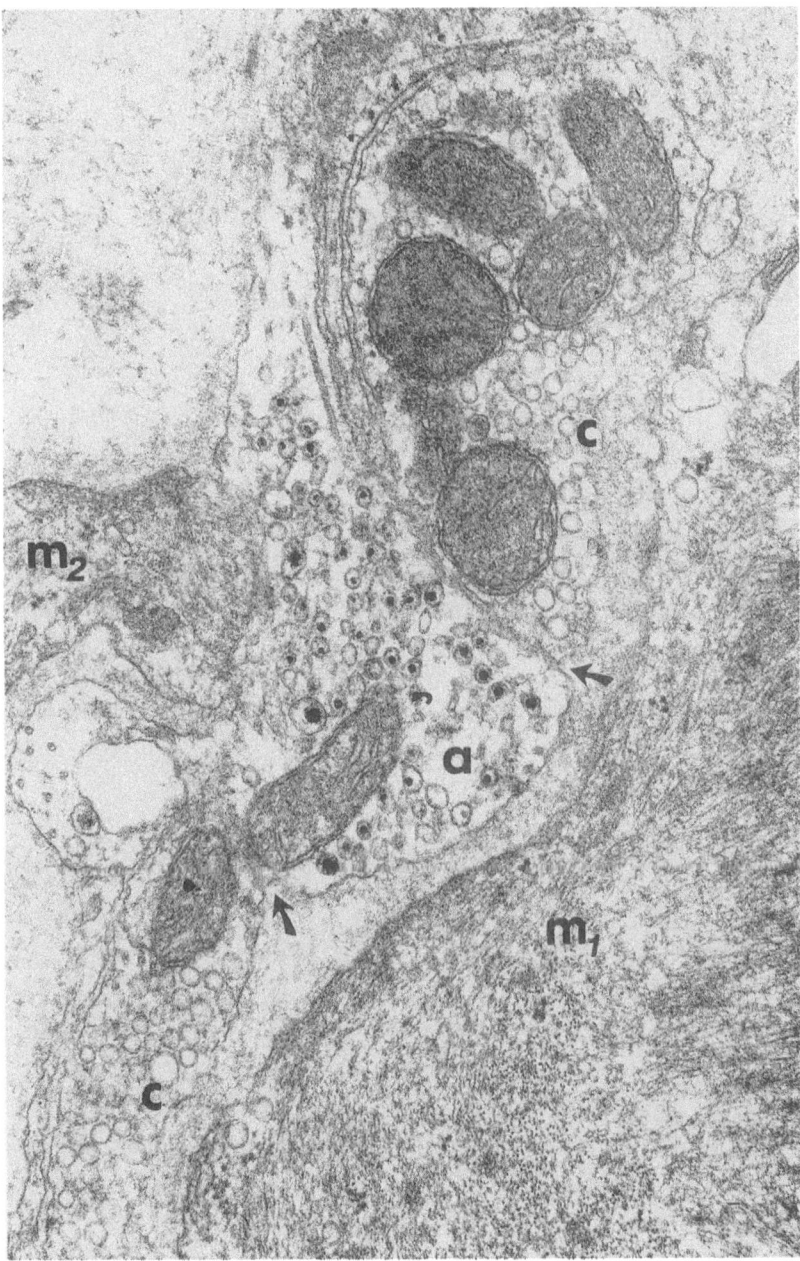

Fig. 2.14. Neuroeffector junctions with axoaxonal synapses in rat bladder base. Two profiles of same cholinergic axon (*c*) form en passant neuroeffector junction with muscle cell m_1; adrenergic axon profile (*a*) forms en passant neureffector junction with m_1 and close junction with process of another muscle cell (m_2); axoaxonal synapse between axon terminals a and c marked with *arrows*. (Elbadawi 1984; © The Williams & Wilkins Co., Baltimore, Maryland.) (× 47 500)

viscera is Pacini's corpuscle, which is relatively large and has a characteristic onionskin-like capsule, with multiple relatively widely spaced lamellae. A cholinergic axon, an adrenergic axon, or both, run vertically in the core of the corpuscle with fine branching between its lamallae. Pacini's corpuscles are not uncommon in the outer muscularis and adventitia of the base detrusor, the vesicourethral junction, the proximal urethra, and the ureterovesical junction, in various mammals—including humans (A. Elbadawi and E.A. Schenk 1967, unpublished work).

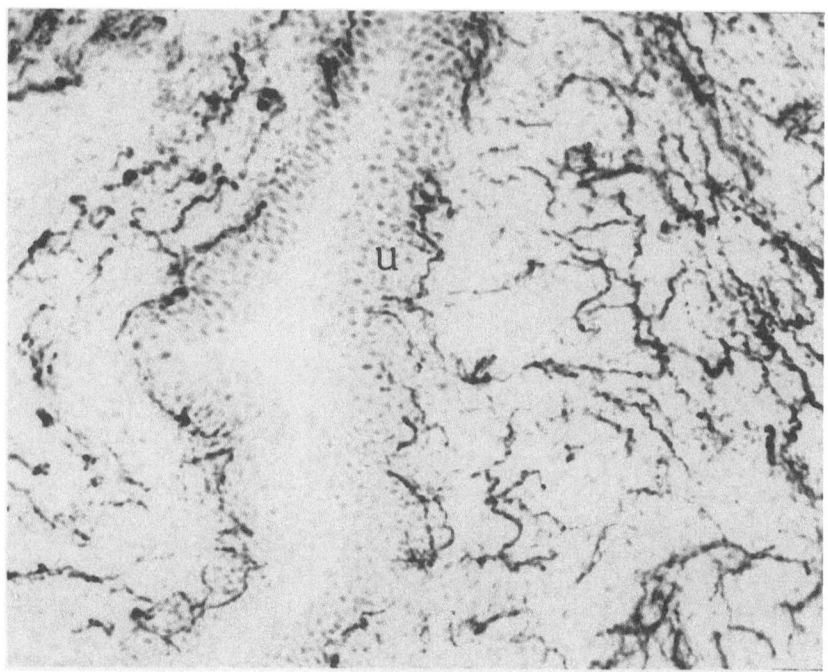

Fig. 2.15. Cholinergic suburothelial plexus, cat bladder base. Cholinergic nerve fibres form rich suburothelial plexus; branches of this plexus penetrate and apparently end within urothelium (*u*).

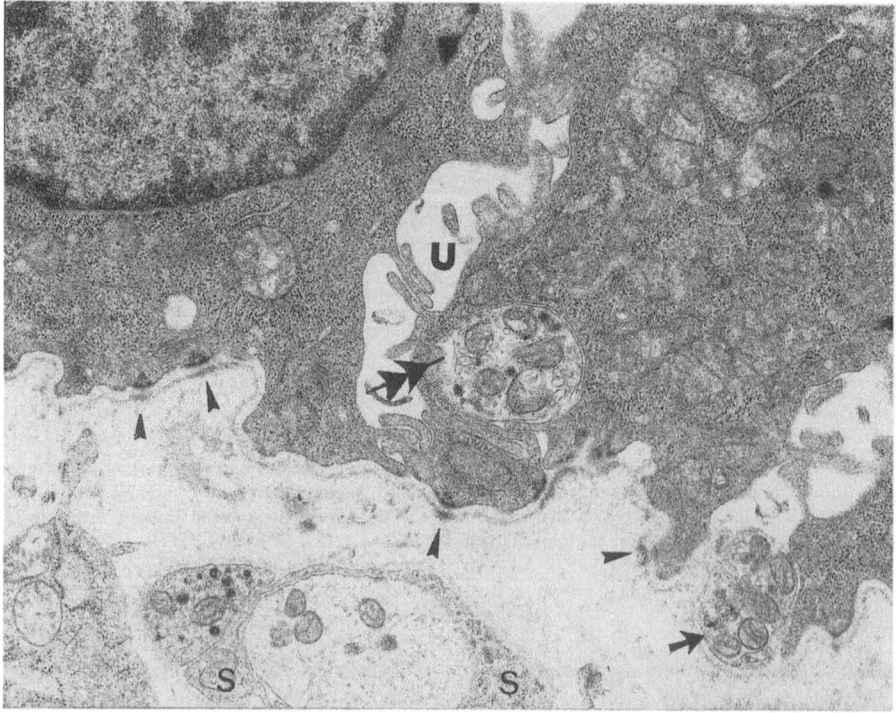

Fig. 2.16. Urothelial innervation, cat bladder base. Schwann cell (*s*) ensheathes adrenergic axon profile and intervaricose segment in suburothelium; basement membrane of urothelium (*u*) with desmosomes marked with *arrowheads*; one adrenergic profile lies superficial to basement membrane in close contact with urothelial cell (*arrow*); another forms invaginated close junction (18 nm cleft) with urothelial cell (*double-headed arrow*). (Elbadawi 1982b.) (× 19 200)

Suburothelial plexuses of cholinergic and adrenergic nerves have been demonstrated histochemically in the bladder and urethra of various nonhuman mammals (Elbadawi and Schenk 1966, 1971a, 1974; Gosling and Dixon 1974; Elbadawi and Goodman 1980; Elbadawi 1982b). These plexuses comprise neuroplexuses of suburothelial vessels, as well as vessel-independent free-ending axons, the branches of which frequently terminate within the urothelium (Figs. 2.15, 2.16) (Elbadawi and Schenk 1966; Elbadawi 1982b). The cholinergic and adrenergic suburothelial plexuses are commensurate in density, which varies regionally. Both are sparse in the dome, progressively become more prominent as the detrusor is followed distally towards the bladder base region, become uniformly rich in the bladder base, and then slightly less prominent in the subvesical portion of the urethra (Elbadawi and Schenk 1966; Elbadawi 1982b). The urethral suburothelial plexuses contain acetylcholinesterase-positive capsulated endings (Elbadawi and Schenk 1974), similar to genital corpuscles (Seto 1963).

Afferent fibres within the cat detrusor appear to be connected to both the lumbar sympathetic and sacral parasympathetic segments of the spinal cord, and to originate as free-ending axons and not in capsulated end-bodies (Uemura et al. 1973, 1974, 1975). According to these authors, sacral afferents are more abundant in the muscularis than the suburothelium, and have a uniform distribution throughout the bladder; lumbar afferents, on the other hand, appear to be localized primarily in the dorsal trigonal and anterior bladder neck regions, and to be more abundant in the suburothelium than the muscularis.

Anatomy of Peripheral Neural Pathways

Pelvic Plexus

The pelvic plexus is orientated in the sagittal plane and has a similar general topography in humans and the other animal species so far studied. The plexus is formed by the hypogastric and pelvic nerves, with caudal contributions from the pudendal nerves (Woźniak and Skowrońska 1967), and incorporates named and/or unnamed ganglia. Branches of the pelvic plexus innervate the lower urinary tract as well as the internal genital organs of both sexes. These branches convey afferent parasympathetic- and sympathetic-associated fibres from, and efferent parasympathetic (cholinergic)

and sympathetic (adrenergic) fibres to the target organs (Kuntz 1945; Mitchell 1953; Elbadawi and Goodman 1980; Elbadawi 1983). Whether afferent or efferent, the fibres contained in branches of the pelvic plexus are distributed to the ipsilateral, and to some extent, the contralateral side of the bladder and urethra across their midline planes, especially dorsally (Uemura et al. 1973, 1974, 1975; Elbadawi et al. 1984).

Four anatomical patterns of the mammalian pelvic plexus can be recognized, depending on the origin of the hypogastric nerves, the location and nature of their junctions with the pelvic nerves, and the topographical arrangement of plexus branches and ganglia (Mitchell 1953; Langworthy 1965; Sjöstrand 1965; Woźniak and Skowrońska 1967; Schnitzlein et al. 1972; Purinton et al. 1975; Baljet and Drukker 1980; Elbadawi and Goodman 1980; Elbadawi 1983; A. Elbadawi, J.W. Downie and M.A. Atta 1984, unpublished work). Ganglia associated with the pelvic plexus include main trunk ganglia (related to the pelvic nerve trunk and the distal portion of the hypogastric nerve), and pelvic plexus ganglia located along its ramifications.

Human Pattern

The human pelvic plexus (inferior hypogastric plexus; plexus of Frankenhäuser) is represented by a three-dimensional network of interconnected nerve strands within the pelvic fascia lateral to the rectum (Mitchell 1953). The right and left plexuses are interconnected freely behind the rectum. Each plexus can roughly be separated by gross dissection into dorsal rectal, cranial ureterovesical and caudal vasourethroprostatic or urethrouterocervical components.

Unlike nonhuman mammals, the inferior mesenteric ganglion is either absent or rudimentary. On each side, the hypogastric nerve arises as a distal subdivision of the superior hypogastric plexus, and is more commonly represented by a narrow elongated nerve meshwork than a single nerve trunk. Visceral branches of the S_{2-4} ventral roots coalesce to form the pelvic nerve, which is represented on each side by 3–4 trunks. The hypogastric and pelvic nerves of each side converge on, and end in a grossly recognizable "plate" of large ganglia that are collectively known as the hypogastric ganglion, the peripheral branches of which form the pelvic plexus.

Feline Pattern

The pelvic plexus of the cat is formed and ramifies in two closely apposed layers of the pelvic fascia, which is cushioned ventrally by a thick pad of fat. The pelvic nerve is formed as a single trunk by visceral branches of the S_2 and S_3 ventral spinal roots and divides into several branches shortly beyond its origin. The right and left hypogastric nerves arise as separate single trunks from a sizable inferior mesenteric ganglion. This ganglion is located in the colonic mesentery, around the origin of the inferior mesenteric artery, and is composed of 3–5 interconnected subunits. As each hypogastric nerve crosses the cranial vesical artery, it divides into a dorsal and a ventral branch. The former courses laterally to join the pelvic nerve or one of its branches close to its origin from the nerve. The ventral branch runs caudally in a more superficial fascial plane to join medially coursing branches of the pelvic nerve. Subsequent branches of the hypogastric and pelvic nerves ramify, mingle and unite to form the pelvic plexus.

The right and left pelvic plexuses are interconnected dorsal to the urogenital organs which they innervate. Two interconnected divisions of each plexus may be discerned by gross dissection. The cranioventral division innervates the uterine cornua and fundus in females, and the urinary bladder, pelvic ureter and ureterovesical junction in both sexes. The dorsocaudal component innervates the urethra in both sexes, and the prostate and penis or the corpus uteri, cervix uteri and vagina.

The pelvic plexus of the cat does not have a discrete hypogastric ganglion that marks the junction of the hypogastric and pelvic nerves, as in humans. Instead, the plexus incorporates many small and microscopic ganglia along its ramifications, especially at the points of their union or bifurcation. A ganglion located at the origin of the pelvic nerve, and another at the termination of the ventral branch of the hypogastric nerve, are constantly present and can be discerned by gross dissection under magnification (M.A. Atta, J.W. Downie and A. Elbadawi 1983, unpublished work).

The pelvic plexus of the rabbit is similar to that of the cat, but occupies a relatively larger and anatomically less defined area of the pelvic fascia. In the rabbit, the hypogastric nerves arise from the inferior mesenteric ganglion as a single trunk that shortly bifurcates into a right and a left branch.

Canine/Primate Pattern

In both the dog and monkey, the pelvic nerve is formed by visceral branches of the S_2 and S_3 ventral roots, as a single trunk in the former and a double trunk in the latter species. In both species, the hypogastric nerves arise as separate right and left trunks from the inferior mesenteric ganglion. Unlike the cat, and similar to the human, the point at which the hypogastric and pelvic nerves meet, away from the urogenital organs, is marked by a sizable and grossly recognizable hypogastric ganglion.

Rodent Pattern

In both the rat and guinea pig, the hypogastric nerves originate as separate trunks from a small, but grossly discernible inferior mesenteric ganglion. The pelvic nerve is formed as a single trunk by visceral branches of S_2 and S_3 ventral spinal roots. In the male rat, the hypogastric and pelvic nerves converge on and join a single ganglion (the major pelvic ganglion), which is adherent to the cranial side of the dorsal surface of the lateral prostatic lobe. The corresponding ganglion in the male guinea pig is represented by an anterior and a posterior pelvic ganglion, which are adherent, respectively to the cranial and caudal parts of the dorsolateral surface of the coagulating gland. The hypogastric ganglion in the female rodent (ganglion of Frankenhäuser) is much smaller than in the male, and is adherent to the dorsolateral surface of the uterine cervix or the uterocervical junction. Because of the presence and location of the hypogastric ganglion, the pelvic plexus of the rat has a very simplified anatomical structure, with very short branches that have limited ramifications and interconnections.

Ganglia

In addition to main trunk ganglia (including the inferior mesenteric ganglion when present), pelvic plexus ganglia, and the hypogastric ganglion when present, smaller ganglia and individual neuronal cell bodies are constantly associated with nerves on the surface of and within the bladder and urethra, including their fine ramifications within the muscularis and suburothelium. These mural ganglia, together with the hypogastric and pelvic plexus ganglia, form the urogenital short neuron system (Elbadawi and Schenk 1970, 1971a, 1974; Elbadawi and Goodman 1980; Elbadawi 1982b, 1983).

Mural ganglia vary in number, distribution and complexity of structure in different species, in different regions of the bladder and urethra, and in different tissue planes in both organs. The ganglia are most abundant, have the widest distribution, and are most complex in arrangement in the cat. In both the bladder and urethra, the ganglia abound in the adventitia, and progressively become less numerous and more restricted in distribution in the outer muscularis, inner muscularis, and suburothelium.

In both the male and female cat, mural ganglia of the bladder and urethra can be grouped as ureterovesical, vesicourethral, vesical, and ureteric ganglion complexes, that are freely interconnected by mural ramifications of branches of the pelvic plexus and have overlapping target areas of innervation (Fig. 2.17; Elbadawi and Schenk 1971a, 1974). The ureterovesical ganglion complex is the most prominent and is located at the ureterovesical junction, both outside and within the periureteric sheath. This complex innervates the base detrusor, the periureteric sheath, the ureterotrigone and the caudal part of the body detrusor. The vesicourethral ganglion complex supplies both the base detrusor and urethral components of the lissosphincter. Ganglia of this complex are more abundant at the vesicourethral junction than in relation to either the base detrusor or urethra. Ganglia of the vesical complex innervate the body detrusor, and become progressively less abundant and more limited in distribution as the bladder body is followed towards the dome, in which virtually no ganglion cells are present. The ureteric complex has a limited number and distribution of ganglia, and innervates the

pelvic ureter beyond the periureteric sheath. In the male cat, numerous mural ganglia are present in relation to, and innervate the terminations of the vasa deferentia, distal portion of the preprostatic urethra, and the prostate, collectively forming the vasourethroprostatic ganglion complex.

Mural vesicourethral ganglia in both the dog and pig are less abundant and more restricted in distribution than in the cat. In both the rat and rabbit, these ganglia are very small and sparse, even at the ureterovesical junction. Because of the immediately subvesical location of the prostate in the dog, pig, rat and rabbit, the ganglia that innervate the prostate and terminal vasa deferentia (plus the seminal vesicles and coagulating glands when present) belong to the vesicourethral ganglion complex.

The human bladder has a modest content of mural ganglia relative to the large size of the organ, as judged in diagnostic biopsies and radical cystectomy specimens submitted for histopathological study. In my experience, and according to a recent study (Dixon et al. 1983), these ganglia, as in nonhuman mammals, are concentrated at the ureterovesical and around the vesicourethral junctions, and are present in decreasing numbers in the adventitia, muscularis and suburothelium of the bladder and urethra.

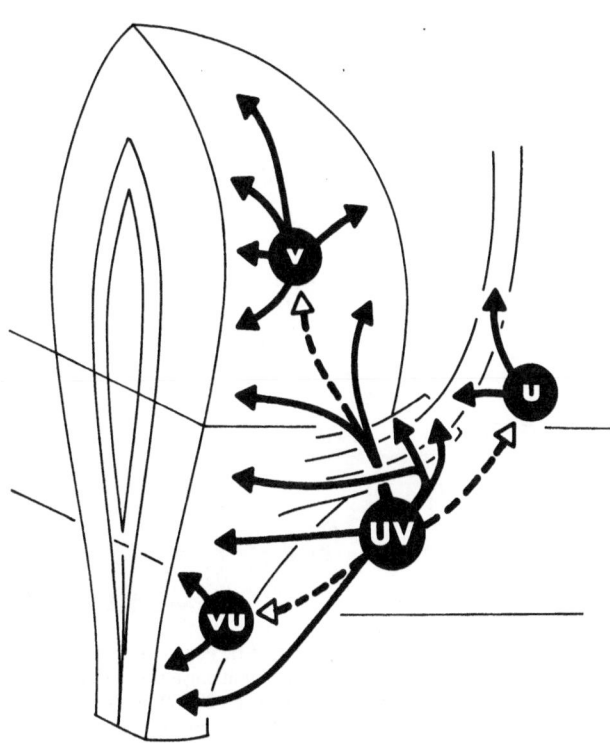

Fig. 2.17. Individual ganglion cells, cell clusters and ganglia innervating cat bladder and urethra can be segregated as four intercommunicating ganglion complexes: *uv*, ureterovesical; *vu*, vesicourethral; *v*, vesical; *u*, ureteric complexes (see text for details). (Elbadawi and Schenk 1971a)

Accessory Efferent Pathways

McCrea and Kimmel (1952) described an accessory efferent pathway to the human bladder that is closely associated with vesical blood vessels, and is derived from the S_{2-4} ventral spinal roots as well as the corresponding segments of the sympathetic chains. This pathway presumably reaches the bladder at the ureterovesical junction, bypassing the pelvic plexus, and was proposed as a source of innervation that can maintain vesical function following interruption of the pelvic plexus by extirpative pelvic surgery. Some authors have described contributions from the sacral sympathetic chain to the human pelvic plexus (Woźniak and Skowrońska 1967), and to the pelvic nerve of the cat (Downie et al. 1984; Kuo et al. 1984).

Microstructure of Peripheral Neural Pathways

Pelvic Plexus

In nonhuman mammals, and probably also in humans, both the hypogastric and pelvic nerve trunks contain cholinergic and adrenergic fibres. The cholinergic fibres in the pelvic nerve are preganglionic efferent parasympathetic, and those in the hypogastric nerve include preganglionic efferent sympathetic as well as postganglionic cholinergic sympathetic fibres; the last probably mediate prostatic secretion (Smith 1975; Elbadawi and Goodman 1980). Adrenergic fibres, which by definition are postganglionic sympathetic in origin, con-

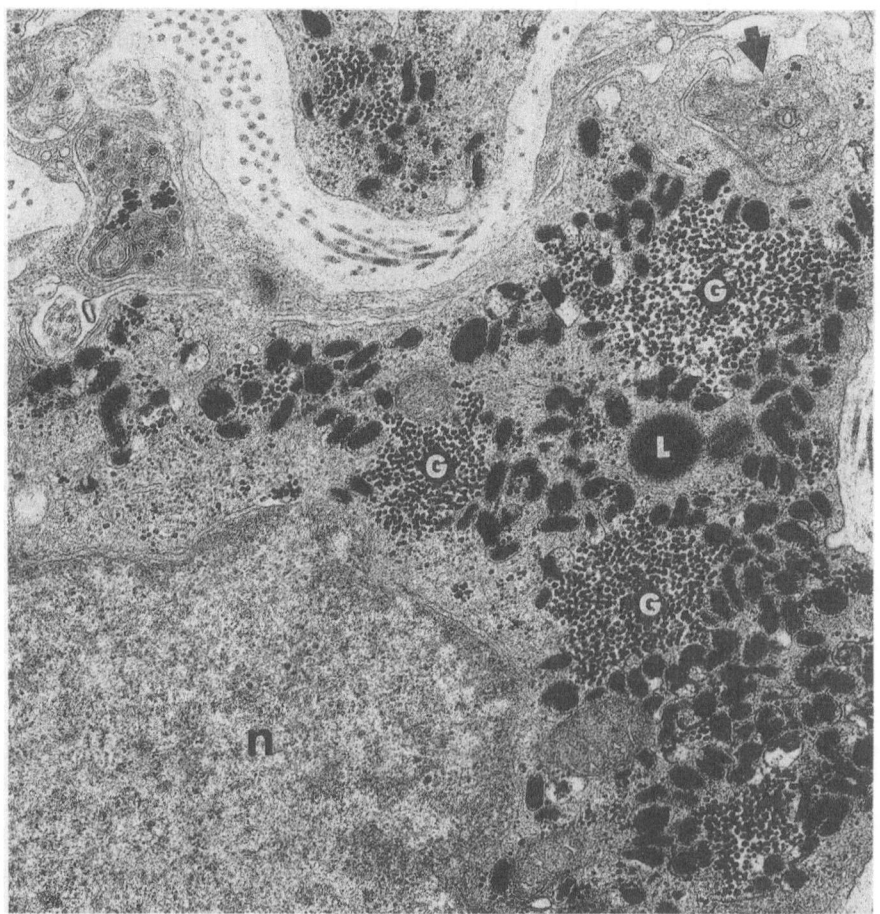

a

Fig. 2.18a, b. Types of neurons, hypogastric ganglion, male rat. **a** Profile of SIF cell is packed with pleomorphic large vesicles with intensely dense cores and contains abundant clusters of glycogen particles (*G*). *Arrow* marks cholinergic preganglionic synapse with cell; *n*, nucleus; *L*, lipofucsin dense body. (× 29 900.) **b** Profile of principal neuron (cholinergic or adrenergic) has abundant sacs of endoplasmic reticulum, extensive number of ribosomes and numerous lipofucsin dense bodies. *n*, nucleus (with prominent nucleolus); *N*, adjacent principal neuron; *s* satellite cells of neurons. (× 4480)

stitute 40%–60% of the hypogastric and 5%–10% of the pelvic nerve in the cat (Sundin and Dahlström 1973). The relative proportions of the different axon types in the hypogastric nerve remain to be studied in other species, including humans.

Based on their connections to the spinal cord, branches of the pelvic plexus contain preganglionic efferent, postganglionic efferent, and afferent, sympathetic and parasympathetic, nerve fibres (Kuntz 1945; Mitchell 1953; Owman and Sjöstrand 1965; Elmér 1975; Elbadawi 1982b, 1983, 1987). Ultrastructurally, these branches are composed of both myelinated and nonmyelinated Schwann cell-ensheathed axons (Elbadawi 1982b, 1983; Donker 1986). It may be assumed that the myelinated fibres are afferent, since efferent autonomic axons are nonmyelinated (Kuntz 1945; Elbadawi 1982b).

Ganglia

The larger ganglia of the urogenital short neuron system, including pelvic plexus ganglia, the hypogastric ganglion (when present) and adventitial mural ganglia, consist of partially segregated clusters of principal (cholinergic and adrenergic) and SIF neuronal cell bodies, that can be distinguished both histochemically and ultrastructurally (Fig. 2.18; Owman and Sjöstrand 1965; Elbadawi and Schenk 1968b, 1970, 1971a, 1974; Elbadawi and Goodman 1980; Elbadawi 1982b, 1983). The relative proportions of the neuronal cell types vary in different ganglia, in the same ganglion in different species, and even between the right and left sides at the same location in the same animal (Elbadawi and Goodman 1980). Intramural vesicourethral ganglia

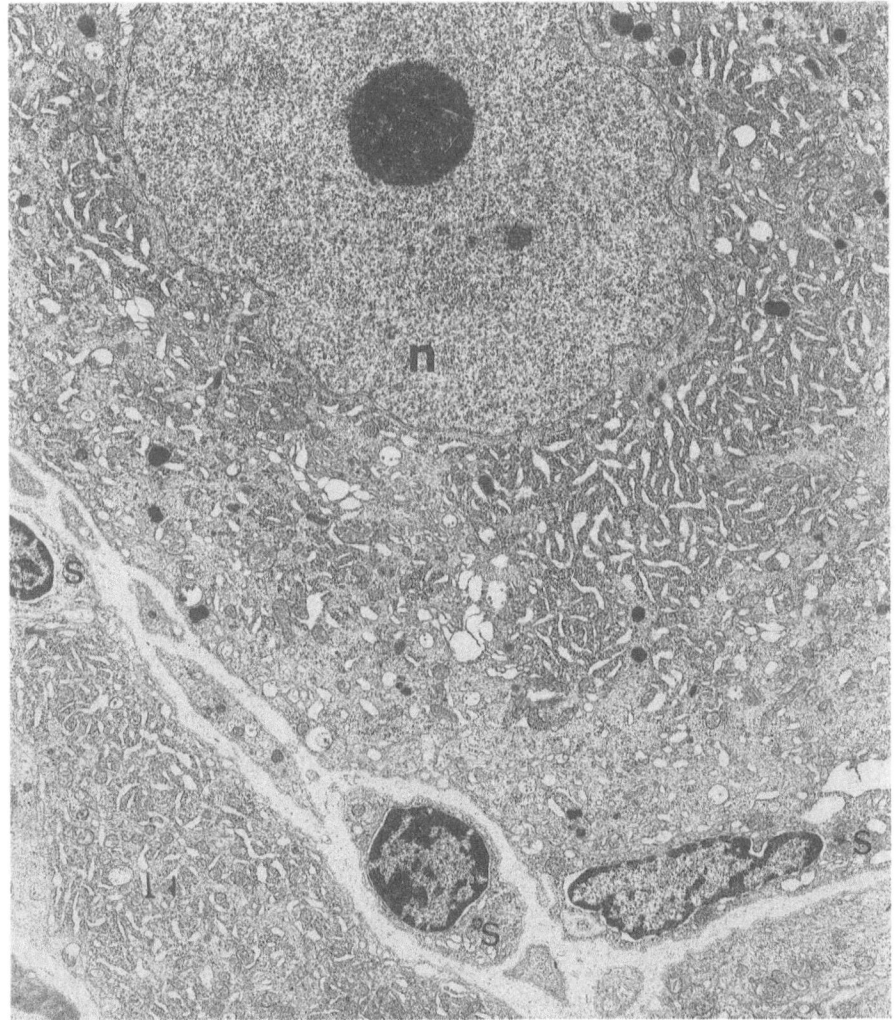

2.18b

are microscopic in size, and usually consist of pure clusters of cholinergic or adrenergic neuronal cell bodies (Elbadawi and Schenk 1968b, 1971a, 1973; Elbadawi 1982b). Rare SIF cells are present in the vesicourethral muscular neuroplexuses of both cat and rat (Elbadawi 1982b, 1985).

The adrenergic and cholinergic cell bodies in peripheral vesicourethral ganglia represent postganglionic neurons that respectively supply effector sympathetic and parasympathetic innervation to the bladder and urethra. The SIF cells in these ganglia are probably true interneurons that originate and end in the same ganglion (Elbadawi 1982b, 1985).

Intraganglionic Synapses

Numerous cholinergic and adrenergic nerve fibres can be identified both histochemically and ultrastructurally within ganglia of the urogenital short neuron system. Many of these fibres are cholinergic preganglionic sympathetic or parasympathetic in nature, and end at preganglionic synapses (Fig. 2.19) with neurons of, or merely traverse each ganglion. Many others originate in each ganglion as cholinergic postganglionic parasympathetic and/or adrenergic postganglionic sympathetic fibres. Still other intraganglionic nerve fibres establish postganglionic synapses with some (possibly all) prin-

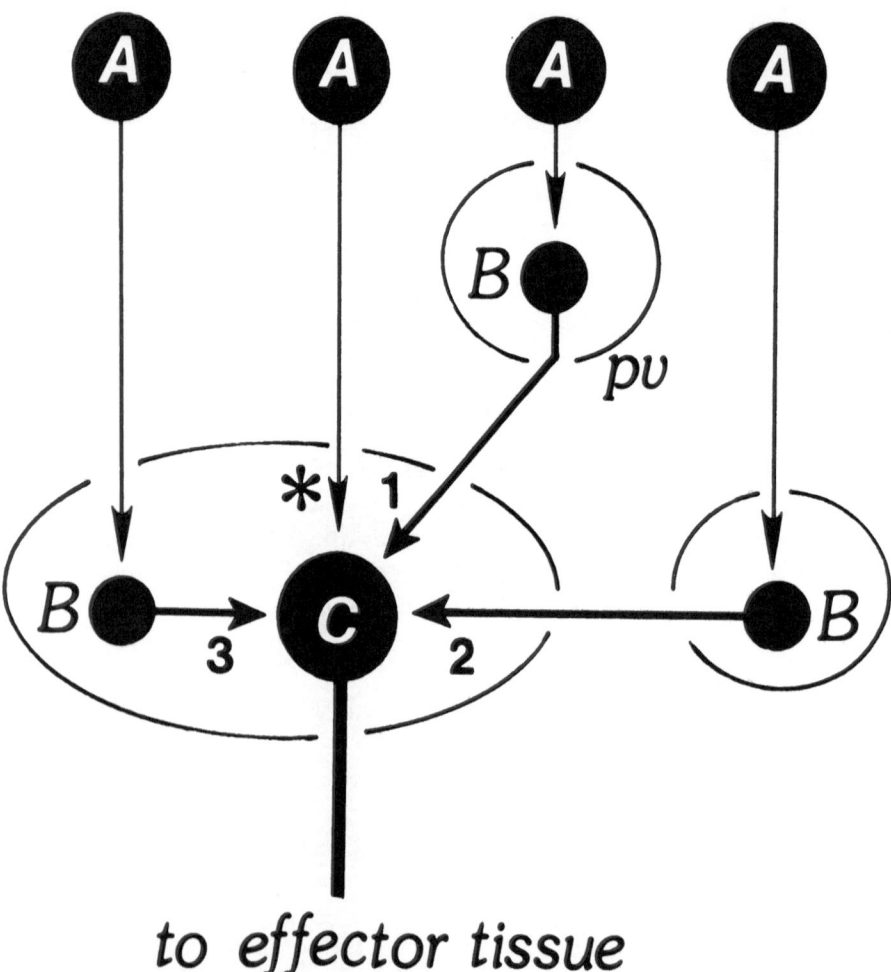

Fig. 2.19. Concept of postganglionic synapses in vesicourethral innervation. *A* represents spinal-cord derived preganglionic neuron; *asterisk* marks preganglionic synapse with effector neuron *C*. *B* represents modulator neuron: *1*, Sympathetic adrenergic origin in paravertebral ganglion (*pv*); *2*, sympathetic adrenergic, parasympathetic cholinergic or ? peptidergic terminal originating in centripetally located ganglion; *3*, sympathetic adrenergic, parasympathetic cholinergic or ? peptidergic terminal originating in same ganglion (interneurons). *C* represents postganglionic effector neuron. (Elbadawi 1983)

cipal cholinergic and/or adrenergic neurons in the same ganglion (Fig. 2.19; Elbadawi and Schenk 1971a, 1973; Buzelin 1984). Postganglionic synapses established by adrenergic fibres have been confirmed histochemically and ultrastructurally in several species (Fig. 2.19; see also Fig. 2.14; Elbadawi and Schenk 1971b, 1973: Elbadawi and Goodman 1980; Elbadawi 1983). On the other hand, cholinergic postganglionic synapses have been described histochemically (Elbadawi and Schenk 1973; Elbadawi 1983) and suspected pharmacologically (Saum and de Groat 1972; Taira 1972; de Groat and Saum 1976), but remain to be proven ultrastructurally. Intraganglionic SIF cells have cholinergic preganglionic (afferent) synapses and establish postganglionic (efferent) synapses with the soma or preganglionic fibre innervating some principal neurons of peripheral autonomic ganglia (Fig. 2.19), including those of the urogenital short neuron system (Elbadawi and Goodman 1980; Elbadawi 1983). In addition to cholinergic and adrenergic fibres, some nerve fibres within peripheral vesicourethral ganglia contain neuropeptides, especially enkephalin (de Groat and Kawatani 1985).

Innervation of Urethral Rhabdosphincter

Histochemistry of Intrinsic Innervation

The male feline rhabdosphincter has triple somatomotor, autonomic cholinergic and autonomic adrenergic innervation of its fibres (Fig. 2.20) (Elbadawi and Schenk 1974). The somatomotor component is in the form of circumscribed, ovoid and strongly acetylcholinesterase-positive motor end-plates. These plates are smaller, but much more densely distributed per unit area than other striated muscles, including the diaphragm, intercostal muscles and extraocular muscles (A. Elbadawi and E.A. Schenk 1973, unpublished work). The autonomic component is represented by intricate cholinergic and adrenergic neuroplexuses that are derived from the corresponding plexuses innervating the urethral smooth muscularis, and are independent of, but are connected with vascular neuroplexuses. Both the cholinergic (Fig. 2.20a) and adrenergic (Fig. 2.20b) neuroplexuses in the rhabdosphincter have ramifications that course between its myofibres, encircling some individually.

Ultrastructure of Intrinsic Innervation

Adrenergic neuroeffector junctions (~ 40 nm separation gaps) with myofibres of the rhabdosphincter have been documented in the rat (Watanabe and Yamamoto 1979). Triple innervation of the rhabdosphincter has recently been confirmed ultrastructurally in the normal and decentralized male cat (Atta and Elbadawi 1985; Elbadawi and Atta 1985, 1986). Similar innervation of the rhabdosphincter in other mammals, including humans, remains to be proven. Proof of triple innervation in humans may not be attainable, even in the distant future, because of the practical impossibility of obtaining fresh samples of the normal rhabdosphincter that are suitable for electron microscopy. Based on recent observations in the cat (Atta and Elbadawi 1985; Elbadawi and Atta 1985), samples of the human rhabdosphincter obtained surgically or transurethrally for therapeutic purposes cannot be regarded as normal.

The male feline rhabdosphincter is innervated by interstitial nerve bundles composed of myelinated and nonmyelinated Schwann cell-ensheathed cholinergic and adrenergic axons, with a less prominent probable copeptidergic component (Elbadawi and Atta 1985). Axons approaching myofibres of the rhabdosphincter establish two ultrastructural types of neuromuscular relationship, namely, motor end-plates and nonspecialized, surface, axon/myofibre junctions (Elbadawi and Atta 1985).

The motor end-plates are considered to be somatomotor in nature, since they have the classical ultrastructural features of neuromuscular junctions in other striated muscles (Fig 2.21; Elbadawi and Atta 1985). These features are a cholinergic axon terminal apposed to a sole plate in the related myofibre, with a 40–80 nm wide gap (primary synaptic cleft). The axon terminal is Schwann cell-ensheathed on the side away from the myofibre, is packed with small clear (agranular) vesicles, and contains occasionally numerous mitochondria, some glycogen particles, and usually a few large vesicles with opaque cores. The sole plate has multiple sarcoplasmic junctional folds that are covered by thickened sarcolemma and are separated from each other by secondary synaptic clefts. In both the primary and secondary synaptic clefts, there is an intermediate electron-dense line, which represents the fusion of the basal laminae of the myofibre and the Schwann cell of its related axon terminal. Sole plates constantly occur adjacent or close to nuclei of the myofibres. The sarcoplasm at the sole plate commonly contains abundant mitochondria, sar-

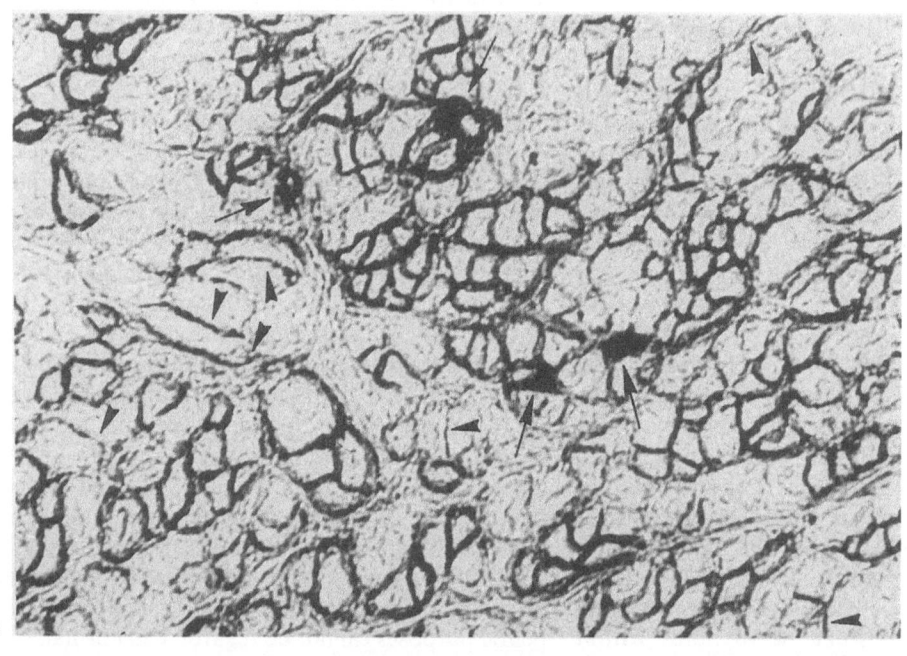

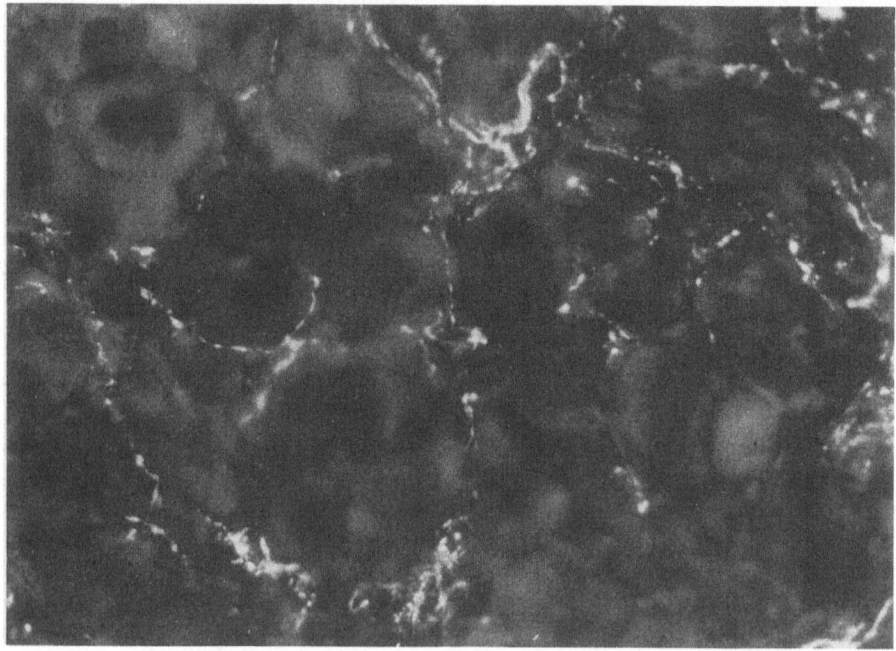

Fig. 2.20. a Cholinergic innervation of cat rhabdosphincter; acetylcholinesterase-positive elements include motor end-plates (*arrows*), individual cholinergic nerve fibres (*arrowheads*) and sarcolemmal staining. Individual cholinergic fibres are independent of blood vessels. **b** Adrenergic innervation of cat rhabdosphincter. Blood vessel-independent adrenergic nerve fibres course between and partially encircle myofibres of rhabdosphincter. (Atta and Elbadawi 1987)

coplasmic reticulum, free ribosomes, and microtubular structures.

Myofibres of the male feline rhabdosphincter have the en plaque (solitary) or less frequently the

en grappe (cluster) type of end-plates (Elbadawi and Atta 1985). These two types have previously been claimed to innervate fast twitch and slow twitch myofibres, respectively, in striated muscle

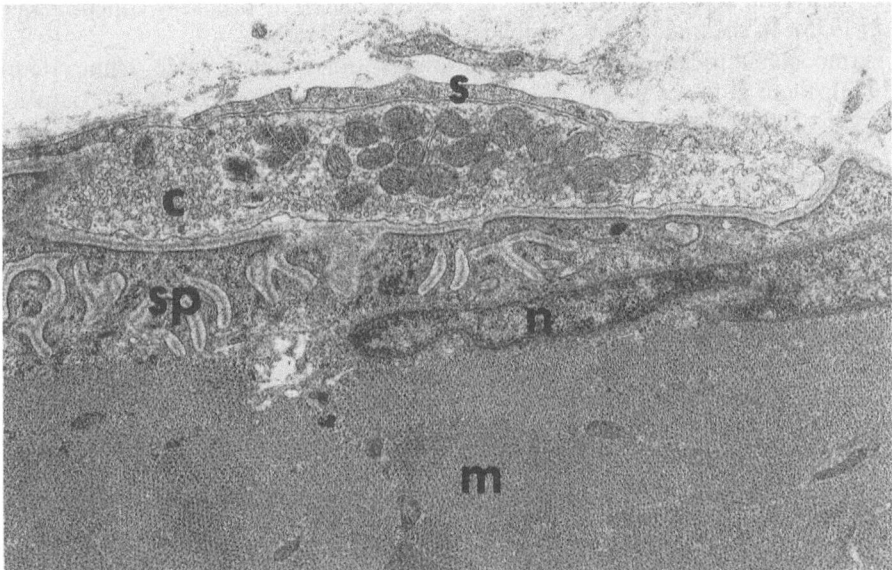

Fig. 2.21. Classical motor end-plate in male cat rhabdosphincter. Cholinergic axon terminal (*c*) is partially ensheathed by Schwann cell (*s*), and its bare side forms neuromuscular junction at sole plate (*sp*) of striated myofibre (*m*); *n*, nucleus of myofibre. (Elbadawi and Atta 1985.) (× 20 000)

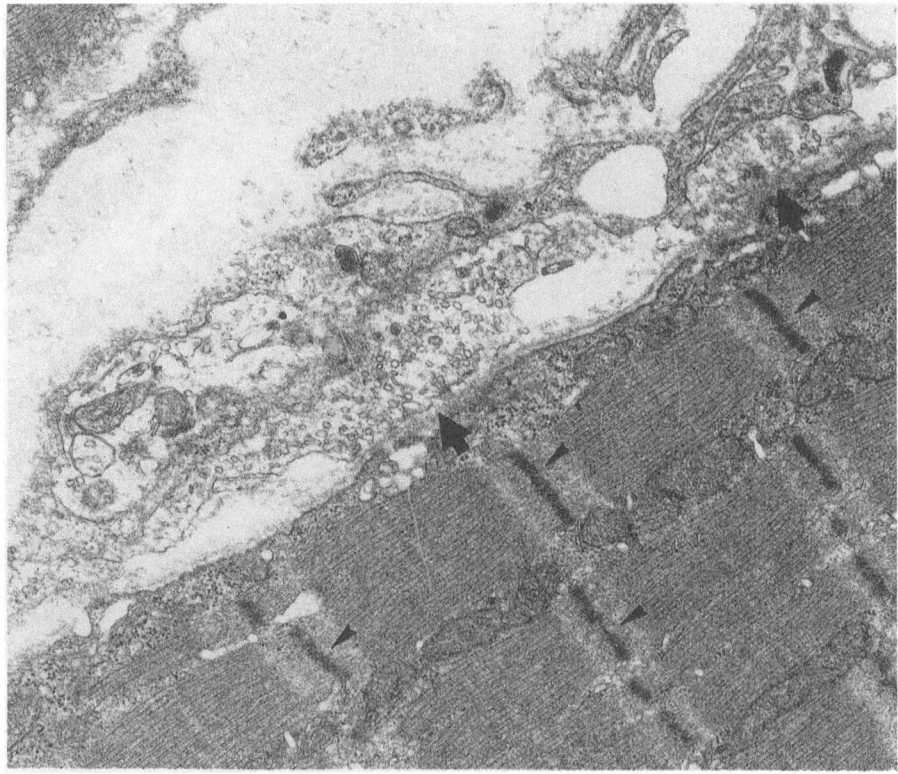

Fig. 2.22. Two close surface neuromuscular junctions between cholinergic axon terminal and myofibre in cat rhabdosphincter with no sole plate specialization (*arrows*). *Arrowheads* mark Z discs of myofibrils. (Elbadawi and Atta 1985.) (× 23 000)

(Murata and Ogata 1969; Padykula and Gauthier 1970; Mitko and Kilarski 1975). In the male feline rhabdosphincter, as in extraocular muscles of the rat (Santa and Engel 1973), there is no constant relationship between the type of end-plate and that of the related myofibre (Elbadawi and Atta 1985).

Surface axon/myofibre junctions in the male feline rhabdosphincter are considered to be autonomic in origin because they lack sarcolemmal specialization in the form of sole plates. The junctions have gaps of 15–120 nm width, and are formed with the flat, slightly raised or focally protuberant surface of the myofibre, away from its nucleus and motor end-plate (Figs. 2.22, 2.23). Cholinergic as well as adrenergic axon terminals form such junctions with both slow and fast myofibres. Some terminals of both axon types contain numerous pleomorphic large dense-core vesicles, and, there-

fore, can be considered as probably copeptidergic (Elbadawi and Atta 1985).

Triple innervation of the male feline rhabdosphincter has been confirmed following complete bilateral sacral ventral rhizotomy, which eliminates its somatomotor innervation. Shortly after such denervation, the axons of *all* motor end-plates degenerate and disappear, but those forming surface junctions persist (Atta and Elbadawi 1985). Eventually, the residual sole plates become reinnervated by sprouts of autonomic cholinergic and adrenergic axons that have a prominent probable copeptidergic component (Elbadawi and Atta 1985).

Acknowledgement. The author's recent studies reviewed in this chapter have been supported by Grants AM-28000, NS-17144 and AG-06202 from the National Institutes of Health, USPHS, USA.

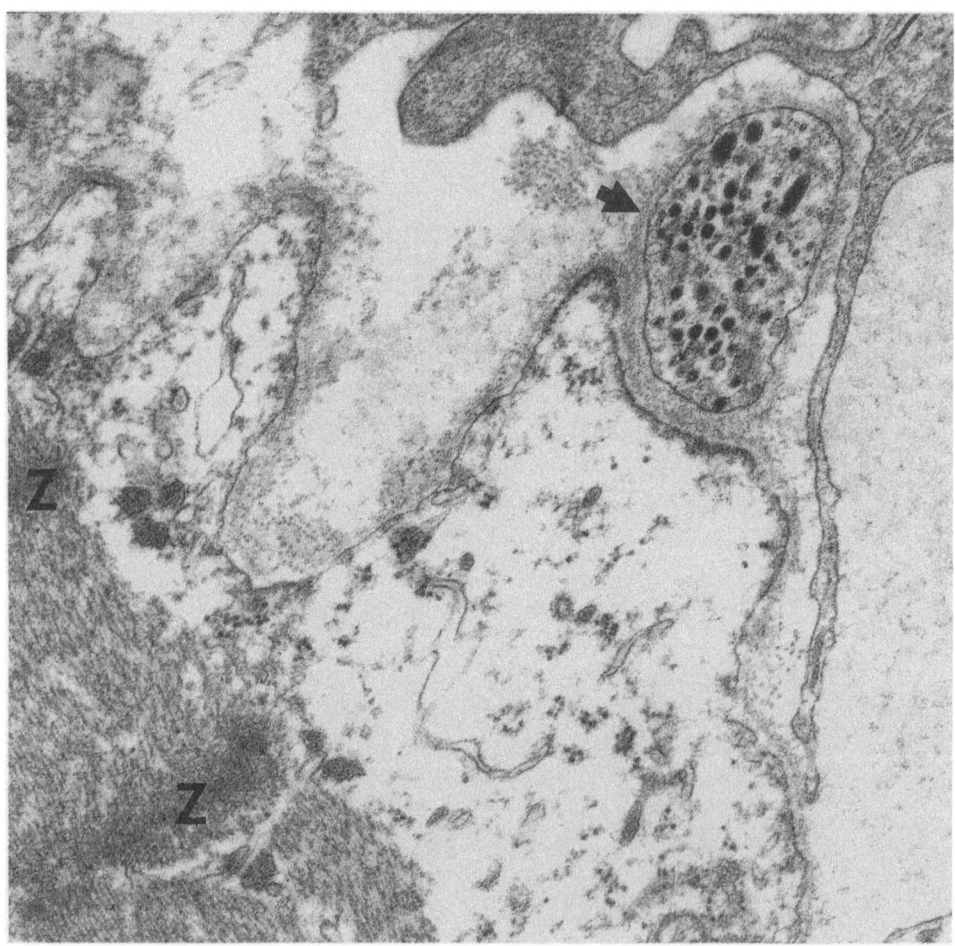

Fig. 2.23. Bare adrenergic axon terminal (*arrow*) forms surface neuromuscular junction with myofibre of cat rhabdosphincter. Note lack of sole plate specialization of myofibre. *Z*, Z disc of myofibril. (Elbadawi and Atta 1985.) (× 46 400)

References

Alm P (1978) Cholinergic innervation of the human urethra and urinary bladder: a histochemical study and review of methodology. Acta Pharmacol Toxicol 43: 56–62

Alm P, Elmér M (1975) Adrenergic and cholinergic innervation of the rat urinary bladder. Acta Physiol Scand 94: 36–45

Alm P, Alumets J, Hakanson R, Sundler F (1977) Peptidergic nerves in the genitourinary tract. Neurosci 2: 751–754

Alumets J, Fahrenkrug R, de Muckadell S, Sundler F, Uddman R (1980) A rich VIP nerve supply is characteristic of sphincters. Nature 280: 155–156

Atta MA, Elbadawi A (1985) Intrinsic neuromuscular defects in the neurogenic bladder. IV. Loss of somatomotor and preservation of autonomic innervation of the male feline rhabdosphincter following bilateral sacral ventral rhizotomy. Neurourol Urodyn 4: 219–230

Atta MA, Elbadawi A (1987) Intrinsic neuromuscular defects in the neurogenic bladder. VII. Neurohistochemistry of the somatically denervated male feline rhabdosphincter. Neurourol Urodyn 6: 47–56

Atta MA, Franck JI, Elbadawi A (1984) Intrinsic neuromuscular defects in the neurogenic bladder. II. Long-term innervation of the unilaterally decentralized feline bladder base by regenerated cholinergic, increased adrenergic and emergent probable "peptidergic" nerves. Neurourol Urodyn 3: 185–200

Baljet B, Drukker J (1980) The extrinsic innervation of the pelvic organs in the female rat. Acta Anat 107: 241–267

Bradley WE, Teague CT (1968) Innervation of vesical detrusor muscle by the ganglia of the pelvic plexus. Invest Urol 6: 251–265

Buzelin JM (1984) Urodynamique. Bas appareil urinaire. Masson, Paris, pp 3–18

Chan-Palay V, Jonsson G, Palay SL (1978) Serotonin and substance P coexist in neurons of the rat central nervous system. Z Zellforsch 135: 287–314

Cuello AC (1982) Co-Transmission. MacMillan, London, pp 1–259

de Groat WC, Kawatani M (1985) Neural control of the urinary bladder: possible relationship between peptidergic inhibitory mechanisms and detrusor instability. Neurourol Urodyn 4: 285–300

de Groat WC, Saum WR (1976) Synaptic transmission in parasympathetic ganglia in the urinary bladder of the cat. J Physiol 256: 137–158

Dixon JS, Gilpin S-A, Gilpin CJ, Gosling JA (1983) Intramural ganglia of the human urinary bladder. Br J Urol 55: 195–198

Donker PJ (1986) A study of the myelinated fibers in the branches of the pelvic plexus. Neurourol Urodyn 5: 185–202

Downie JW, Champion JA, Nance TW (1984) A quantitative analysis of the afferent and extrinsic efferent innervation of specific regions of the bladder and urethra in the cat. Brain Res Bull 12: 735–740

Ek A, Alm P, Andersson K-E, Persson CGA (1977) Adrenergic and cholinergic nerves of the human urethra and urinary bladder. A histochemical study. Acta Physiol Scand 99: 345–352

Elbadawi A (1972) Anatomy and function of the ureteral sheath. J Urol 102: 224–229

Elbadawi A (1982a) Ultrastructure of vesicourethral innervation. I. Neuroeffector and cell junctions in the male internal sphincter. J Urol 128: 180–188

Elbadawi A (1982b) Neuromorphologic basis of vesicourethral function: I. Histochemistry, ultrastructure, and function of intrinsic nerves of the bladder and urethra. Neurourol Urodyn 1: 3–50

Elbadawi A (1983) Autonomic muscular innervation of the vesical outlet and its role in micturition. In: Hinman F Jr (ed) Benign prostatic hypertrophy. Springer, Berlin Heidelberg New York, pp 330–348

Elbadawi A (1984) Ultrastructure of vesicourethral innervation. II. Postganglionic axoaxonal synapses in intrinsic innervation of the vesicourethral lissosphincter: a new structural and functional concept in micturition. J Urol 131: 781–790

Elbadawi A (1985) Ultrastructure of vesicourethral innervation. III. Axoaxonal synapses between postganglionic cholinergic axons and probably SIF-cell derived processes in feline lissosphincter. J Urol 133: 524–528

Elbadawi A (1987) Neuromuscular mechanisms of micturition. In: Yalla SV, McGuire EJ, Elbadawi A, Blaivas JG (eds) Principles and practice of neurourology and urodynamics, McMillan, New York, Chapt 1 (in press)

Elbadawi A, Atta MA (1985) Ultrastructure of vesicourethral innervation. IV. Evidence for somatomotor plus autonomic innervation of the male feline rhabdosphincter. Neurourol Urodyn 4: 23–36

Elbadawi A, Atta MA (1986) Intrinsic neuromuscular defects in the neurogenic bladder. V. Autonomic re-innervation of the male feline rhabdosphincter following somatic denervation by bilateral sacral ventral rhizotomy. Neurourol Urodyn 5: 65–86

Elbadawi A, Goodman DC (1980) Autonomic innervation of accessory male genital glands. In: Spring-Mills E, Hafez ESE (eds) Male accessory sex glands. Elsevier/North Holland, Amsterdam, pp 101–128

Elbadawi A, Schenk EA (1966) Dual innervation of the mammalian urinary bladder. A histochemical study of the distribution of cholinergic and adrenergic nerves. Am J Anat 119: 405–427

Elbadawi A, Schenk EA (1968a) A new theory of the innervation of bladder musculature. 1. Morphology of the intrinsic vesical innervation apparatus. J Urol 99: 585–587

Elbadawi A, Schenk EA (1968b) The peripheral adrenergic innervation apparatus. Intraganglionic and extraganglionic adrenergic ganglion cells. Z Zellforsch 87: 218–225

Elbadawi A, Schenk EA (1969) Innervation of the abdominopelvic ureter in the cat. Am J Anat 126: 103–120

Elbadawi A, Schenk EA (1970) Intra- and extraganglionic peripheral cholinergic neurons in urogenital organs of the cat. Z Zellforsch 103: 26–33

Elbadawi A, Schenk EA (1971a) A new theory of the innervation of bladder musculature. 2. The innervation apparatus of the ureterovesicular junction. J Urol 105: 368–371

Elbadawi A, Schenk EA (1971b) A new theory of the innervation of urinary bladder musculature. 3. Postganglionic synapses in ureterovesicourethral autonomic pathways. J Urol 105: 372–374

Elbadawi A, Schenk EA (1973) Parasympathetic and sympathetic postganglionic synapses in ureterovesical autonomic pathways. Z Zellforsch 146: 147–156

Elbadawi A, Schenk EA (1974) A new theory of the innervation of bladder musculature. 4. Innervation of the vesicourethral junction and external urethral sphincter. J Urol 111: 613–615

Elbadawi A, Amaku EO, Frank IN (1973) Trilaminar musculature of submucosal ureter. Anatomy and functional implications. Urology 2: 409–417

Elbadawi A, Atta MA, Franck JI (1984) Intrinsic neuromuscular defects in the neurogenic bladder. I. Short-term ultrastructural changes in muscular innervation of the decentralized feline bladder base. Neurourol Urodyn 3: 93–113

Elmér M (1975) Autonomic innervation of the rat urinary bladder. Thesis, Lund, pp 1–31

Fletcher TF, Bradley WE (1978) Neuroanatomy of the bladder-urethra. J Urol 119: 153–158

Gil Vernet S (1968) Morphology and function of vesico-prostatourethral musculature. Canova, Treviso

Gosling J (1979) The structure of the bladder and urethra in relation to function. Urol Clin North Am 6: 31–38

Gosling JA, Dixon JS (1974) Sensory nerves in the mammalian urinary tract. An evaluation using light and electron microscopy. J Anat 117: 133–141

Gosling JA, Dixon JS (1979) Light and electron microscopic observations on the human external urethral sphincter. J Anat 129: 216

Gosling JA, Dixon JS, Lendon RG (1977) The autonomic innervation of the human male and female bladder neck and proximal urethra. J Urol 118: 302–305

Gu J, Blank MA, Huang WM, Islam KN, McGregor GP, Christofides N, Allen JM, Bloom SR, Polak JM (1984) Peptide-containing nerves in human urinary bladder. Urology 24: 353–357

Hutch JA (1972) Anatomy and physiology of the bladder, trigone, and urethra. Appleton-Century-Crofts, New York, pp 1–180

Iijima T (1981) Occurrence of uranaffin-positive synaptic vesicles in both adrenergic and non-adrenergic nerves of the rat anococcygeus muscle. Cell Tissue Res 220: 427–433

Islam KN, Gu J, Blank HA, Bloom SR, Polak JM, Shuttleworth KED (1984) Peptide-containing nerves in human urinary bladder. J Urol 31 (4/2): 193A

Klück P (1980) The autonomic innervation of the human urinary bladder, bladder neck and urethra: a histochemical study. Anat Rec 198: 439–448

Kuntz A (1945) The autonomic nervous system. Lee and Febiger, Philadelphia

Kuo DC, Hisamitso T, de Groat WC (1984) The sympathetic projection from sacral paravertebral ganglia to the pelvic nerve and to postganglionic nerves on the surface of the urinary bladder and large intestine of the cat. J Comp Neurol 226: 76–86

Kuru M (1965) Nervous control of micturition. Physiol Rev 45: 425–494

Langworthy OR (1965) Innervation of the pelvic organs of the rat. Invest Urol 2: 491–511

Larsson L-I (1977) Ultrastructural localization of a new peptide (VIP). Histochemistry 54: 170–176

Lentz TL (1971) Cell fine structure. An atlas of drawings of whole-cell structure. Saunders, Philadelphia, pp 84–102

Lundberg JM, Hökfelt T (1983) Coexistence of peptides and classical neurotransmitters. Trends Neurosci 6: 325–333

Lundberg JM, Hedlund B, Anggard A, Fahrenkrug J, Hökfelt T, Tatemoto M, Bartfai T (1982) In: Bloom SR, Polak JM, Lindenlaub E (eds) Systemic role of regulatory peptides, Schattauer, Stuttgart, pp 145–168

McCrea LE, Kimmel DL (1952) The importance of nerve supply of the urinary bladder in surgery of the rectosigmoid. J Int Coll Surg 17: 651–657

Martin WD, Fletcher TF, Bradley WE (1974) Innervation of feline perineal musculature. Anat Rec 180: 15–29

Mattiasson A, Andersson K-E, Sjögren C (1984) Adrenoceptors and cholinoceptors controlling noradrenaline release from adrenergic nerves in the urethra of rabbit and man. J Urol 131: 1190–1195

Mattiasson A, Andersson K, Elbadawi A, Morgan E, Sjögren C (1987) Interaction between adrenergic and cholinergic nerve terminals in the urinary bladder of rabbit, cat and man. J Urol (in press)

Merrillees NCR (1968) The nervous environment of individual smooth muscle cells in the guinea pig vas deferens. J Cell Biol 57: 794–798

Merrillees NCR, Burnstock G, Holman ME (1963) Correlation of fine structure and fine structure and physiology of the

innervation of smooth muscle in the guinea pig vas deferens. J Cell Biol 19: 529–550

Mitchell GAG (1953) Anatomy of the autonomic nervous system. Edinburgh, Livingstone, pp 257–310

Mitko L, Kilarski W (1975) Correlation between nerve terminal size and muscle fiber diameter in urodela extrinsic eye muscle. Z Mikrosk Anat Forsch 89: 1015–1029

Murata F, Ogata T (1969) The ultrastructure of neuromuscular junctions of human red, white and intermediate striated muscle fibers. Tohoku J Exp Med 99: 289–301

Nathanson JA (1977) Cyclic nucleotides and nervous system function. Physiol Rev 57: 157–256

Nordling J, Christensen B (1978) Human sympathetic bladder innervation. Urol Int 33: 359–365

Owman C, Sjöstrand NO (1965) Short adrenergic neurons and catecholamine-containing cells in vas deferens and accessory male genital glands of different mammals. Z Zellforsch 66: 300

Padykula HA, Gauthier GF (1970) The ultrastructure of the neuromuscular junctions of mammalian red, white and intermediate skeletal muscle fibers. J Cell Biol 46: 27–41

Purinton PT, Fletcher TF, Bradley WE (1975) Gross and light microscopic features of the pelvic plexus in the rat. Anat Rec 175: 697–706

Raezer DM, Wein AJ, Jacobwitz D, Corriere JN Jr (1973) Autonomic innervation of canine urinary bladder. Cholinergic and adrenergic contributions and interaction of sympathetic and parasympathetic nervous systems in bladder function. Urology 2: 221

Richards JG, da Prada M (1977) Uranaffin reaction: a new cytochemical technique for the localization of adenine nucleotide in organs storing biogenic amines. J Histochem Cytochem 25: 1322–1336

Santa T, Engel AG (1973) Histometric analysis of neuromuscular junction ultrastructure in rat red, white and intermediate muscle fibres. In: Desmedt JE (ed.) New developments in electromyography and clinical neurophysiology, vol 1. Karger, Basel, pp 202–221

Saum WR, de Groat WC (1972) Parasympathetic ganglia. Activation of an adrenergic inhibitory mechanism by cholinomimetic agents. Science 175: 659–661

Schnitzlein HN, Hofman HH, Tucker CC, Quigley MB (1972) The pelvic splanchnic nerves of the male Rhesus monkey. J Comp Neurol 114: 51–65

Seto H (1963) Studies on the sensory innervation. Human sensibility, 2nd edn. Igaku Shoin, Tokyo, p 250

Sjöstrand NO (1965) The adrenergic innervation of the vas deferens and the accessory male genital glands. Acta Physiol Scand [Suppl] 257: 1–82

Slack BE, Downie JW, Elbadawi A (1982) Paradoxical resistance to adrenolytic agents of field-stimulated bladder base of rabbit. J Pharmacol Exp Ther 220: 216–222

Smith ER (1975) The canine prostate and its secretion. Adv Sex Horm Res 1: 167

Sneddon P, Westfall DP (1984) Pharmacological evidence that adenosine triphosphate and noradrenaline are co-transmitters in the guinea pig vas deferens. J Physiol 347: 561–580

Stach W, Fänghanel J, Schultz E (1970) Afferente terminalorgane in der Wand der Harnblase verscheidner Laboratoriumstiere und ihre Topographie. J Neurovisc Rel 31: 26

Sundin T, Dahlström A (1973) The sympathetic innervation of the urinary bladder and urethra in the normal state and after parasympathetic denervation at the spinal root level. An experimental study in cats. Scand J Urol Nephrol 7: 131–149

Taira N (1972) The autonomic pharmacology of the bladder. Ann Rev Pharmacol 12: 197–208

Tanagho EA, Pugh RCB (1963) The anatomy of the ureterovesical junction. Br J Urol 35: 151–165

References

51

Uemura E, Fletcher TF, Dirks VA, Bradley WE (1973) Distribution of sacral afferent axons in cat urinary bladder. Am J Anat 136: 305–314

Uemura E, Fletcher TF, Bradley WE (1975) Distribution of lumbar and sacral afferent axons in submucosa of cat urinary bladder. Anat Rec 183: 579–588

Uemura E, Fletcher TF, Bradley WE (1974) Distribution of lumbar afferent axons in muscle coat of cat urinary bladder. Am J Anat 139: 389–398

Uhlenhuth E, Hunter DW Jr, Loechel WF (1953) Problems in the anatomy of the pelvis. Lippincott, Philadelphia, pp 1–157

Waldeyer W (1892) Ueber die sogenannte Ureterscheide. Anat Anz 259–260

Waldeyer W (1899) Das Becken. Friederich Cohen, Bonn, p 329

Watanabe H, Yamamoto TY (1979) Autonomic innervation of the muscles in the wall of the bladder and proximal urethra of male rats. J Anat 128: 873–886

Westfall TC (1977) Local regulation of adrenergic neurotransmission. Physiol Rev 57: 659–675

Woodburne RT (1964) Anatomy of the ureterovesical junction. J Urol 92: 431–438

Woodburne RT (1967) Anatomy of the bladder. In: Boyarsky S (ed) The neurogenic bladder. Williams and Wilkins, Philadelphia, pp 1–17

Woźniak W. Skowrońska U (1967) Comparative anatomy of pelvic plexus in cat, dog, rabbit, Macaque and man. Anat Anz 120: 457–473

Yoshida T (1957) A histologic study of sensory nerves in the urinary organs. Arch Jpn Chir 26: 55–65

Young BW (1972) Lower urinary tract obstruction in childhood. Lee and Febiger, Philadelphia

3 · Neural Connections Between the Lower Urinary Tract and the Spinal Cord

John F. B. Morrison

There have been enormous advances in our knowledge of the functional anatomy of the autonomic nervous system in the last decade. Much of the detailed knowledge that has been acquired has been the result of animal studies, and different species will be considered separately in this chapter, as the research worker may find it helpful to refer to the differences between them. In all species there exist two or more segments in the lower lumbar region between the lumbar sympathetic and the sacral parasympathetic outflows that have no afferent or efferent connections with the viscera. Confusion may arise, however, because the nomenclature of spinal segments is defined according to the corresponding vertebrae, which vary in number between the species; thus, some of the common laboratory animals have 13 thoracic, 7 lumbar and 4 sacral segments and, as a result, the parasympathetic outflow may be from the lumbosacral junction rather than the middle of the sacral cord. In other cases, the lower end of the sympathetic outflow may be as caudal as the fifth lumbar segment.

Nervous connections between the sacral cord and the lower urinary tract include pelvic nerve afferent fibres from the bladder and urethra, pudendal nerve afferents from the urethra and perineal skin, pelvic parasympathetic efferents to the vesical ganglia, and pudendal somatic motoneurons to the external sphincters and pelvic floor. Segments of the upper lumbar and lower thoracic cord (down to L-5 in some species) also provide both sensory and motor innervation for the urogenital tract. Sympathetic efferent fibres run via the white rami communicantes to the sympathetic chain, through the lumbar splanchnic nerves to the inferior mesenteric ganglia, the vesical plexus and bladder wall before they synapse with postganglionic elements; postganglionic fibres arise in the inferior mesenteric ganglia, in the lumbar and (mainly) sacral sympathetic chain ganglia, and in the vesical ganglia (see Fig. 3.5). Afferent fibres travel in the hypogastric and lumbar splanchnic nerves and the white rami, and have their cell bodies in the upper lumbar and lower thoracic dorsal root ganglia. The morphology of sensory nerve endings in the lower urinary tract has been considered by Kuru (1965). Many sensory nerve endings have free endings, and in addition there are lamellated corpuscular endings such as Pacini's corpuscle and its smaller relatives.

These lammellated endings have large myelinated afferent fibres and occur in many viscera, but represent only a small proportion of the afferent fibres. The majority of afferents from the bladder have small myelinated or unmyelinated axons, and, unlike Pacini's corpuscles, are slowly adapting receptors. The properties of these endings will be discussed in Chapter 4 (see p. 98). Some of these afferents may have axon collaterals which synapse on the postganglionic cells in the inferior mesenteric ganglia, possibly using substance P as a neurotransmitter. Reflexes from the bladder to the internal anal sphincter are believed to be partly mediated by the prevertebral ganglia (see Chap. 7, p. 218); it is not clear to what extent the collaterals of bladder afferents are involved in this process. There is evidence in other viscera for the existence of enteric (intramural) ganglion cells which have axons that project centripetally towards the prevertebral ganglia; they are not thought to reach the spinal cord, but are mentioned at this point because they are present alongside spinal neurons in some visceral nerves.

Our increased knowledge of the extrinsic innervation of the lower urinary tract in animals has come mainly from two sources: first, the ability to trace all the processes of neurons using substances which are transported along axons, and, second, quantitative electron microscopy. This new data provides opportunities to study the changes in innervation in different conditions which influence the survival of neurons.

Studies on the distribution of afferent terminals and efferent cell bodies have been analysed with reference to a standardized map of the spinal cord, so as to allow comparisons to be made between different laboratories and different experiments. The maps generally used for this purpose are based on those originally published in 1954 by Rexed. Figure 3.1 shows examples of Rexed's laminae. Rexed's maps of the spinal cord of cats were based on the cytoarchitecture of the neurons in different regions of the grey matter. Nissl staining was used to determine the locations, sizes and packing densities of neuronal somata, and this allowed the division of the grey matter into regions which have become known as Rexed's laminae: in the dorsal horn there are six laminae which form a sequence of layers, lamina I being the most superficial dorsal layer, and lamina VI the deepest layer in the dorsal horn. The intermediolateral horn, the ventral horn, and the grey matter adjacent to the central canal are also divided into cell groups by cytoarchitectonics, but the arrangement is not strictly layered; it is arranged as a series of columns of cells. However, some laminations can be found within

these areas if the ventral and intermediolateral horns are examined with other neuroanatomical techniques. Rexed's map has provided a satisfactory reference point for much anatomical and physiological work on the cord of the cat. A similar map (Fig. 3.2) has been published for the rat (Molander et al. 1984). These cytoarchitectonic maps give useful information about the distribution, shapes and sizes of the cell bodies of neurons, but do not provide any information about the distribution of dendrites of the neurons, the sites of much of the synaptic input, or the axonal projections of these cells.

Tracing Neuronal Pathways Using Axonal Transport

During the last 15 years or so, a revolution in neuroanatomical methods has allowed new insights into the connections of neurons. The new methodology has resulted from an understanding of the process of axonal transport. Substances and subcellular particles are constantly travelling up and down axons at rates which can be quite fast, of the order of hundreds of millimetres a day. This transport of chemicals in both directions may be associated with the need to provide nerve endings with essential chemicals that can only be synthesized in the neuronal soma, and to provide the cell soma with some sort of information concerning events at the nerve ending, such as the existence of a functional contact with a postsynaptic cell. Some substances can be taken up by nerve terminals and carried back to the cell body, which can be stained for their presence. Thus, if a suitable substance is injected into a nerve or a discrete area of innervated tissue, it is possible to identify the site of the neuronal perikarya after time has elapsed for axonal transport to occur. The nature of this transport of substances within the axon, from the site of axotomy, or from axon terminals back to the cell body of the neuron of origin, is known as *retrograde transport*.

One substance that has been used commonly to trace neurons is the enzyme horseradish peroxidase (HRP); this enzyme is fairly resistant to the action of histological fixatives, and its presence can be demonstrated by histochemical reactions in which the site of breakdown of hydrogen peroxide is converted to water and oxygen, which reacts with a chromogen to yield a coloured reaction product. Retrograde axonal transport is commonly used to

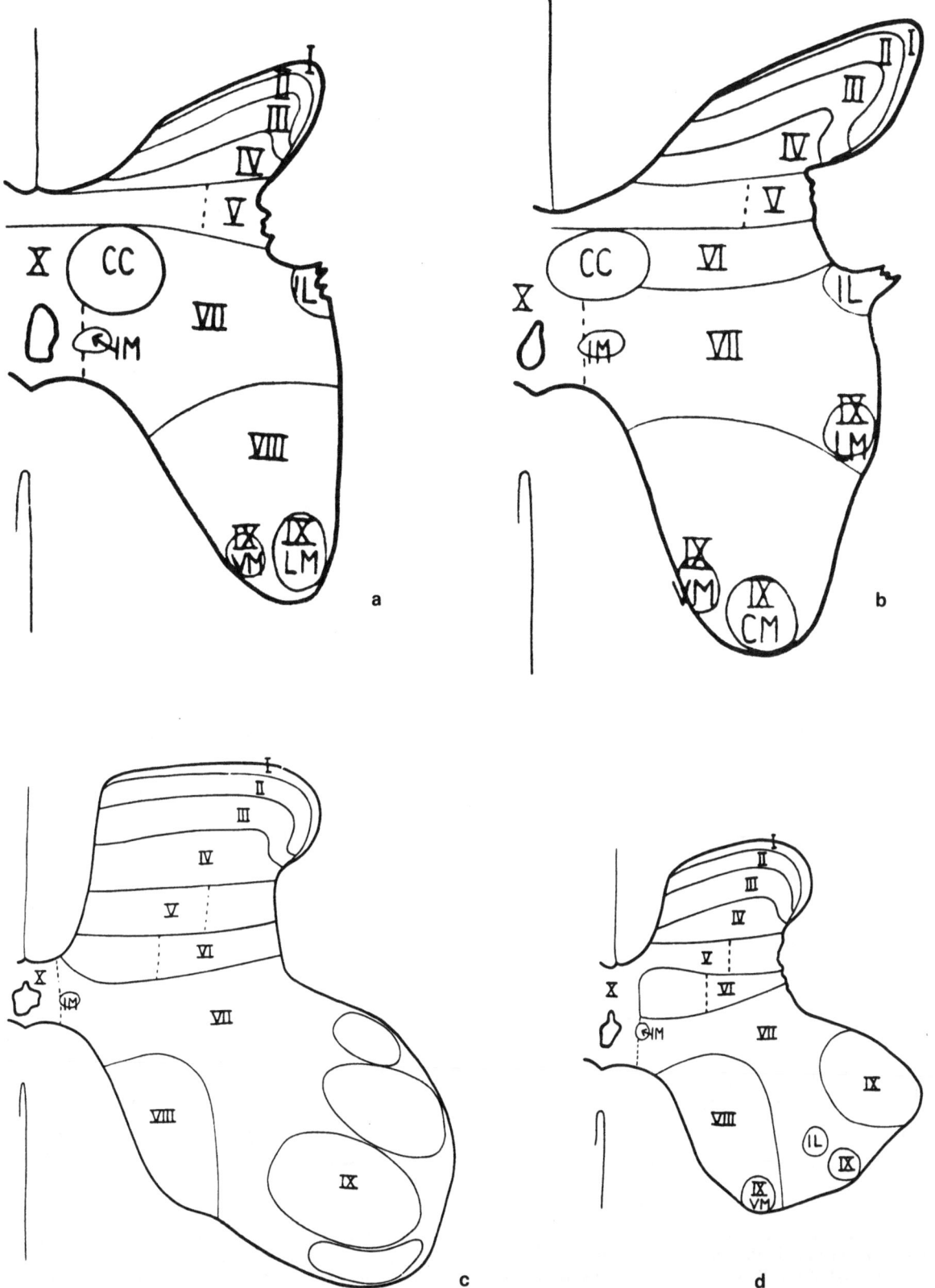

Fig. 3.1a–d. Arrangement of different cytoarchitectonic groups in **a** the second and **b** third lumbar and **c** the first and **d** second sacral segments in the cat. (Rexed 1954)

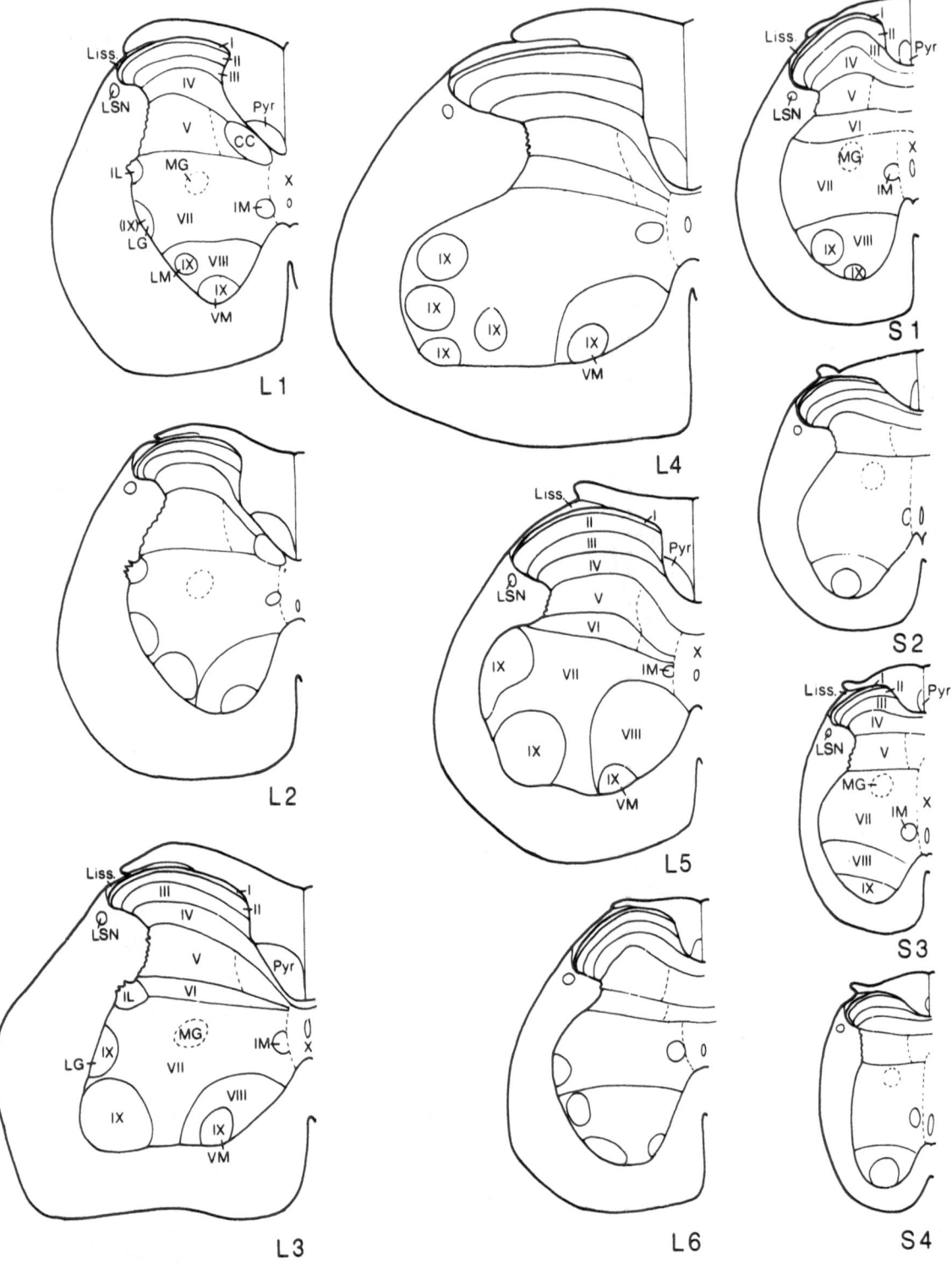

Fig. 3.2. Arrangement of cytoarchitectonic groups in the lumbosacral cord of the rat: *LX,* spinal cord laminae; *CC,* column of Clarke; *IL,* intermediolateral nucleus; *IM,* intermediomedial nucleus; *LSN,* lateral spinal nucleus; *Liss,* Lissauer's tract; *LG,* lateral group of large cells in the dorsolateral part of the ventral horn; *LM,* lateromedial nucleus; *MG,* medial group of large neurons in the intermediate zone. Note that *LG, LM,* and *VM* are parts of lamina *IX.* (Molander et al. 1984)

transport HRP and other markers (such as True Blue, Fast Blue, Nuclear Yellow, Propidium iodide and Bisbenzimide) from the site of projection of the neuronal pathway being studied back to the somata of the cells of origin of the pathway. Most of these other markers fluoresce in ultraviolet light, and can be visualized without much histological processing. However, HRP can show up more detail of the structure of these neurons than some of the other tracers.

In primary afferent neurons, HRP applied to a peripheral nerve is transported to the cell body in the dorsal root ganglion, but some is also transported past the ganglion to the central terminations of the afferent fibres within the spinal cord, and can be stained with sensitive techniques, such as the tetramethylbenzidine (TMB) method (Mesulam and Brushart 1979); this type of transport past dorsal root ganglia is known as *anterograde transport*. Anterograde transport has been used to study the segmental distribution of sensory neurons and the sites of their terminations within the spinal cord. Thus, the combination of both techniques can be used to elucidate the likely sites of connections between sensory and motor neurons involved in integration of function in the lower urinary tract and elsewhere.

HRP can also be introduced into neurons through micropipettes; following intracellular injection, the marker becomes distributed throughout the dendritic system of the neuron; the beautiful pictures of neurons that result have transformed our thinking on their function. The detailed morphology, when combined with the techniques described, provides information about the possible sites of contacts between axon terminals and postsynaptic dendrites; it does not follow, however, that such synaptic contacts are necessarily functional. This is too large a topic to mention in detail, and the reader is directed to one of the major reference works on the application of this technique in the spinal cord, such as that by Brown (1981).

These techniques depend on the tracer being confined to the inside of the cell membrane, and not being passed from cell to cell. HRP is satisfactory in this respect, but some other substances, such as wheat germ agglutinin (WGA) appear to be able to travel from cell to cell and this can be used to advantage in certain situations, to study connections between cells.

Axonal transport can be used in one other way to elucidate connections; tritiated amino acids injected into localized groups of cell bodies will be incorporated into protein which in turn is transported along the axon to its terminals. The presence of radioactive protein in distant terminals can be dem-

onstrated autoradiographically, and establishes axonal projections from the site of injection to the site at which autoradiographic grains can be demonstrated.

Quantitative Estimates of Fibre Numbers in Peripheral Nerves Using Electron Microscopy and Selective Denervations

An exact knowledge of the fibre composition of the different nerves to the urinary tract is of great importance to experimental studies that investigate the effects of changes in peripheral innervation; quantitative electron microscopy, combined with selective lesions that remove certain fibre types, has complemented the neuronal tracing methods. The results of such studies, many of which have been performed by Coggeshall and colleagues in Galveston (see Coggeshall 1980), provide precise information regarding the numbers of axons of different sizes and types in a nerve, and, when correlated with the results of HRP studies, allow deductions to be made as to the degree of peripheral branching of axons. Thus, we know, in several species, the pelvic nerve efferents divide several times before the axons reach the pelvic ganglia, whereas the afferent fibres probably do not: the number of afferent fibres in this nerve is very similar to the number of dorsal root ganglion cells which send axons peripherally in this nerve. Similarly, in the cat, sympathetic postganglionic fibres outnumber parasympathetic efferents in the pelvic nerve, and in several species sex differences in the number of sympathetic preganglionic fibres have been demonstrated.

Segmental Origin and Central Terminations of Visceral Afferents

Pelvic and Pudendal Nerves

In recent years, anterograde transport of HRP has demonstrated beautifully the location and distribution of primary visceral afferent terminals and cell bodies (see de Groat 1986). The majority of the pelvic nerve afferents enter the spinal cord through one or two dorsal roots, and there is some species variation in the level at which these afferents enter

Table 3.1 The spinal segmental origins of afferent fibres of the pelvic, hypogastric and pudendal nerves in the human, rat, cat and monkey

Species segmental differences			Locations of dorsal root ganglion cells of		
H	R	C & M	Pelvic n.	Hypogastric n.	Pudendal n.
T-10	T-11	T-11		R	
T-11	T-12	T-12		R	
T-12	T-13	T-13		R	
L-1	L-1	L-1		R	
L-2	L-2	L-2		R	
L-3	L-3	L-3		C	
L-4	L-4	L-4		C	
L-5	L-5	L-5		C	
S-1	L-6	L-6	R		
S-2	S-1	L-7	RH		C
S-3	S-2	S-1	CHM		CM
S-4	S-3	S-2	CHM		CM
	S-4	S-3	C		CM

From: Applebaum et al. (1980), Baron et al. (1985 a,b), Hulsebosch and Coggeshall (1982), Nadelhaft and Booth (1984), Nadelhaft et al. (1983), Neuhuber (1982), Roppolo et al. (1985), Sharkey et al. (1983), Ueyama et al. (1984).

C, cat; H, human; M, monkey; R, rat.

the cord (Table 3.1). Nevertheless, there is a close correlation between the level of entry of pelvic nerve afferents and the level of the parasympathetic autonomic nucleus in all species; in several species there are numerical estimates of the numbers of cells and axons in these nerves (Table 3.2). The general pattern of the central distribution of the afferents is as follows: they travel centrally in the dorsal rootlets and, after entering the cord, follow a course that is quite distinctive, and partly segregated from the somatic afferents at the lateral border of the dorsal horn. They give off axon collaterals which pass up and down the cord in Lissauer's tract, enter lamina I, and from two collateral pathways that travel laterally (the major pathway) or medially round the dorsal horn to reach the deep layers of the dorsal

horn, the intermediolateral column, the dorsal commissure and the area around the central canal. These collaterals form bundles which travel dorsoventrally at intervals along the longitudinal axis of the cord. Other collaterals pass up the dorsal columns to reach the nucleus gracilis. Although such detailed information is not available for humans, the consistency between the arrangement in other species suggests that a similar type of pattern might also be present in humans.

Cat

The majority of the pelvic nerve afferents enter the spinal cord through the second sacral dorsal root,

Table 3.2. The numbers of afferent fibres in the pelvic, lumbar and pudendal afferent pathways of cat, rat, monkey and guinea pig

Pathway	Cat	Rat	Monkey	Guinea pig	
				Male	Female
Lumbar	2600[a]	840[b]		430[c]	515[c]
		260[d]		2–10[c]*	2–10[c]*
Pelvic n.		1675[d]			
	3675[e]	1540[f]	3000[g]		
			5800[h]		
Pudendal n.		4150[d]	9200[j]		

From: [a] Baron et al. (1985a,b), [b] Neuhuber (1982), [c] McLachlan (1985), [d] Hulsebosch and Coggeshall (1982), [e] Morgan et al. (1981), [f] Nadelhaft and Booth (1984), [g] Nadelhaft et al. (1983), [h] Schnitzlein et al. (1960), [j] Roppolo et al. (1985).

* Sacral afferents in the hypogastric nerve.

with some also travelling in S-1 and in S-3. There are about 3700 dorsal root ganglion cells on each side of the neuraxis concerned with visceral function. Their collaterals pass up and down the cord in Lissauer's tract as far as L-4 rostrally and down as far as the seventh coccygeal segment. Axon collaterals pass into lamina I, and laterally down the edge of the dorsal horn (the "lateral collateral pathway") to reach laminae V, VI and X and the dorsal commissure and area around the central canal. Figure 3.3 shows the distribution of pelvic nerve afferent terminals in the sacral cord, as shown by anterograde transport of HRP. These collaterals form bundles which travel dorsoventrally at intervals of around 215 μm in the longitudinal axis. A few collaterals pass medial to the dorsal horn (the

"medial collateral pathway") and others pass up the dorsal columns to reach the nucleus gracilis (Morgan et al. 1981). Figure 3.5 shows the arrangement of visceral afferent fibres compared with the terminals of pudendal nerve afferents (Roppolo et al. 1985).

Applebaum et al. (1980) investigated the numbers of afferents which projected to the cord from the bladder by giving multiple injections of HRP into the bladder wall, and observing the number of labelled dorsal root ganglion cells found in dorsal root ganglia after sufficient time for axonal transport had been allowed. They concluded that about 800 afferents from the bladder enter the first, second and third sacral segments. The crossover of pelvic nerve afferents was estimated to be approximately

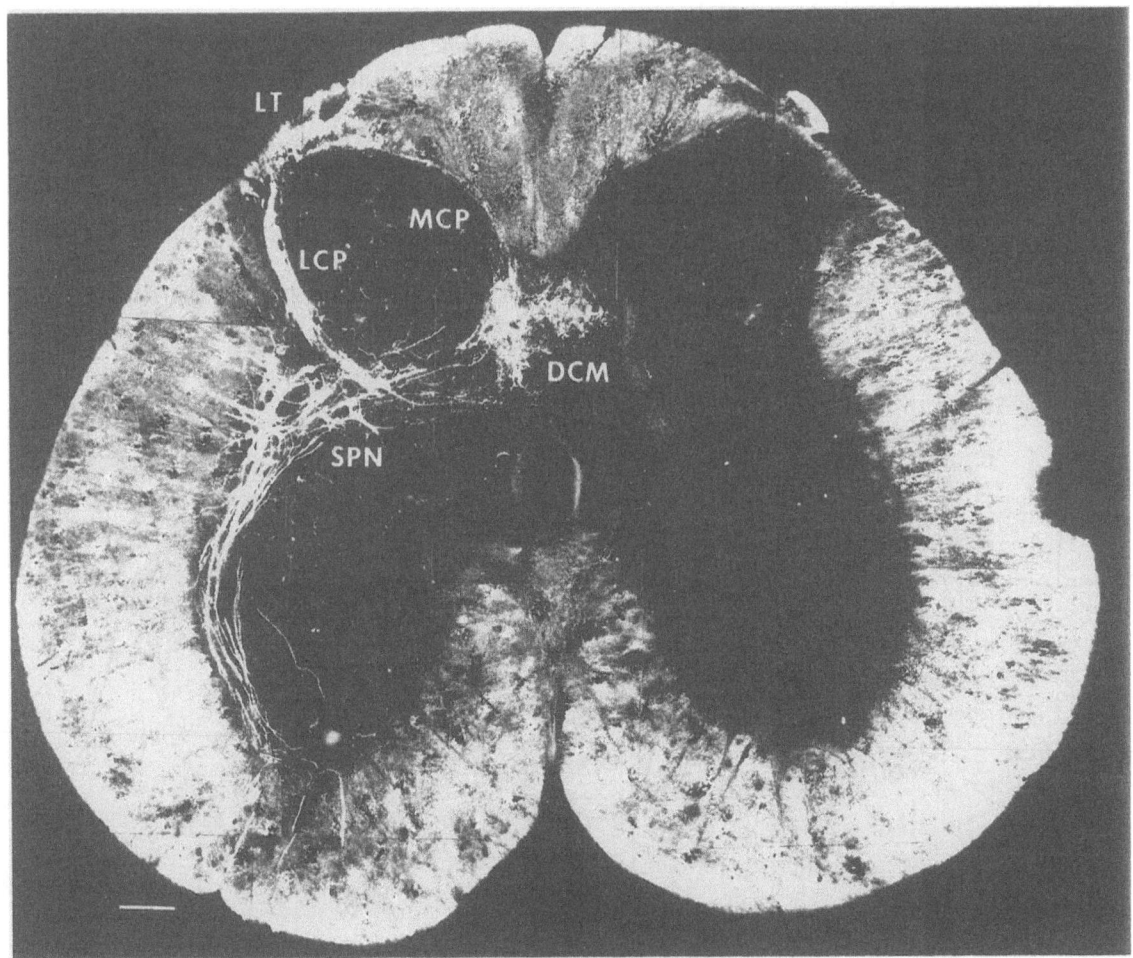

Fig. 3.3. Transverse section of the second sacral segment of the spinal cord showing afferents and efferents stained following the application of HRP to the pelvic nerve. Pelvic afferents enter Lissauer's tract (*LT*) and lamina I, to extend mainly via the lateral collateral pathway (*LCP*) to the area of the sacral parasympathetic nucleus (*SPN*). A smaller bundle of afferents extends to the dorsal grey commissure (*DCM*) via the medial collateral pathway (*MCP*). The cell bodies of parasympathetic efferents can also be seen in the *SPN*. (Morgan et al. 1981)

10%. Uemura et al. (1973) studied the distribution of afferent axon terminals in this species, and found that afferent endings that degenerated following removal of sacral dorsal roots were present at the detrusor and the trigone, with no preferred distribution. They estimated that one-third of afferent axons crossed to the opposite side of the viscus. Downie et al. (1984) found that the second sacral segment was the main one concerned with afferent inflow from the detrusor, bladder base and urethra, and very few neurons innervated more than one of these sites. Thus there is a small degree of crossover of afferents in the periphery, and little evidence that bladder afferents have axon collaterals to other neighbouring structures in the lower urinary tract.

However, there is some evidence for crossover of afferent fibres in the sacral cord. Ueyama et al. (1984) found that terminals of pudendal nerve afferents could be found ipsilaterally in lamina I of the lower lumbar, sacral and upper caudal segments and in the gracile nucleus; terminals could also be visualized in laminae III to VI and in the dorsal commissural grey matter bilaterally, but with ipsilateral predominance. Matsushita and Tanami (1983) have also reported that primary afferents can cross the midline in the dorsal and central commissures of the sacral cord. Figure 3.4 shows the distribution of pelvic afferent terminals compared with the locations of pelvic efferent cell bodies.

In 1974, Coggeshall et al. published a quantitative electron microscopic study in which they showed that the sacral ventral roots of the cat contained afferent fibres, many of which innervated the viscera; it now transpires that these afferent fibres have cell bodies in the dorsal root ganglia, the peripheral portion of the axon makes U-turns in the ventral root and the central axon actually enters the spinal cord through the dorsal root. Thus, the peripheral axon of the bipolar dorsal root ganglion cell has a minor diversion into the ventral loop, but turns back towards the spinal nerve before reaching the spinal cord (Coggeshall 1980; Risling et al. 1984).

Rat

In this species the dorsal root ganglion cells that have axons in the pelvic nerve number 1500–1600 on each side, and are distributed mainly in L-6 and S-1 (Hulsebosch and Coggeshall 1982; Nadelhaft and Booth 1984). Their central terminations can be found between L-3 and S-3 and in Lissauer's tract; the major collateral pathway is the lateral one, and fibres pass ventrally in bundles to become intermingled with the dendrites of cells of the sacral parasympathetic nucleus. The medial fibres reach the dorsal grey commissure (Nadelhaft and Booth

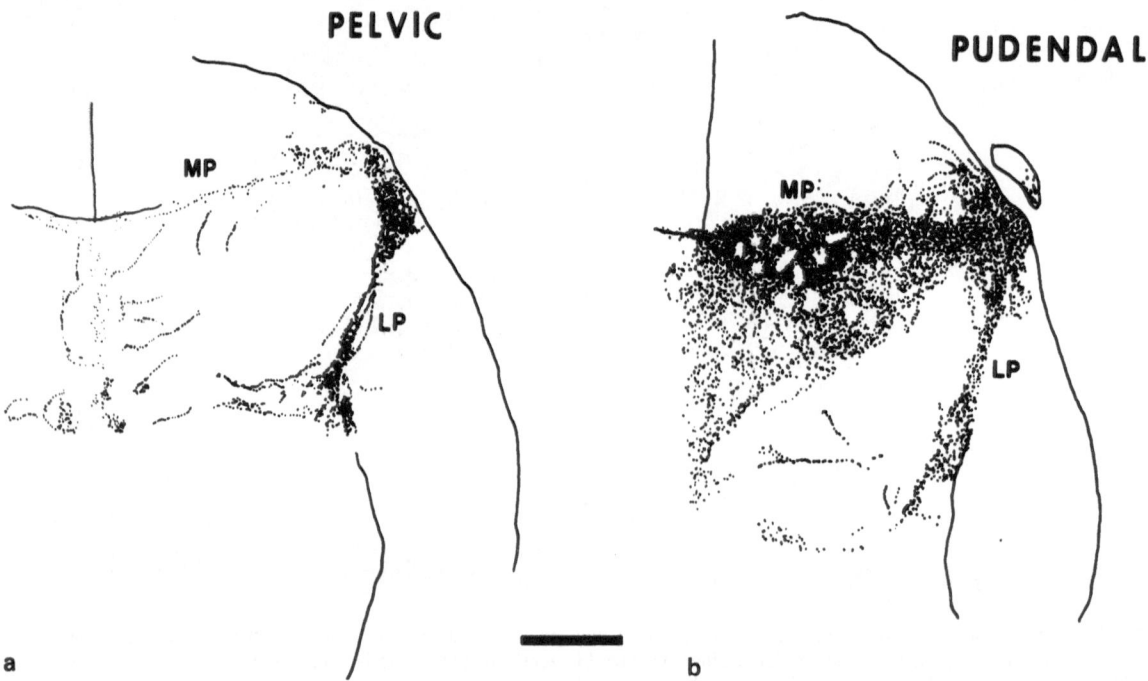

PELVIC

PUDENDAL

a   b

Fig. 3.4. The location of afferent collaterals and terminals of visceral afferents in **a** the pelvic nerve and **b** the pudendal nerve of the monkey as shown by the application of HRP to these nerves. *Bar* = 400 μm. (Roppolo et al. 1985)

1984). Applebaum et al. (1980) concluded that the bladder is innervated by just over 100 afferents in this species. There is electrophysiological evidence that the ventral roots in this species contain afferent fibres that are presumed to loop as in the cat (as discussed previously). The pudendal nerves of the rat contain about 4100 sensory axons, two-thirds of which are unmyelinated (Hulsebosch and Coggeshall 1982).

Monkey

In this species, anterograde transport of HRP shows the presence of about 3000 afferent cell bodies of pelvic nerve fibres in the dorsal root ganglia of segments S-1 to S-3, with the majority in S-2 (Nadelhaft et al. 1983). These axons enter Lissauer's tract and travel for about three segments in each direction. Axon collaterals travel ventrally (in bundles at intervals of around $250\,\mu m$). Schnitzlein et al. (1960) found a total of about 5800 sensory neurons in the pelvic nerves of the monkey; it is not clear why there is a discrepancy between these results and those of Nadelhaft et al. (1983). The possibility that the neurons were incompletely filled with HRP in the latter study seems unlikely given the agreement between HRP and electron microscopic studies in other species and nerves. The estimated error of the light microscopic methods used by Schnitzlein et al. (1960) was about 20%. Many of the large sensory fibres in this study originated from a large branch of the pelvic nerve that innervates the anal musculature; it is possible that HRP might not have been applied to this branch in Nadelhaft's study. This group also use a correction factor in their numerical analysis that may not be fully justified, and may explain the discrepancy (see Baron et al. 1985a–c).

The pudendal nerve afferents enter the cord predominantly through either the first or second sacral dorsal roots (85% of the pudendal afferents are found in one of these ganglia). Afferents in the pudendal nerve considerably outnumber efferents; in the monkey there are 418 efferents compared with 9200 afferents (Roppolo et al. 1985). They can travel in Lissauer's tract from S-3 to L-1, and enter the dorsal columns; another prominent fibre bundle caps the dorsal horn to end medially in the upper four laminae of the dorsal horn, a pattern characteristic of many somatic nerves. Some afferents pass laterally along the edge of the dorsal horn to enter lamina V, and some reach the dorsal commissure, while others terminate just dorsal to the sacral parasympathetic cell column. This is one site where they could contact interneurons concerned with mic-

turition, such as those described by de Groat et al. (1981) and McMahon and Morrison (1982a,b). Ueyama et al. (1985) found that some afferent terminals are distributed contralaterally in laminae I–VI, especially in the medial portions of laminae III–V; evidence was also presented for projections of pudendal afferents to the gracile nuclei. Figure 3.5 shows the distribution of pelvic and pudendal nerve terminals in the monkey.

Guinea Pig

McLachlan (1985) found between two and ten labelled dorsal root ganglion cells in S-1 and S-2 following application of HRP to the distal stump of a transected hypogastric nerve. This confirms older ideas that the peripheral plexuses contain a partially mixed population of axons.

Human

The older anatomical techniques located the pelvic nerve afferents to sacral roots 2–4. This is confirmed by lesioning studies, where spinal roots have been divided for relief of pain (White 1943; White and Sweet 1955), and is supported by studies of the distribution of referred pain, cutaneous hyperalgesia and deep muscle tenderness in visceral disease (see Jänig and Morrison 1986).

Hypogastric Nerves

Cat

The majority of hypogastric nerve afferents enter the cord through the fourth lumbar dorsal root, with smaller numbers in L-3 and L-5. About 1300 afferent neurones were found on each side of the neuroaxis by Baron et al. (1985a,b). However, Morgan et al. (1986) estimated that there are only 700 afferents on each side, located in the T-12 to L-5 dorsal root ganglia with the majority in the third and fourth, or fourth and fifth lumbar segments. The discrepancy between these two results appears to be due to a difference in the type of correction made for counting cells that appear in more than one section, and the effectiveness of filling with HRP. The general arrangement of afferent terminations within the spinal cord is similar to those of the pelvic nerve in the sacral cord (Morgan et al. 1986), and the hypogastric fibres travel in the tract of Lissauer (particularly the medial portion of the tract) from T-13 to L-7, although the density of

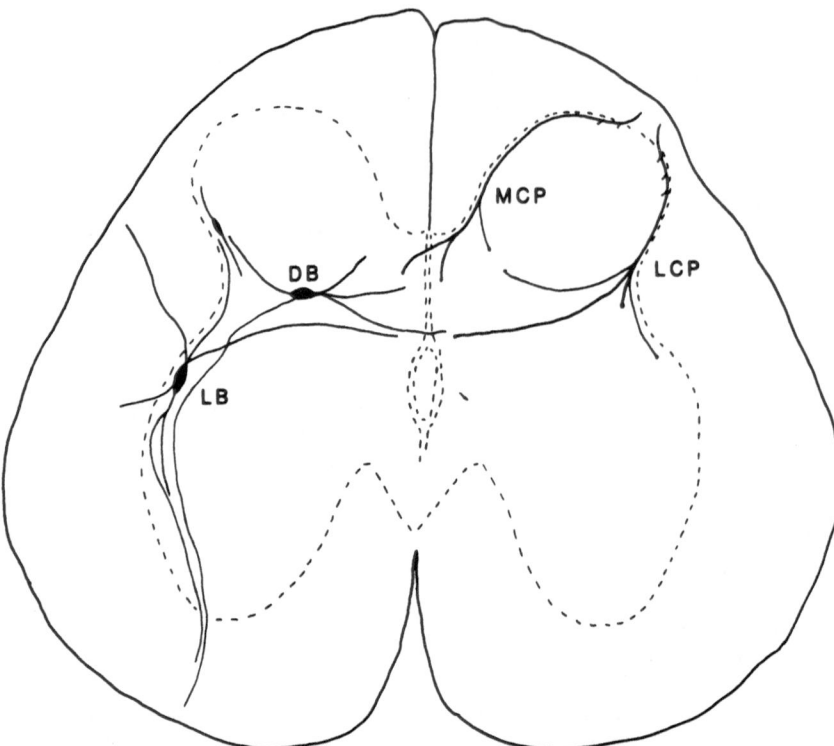

Fig. 3.5. Diagram comparing the pelvic nerve afferent distribution with the location of the preganglionic neurons and their dendrites at the S-2 level. Dendrites from a single dorsal band (*DB*) cell extend to three medial and one lateral locations. These are the same sites of termination for the three collaterals in the medial collateral pathway (*MCP*) and three of the four collaterals in the lateral collateral pathway (*LCP*). Dendrites of lateral band (*LB*) cells extended dorsolaterally into lamina I and dorsomedially into the lower dorsal grey commissure, thereby overlapping the pelvic nerve afferents at three points of the LCP. LB cells also sent dendrites into the lateral and dorsolateral funiculus and into the ventral horn. (Morgan et al. 1986)

staining was always heaviest from L-2 to L-5 (Fig. 3.6). There appears to be no systematic difference in the central sites of termination of the hypogastric afferents (which innervate a variety of pelvic structures) and those of the lumbar colonic nerves or the pelvic nerves. All afferent fibres enter the tract of Lissauer, where bundles of axons from the same nerve may occupy different positions within the tract. Primary afferents and propriospinal neurones are both present in this pathway, and its structure appears to be different at different segmental levels. The terminals of these afferents end in laminae I and V, and overlap the dendrites of neurons whose axons form the spinothalamic and spinoreticular tracts, and the preganglionic sympathetic efferents (Fig. 3.6; see also Fig. 4.19, p. 117). There are reports that monosynaptic connections between primary afferents and preganglionic efferents may occur, although the functions of these is uncertain (Mawe et al. 1984).

Applebaum et al. (1980) found that the cell bodies of afferent fibres labelled following injections of HRP into the wall of the bladder were distributed in dorsal root ganglia L-2 to L-5, and 10% of neurons projected across the midline. They estimated the size of this pathway to be about 160 neurons. Uemura et al. (1974) found that hypogastric afferent terminals are preferentially distributed in the bladder base; most were free nerve terminals, except for a sparse distribution of Pacini's corpuscles.

Rat

The majority of lumbar afferents enter the first and second segments of the lumbar cord, with only a few cells in the lower thoracic ganglia. About 400 neurons are present on each side, and these project into Lissauer's tract; their collaterals travelled mainly medial to the dorsal horn to reach the area around the central canal (Neuhuber 1982). Applebaum et al. (1980) could find only about 20 of these neurons labelled after injection of the bladder with HRP.

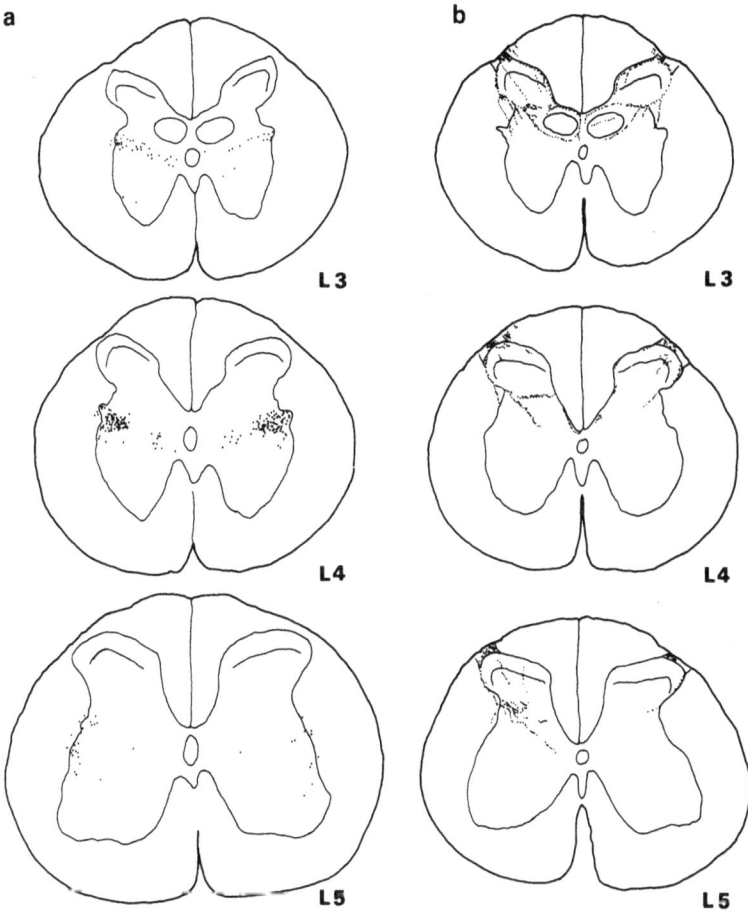

Fig. 3.6a,b. Representative distributions of preganglionic neurons (PGN) and primary afferent axon collaterals after HRP labelling of the hypogastric nerves bilaterally. **a** Composite maps of PGN made from 18 56-μm serial sections (1.1 mm) taken from the caudal ends of L-3 and L-4 and the middle of L-5 in one experiment. The PGN are spread throughout the intermediate zone at the rostral end of the distribution, L-3, but are much more restricted to the lateral edge of the grey matter at the caudal end in L-5. Most of the labelled neurons were located in L-4 and were concentrated in the lateral half of the intermediate zone (lamina VII), **b** Composite maps of primary afferent collaterals made from two or three serial sections each at the L-3 to L-5 segments. HRP was observed in medial LT, the lateral DCL, laminae I, II, V, VI, lateral VII, and X. Most axon collaterals extended in lamina I around the lateral side of the dorsal horn in the lateral collateral pathway or on the medial side in the medial collateral pathway to reach the more ventral laminae. However, a small number of axons travelled directly through laminae II–IV to reach these same areas. *Bar* = 300 μm. (Morgan et al. 1986)

Guinea Pig

The majority of hypogastric afferents enter the third lumbar segment, but can be found in lesser numbers in L-1, L-2 and L-4. About 500 hypogastric neurons are present in these dorsal root ganglia on each side, and around 40% of cells had axons in the contralateral hypogastric nerve (McLachlan 1985); whether or not this indicates that some neurons branch and have axon collaterals in both lumbar splanchnic nerves is not clear, but it seems probable.

There is recent evidence in this species that visceral afferents with cell bodies in L-2 or L-3 have been shown to have axon collaterals with profiles suggestive of axonal terminals within the inferior mesenteric ganglion (Aldskogius et al. 1986). These could be the basis of some of the peripheral reflexes that are known to be mediated by prevertebral ganglia.

Human

Studies of the relief of pain by transections or local anaesthetic block of white rami, and the distribution of referred pain, cutaneous hyperalgesia and deep muscle tenderness in bladder pain suggest that the afferent innervation through sympathetic nerves enters the tenth thoracic to second lumbar dorsal roots (Jänig and Morrison 1986).

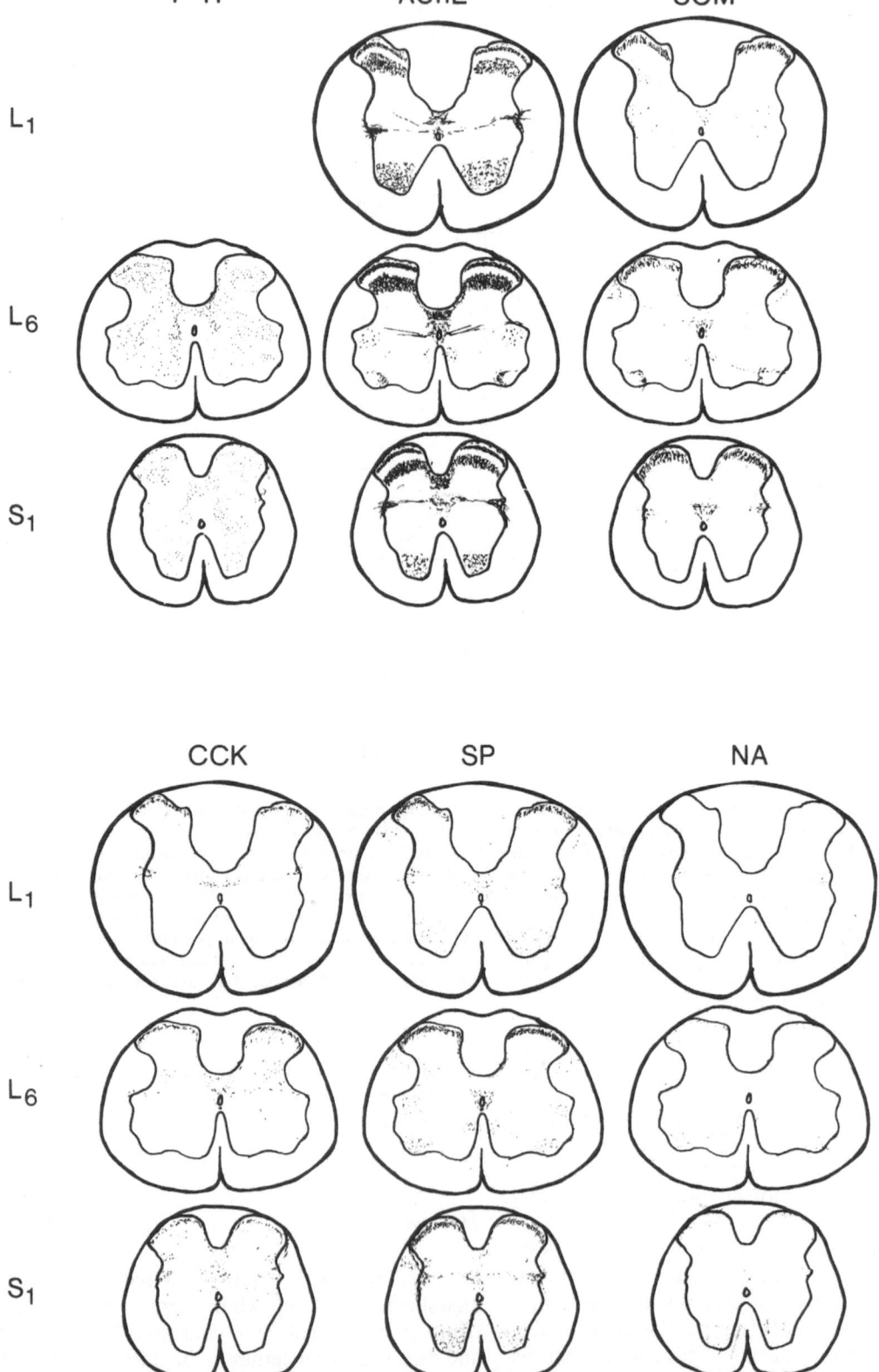

Fig. 3.7. The distribution of acetylcholinesterase (*AChE*), somatostatin (*SOM*), cholecystokinin (*CCK*) and substance P (*SP*) in the rat lumbosacral cord. (Schrøder 1985)

Neurochemistry of Visceral Afferents and the Central Distribution of Neuropeptides

Many peptides can be found in the central nervous system and the viscera, and they form one category of compounds that may act as neurotransmitters or neuromodulators, i.e. substances that can modulate transmission by some other chemical species. Each biologically active peptide may exist in a number of forms and they can be recognized individually using extremely sensitive immunocytochemical techniques (Bloom and Polak 1981; de Groat 1986; Dockray and Sharkey 1986). Neuropeptides of various types are found in small myelinated and unmyelinated primary afferent terminals, both central and peripheral, and afferent as well as efferent, and certain peptides appear to be present in larger quantities in visceral nerves than in somatic ones. The list of chemicals found in visceral afferent terminals is not confined to neuropeptides; an enzyme, fluoride-resistant acid phosphatase, is present in many small nerve terminals, including those of the pelvic nerve (McMahon 1986). The distribution of these neuropeptides in visceral nerves and in the spinal cord is of particular interest as these axons are all of small diameter and regional differences in the distribution of these substances in the cord appear to be due partly to the presence of visceral afferents (Figs. 3.7–3.9).

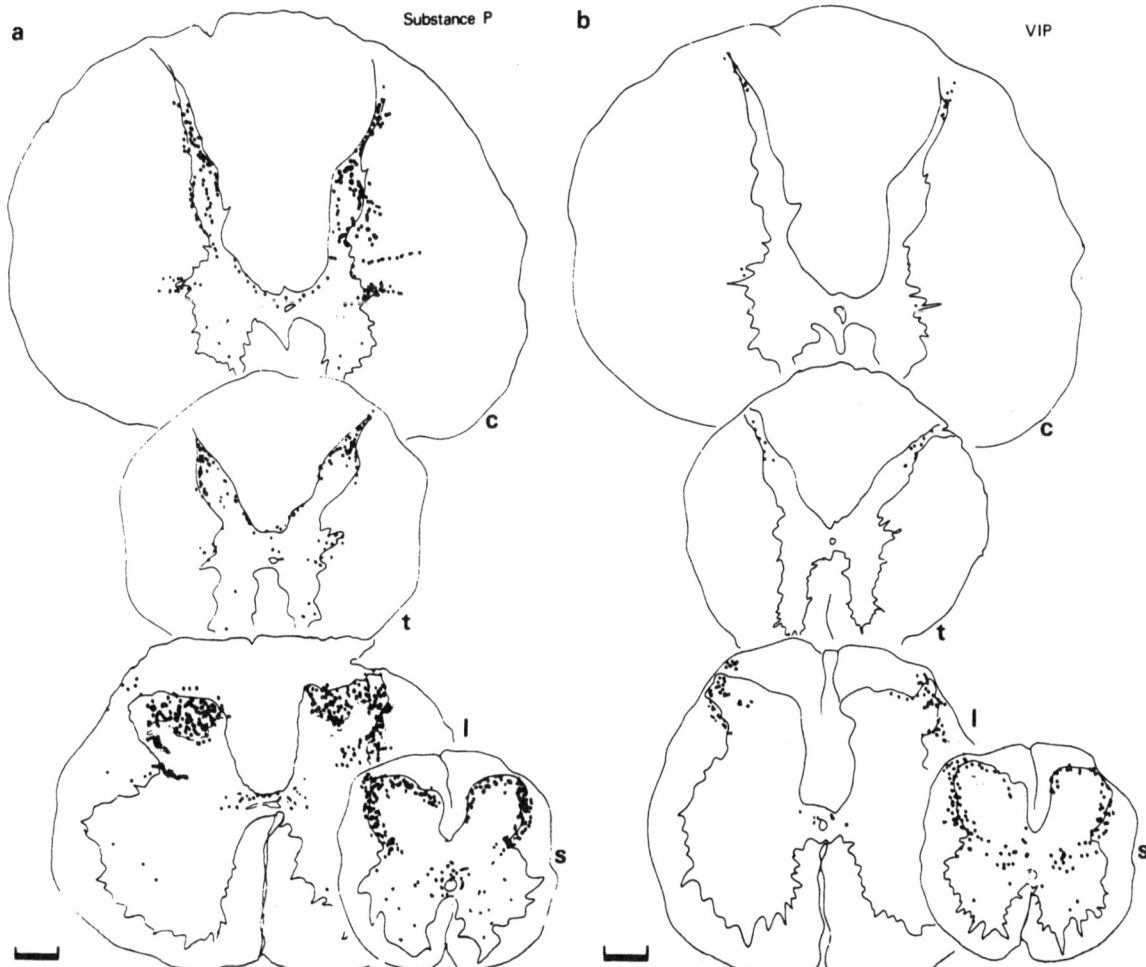

Fig. 3.8a,b. Distribution of substance P (**a**) and VIP (**b**) fibres and terminals in the cervical (*c*), thoracic (*t*), lumbar (*l*) and sacral (*s*) human spinal cord. The maps were constructed using a camera lucida. Substance P-immunoreactive fibres were more numerous in the substantia gelatinosa, around the central canal and in the intermediolateral cell column of the thoracic segments. VIP-containing fibres were less frequently observed, but the distribution followed a similar pattern. There was a remarkable increase of fibres in the dorsal horn of lumbar and especially sacral segments. *Bars* = 1 mm. (Anand et al. 1983)

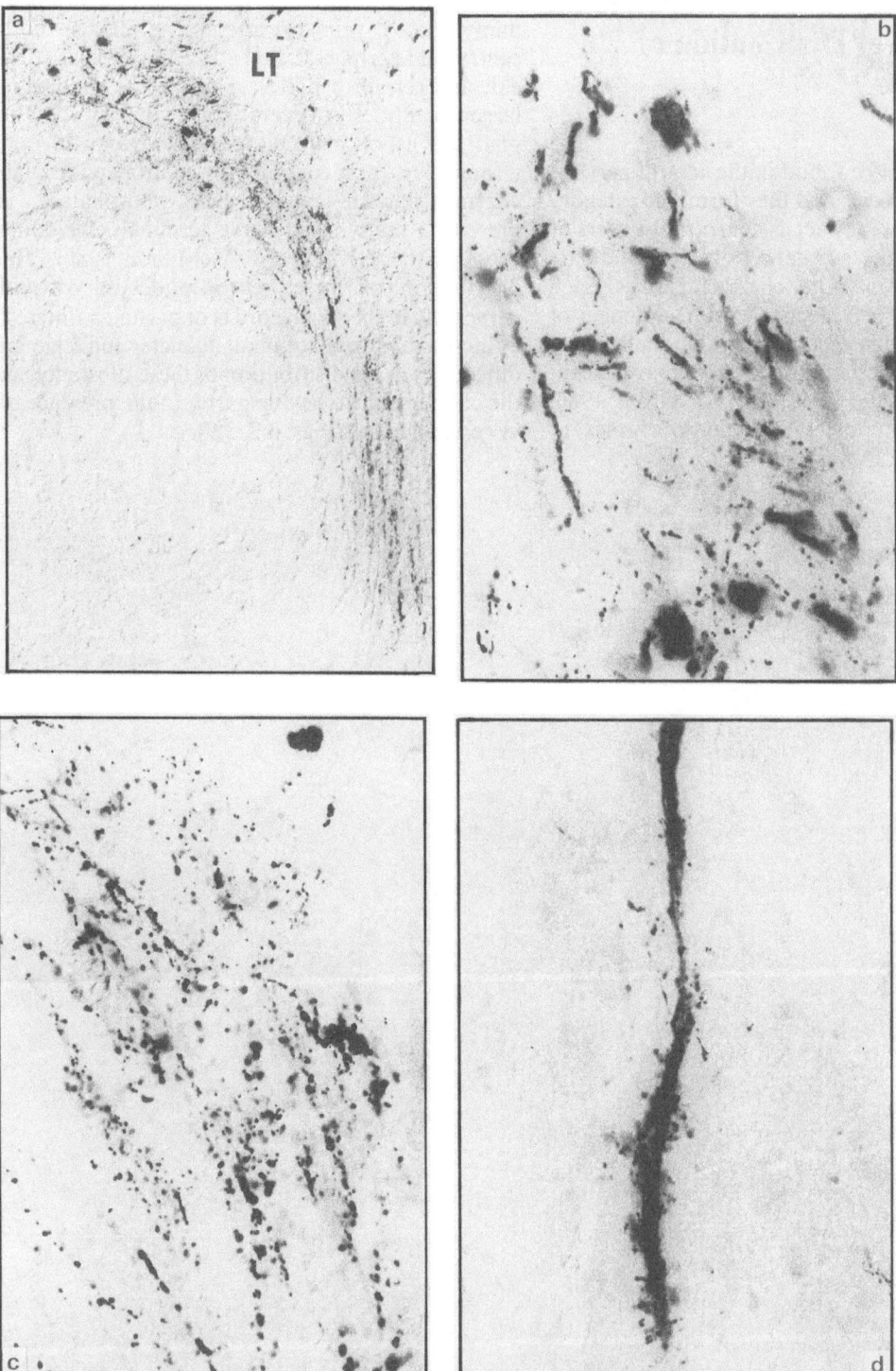

Fig. 3.9. VIP immunoreactivity in the human sacral spinal cord. **a** Low power photomicrograph of the dorsal horn. Immunoreactive fibres are present in Lissauer's tract (*LT*), seen as discrete bundles. In the white matter, some axon collaterals course medially, but for the most part laterally, in the dorsal horn. (×115.) **b** High power photomicrograph of VIP-immunoreactive fibres and bundles of fibres at the outermost region of the dorsal grey matter. (×560.) **c** High power photomicrograph of immunoreactive fibres in the lateral dorsal horn. (×560.) **d** Some fibres aggregate to form tracts which span the area between the dorsal grey commissure and the lateral dorsal horn. (×390.) Formalin-fixed spinal cord obtained 48 h post mortem. Cryostat (20 μm) sections immunostained using the peroxidase/antiperoxidase method. (Courtesy of Sally Gibson and Prof. J. M. Polak, Department of Histochemistry, Royal Postgraduate Medical School, London)

Visceral afferents comprise probably only a few percent of spinal afferents in the cat (Jänig and Morrison 1986), and one of the early studies on the distribution of neuropeptides (Hökfelt et al. 1976) found that about 10%–20% of rat dorsal root ganglion cells contain substance P, and a further separate population, numbering about 10% of the total, contain somatostatin. In the rat, about 10%–20% of dorsal root ganglion cells innervating the bladder contain substance P, and few, if any, contain somatostatin (Sharkey et al. 1983). Thus, substance P seems to be present in a higher proportion of visceral afferents than elsewhere, a fact which seems to be true for other peptides as well (de Groat 1986). The potential importance of these studies to pathophysiology can be seen from studies on rats treated with a neurotoxin, capsaicin, which, when administered on the second day of postnatal life, causes a failure of development of small afferent axons (Fitzgerald 1983). Capsaicin treatment causes the substance P neurons innervating the bladder to disappear; these animals develop large distended bladders, which indicates the importance of afferent neurons in the regulation of bladder capacity and motility.

Substance P neurons are in fact not the the the only peptidergic neurones affected by capsaicin, and a variety of other peptides have been found in visceral afferents that innervate pelvic viscera. Vasoactive intestinal peptide (VIP), cholecystokinin (CCK), calcitonin gene-related peptide (CGRP), PHI (a peptide derived from the same precursor molecule with VIP) and enkephalins (ENK) are amongst the peptides which have been demonstrated in visceral afferent fibres. VIP particularly seems to be associated with visceral afferent fibres, and some of the substance P-containing fibres in the viscera also contain CGRP (de Groat 1986; Su et al. 1986).

Many, if not all, pelvic nerve afferent neurons also contain peptides in their central nerve terminals; VIP, substance P, cholecystokinin and the enkephalins can be found in the areas where these nerve fibres terminate, and some of the neurons contain more than one peptide (de Groat 1986). The neuropeptides present in the pelvic nerve are found less commonly in pudendal nerve afferents. VIP is especially associated with visceral afferent fibres in the lateral collateral pathway (Gibson et al. 1984a; de Groat 1986), and can be released into the cerebrospinal fluid during stimulation of the vesical branches of the pelvic nerve (Blank et al. 1984). It is believed that VIP is released only from unmyelinated fibres (Honda et al. 1983, Morgan and O'Hara 1984), and the VIP present in them accounts for about 90% of the VIP in the sacral cord of the cat, a species which, like humans, has

particularly high concentrations of this peptide in the sacral segments (Anand et al. 1983; Gibson et al. 1984a). Figures 3.8–3.10 show the distribution of substance P and VIP in the sacral cord. Approximately a quarter of the pelvic afferent fibres from the bladder contain VIP, and in over half of them there is evidence for coexistence of substance P and VIP. However some substance P-containing terminals are found in different parts of the spinal cord from those containing VIP. Coexistence of other peptides can also occur, e.g. primary afferents containing Leu-enkephalin may also contain substance P, VIP or cholecystokinin, but not somatostatin. This is particularly interesting, as VIP and substance P are usually regarded as excitatory neurotransmitters, whereas the actions of the enkephalins are usually inhibitory. Enkephalins can be released into the cerebrospinal fluid during stimulation of afferents; the presence of opiate receptors on primary afferent terminals, and the effects of opiates in depressing the release of substance P suggest that there may be interactions between these peptides in the dorsal horn. It has been proposed that corelease of peptides in the spinal cord may be involved in feedback inhibitory mechanisms at primary afferent terminals (de Groat 1986).

Summary

Recent experiments using axonal transport of chemicals such as HRP have shown a considerable amount of consistency between the arrangement of afferents from the lower urinary tract. Although there are differences in segmental origin in laboratory animals and humans, the arrangement of axonal terminals of the bladder and urethral afferents is remarkably consistent. The majority of afferent fibres from the lower urinary tract enter a couple of spinal segments in the sacral region and several segments at the thoracolumbar junction and upper lumbar cord; their central axon collaterals pass up and down the tract of Lissauer for several segments on either side of the entry point; bundles of afferents descend ventrally from this tract at intervals of 200–300 μm, taking a predominantly lateral path round the dorsal horn to reach the deeper laminae and the sacral autonomic nucleus. VIP, the most common peptide in the pelvic nerve afferents, has a similar distribution, and is found in larger concentrations in the sacral cord than in other sites; in this the cat is the closest of the animal species to humans. The other peptides that are present in visceral afferent terminals include substance P, calcitonin gene-related peptide, cholecystokinin and enkephalin; some of these coexist in the same neuron.

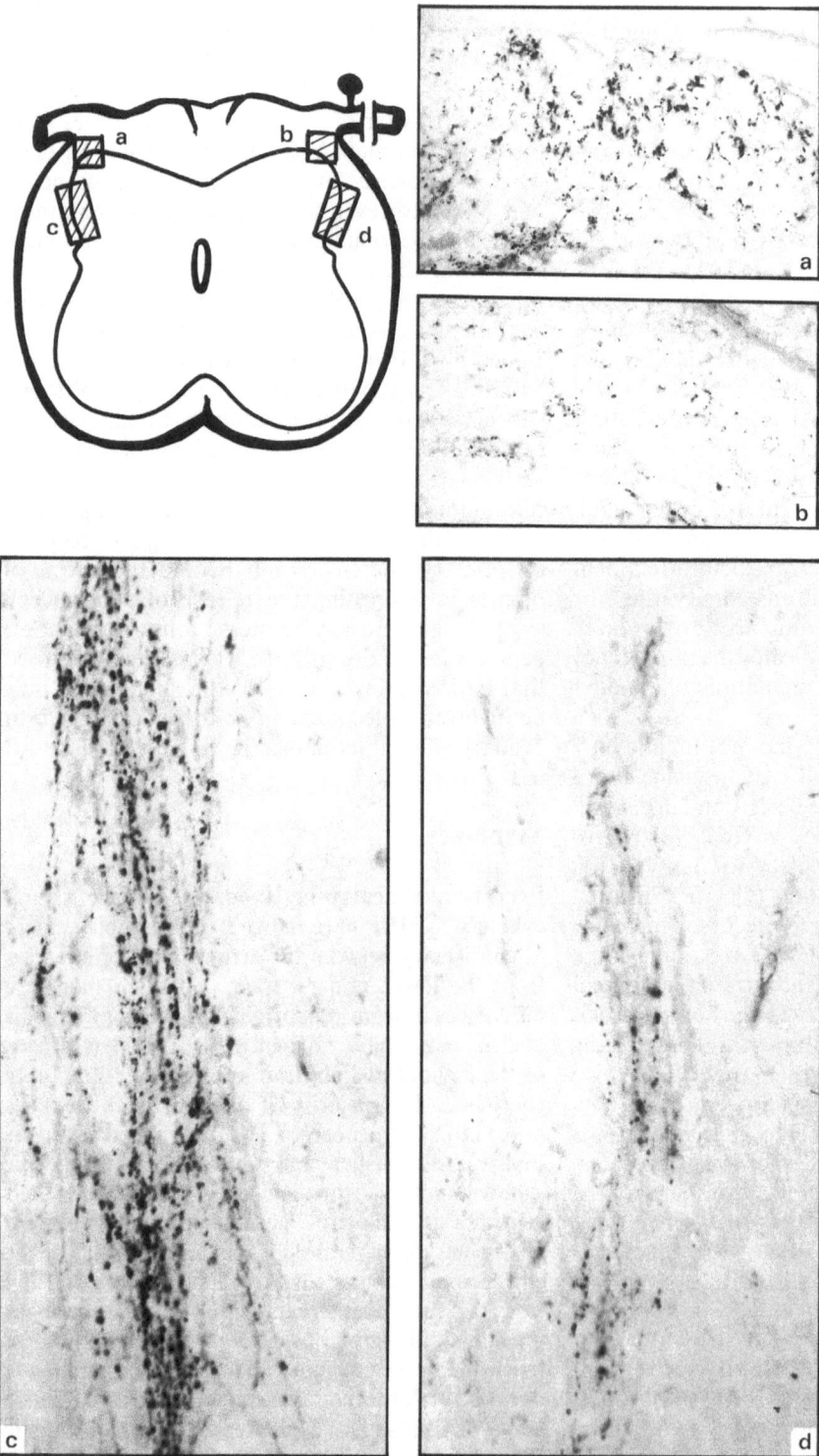

Fig. 3.10a–d. High power photomicrographs of VIP-immunoreactive fibres in the second sacral segment of the cat spinal cord. A diagrammatic representation of the areas shown is provided in the drawing. Four weeks after section of the pelvic nerve there is a marked loss of VIP-immunoreactive fibres from Lissauer's tract (*b*) and the lateral dorsal horn (*d*) compared with the intact side (*a* and *c* respectively). Cryostat sections (20 μm), bezoquinone immersion fixation. Immunostained with the peroxidase/antiperoxidase method. (Courtesy of Sally Gibson and Prof. J. M. Polak, Department of Histochemistry, Royal Postgraduate Medical School, London.) Magnification: **a,b** × 560; **c,d** × 780

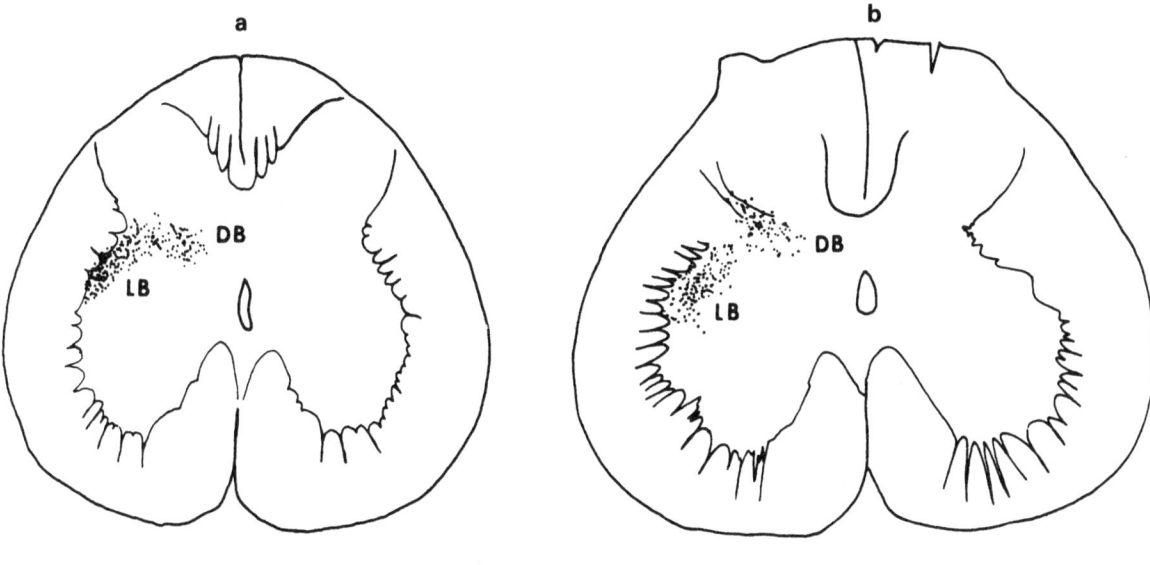

Fig. 3.11a,b. The distribution of the cell bodies of pelvic parasympathetic efferents. Composites (1 mm in length) from the middle portion of the sacral parasympathetic nucleus (SPN) from two animals. **a** Cells from dorsal band (*DB*) and the lateral band (*LB*) overlap and the SPN has an inverted L shape. **b** Cells from these two portions of the SPN are separated and the dorsal band extends laterally beneath the dorsal horn, resulting in a nucleus with a T shape. *Bar* = 3 mm. (Nadelhaft et al. 1980)

Efferent Pathways from the Cord to the Lower Urinary Tract

Pelvic Nerves

The portion of the pelvic nerves central to the pelvic ganglia contain parasympathetic preganglionic efferents that originate from the spinal cord, and postganglionic sympathetic efferents that originate from the sympathetic chain. This section will deal mainly with the former, and the sympathetic neurons will be considered in the review of sympathetic innervation of the lower urinary tract.

Parasympathetic preganglionic efferent cell bodies form a continuous column of cells that are located in the deep layers of the dorsal horn and in the intermediolateral column, at segmental levels of the cord that vary with species (Fig. 3.11 and Table 3.3). The efferent axons generally pass towards the ventral roots along the lateral edge of the ventral horn. Some of these neurons give rise to more than one axon, probably during the peripheral course of the nerve, and in several species the ratio of fibres in the pelvic nerve to the number of parasympathetic cell bodies is of the order of 3:1, 4:1, or more. There is generally some degree of spatial separation of neuronal cell bodies concerned with different functions, and those concerned with bladder function have larger cells and axons than those concerned with the regulation of large intestinal function. Little is known of the parasympathetic efferents that innervate the genitalia. In some species, there is evidence that the urethra is innervated mainly from the nucleus intercalatus whereas the bladder is innervated mainly from the intermediolateral nucleus; the colon is innervated from a more dorsal band of cells at the base of the dorsal horn. In addition, there is evidence in some species of pelvic nerve efferents projecting to the urethra from the ventral horn; these neurons, which originate from the nucleus torsade (Onuf's nucleus), innervate the skeletal muscle of the external sphincters. In humans, parts of this muscle are innervated by the pelvic nerve (see Chap. 1). There is evidence of considerable mixing of parasympathetic and sympathetic components in the pelvic and hypogastric nerves of many species. The ganglia of the pelvic plexus are known to contain not only postganglionic parasympathetic neurons, but also catecholamine-containing neurons, afferent neurons and chromaffin cells; Gabella's review (1976) provides an informative account of this topic. The details of individual species will be dealt with separately; the arrangement in the pelvic plexus of different species has been considered in Chap. 2 (see p. 39). Table 3.3 summarizes the segmental

Table 3.3. The spinal segmental origins of efferent fibres of the pelvic, hypogastric and pudendal nerves in the human, rat, cat and monkey

Species segmental differences			Locations of efferent cell bodies of		
H	R	C & M	Pelvic n.	Hypogastric n.	Pudendal n.
T-10	T-11	T-11		R	
T-11	T-12	T-12		R	
T-12	T-13	T-13		R	
L-1	L-1	L-1		R	
L-2	L-2	L-2		R	
L-3	L-3	L-3		C	
L-4	L-4	L-4		C	
L-5	L-5	L-5		C	
S-1	L-6	L-6	R		R
S-2	S-1	L-7	RH		HM
S-3	S-2	S-1	CHM		CHM
S-4	S-3	S-2	CHM		C
	S-4	S-3	CM		

From: Applebaum et al. (1980), Baron et al. (1985a,b), Hulsebosch and Coggeshall (1982), Nadelhaft and Booth (1984), Nadelhaft et al. (1983), Neuhuber (1982), Roppolo et al. (1985), Sharkey et al. (1983), Sherrington (1892), Ueyama et al. (1984).

C, cat; H, human; M, monkey; R, rat.

levels from which the efferent nerves to the viscera originate in the different species.

Cat

In the cat, the sacral parasympathetic nucleus is found mainly in the second sacral segment with some cells in S-1 and/or S-3; retrograde tracing studies using HRP show a division of these cells, which number about 1100, into two main groups: (a) a lateral band in lamina VII at the lateral border of the intermediolateral grey matter; and (b) a dorsal band in laminae V and VI at the base of the dorsal horn. These two populations are often separated by a gap or "interband" region (Fig. 3.11), which are presumed to be interneurons and spinal tract cells. In addition a few preganglionic neurons exist in lamina I on the lateral edge of the dorsal horn, and in the ventral horn. The cells in the lateral band mainly innervate the bladder, whereas those in the dorsal band, which are smaller and about half as numerous, innervate the colon (de Groat et al. 1981). Figure 3.12 shows the dendritic fields of some of these neurons (Nadelhaft et al. 1980). This region of the sacral cord is richly supplied with terminals containing serotonin (Dahlstrom and Fuxe 1965), and enkephalin (see Chap. 5, p. 140).

About 1080 neurons are present in the sacral parasympathetic nuclei of this species and these give rise to about 3160 efferent fibres in the pelvic nerve; about 30% of these are unmyelinated (Nadelhaft et al. 1980). This is in keeping with the data of Schitzlein et al. (1963), who found the mean diameter of these fibres to be 2–3 μm, with a range between 8 μm and less than 1 μm; the conduction velocities varied between 1 and 30 m.s^{-1}. Nadelhaft et al. (1980) found no evidence of recurrent collaterals of parasympathetic efferents within the spinal cord. These fibres communicate with postganglionic neurons in the pelvic plexus; the paracervical ganglion alone contains about 5000 neurons, some of which are surrounded by endings which fluoresce for catecholamines (Gabella 1976). The postganglionic fibres make contact with 6–7 smooth muscle cells at the apex of the bladder, and 1–2 in the trigone (Fehér et al. 1980); these recent figures have been obtained using quantitative methods, and indicate that the density of innervation at the trigone is about three times that at the apex. These figures contrast with a previous estimate, obtained using less quantitative methods, of a 1:1 ratio of nerve fibres to smooth muscle cells throughout the bladder (Elbadawi and Schenk 1968). The bladder ganglia contain adrenergic terminals and adrenergic postganglionic neurons (Hamberger and Norberg 1965).

Some pelvic preganglionic neurons are known to contain Leu-enkephalin, and are believed to give rise to the enkephalin-containing nerve terminals in the bladder ganglia; enkephalins are inhibitory to

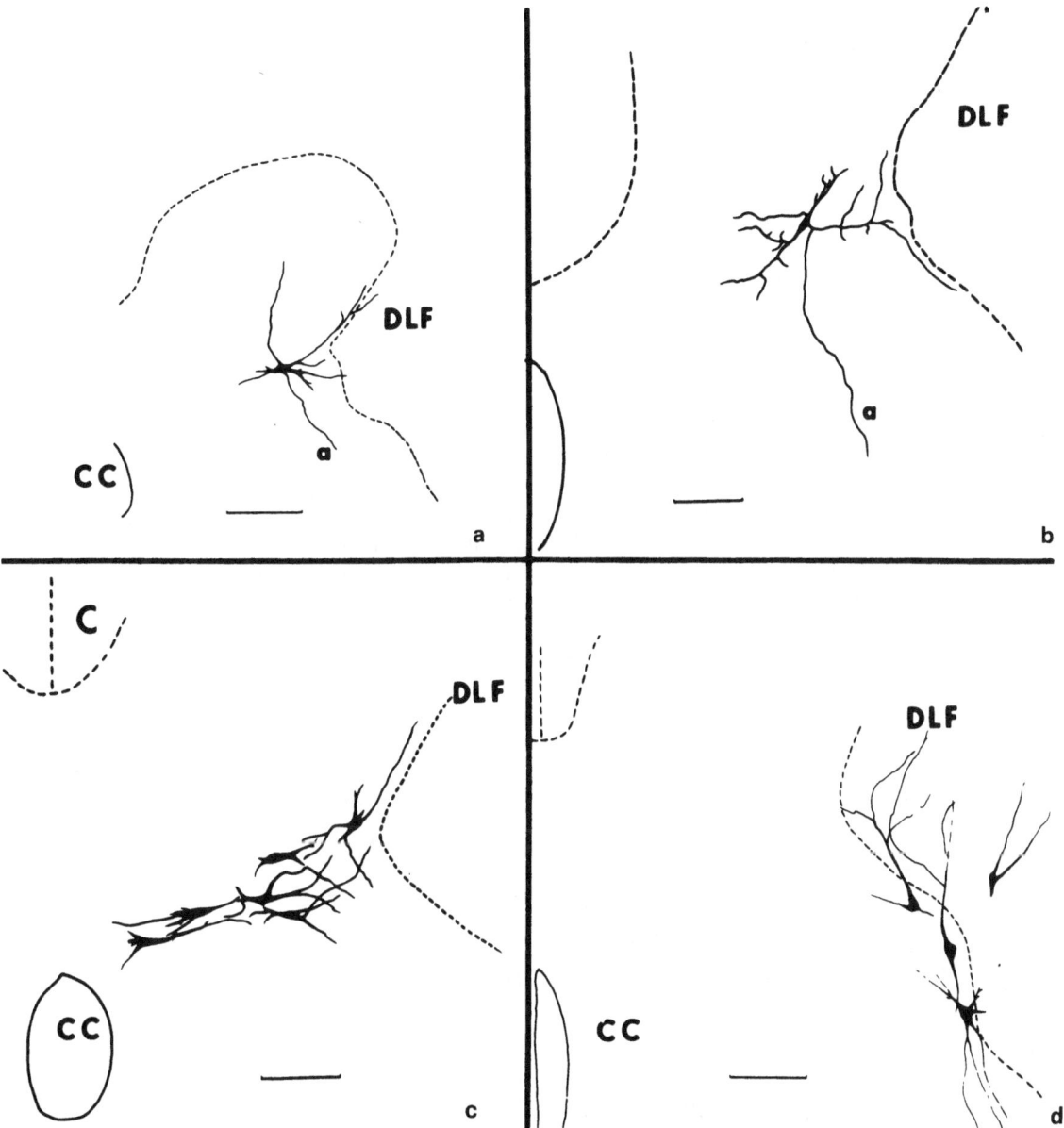

Fig. 3.12a–d. The distribution of the cell bodies and dendrites of pelvic nerve efferents in the deep laminae of the dorsal horn and the intermediolateral column of the sacral cord of the cat. **a, b,** and **c** are dorsal band cells which exhibit the characteristic horizontal orientation. Dendrites are seen to extend horizontally towards the midline and laterally into the marginal layer and into the lateral band. In rare instances dendrites extend into the dorsal horn **(a)**. Axons (*a*) are usually observed at the lateral edge of the grey matter, but occasionally are seen to run more medially **(b)**. **d.** Several lateral band cells are shown. Note the lateral dendritic extensions and one cell which lies in the lateral funiculus. Note also, the three axon-like processes extending ventrally from the ventral-most cell. The *dashed line* approximates the border between white and grey matter. *CC*, central canal; *DLF*, dorsolateral funiculus; *a*, axon. *Bars* = 200 μm. (Nadelhaft et al. 1980)

transmission at this site (de Groat et al. 1983). In the pelvic nerve of the cat, the parasympathetic preganglionic neurons are accompanied by about 2100 sympathetic postganglionic fibres that have their cell bodies in the lower lumbar and upper sacral sympathetic ganglia (Downie et al. 1984; Kuo et al. 1984).

Rat

In the rat, the sacral parasympathetic nucleus exists in L-6 and S-1 (Hancock and Peveto 1979), except for a few cells in animals where it was either pre- or post-fixed. It contains about 550 preganglionic efferent neurons which are located within the intermediolateral grey matter at the base of the dorsal horn. No subdivisions of the nucleus are apparent (Nadelhaft and Booth 1984) as have been reported in the cat (de Groat et al. 1981). The dendrites of these cells enter the dorsolateral funiculus, and extend along the lateral edge of the dorsal horn almost to its apex; in addition they extend medially into the dorsal grey commissure at least as far as the midline. The axons of these efferents generally pass down the lateral border of the ventral horn, but sometimes pass medially to some of the lateral motoneuron cell columns. These efferent axons appear to divide to give rise to about 2400 preganglionic axons in the pelvic nerve, of which about 90% are unmyelinated (Hulsebosch and Coggeshall 1982; Nadelhaft and Booth 1984); the paracervical ganglion of this species contains about 5700 neurons, of which about one-third show specific fluorescence for catecholamines (Gabella 1976). Greenwood et al. (1985) report that there is a sex difference in the numbers of pelvic ganglion cells in the rat; in the female they counted 5892 ± 797, and in the male, 14654 ± 963, and preliminary studies suggest that there may be 50% more preganglionic pelvic nerve axons in the male than in the female. Puriton et al. (1971) have suggested that some of the perikarya in the pelvic ganglia are afferent.

Hancock and Peveto (1979) found 350 preganglionic neurons in this species; some were present in the ventral horn in association with somatic nuclei. The latter observation may fit with reports in some species that the pelvic nerve also innervates the skeletal muscle of the urethral sphincter (see Chap. 1, p. 15).

Monkey

In this species, the sacral parasympathetic nucleus can be found in S-1 to S-3, with the majority of cells in S-2. It consists of about 820 cells, whose dendrites extend deep into the lateral funiculus, into the dorsal grey commissure, and dorsolaterally along the lateral edge of the dorsal horn. As a rule, the axons travel laterally round the edge of the ventral horn, and probably divide before they reach the pelvic ganglia. Nadelhaft et al. (1983) report that a split into two groups of cells can be seen in some sections, as had been reported in the cat, and in both species the cells form a continuous column, of maximum breadth near the centre of the nucleus. These authors also report the existence of few neurons located deep in the ventral horn; whether or not these neurons project to the urethra is unknown. Schnitzlein et al. (1960) report that there are nearly 11 000 parasympathetic (and possibly somatic

motor) fibres in the pelvic nerves; this would be in keeping with the view that parasympathetic neurons give off axon collaterals along their peripheral course (as already discussed). The pelvic nerve in this species contains considerable numbers of sympathetic postganglionic neurons (Schnitzlein et al. 1960); the number of neurons in the paracervical ganglion is around 14 400 (Gabella 1976).

Guinea Pig

McLachlan (1985) found small numbers (about 40 on average) of parasympathetic efferents in the distal stump of a transected hypogastric nerve: these were identified as cell bodies in the sacral cord following application of HRP to the distal stump of the sectioned hypogastric nerve. This represents about 2% of the fibres in this nerve, and illustrates the mixing of autonomic components known to be present in these nerves.

The pelvic ganglia of this species are known to contain monopolar neurons with few dendrites and supplied by a few adrenergic axons; the ganglion cells occur in clusters and some are intensely fluorescent when examined for the presence of catecholamines (Sjöstrand 1965; Costa and Furness 1973). Structurally, these ganglia are less complex than the sympathetic ganglia (Blackman et al. 1969), and appear to function more as relay stations than integrating centres (see Chap. 5, p. 135).

Dog

Petras and Cummings (1978) searched for efferent cell bodies in the sacral cord following the injection of HRP into the bladder wall or the urethra. These workers concluded that parasympathetic preganglionic neurons innervating the bladder originate mainly from the intermediolateral nucleus of the S-2, S-3 and Cx-1 segments, whereas the urethra is innervated mainly by the nucleus intercalatus in the sacral cord. In addition the nucleus torsade (Onuf's nucleus) in the ventral horn was found to contain motoneurons which stained for HRP after the tracer was injected into the urethra. Presumably these motoneurons projected to the rhabdosphincter via the pudendal nerve.

Human

Sheehan (1941) and Pick (1970) found the parasympathetic outflow to be concentrated in S-3 and

S-4, with occasional extensions into S-2 or S-5. Laruelle (1937, 1948) and Laurelle and Reumont (1958) located the parasympathetic preganglionic cell column in humans to the anteromedial border of the ventral horn, dorsal to the somatic motoneurons.

The parasympathetic efferent cell bodies are grouped together at the base of the dorsal horn and the intermediolateral cell column of a couple of sacral segments. There is some spatial separation of neurons concerned with the control of the bladder and the colon. The neurons innervating the bladder are larger than those innervating the colon and are situated laterally within the sacral autonomic nucleus; those innervating the urethra are located mainly in the nucleus intercalatus. The presence of stained motoneurons in Onuf's nucleus following injection of HRP into the urethral sphincter is probably due to the pudendal innervation of the urethral sphincter. However, some studies report that ventral horn cells can be stained following the application of HRP to the pelvic nerve.

The parasympathetic efferent fibres leave the cord by passing laterally round the ventral horn, and appear to divide into a series of axon collaterals in their peripheral course to the pelvic ganglia. In some species it has been suggested that some of the parasympathetic efferents travel in the pudendal nerve, but their destination is unknown.

There is evidence for considerable numbers of postganglionic sympathetic fibres mixed in with the parasympathetic efferents in the pelvic nerve. Table 3.4 summarizes the numbers of sympathetic and parasympathetic efferents in the hypogastric, pelvic and pudendal nerves of several species.

Pudendal Nerves

Sphincteric motoneurons are present in the pudendal nerves of most species and in the pelvic nerves of some, including humans; indeed, the pelvic nerve is regarded by some authors as the main pathway to the urethral sphincter in humans (see the following discussion). The segmental origin of pudendal motoneurons shows some species variation; in humans, Onuf's nucleus (the cell column that contains the cell bodies of motoneurons innervating the urethral and anal sphincters and pelvic floor muscles) is found in the second and third sacral segments, but occasionally extends into S-1 (see Schrøder 1981, 1985). In the rat, the pudendal motoneurons are present in the sixth lumbar

segment; in the cat the first and second sacral segments contain most of the neurons, although the nucleus can extend into L-7 and S-3. (The rostral position of the nucleus in the rat is in keeping with the position of the parasympathetic outflow—L-6 and S-1—in this species.) In the monkey, pudendal motoneurons originate from L-7 and S-1 (see Fig. 3.15 and Table 3.3). The pelvic nerve efferents innervating the external sphincter have been considered in the previous section.

Cat

The analogue of Onuf's nucleus in the cat was described by Romanes (1951) as "cell group Y", and the vast majority of these motoneurons reach the skeletal muscle of the sphincters and perineum via the pudendal nerve; only a few travel in the pelvic nerve (Sato et al. 1978; Ueyama et al. 1984). Motoneurons which innervate the anal and urethral rhabdosphincters are separated into dorsomedial and ventrolateral columns of cells, respectively, in the sacral cord (Kuzukara et al. 1980) and are present in the first and second sacral segments of most animals (Ueyama et al. 1984); there was no evidence for neurons that innervated both sphincters, so similarities of electromyographic responses in these sphincters cannot be due to motoneurons that innervate both structures. Figure 3.13 shows the sites at which Mackel (1979) could record antidromic action potentials in pudendal motoneuron cell bodies, having identified their location by microiontophoesis of fast green. Large neurons such as motoneurons do not contain most of the peptides found in the terminals of small primary afferent fibres; however Gibson et al. (1984a) have found CGRP in the cell bodies of pudendal, and other, motoneurons (Fig. 3.14). The role of this peptide in motoneurons is uncertain, as neurotransmission at the neuromuscular junction is mediated by acetylcholine; the possibility that the peptide may be concerned with trophic functions has not been fully investigated.

Rat

The organization of spinal nuclei concerned with control of the rhabdosphincter has recently received considerable attention, particularly by Schrøder (see Schrøder 1985 for a recent review, and Figs. 3.15 and 3.16). In the rat the neurons which innervate the urethral sphincter also lie lateral to those supplying the external anal sphincter, but a clear

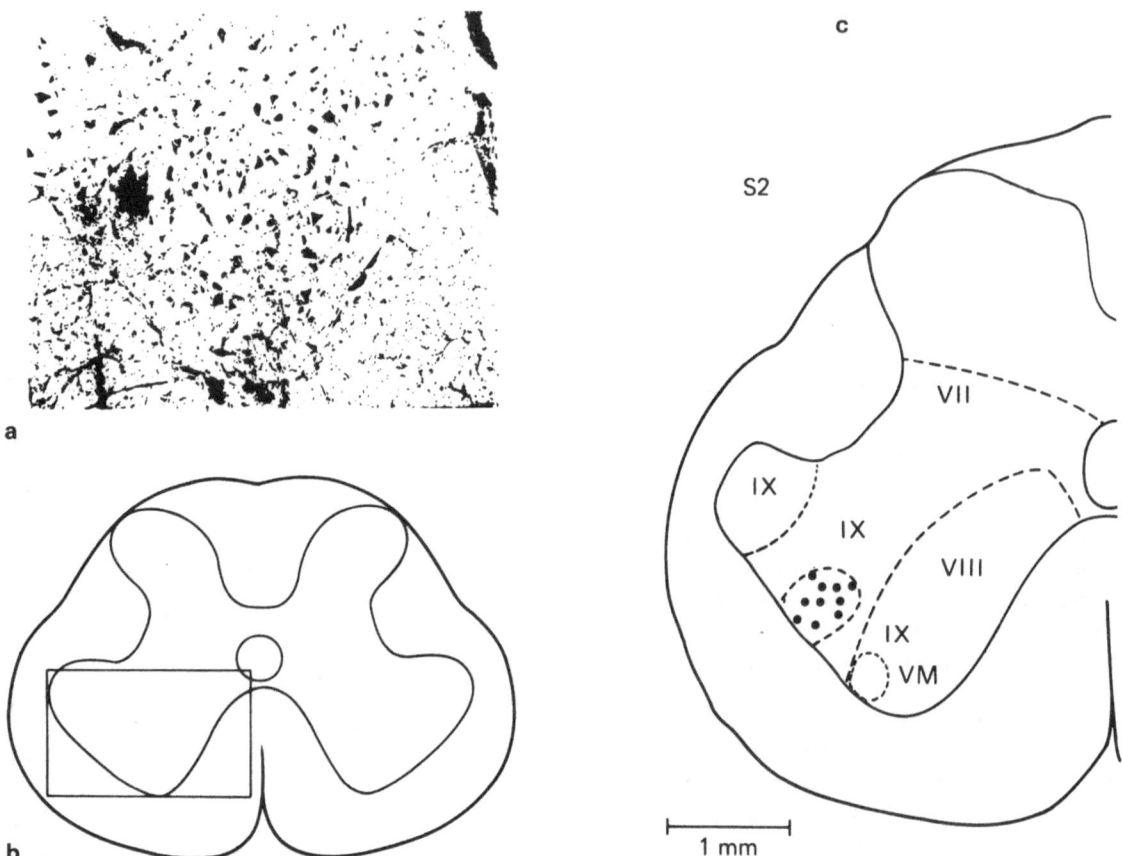

Fig. 3.13a–c. The location of pudendal motoneurons in the cat, as identified using electrophysiological techniques. **a** Photomicrograph of a fast green dye spot in the ventrolateral horn of the S-2 spinal cord. **b** Section through the S-2 segment of the spinal cord from which the photomicrograph in **a** was taken. *Rectangle* indicates the area shown in **a**. **c** Reconstruction of the area where several dye spots were located. This was done by superimposing several sections containing a dye spot. Each *dot* corresponds to a dye spot marking the location of a sphincter motoneuron. (Mackel 1979)

split was only present in one of the four animals studied by Roppolo et al. (1985). Sexual dimorphism has been reported in the rat spinal nucleus that innervates levator ani and bulbocavernosus, with larger numbers and sizes of motoneurons in the male; testosterone binds to these neurons in the male rat (Breedlove and Arnold 1980). Micevych et al. (1986) found differences in the peptide content of neurons and terminals in these nuclei; the nuclei that innervate the ischiocavernosus had cells surrounded by terminals that contained enkephalin or 5-hydroxytryptamine, whereas the nuclei that innervated the sphincter urethrae muscle did not.

In the male rat the pudendal nerve consists of about 4000 axons, two-thirds of which are unmyelinated: 84% are sensory, and disappear when the dorsal root ganglia of segments T-12 to S-2 are ablated; 12% are motor axons and over 80% of

these are unmyelinated (a rather surprising finding which may indicate the presence of parasympathetic fibres in this nerve), and 4% are postganglionic sympathetic axons—one-quarter of them are myelinated (Hulsebosch and Coggeshall 1982). The presence of myelinated postganglionic axons is well documented in visceral nerves and was first reported by Langley and Anderson (1896), but has not been reported in other "somatic" nerves. These authors indicate that the fibre composition of the pudendal nerve, in the rat at least, is unlike that of most somatic nerves; in particular they believe that many of the unmyelinated efferent fibres have a parasympathetic origin; however, this is in contrast to the results of Ueyama et al. (1984) in the cat, where no mention was made of cells in the intermediolateral column being stained after application of HRP to the central cut end of the pudendal nerve.

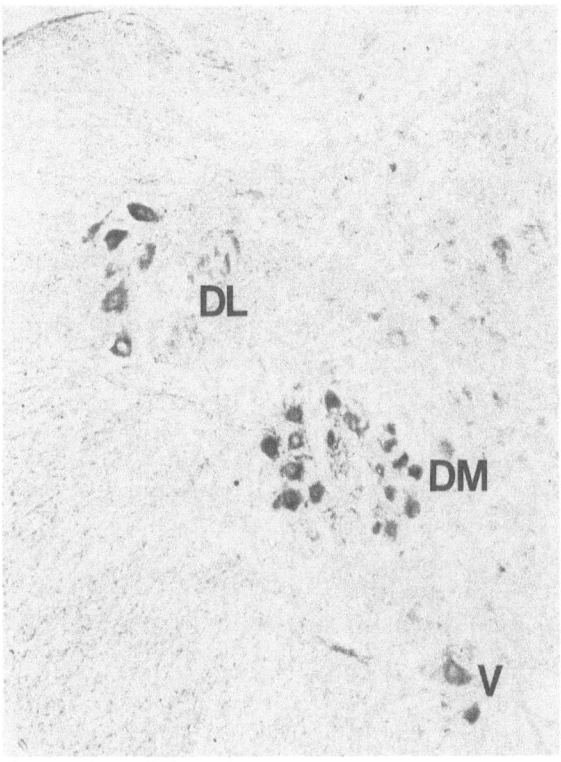

Fig. 3.14. The distribution of calcitonin gene-related peptide (CGRP) in the ventral horn of the dog. *DL*, dorsolateral; *DM*, dorsomedial; *V*, ventral motoneuron cell groups. Many, but not all, motoneurons in the *DL* and *DM* groups, which project in the pudendal nerve, are immunoreactive for CGRP. Cryostat (20 μm) sections; benzoquinone immersion fixation. Immunostaining with the peroxidase/antiperoxidase method. (Courtesy of Sally Gibson and Prof. J. M. Polak, Department of Histochemistry, Royal Postgraduate Medical School, London) (× 80)

Monkey

In this species there are only about 400 somatic motoneurons, as compared with 9200 afferent cell bodies; they are present mainly in the seventh lumbar and first sacral segments of the cord in a small area (Onuf's nucleus), medial to the lateral motor cell columns (Fig. 3.15). This segmental origin is slightly higher than the main pudendal afferent input, which enters the cord mainly through the first and second dorsal roots. The dendrites of pudendal motoneurons extend towards the central canal and into the intermediolateral column (Roppolo et al. 1985). Ueyama et al. (1985) report that male monkeys have more pudendal motoneurons than females, but the data in Roppolo et al. (1985) does not fully support that conclusion. Roppolo et al. (1985) report that the pudendal motoneurons in the monkey were medium-sized, measuring 44 × 26 μm on average, which suggests

that the axonal conduction velocity and the motor units might be small in size.

Dog

In the dog, like the cat, the motoneurons which innervate the anal and urethral sphincters are separated into dorsomedial and ventrolateral columns of cells, respectively (Fig. 3.15). There is no evidence for neurons innervating both sphincters (Kuzuhara et al. 1980).

Human

The motoneurons that innervate the striated muscle of the external urethral and anal sphincters and perineum originate from a localized column of cells in the sacral cord in humans called Onuf's nucleus (Fig. 3.16), first described by Onufrowicz (Onuf 1899, 1900; Schrøder 1981). Recent neuropathological evidence suggests that Onuf's nucleus is concerned with the function of the rhabdosphincter (Mannen et al. 1982). The axons of these cells travel peripherally through Alcock's canal, a site where there is evidence that they may be subjected to mechanical trauma. There is evidence that some controlling the urethral sphincter travel in the pelvic nerves rather than the pudendal nerves; the latter innervate the pelvic floor in humans (Donker et al. 1976). Despite this, there is physiological evidence that stimulation of the pudendal nerves in humans increases the resistance to urine flow (Brindley et al. 1974).

Pudendal nerve motoneurons originate from Onuf's nucleus in the ventral horn of one or two segments of the sacral or lower lumbar cord (depending on species). These neurons are smaller than most α-motoneurons, which would be in keeping with them innervating the small-diameter skeletal muscle fibres of the external anal sphincter, pelvic floor, and possibly parts of the urethral sphincter. The muscle fibres of the rhabdosphincter are type I fibres; these have a small diameter, and are thought to produce sustained contractions over long periods. There is evidence in some species for a spatial separation of neurons concerned with the innervation of the urethra, and those which control the external anal sphincter. Sexual dimorphism has been observed in the Onuf's nucleus of some species, but it is not certain that this affects urethral motoneurons. The pudendal nerves also contain some sympathetic efferent axons, and possibly some parasympathetic preganglionic efferents (in some species).

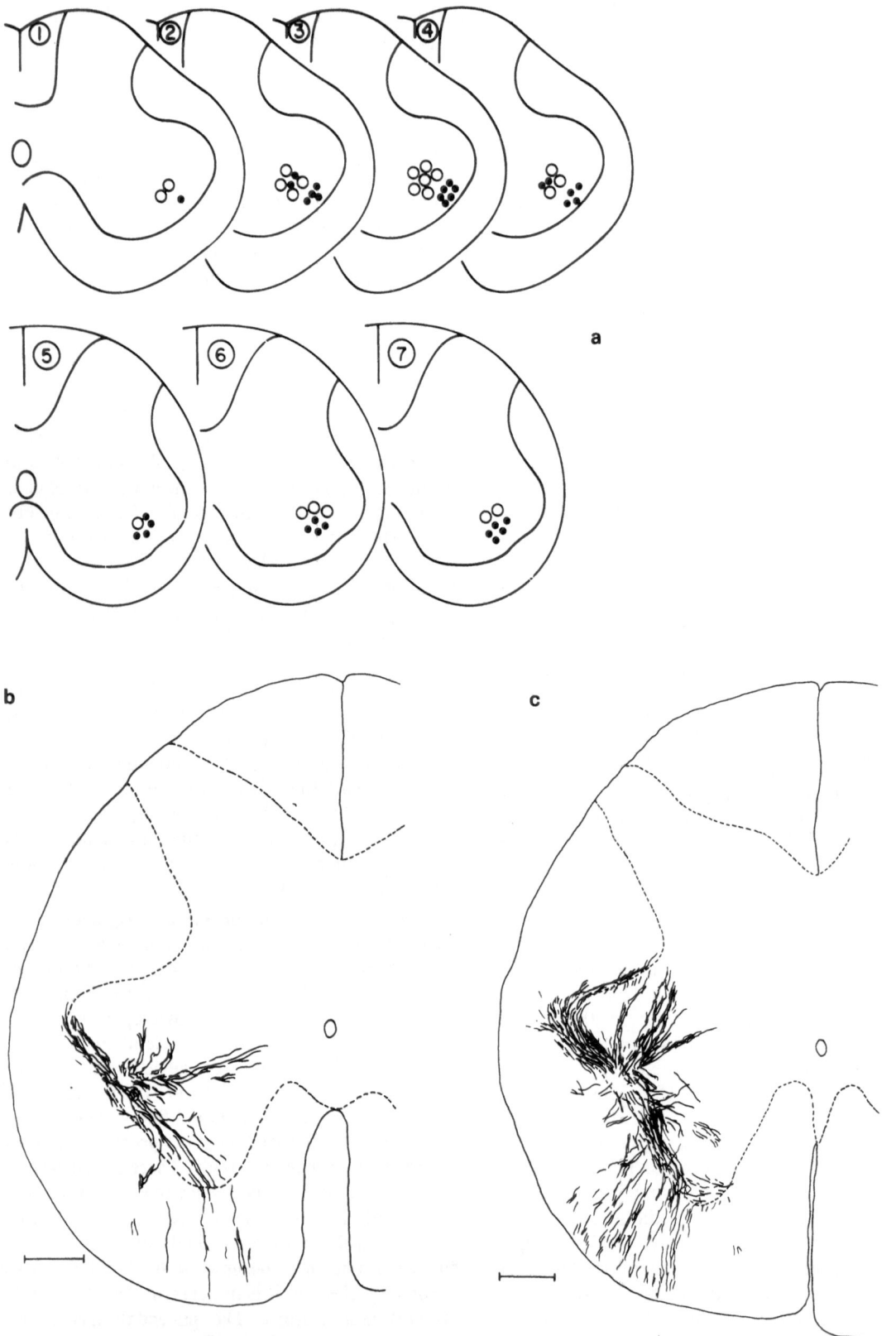

Fig. 3.15a. The location of pudendal motoneurons innervating the urethral (*solid circles*) and anal sphincters (*open circles*) in the cat (diagrams *1–4*) and dog (diagrams *5–7*) (Kuzuhara et al. 1980.) **b,c** The dendritic fields of and axons of pudendal motoneurones in the seventh lumbar segment of the monkey. (Roppolo et al. 1985)

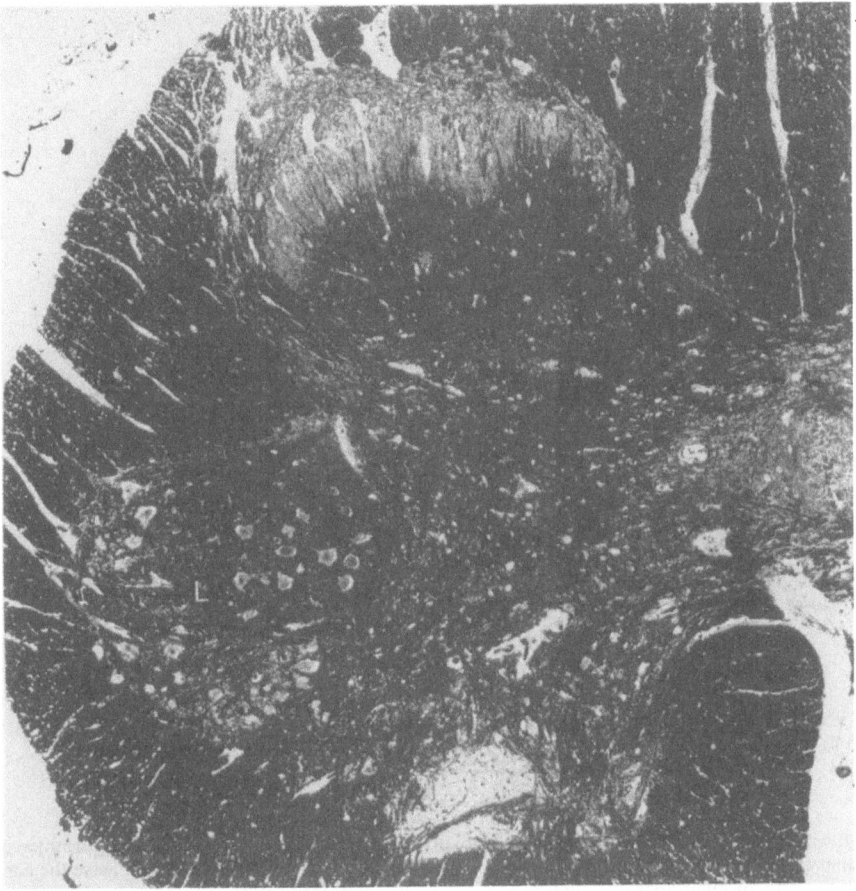

Fig. 3.16. Onuf's nucleus in the human. Transverse section of S-2 of the human spinal cord. Note the difference in size between the motoneurons in the lateral nucleus (*L*) and in Onuf's nucleus. (Schrøder 1985.) (Mahon, × 44)

Sympathetic Pathways to the Lower Urinary Tract

The sympathetic postganglionic neurons that innervate the lower urinary tract (Fig. 3.17) are known to arise from the inferior mesenteric ganglia, the sacral sympathetic chain ganglia and the vesical ganglia. The neurons in the prevertebral ganglia probably take part in local reflex activities, as they can be excited by afferent stimulation following decentralization. However, the main input to these cells comes from the preganglionic fibres of the lumbar splanchnic nerves which arise from the upper lumbar and lower thoracic cord (Fig. 3.18).

In addition to the preganglionic fibres that make contact with cells in the inferior mesenteric ganglion, the sympathetic outflow contains fibres that originate in the spinal cord and project through the prevertebral ganglia to the vesical ganglia. The cells of the pelvic plexus that receive this sym-pathetic preganglionic input have been studied by Blackman et al. (1969), and have a structure that is similar to other parasympathetic ganglia (such as the avian ciliary ganglion) rather than the inferior mesenteric ganglion.

In some species, preganglionic fibres that project to the pelvic plexus appear to originate mainly from the nucleus intercalatus in the intermediate grey matter of the cord (Fig. 3.19). The sympathetic pathways show a degree of crossover, possibly 20% or more in some species. There is a suggestion that there are more preganglionic fibres in the hypo-gastric nerves of male animals (Table 3.4).

There is a rich sympathetic supply to the lower urinary tract, and in some species it is known that a proportion of the postganglionic efferents (possibly 10%) arise from the sacral sympathetic ganglia and travel in the pelvic nerves (Schnitzlein et al. 1960; Downie et al. 1984; Kuo et al. 1984; Baron et al. 1985a). There does not appear to be any sex differ-ence between the numbers of postganglionic sym-

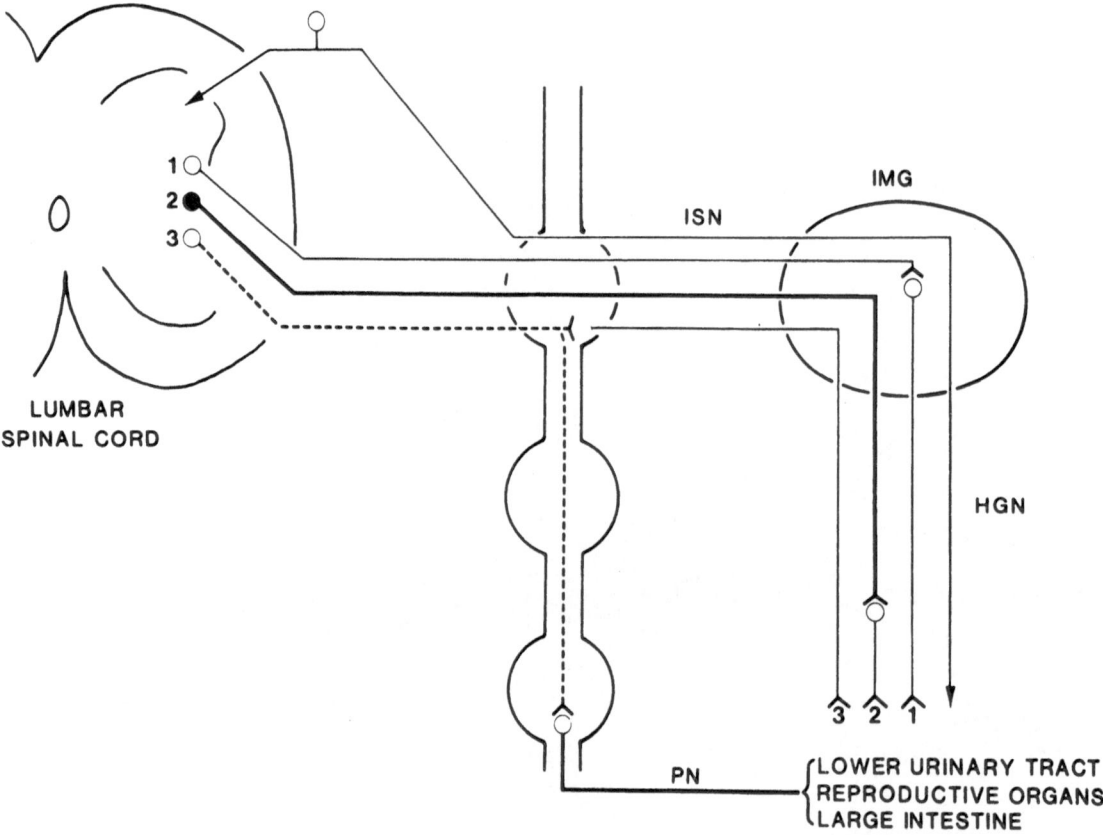

Fig. 3.17. Schematic diagram of the sympathetic innervation of the pelvic organs in segments L-2 to L-5 of the cat. Four different pathways are illustrated: one primary afferent (*arrows*) and three preganglionic (*1, 2,* and *3*). Axons from all PGN leave the cord in the ventral roots and enter the sympathetic chain (paravertebral ganglia) as white rami. Here some of these axons (*3*) synapse upon ganglion cells whose axons reach their targets via the pelvic nerve (*PN*) or hypogastric nerve (*HGN*). Other axons (*1*) pass through the sympathetic chain and inferior splanchnic nerves (*ISN*) and synapse within the inferior mesenteric ganglion (*IMG*) upon neurons whose postganglionic axons follow the HGN to the pelvic organs. Finally, some PGN (*2*) (*solid circle* in spinal cord) send axons through the sympathetic chain, IMG and HGN to synapse upon ganglion cells near or within the pelvic organs. The location and morphology of the PGN in this last pathway and the visceral primary afferents carried in the HGN are discussed by Morgan et al. (1986). (Morgan et al. 1986)

pathetic efferent fibres in the main nerve trunks, but there certainly appears to be a difference in the distribution of sympathetic endings to the posterior urethra. In men, the posterior urethra receives a rich sympathetic innervation, whereas such endings are sparse in women.

Cat

The detailed anatomy of the sympathetic outflow in this species has recently been re-examined by Baron et al. (1985c). Baron et al. (1985a) found about 1700 preganglionic efferents in each hypogastric nerve, the majority of which originated from the fourth lumbar segment; these are accompanied by about 17 000 postganglionic sympathetic efferents, the majority of which originated from the

inferior mesenteric ganglion (IMG). About 20% of the preganglionic axons in the hypogastric nerve have cell bodies in the opposite side of the cord, and a similar proportion of the postganglionic fibres arising from the lumbar sympathetic trunk originate contralaterally. In contrast, Morgan et al. (1986) estimated the number of preganglionic neurons in the hypogastric nerve to be about 400; the difference is remarkable, and it seems doubtful whether it can be explained simply by a difference in the use of correction factors for double counting. The majority of these preganglionic neurons were in the fourth lumbar segment. Most hypogastric preganglionic neurons have cell bodies in the principal intermediolateral nucleus or between it and the central canal, in a position that corresponds to the nucleus intercalatus spinalis (IC nucleus), where Petras and Cummings (1978) found labelled cell

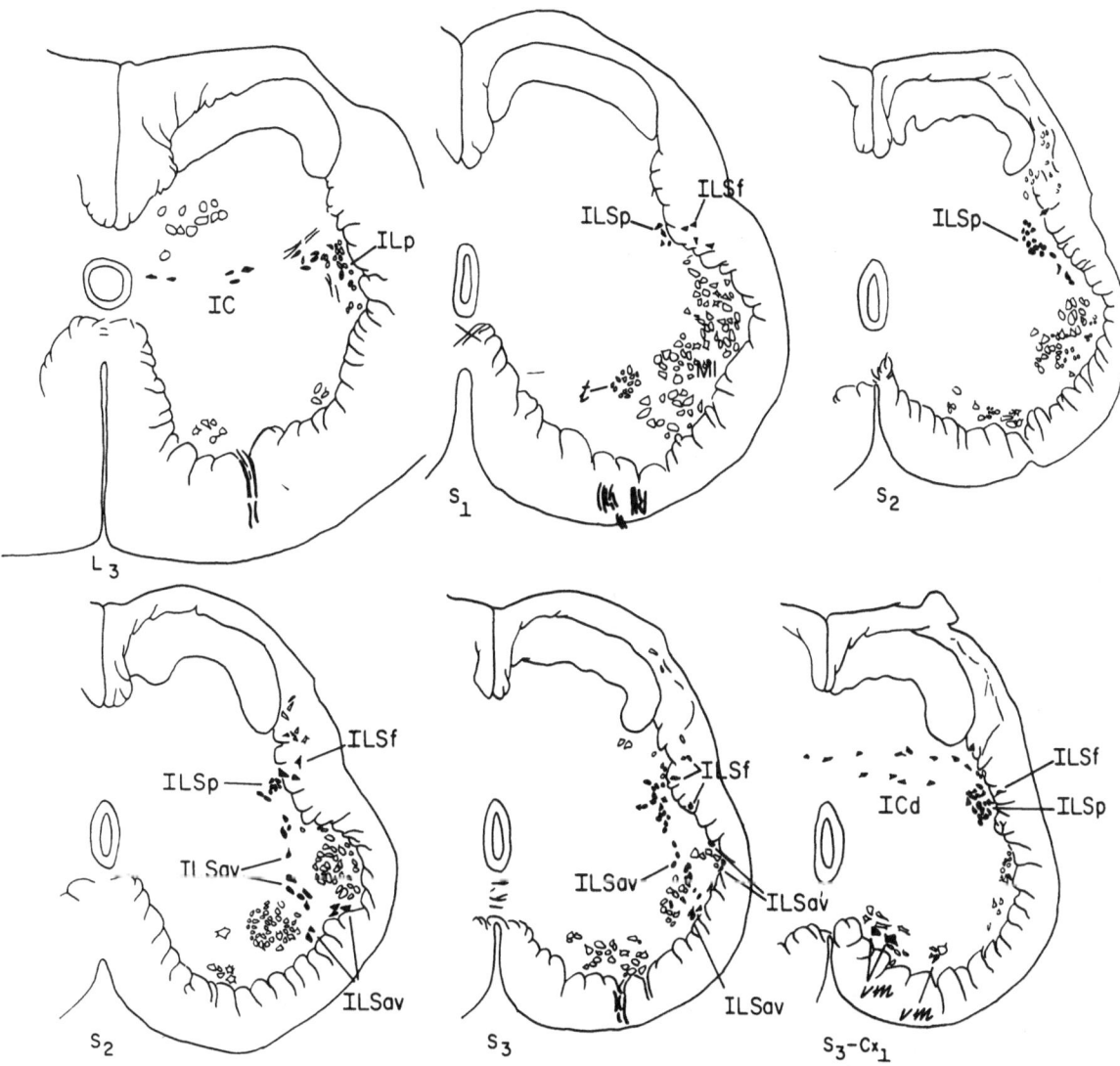

Fig. 3.18. The arrangement of autonomic nuclei in the third lumbar segment and in the sacral cord of neonatal dogs. HRP was injected into the bladder or the urethra and drawings were made from transverse sections of cranial lumbar and from sacral and coccygeal spinal cord segments. The preganglionic sympathetic and parasympathetic nuclear groups described here are indicated by *blackened circular, triangular or polyhedral figures* which represent cell bodies. The bladder and urethra are innervated by several nuclear groups. The bladder receives fibres from two lumbar (*ILp, IC*) and from three sacral (*ILSp, ILSF* and *ILSav*) nuclei, while the urethra receives contributions from four sacrococcygeal nuclei (*ILSp, ILSF, ILSav* and *ICd*), *ILSp*, nucleus intermediolateralis sacralis pars principalis; *ILSf*, nucleus intermediolateralis sacralis pars funicularis; *ILSav*, nucleus intermediolateralis sacralis pars ala ventralis; *ICd*, nucleus intercalatus disseminata; *t*, nucleus torsade (Onuf's nucleus); *Ml*, n. motorius lateralis. (Petras and Cummings 1978)

bodies, following the injection of HRP into the bladder wall in dogs (see Fig. 3.18). Morgan et al. (1986) also found smaller numbers of preganglionic neurons in the ventral horn nucleus and the funicular nucleus.

Kuo et al. (1984) have found the pelvic nerve to contain about 2100 postganglionic efferent fibres which arise from sacral paravertebral ganglia S-1 to S-3. This number is approximately 10% of the number of postganglionic neurons in the hypo-

gastric nerve (Baron et al. 1985a). A further 2900 postganglionic sympathetic cells project in the pudendal nerve, and originate from paravertebral ganglia L-6 to S-2.

Downie et al. (1984), who studied the distribution of fluorescent tracers injected into the walls of the bladder and urethra of male cats, concluded that the bladder and urethra received significant sympathetic postganglionic innervation from the inferior mesenteric and paravertebral (L-7 to S-2)

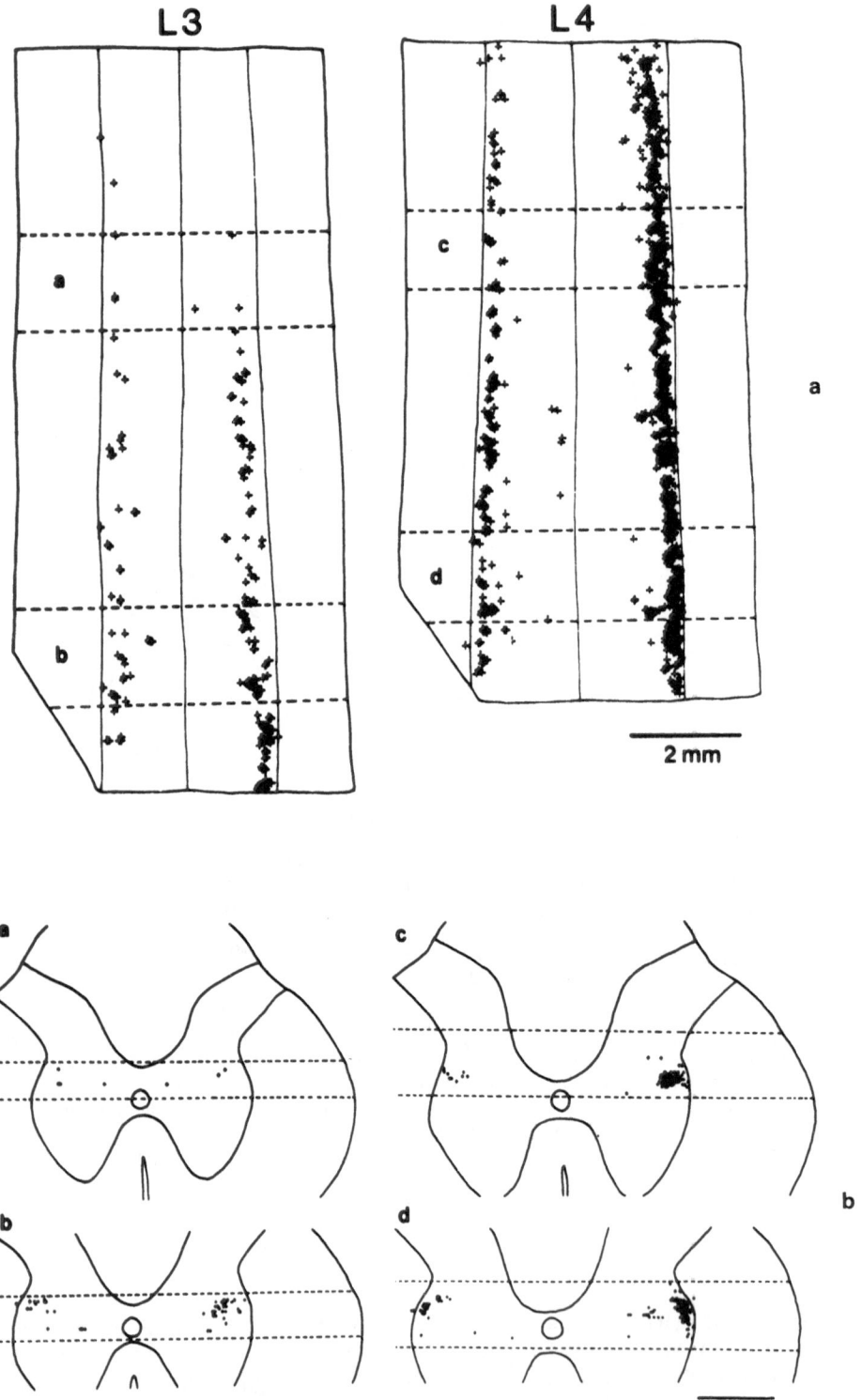

Fig. 3.19a,b. The distribution of sympathetic preganglionic neurons in L-3 and L-4 that project in the hypogastric nerve of the cat. **a** Superimposition of horizontal sections through the intermediate zone showing the rostrocaudal pattern of cell location on both sides after HRP application to one hypogastric nerve. Only a few cells were present in L-5 and these were restricted to the lateral part of the intermediate zone like those at the caudal end of L-4. The segments were trimmed at the caudal end on the contralateral side for purposes of orientation. **b** The lengths of segment L-3 (1.6 mm) and L-4 (1.4 mm) marked by the *broken lines* in **a** have been rotated and displayed as parts of transverse sections. The *broken lines* in **b** indicate the limit of the sections included in the analysis. (Baron et al. 1985a)

Table 3.4. Numbers of sympathetic pre- and postganglionic fibres and parasympathetic preganglionic fibres in the hypogastric, pelvic and pudendal nerves of the cat, rat, monkey and guinea pig

	Cat	Rat	Monkey	Guinea pig
A. Sympathetic efferents				
Hypogastric n.				
Preganglionic fibres	1492C (male)	550F		1270C (male)
	833C (female)			590C (female)
Postganglionic fibres	14 000C (male)	920F		430C (male)
	9250C (female)			580C (female)
Pelvic n.				
Postganglionic fibres	2100C	820F	5500F	
Pudendal n.				
Postganglionic fibres	2900C	200F		
B. Parasympathetic efferents				
Hypogastric n.				40C
Pelvic n.	1080C	560C	820C	
		2375F	11 000F	
Pudendal n.		?500F		
C. Pudendal motoneurons				
Pudendal n.		100F	420C	

Derived from data in: Baron et al. (1985a), Hulsebosch and Coggeshall (1982), Kuo et al. (1984), McLachlan (1985), Nadelhaft et al. (1983), Roppolo et al. (1985), and Schnitzlein et al. (1960). The 500 or so unmyelinated efferents in the pudendal nerve were presumed to be parasympathetic (see Hulsebosch and Coggeshall 1982). The numbers apply to the total (bilateral) innervation.

C, cell bodies by HRP; F, numbers of fibres in the peripheral nerve, by electron microscopy.

ganglia, and that the pelvic plexus made only a small contribution to the innervation of these areas. The detrusor and bladder base received equal contributions from the IMG and paravertebral ganglia, but the urethra received twice as much innervation from the sympathetic chain as from the IMG.

Henry and Calaresu (1972) found significantly higher numbers of preganglionic neurons in the intermediolateral columns of male compared with female cats. McLachlan (1985) found increased numbers of preganglionic fibres in the hypogastric nerves of male guinea pigs (as discussed later).

Rat

Neuhuber (1982) found the majority of the preganglionic cells that contributed to the inferior mesenteric plexus and hypogastric nerve to be located in L-1 and L-2 bilaterally. Preganglionic neurons were located in the intermediolateral nucleus, in the nucleus intercalatus, and in nucleus commisuralis dorsalis. Dendrites of the first of these nuclei ramify in all directions, but mainly into the dorsal horn and dorsolateral funiculus, whereas the dendrites of cells in the other nuclei are distributed medio-laterally, and ventral to the central canal.

Hancock and Peveto (1980) have also found that the hypogastric preganglionic cells in this species are present in L-1 and L-2, particularly along the midline in the dorsal grey commissure. This is in contrast to the distribution of sacral preganglionic cells, which are lateral in the intermediolateral column (Hancock and Peveto 1979).

Monkey

In this species, the lower level of the sympathetic outflow is usually L-3 or L-4, with occasional extensions into L-5. Petras and Cummings (1972) defined the intermediolateral and intercalated nuclei in the lumbar cord as the origin of cells that project through sympathetic nerves; they found no evidence for direct contacts between primary afferent and visceral efferent fibres. They also identified the n. intermediolateralis, n. proprius and central basal area of the dorsal horn as the most likely sites for interneurons that mediated visceral reflexes and visceral sensory pathways. With regard to the peripheral pathways of the postganglionic fibres, Schnitzlein et al. (1960) found that the pelvic nerves contained nearly 5500 sympathetic axons, approximately one-quarter to one-third of the total.

Guinea Pig

McLachlan (1985) found about twice as many pre-ganglionic sympathetic neurons in male (1250 neurons) as in female guinea pigs (600 neurons). In both sexes, they originated mainly from L-3. The hypogastric preganglionic neurons originated medial and ventral to the majority of the sympathetic efferent neurons in the intermediolateral column, and spread ventrally around the border of the ventral horn. Surprisingly, the number of postganglionic sympathetic efferents in the hypogastric nerve in this study was only about 400–500, with no significant difference between the sexes.

The guinea pig appears to have a visible smooth muscle sphincter round the bladder neck, innervated by a dense plexus of adrenergic fibres (Ulmsten et al. 1977).

Dog

Petras and Cummings (1978) studied the distribution of labelled neurons following the injection of HRP into the walls of the bladder and urethra of neonatal dogs. They concluded that the intramural neurons of the bladder are more abundantly innervated by nucleus intercalatus than by the main intermediolateral column, and that these neurons are distributed mainly in spinal segments L-1 to L-4.

Human

The sympathetic innervation of the human urethra has been considered in detail in Chapter 1.

Sympathetic noradrenergic postganglionic fibres that innervate the lower urinary tract originate from the inferior mesenteric ganglion, the sympathetic chain, and the vesical plexus. These neurons receive preganglionic inputs from the upper lumbar and lower thoracic segments, and those in the prevertebral ganglia may also receive excitatory inputs from afferent collaterals. The preganglionic neurons that traverse the inferior mesenteric ganglion on their way to the vesical ganglia originate mainly from the nucleus intercalatus; others originate mainly from the intermediolateral nucleus. The peripheral distribution of sympathetic fibres in the lower urinary tract has been dealt with in Chapters 1 and 2.

Conclusion: Afferent and Efferent Innervation of the Lower Urinary Tract

All the nerves which influence the function of the lower urinary tract contain afferent and efferent fibres. The segments which receive visceral afferents are closely associated with those giving rise to the autonomic preganglionic efferents to the viscera. There are at least two segments of lower lumbar cord that have no afferent or efferent connections with the viscera; the precise levels of the segments concerned are different in humans and the common laboratory animals.

The "somatic" pudendal nerve contains large- and small-diameter afferents, and myelinated and unmyelinated efferent axons. The large myelinated motoneurons innervate the skeletal muscle of the sphincters and pelvic floor, and the unmyelinated fibres are largely sympathetic efferent fibres, although in some species, e.g. the rat, some of the unmyelinated efferent fibres may be of parasympathetic origin. In addition, some of the small myelinated efferent axons may be γ-motoneurons, innervating spindle intrafusal fibres. Pudendal motoneurons originate from Onuf's nucleus and there is some spatial separation between motoneuron cell bodies concerned with the control of the urethra and of the anal sphincter. Sex differences have been found in some nuclei within the Onuf's complex of the rat.

The pelvic parasympathetic nerve contains visceral afferents, parasympathetic efferents, sympathetic efferents, and, in some species, some larger motoneurons that innervate the pelvic floor muscles and the skeletal muscle of the urethral sphincter (see Chap. 1, p. 15). The parasympathetic efferent cell bodies that innervate the bladder or colon occur at the base of the dorsal horn and the intermediolateral cell columns; the neurons controlling the bladder are larger and more lateral than those innervating the colon. The parasympathetic neurons that innervate the urethra are situated in the nucleus intercalatus, and, in some species, in the nucleus torsade (Onuf's nucleus) of the ventral horn. The efferent axons enter the ventral roots by passing laterally round the ventral horn. In the pelvic ganglia they make contact with postganglionic parasympathetic neurons, and these cells do not have inputs from large numbers of proganglionic fibres; their function therefore is not to integrate preganglionic inputs, unlike the inferior mesenteric (sympathetic prevertebral) ganglion (see Chap. 5, p. 135). Some neurons in pelvic ganglia contain catecholamines, and VIP is also present. The intramural plexuses are dealt with in detail in Chapters 1 and 2 (see p. 8 and p. 40).

The hypogastric nerve contains visceral afferents, pre- and postganglionic sympathetic efferents, and in some species a few parasympathetic efferents. The preganglionic fibres have cell bodies in the intermediolateral column or in the nucleus intercalatus (between the intermediolateral column and the central canal below the thoracolumbar junction). Some preganglionic neurons pass through the sympathetic chain and prevertebral ganglia to make contact with intramural ganglion cells within the bladder wall; these preganglionic neurons originate mainly from the nucleus intercalatus. There appear to be more preganglionic neurons in male animals than in the females of some species. Postganglionic fibres in the hypogastric nerve arise from the inferior mesenteric ganglion and the sympathetic chain. The cell bodies of the postganglionic neurons can receive synapses from afferent fibres in the same nerve.

Preganglionic sympathetic and parasympathetic neurons release acetylcholine at their terminals, but there are a variety of other putative neurotransmitters or neuromodulators present in these axons. The pelvic preganglionic terminals are known to contain enkephalins, and the axon collaterals of hypogastric axons that make contacts on inferior mesenteric ganglion cells may contain substance P and possibly other peptides.

References

Aldskogius H, Elfin L-G, Andersson-Forsman C (1986) Primary sensory afferents in the inferior mesenteric ganglion and related nerves of the guinea pig. J Auton Nerv Syst 15:179–190

Applebaum AE, Vance WH, Coggeshall RE (1980) Segmental localisation of sensory cells that innervate the bladder. J Comp Neurol 192:203–209

Anand P, Gibson SJ, McGregor GP, Blank MA, Ghatei MA, Bacarese-Hamilton AJ, Polak JM, Bloom SR (1983) A VIP-containing system concentrated in the lumbosacral region of the human spinal cord. Nature 305:143–145

Baron R, Jänig W, McLachlan EM (1985a) The afferent and sympathetic components of the lumbar spinal outflow to the colon and pelvic organs in the cat. I. The hypogastric nerve. J Comp Neurol 238:135–146

Baron R, Jänig W, McLachlan EM (1985b) The afferent and sympathetic components of the lumbar spinal outflow to the colon and pelvic organs in the cat. I. The lumbar splanchnic nerves. J Comp Neurol 238:147–157

Baron R, Jänig W, McLachlan EM (1985c) On the anatomical organisation of the lumbosacral sympathetic chain and the lumbar splanchnic nerves of the cat—Langley revisited. J Auton Nerv Syst 12:289–300

Blackman JG, Crowcroft PJ, Devine EE, Holman ME, Yonemura K (1969) Transmission from preganglionic fibres in the hypogastric nerve to peripheral ganglia of male guinea-pigs. J Physiol 201:723–743

Blank MA, Anand P, Lumb BM, Morrison JFB, Bloom SR (1984) Release of VIP-like immunoreactivity from cat urinary bladder and sacral spinal cord during pelvic nerve stimulation. Dig Dis Sci 29 (8):115P

Bloom SR, Polak JM (eds) (1981) Gut hormones, 2nd edn. Churchill Livingstone, Edinburgh, p 605

Breedlove SM, Arnold AP (1980) Hormone accumulation in a sexually dimorphic nucleus of the rat spinal cord. Science 210:564–566

Brindley GS, Rushton DN, Craggs MD (1974) The pressure exerted by the external sphincter of the urethra when its motor nerve fibres are stimulated electrically. Br J Urol 46:453–462

Brown AG (1981) Organisation in the spinal cord: the anatomy and physiology of identified neurones. Springer, Berlin Heidelberg New York

Coggeshall RE (1980) Law of separation of function of the spinal roots. Physiol Rev 60:716–755

Coggeshall RE, Coulter JD, Willis WD (1974) Unmyelinated axons in the ventral roots of the cat lumbosacral enlargement. J Comp Neurol 153:39–58

Costa M, Furness JB (1973) Observations on the anatomy and amine histochemistry of the nerves and ganglia which supply the pelvic viscera and on the associated chromaffin tissue in the guinea-pig. Z Anat EntwGech 140:85–108

Dahlstrom A, Fuxe K (1965) Evidence for the existence of monoamine neurones in the central nervous system. II. Experimentally induced changes in the interneuronal amine levels of bulbospinal neuron systems. Acta Physiol Scand 64 (suppl 247):7–34

de Groat WC (1986) Spinal cord projections and neuropeptides in visceral afferent neurons. In: Cervero F, Morrison JFB (eds) Visceral sensation. Elsevier, Amsterdam, pp 165–187 (Progress in brain research, vol 67)

de Groat WC, Nadelhaft I, Milne RJ, Booth AM, Morgan C, Thor K (1981) Organisation of the sacral parasympathetic reflex pathways to the urinary bladder and large intestine. J Auton Nerv Syst 3:135–160

de Groat WC, Kawatani M, Hisamitsu T, Lowe I, Morgan C, Roppolo J, Booth AM, Nadelhaft I, Kuo D, Thor K (1983) the role of neuropeptides in the sacral autonomic reflex pathways of the cat. J Auton Nerv Syst 7:339–350

Dockray GJ, Sharkey KA (1986) Neurochemistry of visceral afferent neurones. In: Cervero F, Morrison JFB (eds) Visceral sensation. Elsevier, Amsterdam, pp 133–148 (Progress in brain research, vol 67)

Donker PJ, Droes JThPM, Ulden BM (1976) Anatomy of the musculature and innervation of the bladder and the urethra. In: Williams DI, Chisholm GD (eds) Scientific foundations of urology, vol II, Heinemann, London, pp 32–39

Downie JW, Champion JA, Nance DM (1984) A quantitative analysis of the afferent and extrinsic efferent innervation of specific regions of the bladder and urethra in the cat. Brain Res Bull 12:735–740

Elbadawi A, Schenk EA (1968) A new theory of the innervation of bladder musculature. Part 1. Morphology of the intrinsic vesical innervation apparatus. J Urol 99:585–587

Fehér E, Vajda J, Csa'nyi K (1980) Quantitative analysis for innervation of smooth muscle cells in the wall of the urinary bladder. J Auton Nerv Syst 2:71–80

Fitzgerald M (1983) Capsaicin and sensory neurones—a review. Pain 15:109–130

Gabella G (1976) Structure of the autonomic nervous system. Chapman and Hall, London, p 214

Gibson SJ, Polak JM, Anand P, Blank MA, Morrison JFB, Kelly JS, Bloom SR (1984a) The distribution and origin of

VIP in the spinal cord of six mammalian species. Peptides 5:201–207

Gibson SJ, Polak JM, Bloom SR, Sabate IM, Mulderry PM, Ghatei MA, McGregor GP, Morrison JFB, Kelly JS, Evans RM, Rosenfield MG (1984b) Calcitonin gene-related peptide immunoreactivity in the spinal cord of man and eight other species. J Neurosci 4:3101–3111

Greenwood D, Coggeshall RE, Hulsebosch CE (1985) Sexual dimorphism in the numbers of neurones in the pelvic ganglia of adult rats. Brain Res 340:160–162

Hamberger B, Norberg KA (1965) Adrenergic synaptic terminals and nerve cells in bladder ganglia of the cat. Int J Neuropharmacol 4:41–45

Hancock MB, Peveto CA (1979) Preganglionic neurones in the sacral spinal cord of the rat: an HRP study. Neurosci Lett 11:1–5

Hancock MB, Peveto CA (1980) A preganglionic autonomic nucleus in the dorsal gray commissure of the lumbar spinal cord in the rat. J Comp Neurol 183:65–72

Henry JL, Calaresu FR (1972) Topography and numerical distribution of neurons of the thoraco-lumbar intermediolateral nucleus in the cat. J Comp Neurol 144:205–214

Hökfelt T, Elde R, Johansson O, Luft R, Nilsson G, Arimura A (1976) Immunohistochemical evidence for separate populations of somatostatin-containing and substance P-containing primary afferent neurones in the rat. Neuroscience 1:131–136

Honda CN, Rethelyi M, Petrusz P (1983) Preferential histochemical location of vasoactive intestinal polypeptide (VIP) in the sacral spinal cord of the cat: light and electron microscopic observations. J Neurosci 3:2183–2196

Hulsebosch CE, Coggeshall RE (1982) An analysis of the axon populations in the nerves to the pelvic viscera in the rat. J Comp Neurol 211:1–10

Jänig W, Morrison JFB (1986) Functional properties of spinal visceral afferents supplying abdominal and pelvic organs, with special emphasis on visceral nociception. In: Cervero F, Morrison JFB (eds) Visceral sensation. Elsevier, Amsterdam, pp 87–114 (Progress in brain research, vol 67)

Kuo DC, Hisamitsu T, de Groat WC (1984) A sympathetic projection from sacral paravertebral ganglia to the pelvic nerve and to postganglionic nerves on the surface of the urinary bladder and large intestine of the cat. J Comp Neurol 226:76–86

Kuru M (1965) Nervous control of micturition. Physiol Rev 45:425–494

Kuzuhara S, Kanayawa I, Nkanaishi T (1980) Topographical localisation of the Onuf's nuclear neurones innervating the rectal and vesical striated muscle sphincters: a retrograde fluorescent double labelling in cat and dog. Neurosci Lett 16:125–130

Langley JN, Anderson HK (1896) The innervation of the pelvic and adjoining viscera. VII. Anatomical observations. J Physiol 20:372–406

Laruelle L (1937) La structure de la moelle épinière en coupes longitudinales. Rev Neurol 67:695–725

Laruelle L (1948) Étude d'anatomie microscopique du névraxe sur coupes longitudinales. Acta Neurol Psychiatr Belg 48:138–280

Laruelle L, Reumont M (1958) Histologie de quelques aspects physiopathologiques de la poliomyelite humaine. Acta Neurol Psychiatr Belg 58:297

Mackel R (1979) Segmental and descending control of the external urethral and anal sphincters in the cat. J Physiol 294:105–122

Mannen T, Iwata M, Toyokura Y, Nagashima K (1982) The Onuf's nucleus and the external anal sphincter muscles in amyotrophic lateral sclerosis and Shy-Drager syndrome. Acta Neuropathol 58:255–260

Mawe GM, Bresnahan JC, Beattie MS (1984) Primary afferent projections from dorsal and ventral roots to autonomic preganglionic neurons in the cat sacral spinal cord: light and microscopic observations. Brain Res 290:152–157

McLachlan Elspeth M (1985) The components of the hypogastric nerve in male and female guinea-pigs. J Auton Nerv Syst 13:327–342

McMahon SB (1986) The localisation of fluoride-resistant acid phosphatase (FRAP) in the pelvic nerves and sacral spinal cord of rats. Neurosci Lett 64:305–310

McMahon SB, Morrison JFB (1982a) Spinal neurones with long projections activated from the abdominal viscera of the cat. J Physiol 332:1–20

McMahon SB, Morrison JFB (1982b) Two groups of spinal interneurones that respond to stimulation of the abdominal viscera of the cat. J Physiol 332:21–34

Matsushita M, Tanami T (1983) Contralateral terminations of primary afferent axons in the sacral and caudal segments of the cat, as studied by anterograde transport of horseradish peroxidase. J Comp Neurol 220:206–218

Mesulam MM, Brushart TM (1979) Transganglionic and anterograde transport of horseradish peroxidase across dorsal root ganglia: a tetramethyl benzidine method for tracing central sensory connections of muscles and peripheral nerves. Neuroscience 4:1107–1117

Micevych PE, Coquelin A, Arnold AP (1986) Immunochemical distribution of substance P, serotonin and methionine-enkephalin in sexually dimorphic nuclei of the rat lumbar cord. J Comp Neurol 248:235–244

Molander C, Xu Q, Grant G (1984) The cytoarchitectonic organisation of the spinal cord in the rat. I. The lower thoracic and lumbosacral cord. J Comp Neurol 230:133–141

Morgan C, O'Hara P (1984) Electronmicroscopic identification of vasoactive intestinal polypeptide (VIP) in visceral primary afferent axons in the sacral spinal cord of the cat. Anat Rec 208:121A

Morgan C, Nadelhaft I, de Groat WC (1981) The distribution of visceral primary afferents from the pelvic nerve within Lissauer's tract and the spinal gray matter and its relationship to the sacral parasympathetic mucleus. J Comp Neurol 201:415–440

Morgan C, de Groat WC, Nadelhaft I (1986) The spinal distribution of sympathetic preganglionic and visceral afferent neurons that send axons into the hypogastric nerve of the cat. J Comp Neurol 243:23–40

Nadelhaft I, Booth AM (1984) The location and morphology of preganglionic neurons and the distribution of visceral afferents from the rat pelvic nerve: a horseradish peroxidase study. J Comp Neurol 226:238–245

Nadelhaft I, de Groat WC, Morgan C (1980) Location and morphology of parasympathetic preganglionic neurons in the sacral spinal cord of the cat revealed by retrograde axonal transport of horseradish peroxidase. J Comp Neurol 193:265–281

Nadelhaft I, Roppolo J, Morgan C, de Groat WC (1983) Parasympathetic preganglionic neurons and visceral primary afferents in the monkey sacral spinal cord revealed following the application of horseradish peroxidase to pelvic nerve. J Comp Neurol 216:36–52

Neuhuber W (1982) The central projections of visceral afferent neurons of the inferior mesenteric plexus and hypogastric nerve and the location of the related sensory and preganglionic sympathetic cell bodies in the rat. Anat Embryol 164:413–425

Onuf (Onufrowicz) B (1899) Notes on the arrangement and function of the cell groups in the sacral region of the spinal cord. J Nerv Ment Dis 26:498–504

Onuf (Onufrowicz) B (1900) On the arrangement and function of the cell groups in the sacral region of the spinal cord. Arch Neurol Psychopathol 3:387–411

Petras JM, Cummings JF (172) Autonomic neurons in the spinal cord of the rhesus monkey: a correlation of the findings of cytoarchitectonics and sympathectomy with fiber degeneration following dorsal rhizotomy. J Comp Neurol 146:189–218

Petras JM, Cummings JF (1978) Sympathetic and parasympathetic innervation of the urinary bladder and urethra. Brain Res 153:363–369

Pick J (1970) The autonomic nervous system: morphological, comparative, clinical and surgical aspects. Lippincott, Philadelphia

Puriton T, Fletcher T, Bradley W (1971) Sensory perikarya in autonomic ganglia. Nature 231:63–64

Rexed B (1954) A cytoarchitectonic atlas of the spinal cord in the cat. J Comp Neurol 100:297–379

Risling M, Dahlsgaard C-J, Cukierman A, Cuello AC (1984) Electronmicroscopic and immunohistochemical evidence that unmyelinated ventral root axons make U-turns or enter the spinal pia mater. J Comp Neurol 225:53–63

Romanes GJ (1951) The motor cell columns of the lumbosacral spinal cord of the cat. J Comp Neurol 94:313–363

Roppolo JR, Nadelhaft I, de Groat WC (1985) The organisation of pudendal motoneurones and primary afferent projections in the spinal cord of the Rhesus monkey revealed by horseradish peroxidase. J Comp Neurol 234:475–488

Sato M, Mizuno M, Konishi A (1978) Localisation of motoneurones innervating perineal muscles: a HRP study in the cat. Brain Res 140:149–154

Schnitzlein HN, Hoffman HH, Tucker CC, Quigley MB (1960) The pelvic splanchnic nerves of the male rhesus monkey. J Comp Neurol 114:51–65

Schnitzlein HN, Hoffman HH, Hamlett DM, Howell EM (1963) A study of the sacral parasympathetic nucleus. J Comp Neurol 120:477–493

Schrøder HD (1981) Onuf's nucleus X: morphological study of a human spinal nucleus. Anat Embryol 162:443–453

Schrøder HD (1985) Anatomical and pathoanatomical studies on the spinal efferent systems innervating pelvic structures. J Auton Nerv Syst 14:23–48

Sharkey KA, Williams RG, Schultzberg M, Dockray GJ (1983) Sensory substance P-innervation of the urinary bladder: possible site of action of capsaicin in causing urine retention in rats. Neuroscience 10:861–868

Sheehan D (1941) Spinal autonomic outflows in man and monkey. J Comp Neurol 75:341–370

Sherrington CS (1892) Notes on the arrangement of some motor fibres in the lumbo-sacral plexus. J Physiol 13:621–772

Sjöstrand NO (1965) The adrenergic innervation of the vas deferens and of the accessory male genital organs. Acta Physiol Scand [Suppl] 257:1–82

Su HC, Wharton J, Polak JM, Mulderry PK, Ghatei MA, Gibson SJ, Terenghi G, Morrison JFB, Ballesta J, Bloom SR (1986) Calcitonin gene-related peptide immunoreactivity in afferent neurons supplying the urinary tract: combined retrograde tracing and immunohistochemistry. Neuroscience 18:737–747

Uemura E, Fletcher TF, Dirks VA, Bradley WE (1973) Distribution of sacral afferent axons in cat urinary bladder. Am J Anat 136:305–314

Uemura E, Fletcher TF, Bradley WE (1974) Distribution of lumbar afferent axons in muscle coat of cat urinary bladder. Am J Anat 139:389–398

Ueyama T, Mizuno N, Nomura S, Konishi A, Itoh K, Arakawa H (1984) Central distribution of afferent and efferent components of the pudendal nerve in the cat. J Comp Neurol 222:38–46

Ueyama T, Mizuno N, Takahashi U, Nomura S, Arakawa H, Matsushima R (1985) Central distribution of efferent and afferent components of the pudendal nerve in Macaque monkeys. J Comp Neurol 232:548–556

Ulmsten U, Sjöberg NO, Alm P, Andersson KE, Owman C, Walles B (1977) Functional role of an adrenergic sphincter in the female urethra of the guinea-pig. Acta Obstet Gynecol Scand 56:387–390

White JC (1943) Sensory innervation of the viscera: studies on visceral afferent neurones in man based on neurosurgical procedures for the relief of intractable pain. Res Publ Assoc Nerv Ment Dis 23:373–390

White JC, Sweet WH (1955) Pain. Its mechanisms and neurosurgical control. Thomas, Springfield, Ill

SENSATION AND MOTOR CONTROL

Introduction

This section concerns itself with experimental physiology, mainly in animals. As much information on human function as possible is discussed and compared. Chapter 4 describes the different sensations that can arise from the lower urinary tract and attempts to explain them in terms of the function of primary afferent fibres and central pathways. There have been a number of advances in this area in recent years, and these are outlined. One of the important concepts is the existence of somatovisceral convergence and viscerovisceral convergence in the spinal cord. There is no evidence for a specialized ascending tract that carries only sensory information from the bladder; instead, afferent information from one or more viscera is combined with afferent information from skin and/or muscle in the ascending pathways in the spinal cord. These observations provide a basis for understanding the mechanism of referred pain and other sensory phenomena.

Chapter 5 deals with functional aspects of the efferent innervation to the bladder and urethra. In addition to the effects these nerves have on the target organs, the mechanisms of action are considered, with a review of the morphological, electrophysiological and pharmacological data.

In Chapter 6, Alison Brading discusses mechanical, electrical and contractile properties of smooth muscle in the lower urinary tract. The role of electrical potentials, ion channels and exchanges and of different transmitters in the control of muscle contraction are considered.

The central pathways concerned with reflex control in the bladder and urethra are outlined in Chapter 7. Barrington's reflexes are reconsidered, and a more modern evaluation of these pathways, particularly in relation to the coordinated processes of filling and voiding, are discussed. Electrophysiological and neuropharmacological techniques have been essential to the developments in understanding in this area, and the new information that has given rise to concepts such as "gating" is outlined.

In Chapter 8, a number of different ideas about the central control of the lower urinary tract have been taken from the literature and discussed in relation to each other. Some of the ideas are mainly anatomical, and emphasis has been put on a more functional approach, considering in particular the neuroscientific correlates of observations made during urodynamic procedures. For instance, "unstable bladder" is considered as a defect in the neural control of the bladder, and some of the points in the control system at which changes might arise and produce the observed cystometric abnormalities are discussed. An attempt is made to integrate the anatomical and functional aspects of control systems.

4 · Sensations Arising from the Lower Urinary Tract

John F. B. Morrison

Conditions that Elicit Sensations from the Lower Urinary Tract

Visceral sensations are often vague, and sensations arising from the lower urinary tract are no exception, with the result that a surprisingly large number of sensations associated with micturition have been described (see Denny-Brown and Robertson 1933; Nathan 1956). Some of these sensations that can be induced by mechanical stimuli have been studied in relation to the pressure–volume relationship for the viscus; other, more intense and less tolerable sensations occur at high intravesical pressures or during inflammatory states, and less is known about the precise circumstances which elicit them. Urgency and pain can be regarded as separate sensations (Denny-Brown and Robertson 1933), and Procacci et al (1986) distinguish between true visceral pain and two types of referred pain. True visceral pain is said to be "deep, dull, not well defined, but described by the patient in especial terms". It can radiate and occur in parts of the body far from the affected organ. The patterns and causes of pain in disease states, however, may be complex, but even in experimental environments, where pain can be induced and relieved by distension and emptying of a viscus, the description of true visceral pain would have the same characteristics as the description given here. It is clear from the observations of Denny-Brown and Robertson (1933) that a similar description could also be applied to non-noxious sensations arising from viscera, except that radiation may be less widespread.

Sensations During Cystometry

The point on the pressure–volume curve at which sensations of different characters occur during controlled filling is commonly used in the assessment of urinary tract sensation (Fig. 4.1; Andrew and Nathan 1964); thus, the first sensation, and the desire to micturate in normal people can be related to the pressure–volume curve of the bladder. Some authors refer to these sensations as proprioceptive. The rate of filling is one variable which can influence both the pressure–volume curves, and the volume at which sensations are preceived; thus, the slope of the pressure–volume curve during cystometry is increased if the filling is done quickly, as is usual, compared with the normal physiological rate of urine production (Klevmark 1977). During fast cystometry, Torrens and Abrams (1979) noted that the volumes at which patients reported sensations were generally lower than those occurring naturally. Presumably this is due to differences in the accommodation of the bladder, which is known to be affected by the rate of filling. Furthermore, the sensation of desire to void during cystometry is also influenced by the physical state of the viscus, and can be delayed by factors that influence the tone of the smooth muscle. Thus, the sensation of desire to void occurs at increased bladder volume following the intravenous administration of thymoxamine, an α-adrenoceptor antagonist which also raises the

threshold of the first uninhibited bladder contraction and reduces its size (Pedersen 1983). In contrast, naloxone, an opiate antagonist that may act on the pelvic ganglia (see Chap. 5, p. 139 and Chap. 7, p. 227) to block the inhibitory action of enkephalins on ganglionic transmission, reduces the bladder volumes at which the first sensation of filling and the urge to micturate occur (Murray and Feneley 1982). These observations all suggest that the bladder volume at which innocuous sensations occur is influenced by the resting activity of bladder smooth muscle, and transmission of excitatory efferent signals through pelvic ganglia. In the case of naloxone, however, central interactions cannot be ruled out.

The volume in the bladder at the time when sensations can be elicited by filling nevertheless appears to be a useful assessment of bladder sensation, and can be easily monitored during routine cystometry. Other sensations have also been described in association with micturition; although these descriptions are not usually accompanied by any physical or physiological measurements, they are useful as they can indicate a number of factors that influence the character, the site of origin, or the pathways concerned in sensation from the lower urinary tract. Thus, the sense of filling and the desire to micturate probably depend on afferent endings in the bladder itself, whereas the sensation that micturition is about to happen probably originates from the urethra, as does the sensation that urine is being passed (Nathan 1956). Brindley et al. (1974), however, believe that the sensation of impending micturition can be a reliable index of opening of the bladder neck and contraction of the detrusor. Reports of intelligent introspective observers can provide useful opinions, but reports about sensations by members of the general public can be vague, and the sensations only identifiable after their attention has been drawn to possible differences in the character of different sensations.

The vagueness of the first sensation during distension of the bladder can be illustrated by reference to the classical paper by Denny-Brown and Robertson (1933). The first sensation they described occurred at a bladder volume of 250–300 ml (although other authors report 150 ml as the lower end of the normal range), and could be felt as "slight pressure deep in the mid-perineum" or "a faint pressure from within" sometimes with movement "like rivulets running down the urethra". The sensation was not constant, but could come and go. It was not comparable with any sensation in normal experience, and may have been related to the presence of a catheter in the urethra and bladder. At higher volumes of around 400 ml

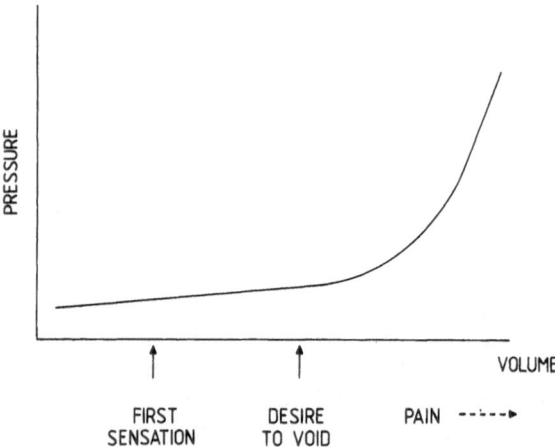

Fig. 4.1. Pressure–volume curves for the human bladder, showing the conditions in which the first sensation of filling and the desire to micturate are felt.

a sense of distension was reported: it is known that both of these sensations can occur without an increase in intravesical pressure, a fact which fits with the known sensitivity of the bladder afferents to volume and to pressure.

Desire to Micturate: Urgency, Pain

In the subjects studied by Denny-Brown and Robertson (1933), the sensation of desire to micturate occurred at about 500 ml and 22 cm H_2O, and followed a sense of distension, neither of which was described as painful; discomfort is generally reported by the time a bladder volume of 600 ml is achieved. Individual subjects, however, do show variations in the response to bladder distension, as noted by Denny-Brown and Robertson (1933). Klevmark (1980) has also pointed out the variation in response seen in patients with symptoms of urgency at low bladder volumes: he noted that cystometric capacity in some of these patients could be normal when investigated under anaesthetic, but was markedly reduced when awake. These patients could never be distracted fully from their symptoms during cystometry, whereas another group of patients with the urge syndrome could be persuaded to accept a normal bladder capacity. In patients with symptoms of urgency at low bladder volumes, treatment with muscle relaxants such as oxybutynin, terbutaline, terodiline and dicyclomine can relieve or improve the symptoms (i.e. increase the bladder volume at which symptoms of urgency are experienced), and increase the bladder capacity as well as the bladder volume at which the first contraction occurs during cystometry (Wein 1985). In contrast, the volume at which the desire to micturate is experienced is lowered by naloxone, possibly through its action in blocking inhibitory mechanisms in pelvic ganglia. This suggests that the bladder volume at which these symptoms occur is influenced by the contractile state of the bladder muscle.

Increased volume and contractions both give rise to increased tension in the wall of the bladder, and in its supporting structures; sensory receptors are present in all of these sites, and appear to respond to changes in local tension set up by distensions and contractions (Talaat 1937; Iggo 1955; Floyd et al 1976). Furthermore, these effects summate to a certain extent. The sensitivity to these two factors has been studied only in acute conditions, when distensions are produced by injection or infusion over tens of seconds. The properties of these receptors will be considered later, but it is clear that the

mechanoreceptors that have been described in the pelvic and hypogastric nerves are able to respond to changes in bladder volume and in the tone of the detrusor; the sensations described during bladder filling can therefore be mediated by these receptors.

Overt pain has generally been associated with reflex contractions at high bladder volume in experimental studies, particularly when there is outflow obstruction or near-isometric conditions (Lewis 1942); Nathan (1956) quotes pressures of about 90 cm H_2O, achieved by contraction, as a level associated with pain. The intravesical pressure level associated with pain in subjects with a normal lower urinary tract, however, can be quite variable: M. J. Torrens (1986 personal communication) has seen normal subjects who can tolerate pressures of > 300 cm H_2O without pain. Procacci et al. (1986) state that pain is referred when the algogenic process in the viscus becomes more intense and prolonged, and the location of the pain becomes more exact and is felt more superficially. These authors state that referred pain can occur with or without superficial and-or deep hyperalgesia. If superficial and/or deep hyperalgesia is present, there may also be myalgic spots or trigger points. These algogenic points are generally due to involvement of muscular or cutaneous structures and can often disappear after injection of local anaesthetic into them. They appear to be the result of reflex phenomena (these are discussed later in the section on the reflex basis of some referred sensory phenomena). In some cases this hyperalgesia inhibits the perception of true visceral pain, and local anaesthesia of the algogenic points changes the character of the visceral pain, causing it to increase in intensity. The possible mechanisms underlying this phenomenon will be dealt with later under the heading of the mediation of referred pain and counterirritation phenomena. Referred pain without the associated phenomena occurs with a similar distribution, but is not accompanied by the reflexly induced trigger points; one result of the absence of cutaneous hyperalgesia or of deep muscle tenderness is that the visceral pain is severe. The reason for this would appear to be along the lines described by Cadden (1985) and Lumb (1986) for counterirritation phenomena in general (see later). As far as we know, the receptors that mediate the innocuous and the noxious, painful, sensations induced by distension cannot be divided into two separate categories with different sensitivity or threshold to intravesical pressure (Morrison 1981): the conclusion that one type of sensory receptor can mediate more than one sensation is embodied in the intensity theory of visceral sensation (see Jänig and Morrison 1986), which will be considered later alongside other ideas on the mediation of noxious sensations.

Sensations Associated with Bladder Contractions

During micturition, intravesical pressures are achieved which are well above the level that would give rise to unpleasant sensations if they were achieved by distension. Thus, pressures of 30 cm H_2O are produced during normal micturition (Scott et al. 1964), and pressures much higher than this can be reached if partial outlet obstruction is present. Nevertheless, pain is not a dominant feature of these situations, unless isovolumetric conditions are present. Table 4.1 shows the ranges of intravesical pressures during micturition and the relationship with sensation.

Table 4.1. Bladder pressure at different volumes

	Pressure (cmH₂O)
Typical normal values in the accommodated state (Abrams et al. 1983)	
Zero volume	<10
300 ml	<10
Full	<15
Pressure in the bladder during normal painless micturition	
Shah (1984)	48.4 ± 9.3
Scott et al. (1964)	49.5 ± 10
Murphy and Schoenberg (1960)	29 (21–35)
Abrams et al. (1983)	72–78 (males)
	55–64 (females)
Pressure in the bladder giving rise to urgency or pain	
Denny-Brown and Robertson (1933)	~40 (urgency)
Nathan (1956)	~90 (pain)

Sensations Induced by Touch and Temperature Change in the Lower Urinary Tract

Touch sensation appears to be confined to the urethra and the bladder base: a number of studies report that the bladder dome can be touched without giving rise to conscious sensation (Morley 1931), but mechanical stimulation applied to the bladder mucosa with sufficient intensity is perceived as pain (Bors et al. 1956). The trigone region is more sensitive to touch than the rest of the bladder, and receives fibres from the thoracolumbar junction as well as the sacral cord. Mechanical stimulation of the bladder mucosa can be lateralized

if the sympathetic pathway is intact (Bors et al. 1956).

Thermal sensation has also been shown to be confined to the urethra (Nathan 1952; Bors et al. 1956), and temperatures of greater than 37°C give rise to a sensation of warmth, whereas a cool sensation is reported at temperatures of 30°C and below (McDonald and Murphy 1959). The pathways for temperature sensation run in nerves that entered the second to fourth sacral segments (McDonald and Murphy 1959). Raising the temperature of the bladder mucosa to 47°C or over induces pain; it is not clear whether the pathways for these sensations travel in the pelvic or the pudendal nerves, or both.

Sensations Induced by Electrical Stimulation of the Lower Urinary Tract

Powell and Feneley (1980) have attempted to assess the sensitivity of the bladder neck to constant current electrical stimulation, using electrodes mounted on a catheter; they report a good correlation between the threshold current that induces a sensation and the volume threshold for the first sensation of filling during cystometry. Patients with bladder instability and a number of other abnormalities in bladder physiology also had raised thresholds to electrical stimulation. Whether this is due to a deficiency of the sensory innervation, or other changes in the tissues at the bladder base is uncertain; nevertheless, posterior urethral sensation, assessed by this method, is measurably reduced in some individuals who never develop proper control over their bladder contractions. Frimødt-Moller (1972) has attempted similar assessments on the bladder using a constant current stimulator; the current which gave rise to sensation was monitored and was greater in subjects with neurological disease (see Chap. 9, p. 302).

Summary

In normal subjects, sensations of filling, fullness, the desire to micturate, urgency and pain can be elicited from the lower urinary tract at different degrees of bladder filling. The volume at which the first sensation of filling occurs during cystometry is a useful measurement in clinical practice and is dependent on the tone of the bladder, and the rate of filling. Bladder contractions can also elicit sensations, which may be nonpainful or painful, depending on the bladder volume, the size of the contractions, and whether they are accompanied by voiding or not. A sensation of

fullness at low bladder volumes may be associated with the occurrence of bladder contractions, or may be due to the presence of pathological conditions, such as inflammation. Pain occurs at high pressures, produced by distension or contraction or both, and by inflammatory or other pathological conditions. The sensation that micturition is imminent or has started probably arises from the bladder neck and the urethra. Touch and temperature sensations appear to be confined to the urethra; warmth and cold can be distinguished, and high temperatures give rise to a sensation of pain.

Peripheral Pathways

The pelvic nerves appear to be the major pathway for reflexes and for sensation from the lower urinary tract; animal studies (see p. 58) suggest that the number of afferent neurons from the bladder is much greater in this pathway than in the hypogastric nerve, and the dominance of the pelvic innervation in pain arising from the bladder is well recognized by neurosurgeons. White (1943) and White and Sweet (1955) realized that it was necessary to section pelvic as well as hypogastric afferent pathways in order to produce relief of bladder pain. More recently Torrens and Hald (1979) found that selective neurectomy of the third sacral segment could relieve painful frequency and urgency. Furthermore, Gunterberg et al. (1975) found that there was a complete loss of bladder mucosal pain (elicited by electrocoagulation of the mucosa, or pinprick in the urethra) following bilateral resection of sacral roots 3–5. The importance of the sacral pathways was also emphasized by Ruch (1979) who felt that the role of the pelvic nerve was sufficient to delineate a "pelvic pain line" separating visceral sensation mediated by this pathway from the lumbar splanchnic innervation. This separation of the pathways is possible because sensory information from the viscera does not enter spinal segments between the lumbar and sacral autonomic outflows (for details see Chap. 3, p. 58).

The hypogastric nerves appear to be able to transmit information about the degree of filling, as patients with spinal transections between the pelvic and hypogastric inputs can sense bladder distension; furthermore, the character of the sensation is similar to that occurring before the spinal injury (Personal observations; Riddoch 1921). However, an extravesical source of sensory input, such as the lower abdominal wall cannot be excluded in this situation. Gunterberg et al. (1975) found that

stretch of the detrusor could be sensed via the hypogastric pathway. The role of afferent endings outside the lower urinary tract, e.g. in the muscle of the abdominal wall or pelvic floor, in transmitting sensory information about the state of these viscera in normal individuals is unknown; however, some patients with denervated bladders may experience a vague, poorly localized sensation in the abdomen when the bladder is full. Similarly, patients with complete spinal transection above the afferent inflow from the bladder can tell when the bladder is overfull because of reflex changes that can be sensed by receptors connected to segments of the neuraxis above the transection. In this instance the sensation is not referred to the lower urinary tract, but its origin is inferred from previous experience of lower urinary tract problems, such as blocked catheters, that can elicit it. Details of these sensations can be found in the paper by Nathan (1956).

In addition to an involvement in sensations associated with filling, it is likely that the hypogastric nerves mediate pain sensation in some circumstances. There has been a view, based on scant evidence, that the hypogastric nerve is mainly responsible for pain sensation, whereas the pelvic nerve is involved mainly with reflexes. There is no doubt that the pelvic nerve is essential for reflexes from the bladder, but sensory information is certainly carried as well (as discussed previously). The view that the hypogastric nerve is mainly concerned with pain sensation is based largely on some recordings by Evans (1936), in the early days of electrophysiology, who stated that hypogastric afferents responded only to overdistension of the bladder, and not to physiological changes in volume. The result has been refuted in the modern literature (see Jänig and Morrison 1986); the properties of the mechanoreceptors in the pelvic and hypogastric nerves are very similar, and the former outnumber the latter. Table 4.2 shows the pathways that are associated with different sensations from the lower urinary tract.

Location of Sensations Arising from the Lower Urinary Tract

The site of the first sensation during filling of the bladder is often within the mid-perineum and penis, whereas the sensations of distension and fullness tend to be suprapubic, i.e. abdominal. At high bladder volumes, the sensation of desire to pass urine merges into pain referred to the mid-

Table 4.2. Pathways responsible for different sensations from the lower urinary tract

Sensation	Afferent pathway			
	Pelvic	Lumbar	Pudendal	Abdominal wall
Fullness	Yes	Probably	No	Possibly
Desire to micturate	Yes	Probably	No	No
Pain, bladder	Yes	Yes	No	No
Imminence of micturition	Possibly	Probably not	Yes	No
Pain, urethral	Probably	Probably not	Yes	No

perineum. Suprapubic pain in the midline can also be elicited by distension of the lower third of the ureter if ureteral occlusion is absent; if it is present, then pain will also occur in the loin (Risholm 1954). According to Jänig and Morrison (1986), the site of visceral pain, true or referred, depends on the segmental innervation of the viscus, the degree of lateralization of that pathway, and the density of innervation of the viscus. In the case of the urinary tract, distension of the renal pelvis produces pain in the ipsilateral loin; the middle third of the ureter is insensitive to distension, provided obstruction that raises renal pelvic pressure is avoided, and distension of the lower ureter produces midline suprapubic pain. Pain produced from the renal pelvis is ipsilateral, presumably because the afferent innervation is entirely from one side of the cord. Viscera with afferent innervation from both sides of the cord generally give rise to sensations in the midline, and the intravesical portions of the ureter, the bladder and the urethra fall into this category. The insensitivity of the middle third of the ureter is attributed to the low density of innervation at this site. Distension of the bladder can induce suprapubic and perineal pain; events in the urethra produce sensation in the perineum and penis. The midline suprapubic sensation associated with overdistension of the bladder is almost certainly mediated by the hypogastric afferent innervation, which enters the upper lumbar and lower thoracic cord; these segments innervate suprapubic dermatomes and myotomes. The perineal sensations are believed to be mediated by the pelvic nerve which enters the sacral segments that also innervate the perineal skin and musculature in humans. This difference in pathway was referred to by Ruch (1979) as the pelvic pain line, a hypothetical division between the sensory territories of the sacral and upper lumbar cord. The reason for the difference would appear to be the fact that visceral sensation is carried by ascending spinal pathways that also transmit somatic sensory information; the brain does not recognize visceral impulses as coming from the viscera, and "sees" the sensory input as having arisen from the somatic domain innervated from the same segmental levels. Thus, pelvic afferent inputs give rise to sensation in the sacral segments in humans, whereas the hypogastric afferents, which project to the thoracolumbar junction in humans, give rise to sensation in the dermatomes and myotomes supplied by these segments. One corollary of this is that somatic structures that are innervated by the segments of cord between the lumbar and sacral autonomic outflows (which do not participate in visceral sensation) should not be the locations of referred pain from the viscera. In practice, referred pain from the lower urinary tract does not occur in these segments. Figure 4.2 shows the distribution of pain arising from the bladder.

Summary

Sensory nerves from the bladder travel with the sympathetic nerves to enter the thoracolumbar junction, and with the parasympathetic nerves to enter the sacral cord in humans. Sensory information from the urethra also reaches the sacral cord via the pudendal nerves. The hypogastric and pelvic pathways can be implicated in the sensations associated with normal bladder filling and with bladder pain; the pelvic and pudendal pathways are concerned also with the sensation that micturition is imminent, and thermal sensation from the urethra.

Sensations arising from the lower urinary tract are usually located deep in the skin, and in the distribution of sensory nerves that arise from the cord segments which receive the visceral afferent input from the lower urinary tract; the sensations are therefore felt in the suprapubic and perineal regions, and sometimes

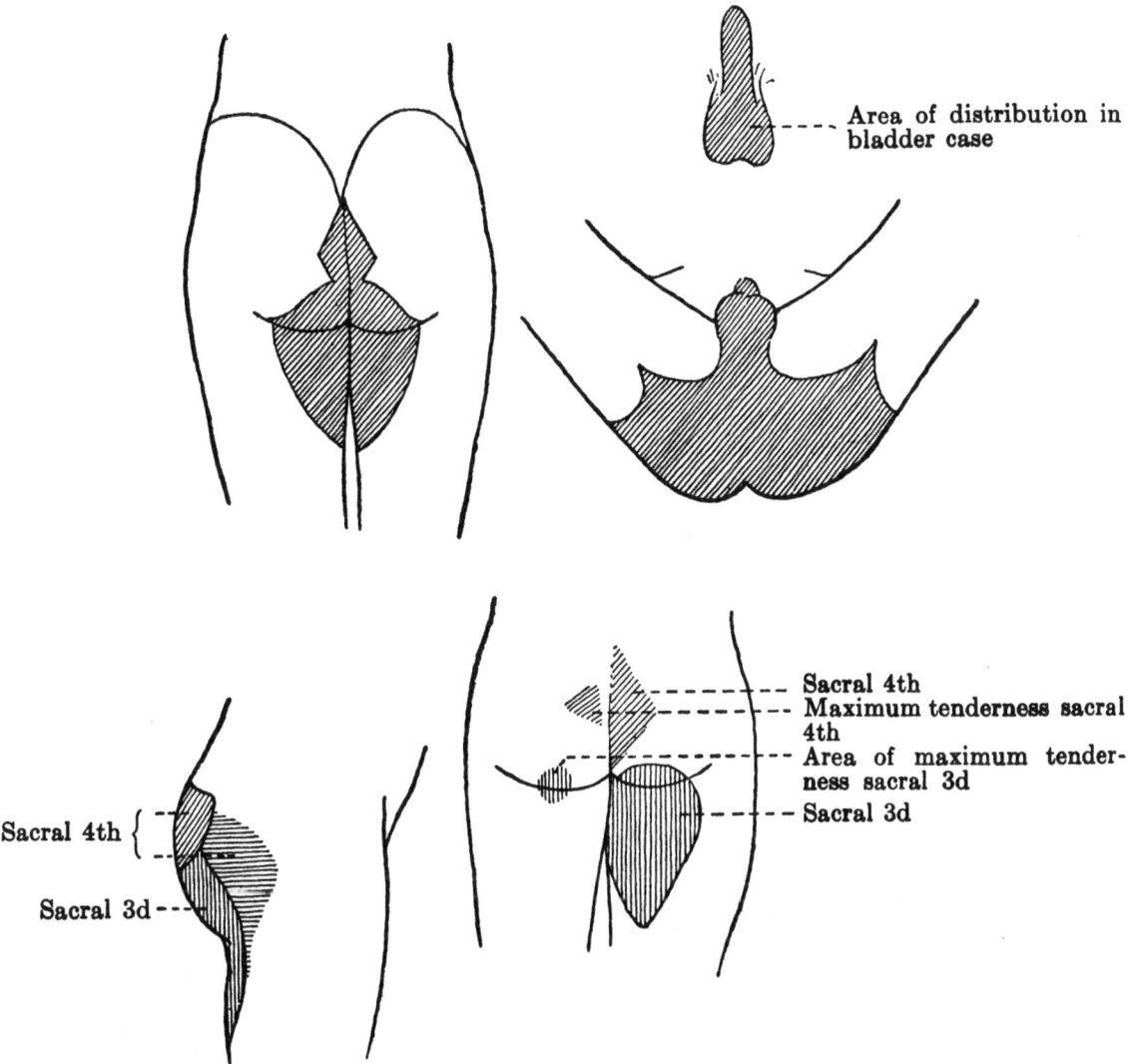

Area of distribution in bladder case

Sacral 4th
Maximum tenderness sacral 4th
Area of maximum tenderness sacral 3d
Sacral 3d

Sacral 4th {
Sacral 3d

Fig. 4.2. The distribution of referred pain arising from the bladder. (Behan 1914)

in the thighs, but not in the leg or foot, which are innervated by cord segments that do not have visceral afferent inputs. It is thought that the sensations evoked occur in the midline because the innervation of these viscera is bilateral.

Abnormal Sensations Induced or Influenced by Activity in the Lower Urinary Tract

Some abnormal sensations, such as the urge to pass urine at low bladder volume, are not usually described as painful, but are likened to the normal sensations of filling or fullness occurring repetitively. Other abnormal sensations, painful or innocuous, may be induced or influenced by bladder activity (see Nathan 1956). Abnormal sensations associated with peripheral nerve lesions, such as pain or paraesthesiae can be increased in intensity during micturition, so much so that some of these people dread having to empty their bladders. Such sensations are not uniformly painful, however, and may consist of warm, or other pleasant sensations in the thighs. A possible explanation for these phenomena is that, when visceral and somatic information converge and become mixed in ascending pathways of the cord, the mixture may arise not only from visceral plus somatic nociceptive pathways, but from visceral plus low threshold somatic pathways (McMahon and Morrison 1982a,b). Thus, the brain may not only misinterpret visceral impulses as having arisen by activation of somatic noxious inputs (giving rise to referred pain), but may confuse visceral impulses as having originated from low threshold somatic receptors (giving rise to referred innocuous sensations); this may be the explanation of paraesthesiae and innocuous sensations in these circumstances (Fig. 4.3). The electrophysiology of neurons which may mediate these and other referred sensory phenomena will be discussed later (see p. 118)

Sensory changes in the lower urinary tract can be caused by a variety of neurological conditions, such as multiple sclerosis, meningeal irritation, tabes dorsalis, diabetes, inflammation of the sacral dorsal roots, herpes zoster, and lesions of the spinal cord and peripheral nerves; the reader is referred to the review by Bors and Comarr (1971) for details.

Phantom Pain

Nathan (1956) also describes a patient with a phantom limb following amputation; shooting pains in the phantom limb were invariably accompanied by the desire to empty his bladder, even though his bladder was empty. Phantom urinary tract pain has also been reported following

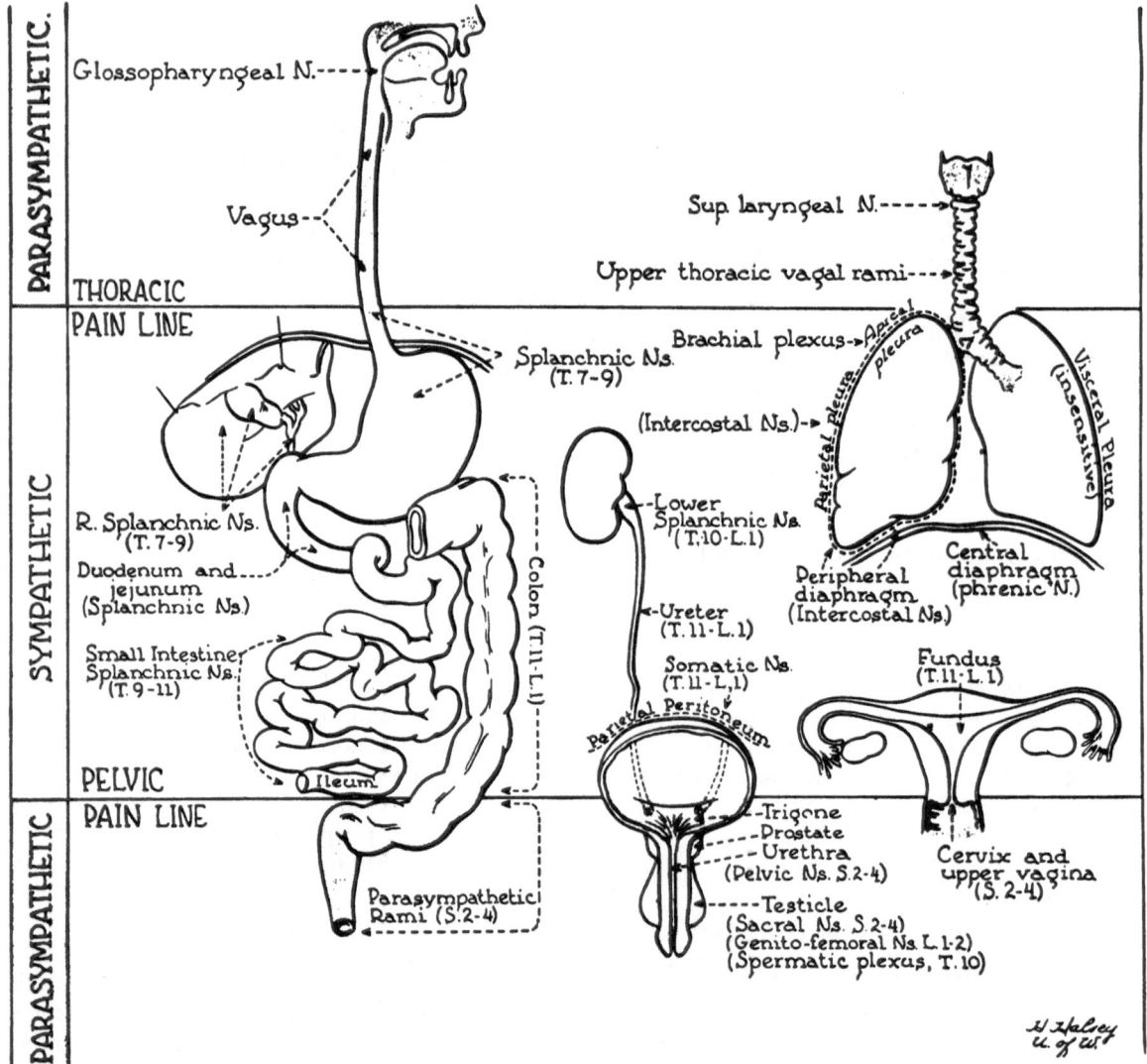

Fig. 4.3. The peripheral pathways by which pain arising from the lower urinary tract and other viscera is mediated. (White 1943; Ruch 1979)

cystectomy, and a recent case was successfully managed using lumbar sympathetic block (which affects the hypogastric afferent pathway) and transcutaneous electrical stimulation in a paravertebral site (Brena and Sammons 1979). This electrically induced analgesia is a consequence of stimulation of somatic afferent fibres, and mechanisms by which this may give rise to pain relief are discussed in the section on counterirritation phenomena.

Reflex Basis of Some Referred Sensory Phenomena

In this section, some of the reflex pathways that appear to be responsible for phenomena associated with referred pain will be discussed. Procacci et al. (1986) describe deep muscle tenderness and cutaneous hyperalgesia with their accompanying myalgic spots, algogenic points and trigger points. These local cutaneous and muscular foci of noxious sensation disappear when the area is anaesthetized with local anaesthetic and are thought therefore to be mediated by reflex pathways from the viscera to the skin or muscle (Fig. 4.3). The changes in skeletal muscle tone during noxious stimulation of the ureter were described by MacLennan and Goodell (1943). A sustained contraction of skeletal muscle occurred in the loin, and a severe deep dull aching sensation developed and persisted for about 6 h, leaving residual tenderness. Evans and McPherson (1959) describe some of the somatic reflex effects induced by distensions and constrictions of the bladder, and these pathways, akin to those responsible for the rigidity of abdominal muscles in the acute abdomen, can probably explain the changes in tone seen in these situations. Woolf (1984) has described long-term alterations in the excitability of the flexion reflex following a short exposure to a noxious thermal or chemical stimulus to the skin. Woolf proposed that these reductions in flexor reflex threshold were analagous to the prolonged hyperalgesia and tenderness seen after tissue injury.

The cutaneous hyperalgesia is more difficult to explain. Wolff (1963) suggested that it may be due to antidromic activation of sensory neurons which then release kinins into the tissues. The ways in which visceral stimuli should give rise to antidromic impulses in cutaneous nerve fibres is unclear. A more likely explanation is that autonomic reflexes to cutaneous structures are responsible (Procacci 1969; Procacci et al 1975). Local anaesthetic block of sympathetic ganglia decreased referred pain and hyperalgesia and altered the cutaneous pain threshold, as measured with a thermal algometer, and the skin potential responses. Jänig (1986) reports

that one of the most powerful inputs to cutaneous vasoconstrictor and sudomotor, and to muscle vasoconstrictor sympathetic efferent units is from the viscera, and in the case of the bladder the pelvic nerves provide the main afferent pathway. It is not clear how activation of the sympathetic efferent pathways could give rise to pain in normal tissues, although there is evidence that sympathetic efferents can take part in pain generation in animals with nerve injury (Wall et al. 1979). A further explanation might be that information from low threshold receptors in certain sites might be misinterpreted as being of noxious origin.

Information concerning the presence of innocuous and of noxious stimuli converges on wide dynamic range neurons in the cord (see p. 120). One cannot exclude the possibility that the local anaesthetic blockade of algogenic points already referred to did not act by blocking low threshold rather than high threshold afferents, and that the existence of algogenic spots may be attributable to changes induced in the central nervous sensory pathways.

Summary

Abnormal sensations, which include pain and innocuous sensations, can occur following lesions of peripheral (somatic or visceral) nerves or the spinal cord, and accompany bladder distension or contractions; they may be explained by the way in which visceral sensations are processed in the spinal cord. The neurons which transmit sensory information from the bladder from the spinal cord to the brain also carry information from muscles and/or skin innervated by the same spinal segment; sensations arising from viscera can be referred to these somatic structures because of this arrangement. This explanation is embodied in Ruch's convergence–projection theory of referred pain. Some of these ascending neurons carry information concerning noxious events, whereas others are concerned with innocuous activities, or a mixture of the two. This may explain the wide variety of sensations that can be elicited when the brain misinterprets visceral afferent activity as having arisen from somatic structures. Referred sensations elicited by these mechanisms are not alleviated by local anaesthesia of the somatic site.

Another source of sensation during excitation of visceral afferents is skeletal muscle spasm or vascular changes in the skin or muscle; these occur during excitation of visceral afferents because of reflex contractions of skeletal or reflex vascular reactions mediated by the autonomic nervous system. Such sensations are localized in myalgic spots or trigger

points, and the sensation arising from these sites can be removed by local anaesthesia of the painful spot.

Natural Activation of Sensory Endings in the Lower Urinary Tract

All of the sensory or reflex functions discussed up until now must be signalled by trains of impulses in the axons that transmit information concerning bladder function to the brain; in the cat there are 800 or so axons that operate as independent channels of communication in this manner. The technique of recording nerve impulses from single primary afferent fibres has provided most of our knowledge of what information is signalled. There is no evidence of crosstalk between adjacent fibres in normal nerves, and each neuron can be considered as a private pathway of communication from a receptor ending; studies of single afferent fibres provide information about the transducer function of that ending. The subsequent sections are concerned with the analysis of how sensory receptors transmit information, and in particular, how sensory endings in the bladder mediate the sensations that can be elicited from this viscus. Attempts will be made to understand the nature of the messages that give rise to innocuous and noxious sensations, and whether they can be transmitted along one or more types of primary afferent neurons.

Properties of Sensory Receptors

When a sensory receptor is subjected to an appropriate natural stimulus it is excited and produces a train of action potentials, the frequency of which is a function of the intensity of the stimulus. The most commonly studied sensory receptors are found in skin and muscle (see Schmidt 1978; Kandel and Schwartz 1981), and sensory receptors in the bladder behave in a similar manner. The frequency of the action potentials is used as a code that can be recognized by the brain, which interprets it as an index of the intensity of the stimulus. In skin and muscle, there is considerable specialization of sensory endings, in terms of their structure and their function. Thus, a sensory ending might be selectively responsive to only one type of natural stimulus, such as touch, temperature or vibration (Table 4.3). The receptive properties of small sensory endings are generally less specialized than those of the large myelinated axons, and one or more types of stimulus (mechanical. chemical or thermal) may excite these endings to produce trains of action potentials. They may also be specialized in other ways; different sensory receptors may respond to one type of stimulus in different ways, that can be defined using quantitative techniques. Some mechanoreceptors produce action potentials for the duration of a maintained mechanical stimulus, while others respond only to the sudden change in position at the start or the end of the event. The former behaviour is that of a slowly adapting receptor, and this type of ending is good at signalling the intensity of a stimulus. The latter type of response is that of a rapidly adapting ending, and is specialized to respond to rapid changes in position, such as occur during vibration (Fig. 4.4). Another type of specialization is that some receptors are extremely sensitive to mechanical stimuli, while others only respond to intense stimulation; the latter are generally believed to be implicated in signalling the presence of a noxious, i.e. damaging, mechanical stimulus, and the brain interprets their messages as pain. The terms low and high threshold mechanoreceptor are used to describe these two types of response (Fig 4.5). Thus receptors may be specialized in three ways: (a) they may respond

Table 4.3. Properties of some cutaneous receptors

Structure	Adaptation	Adequate stimulus	Threshold	Conduction velocity
Merkel disc	Slow	Indentation	Few milligrams	Fast
Pacinian corpuscle	Rapid	Vibration	Optimally 200–250 Hz	Fast
Free nerve endings (nociceptor)	Slow	Damaging thermal and mechanical stimuli	>45°C hundreds of grams	Slow

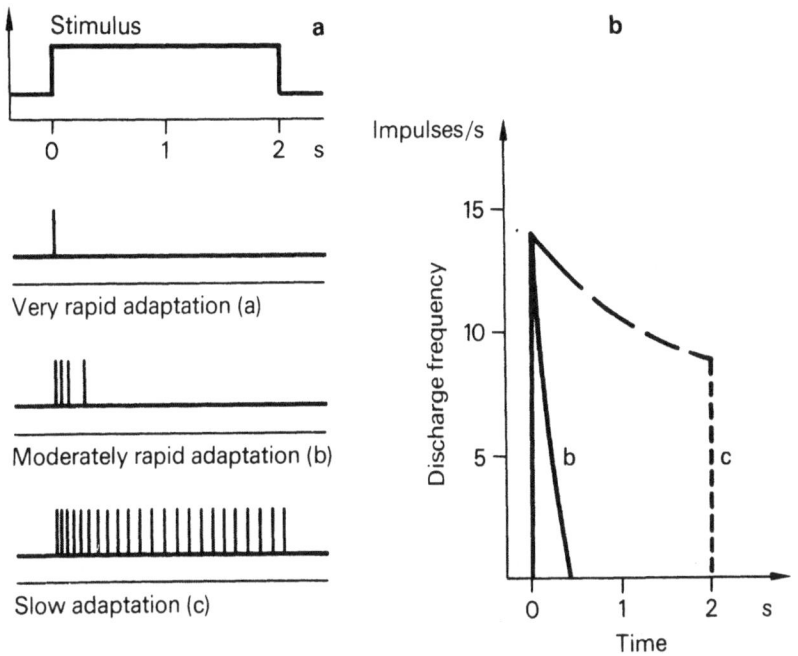

Fig. 4.4a,b. Schematic diagram of the discharges of receptors adapting at different rates. **a** Action potentials of a very rapidly (*a*), moderately rapidly (*b*) and slowly adapting (*c*) receptor during a prolonged stimulus held constant in time. **b** Graph of the time course of the instantaneous discharge rate during the stimulus, in the receptors (*b*) and (*c*). Note that the discharge in (*c*) incorporates both phasic and tonic response components. (Schmidt 1978)

specifically to only one type of physical or chemical stimulus; (b) they may respond to rapidly changing aspects of the stimulus, or to its continued presence; and (c) they may be very sensitive (i.e. have low thresholds) to that form of natural stimulation, or be quite insensitive (i.e. have a high threshold). As stated previously, certain sensory endings are relatively unspecialized, and may be excited by more than one type of stimulus, e.g. mechanical and thermal stimuli; these receptors are often called polymodal receptors, and the information they signal is therefore less specific. Nevertheless, this sort of information can be used by the brain, e.g. in signalling the presence of stimuli that cause tissue damage. The term nociceptor is applied to a sensory ending that responds only to intense, damaging stimuli such as pinch and high temperatures (Fig. 4.6).

The idea that innocuous and noxious sensations are mediated by separate populations of nerve fibres is embodied in the "Specificity theory" of sensation. In skin, the low threshold fibres can be maximally activated by innocuous stimuli, and the thresholds of the high threshold mechanical nociceptors may be several orders of magnitude higher than the low threshold fibres; indeed they respond to injurious stimuli only.

An Approach to the Study of Bladder Receptors

In the case of the bladder, an increase in intravesical pressure, produced by distensions or contractions will increase the firing rate of the afferent fibres. Certain pressures are too low to excite the sensory endings, and the terms "pressure threshold" or "volume threshold" indicate the pressure or the volume which just causes the sensory ending to discharge repetitively. Thus, any given level of pressure in the bladder may be subthreshold or suprathreshold for a particular sensory ending. The measurement of such a threshold is a useful property of the ending since it defines whether the ending is capable of responding to natural or supranormal levels of pressure. If there were a population of receptors that responded only to overdistension, or to excessive intravesical pressure, one could argue that these were specifically concerned with the transmission of noxious information from the bladder. Painful sensations could then be mediated by a separate population of afferent nerve fibres, which, as in skin, would be called nociceptors.

Fibres with a pressure or volume threshold in the normal range are clearly capable of sensing natural

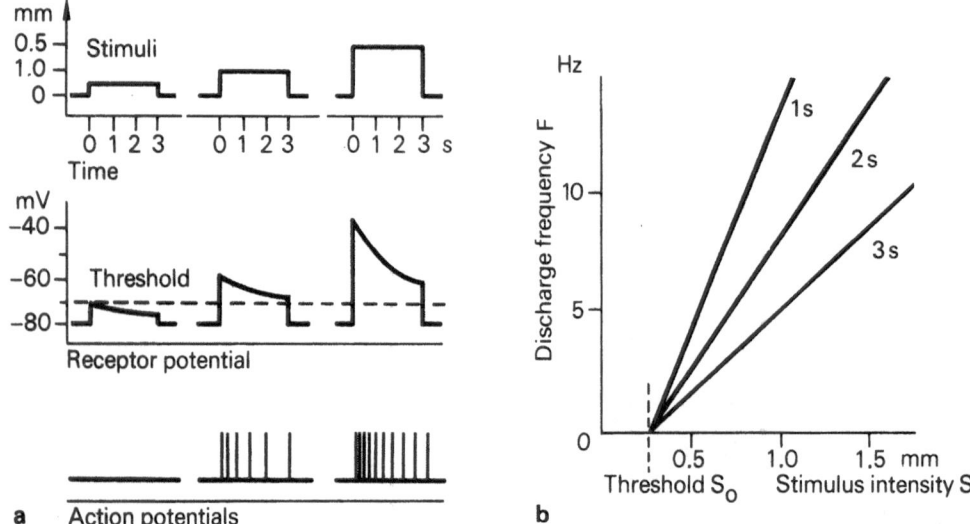

Fig. 4.5a,b.. Schematized relationship between stimulus intensity, receptor potential and discharge frequency. **a** Receptor potentials and action potentials at three different intensities of the stimulus. Action potentials are elicited as soon as the receptor potential exceeds the threshold (*dashed line*). **b** Graph of the instantaneous action potential frequency F at three different times (1, 2, and 3 s) after the onset of the stimulus, as a function of stimulus intensity S/S_0 is the threshold intensity. (Schmidt 1978)

events; they may continue to increase their frequency of discharge when the intravesical pressure is raised into the supraphysiological range, and this would be evident from the stimulus–response curve, i.e. the relationship between the pressure (or volume) in the viscus and the response (spike frequency) of the sensory ending. Sensory endings with these properties can signal normal and supranormal pressures and may be involved in sensory experiences such as the normal sensation of filling and, in addition, in visceral pain. If this were the case, those sensory pathways would behave rather differently from pathways that carry sensation from skin, and the function of that pathway would fit with the "intensity theory" or "summation hypothesis" of Goldscheider (1920). According to this theory, one set of afferent fibres may encode information that gives rise to innocuous and noxious sensations. In our consideration of the properties of bladder afferents, it is necessary therefore to ask how many different functional types of sensory endings are present in the bladder, and in particular to seek evidence for high threshold afferents that might only signal events that give rise to pain. In considering this question in respect of distending pressures, it has to be remembered that, to date, all the work on bladder afferents has been done on animals with a normal lower urinary tract; while there is no doubt that pain is normally generated by overdistension of the bladder, it can commonly be produced at low volumes, if the bladder is inflamed.

Recently, there have been reports of change in reflex excitabilities during bladder inflammation, but no information is currently available concerning the afferents in this condition. In inflammation, pain could be induced by low levels of distension either by an increased response of the sensory endings that normally sense bladder distension, or it could be that a different population of afferents sense distension only during inflammation. The answers to this question should become available in the next few years, and the answer will be quite significant, as it will determine whether we think in terms of specificity or intensity theory.

Summary

Sensory receptors act as transducers which transform natural energy into a train of action potentials. Many receptors are specific in that they mainly respond to one type of natural energy (mechanical, thermal, chemical), which is known as the adequate stimulus. Mechanoreceptors can be subdivided into rapidly adapting receptors, that respond to a change in position of the sensory ending, and slowly adapting receptors, that respond to a maintained displacement or force. Rapidly adapting endings are optimally sensitive to vibration, whereas slowly adapting receptors monitor position. The frequency of the discharge of the receptors is a function of the

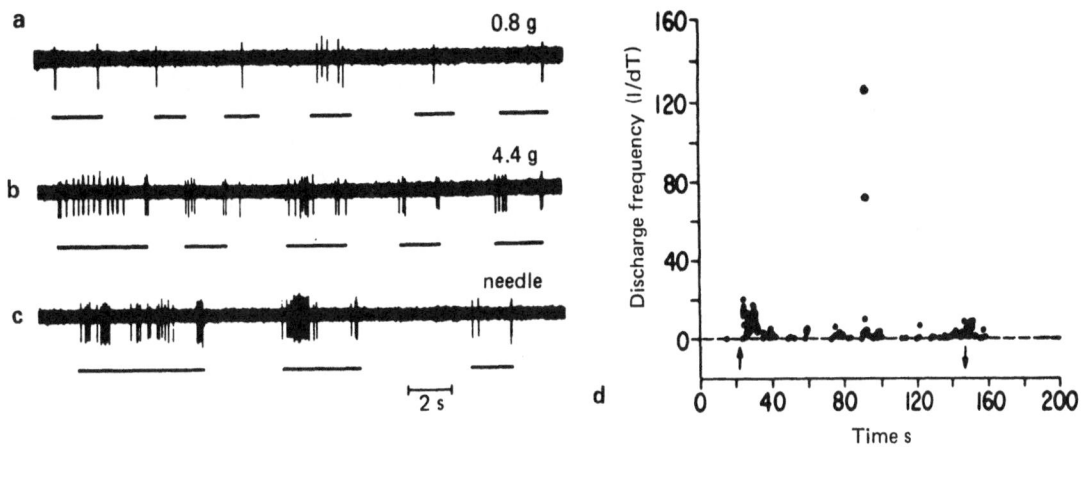

Fiber 2572-C14

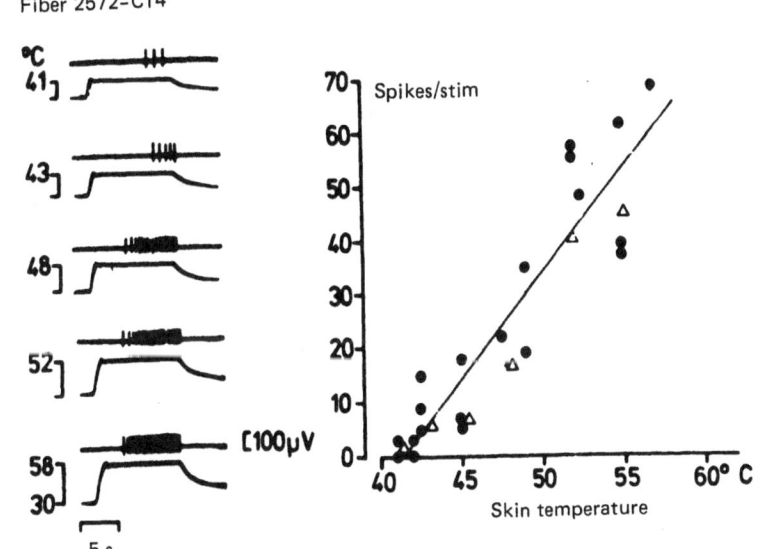

Fig. 4.6a–e. Properties of a cutaneous polymodal nociceptor. In **a** are shown the responses of a C polymodal nociceptor to stimulation with a von Frey filament bending with a force of 0.8 g. A greater response is shown in **b**, when the stimulus was a 4.4 g von Frey filament. A needle that penetrated the skin was used for **c**. The discharge produced in a C polymodal nociceptor by the application of dilute acid to the skin is shown in **d** (*a–d* are from Bessou and Perl 1969). In **e** are shown the responses of a C polymodal nociceptor to graded thermal stimuli. The relationship between discharges produced by each stimulus and skin temperature is shown by the graph at the *right* (Beck et al. 1974; quoted in Willis and Coggeshall 1978)

intensity of the natural stimulus, and the intensity of stimulus which just causes the receptor to be excited is known as the threshold. Mechanoreceptors can have low thresholds (and these respond to a few milligrams force or a few micrometres displacement) or high thresholds (these may only respond when the force applied is over 100 G). The relationship between spike frequency and the intensity of the natural stimulus is often referred to as the stimulus–response curve, and it describes the range of transduction properties of the ending. In skin, low threshold mechanoreceptors can be maximally activated by innocu-

ous events (i.e. stimuli that do not cause injury or pain); high threshold mechanoreceptors do not begin to respond to a mechanical stimulus until its intensity is high, and often in the noxious range.

Not all sensory receptors are absolutely specific, and some are quite nonspecific, being capable of excitation in response to mechanical, thermal and chemical stimuli; such endings are polymodal. The term polymodal nociceptor is applied to sensory endings that are activated by damaging levels of mechanical and thermal, and possibly chemical stimuli. Endings of this type usually have unmyelinated axons.

In the bladder and its ligaments, receptors that respond to distension and contraction and to the intra-arterial administration of bradykinin have been found, and they have small myelinated and unmyelinated axons. They travel to the spinal cord in the pelvic and hypogastric nerves; hypogastric nerve endings are found in greater numbers at the bladder base, but pelvic endings are distributed more evenly throughout the viscus. The endings respond to dis-tension by giving a slowly adapting discharge, and the thresholds of both groups of endings are in the range of pressure that is considered physiological. The threshold of the most sensitive endings is at a point on the pressure–volume curve that corresponds to the point in the cystometrogram when the first sensation of filling is normally reported. No high threshold endings have been found in afferent axons coursing centrally in the hypogastric or pelvic nerves.

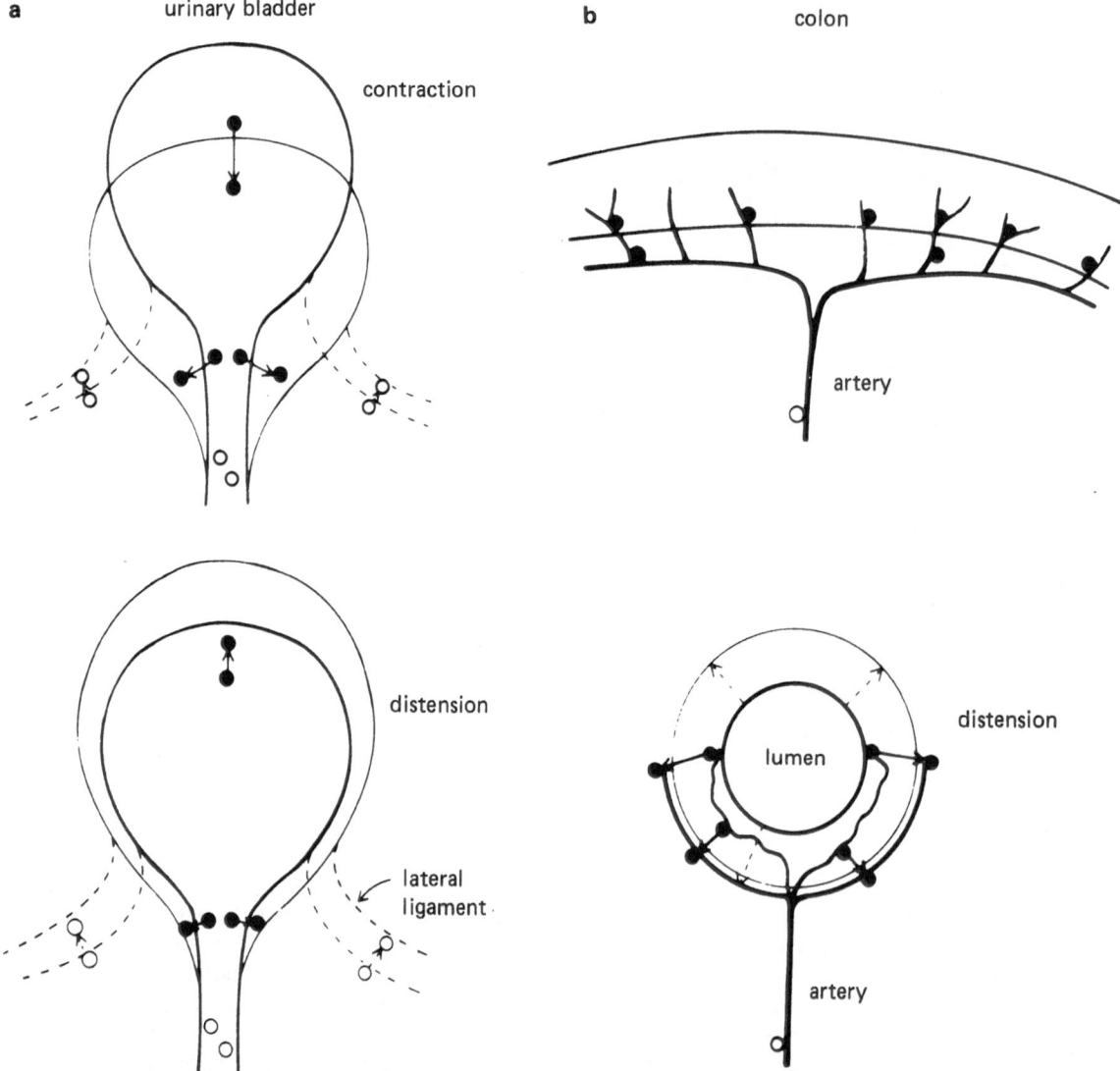

Fig. 4.7(a,b). The receptive fields of pelvic and hypogastric nerve endings in the bladder (**a**) and colon (**b**), showing the distortion produced by distension and contraction of the viscera: the bladder neck and proximal urethra become distended during detrusor contraction, producing greater distortion of these structures than occurs during distension. Similar diagrams show the effects of colonic distension. Positions of receptors which are excited by these procedures are indicated by *dots*. Positions of receptors which are not excited or only weakly excited with a few spikes at the beginning of a distension or a contraction are indicated by *circles*. (Jänig and Morrison 1986)

The low threshold endings can, however, signal the existence of high pressures, as demonstrated by their stimulus–response curves.

Sensory Receptors in the Lower Urinary Tract: Results

Sensory endings that respond to mechanical changes in the lower urinary tract of mammals have been described by Evans (1936), Talaat (1937), Iggo (1955), Arlhac (1971), Winter (1971), Clifton et al. (1976), Floyd et al. (1976), Floyd and Lawrenson (1979), Morrison (1981) and Bahns et al (1986, 1987). Most workers have studied cats.

Apart from the occasional presence of a few pacinian or paciniform copuscles projecting centrally in the pelvic and hypogastric nerves in cats, the main afferent fibres in these nerves are finely myelinated or unmyelinated. The pacinian corpuscles are rapidly adapting receptors with large myelinated axons, and respond to vibrations, usually transmitted from the exterior. They are common in some members of the cat family, but are rare in human viscera. The smaller fibres include mechanoreceptors which respond to bladder distention and contraction, and to certain chemical stimuli. The adequate stimulus is tension at the site of the sensory endings, which may be within the bladder wall or outside it (Talaat 1937; Winter 1971; Floyd et al 1976). There is no doubt that some sensory endings are in the walls of the viscera in series with smooth muscle, but others are outside the muscle layers, and can respond to forces transmitted through these stabilizing ligaments during contraction and distension. The concept of the "in series" tension receptor, defined as a receptor that responds to distension and contraction (Iggo 1955) includes receptors in both situations, and may group under one heading a rather heterogeneous bunch of endings. Nevertheless, reports that receptors respond to only one of the combination of contraction and distension are rare (see Clifton et al. 1976), and there do not appear to be any major differences in the thresholds or sensitivities of these endings, when studied with quantitative techniques. The sensory endings can be localized within the tissues, usually as punctate (spot-like) sites, and sometimes as small areas of mechanosensitivity (Fig. 4.7).

Responses to Bladder Distension

Pelvic Nerve Afferents Several groups of investigators have examined the range of pressure thresholds in the bladder afferents during disten-

Table 4.4. Threshold intravesical pressures necessary to excite bladder afferents

	Intravesical pressure (mmHg)
A. Pelvic nerve	
Distension:	
Talaat (1937)	5–20
Floyd and Lawrenson (1979)	4–25
Bahns et al. (1985)	5–18
Contraction:	
Bahns et al. (1985)	5–12
B. Hypogastric nerve	
Distension:	
Floyd et al. (1976)	4–20
Bahns et al. (1986)	5–30
Contraction:	
Bahns et al. (1986) (see also Jänig and Morrison 1986)	4–25

sions, and are agreed that these afferents form a unimodal population of nerve fibres with thresholds within the physiological range of pressures, i.e. 4–30 mm Hg (see Table 4.4). The threshold pressures are usually on the compliant part of the pressure-volume curve, between 25% and 75% of the pressure at which the curve becomes steep (G. Lawrenson, personal communication). Thus, the position on the pressure–volume curve at which the first discharge of vesical afferents occurs is close to that at which the first sensation of filling is reported during cystometry (Fig. 4.8). The stimulus–response curve obtained by gradual filling differs

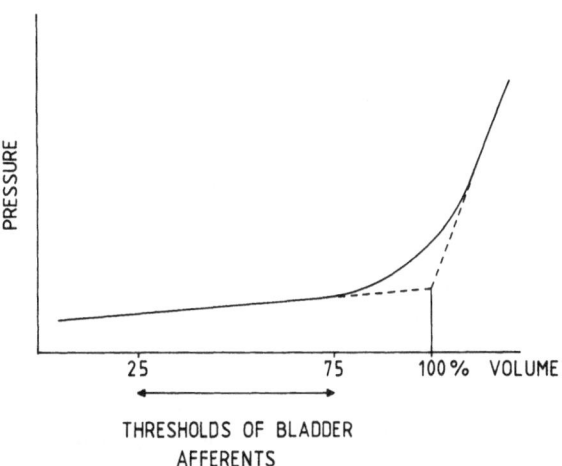

Fig. 4.8. The thresholds of pelvic afferent fibres relative to a standardized pressure–volume curve in cats. 100% indicates the pressure at which the flat and steep parts of the pressure–volume curve intercept. (Courtesy of G. Lawrenson)

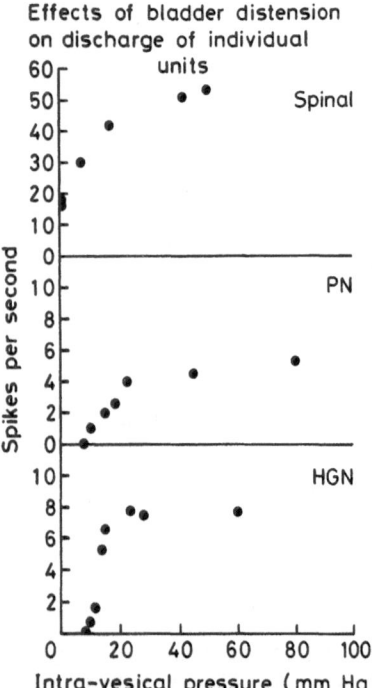

Fig. 4.9. The stimulus–response curves of single pelvic nerve and hypogastric nerve afferents, and of a neuron within the spinal cord; the bladder was distended 5–10 ml at a time to reach the higher volumes. Adaption could therefore have contributed to the responses obtained at high volumes. (Morrison 1981)

nerve afferents in the literature have been from finely myelinated fibres.

Responses to Bladder Contraction

Pelvic Nerve Afferents Bladder afferents can monitor not only the state of filling of the bladder, but also the contractile state of the detrusor. Thus, when the bladder contracts, the discharge of the afferent endings increases, and the rate of discharge is a function of the size of the contraction. Some endings are so sensitive that they can respond to slow changes in the tension of the vesical smooth muscle. This ability of the nerve endings to respond to bladder distension and contraction may explain how the sensory experience during bladder filling is dependent on the mechanical state of the viscus. The pressures at which contractions elicit activity is close to the pressure threshold during distensions.

The responses of bladder afferents during contractions is greater if the conditions are isovolumetric. This is not just a situation which can occur in the laboratory, but is achieved in humans during a sudden voluntary interruption of micturition; in these conditions flow ceases because of the contraction of the skeletal muscle of the urethral sphincter, and the subject reports a desire to void until the smooth muscle of the detrusor becomes relaxed (Fig. 4.11).

Hypogastric Nerve Afferents The hypogastric nerve afferents respond to bladder distension and contraction in a manner similar to that described for the pelvic nerve afferents. There is no basis for the belief that the hypogastric nerve contains high threshold sensory endings as had been suggested by the early electrophysiological work of Evans (1936). The literature on this topic has recently been reviewed by Jänig and Morrison (1986). Figure 4.12 shows the response of a single hypogastric afferent unit during a small spontaneous bladder contraction.

Responses to Chemicals and Inflammatory Agents

The hypogastric slowly adapting mechanoreceptors are able to respond to intra-arterial injection of chemicals that are known algesic agents, such as bradykinin. In the cat, Floyd et al. (1977) found that the excitation of these endings was mainly related to contractions induced by the bradykinin, and if reflex activity of the bladder was prevented,

from that during sudden distensions, in that the former plateaus at high pressures (Fig. 4.9), presumably owing to adaptation of the nerve endings (Jänig and Morrison 1986).

Thus the first sensation of filling can be related to the onset of activity in mechanosensitive afferents from the bladder, and the differences in response to sudden and gradual filling may be explained partly by the adaptation of sensory nerve endings. Little information is available about the effects of distensions of the proximal urethra on sensory endings.

Hypogastric Nerve Afferents The response of hypogastric afferent fibres to bladder distension is essentially the same as that of the pelvic nerve afferents. These afferents differ in a number of respects, however. First, the sensory endings are not all in the bladder wall, but can be in the vesical ligaments; second, more than half of them exhibit some resting discharge; third, the hypogastric afferent fibres that respond in the manner described have conduction velocities in the A-delta and C-fibre ranges (Fig. 4.10). In contrast, all of the recordings of pelvic

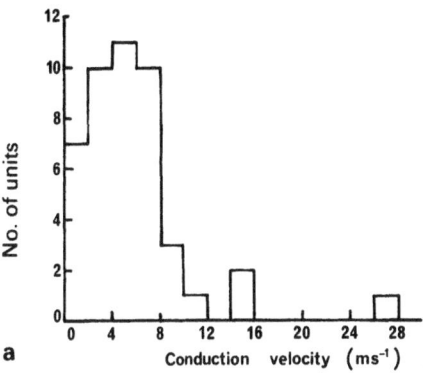

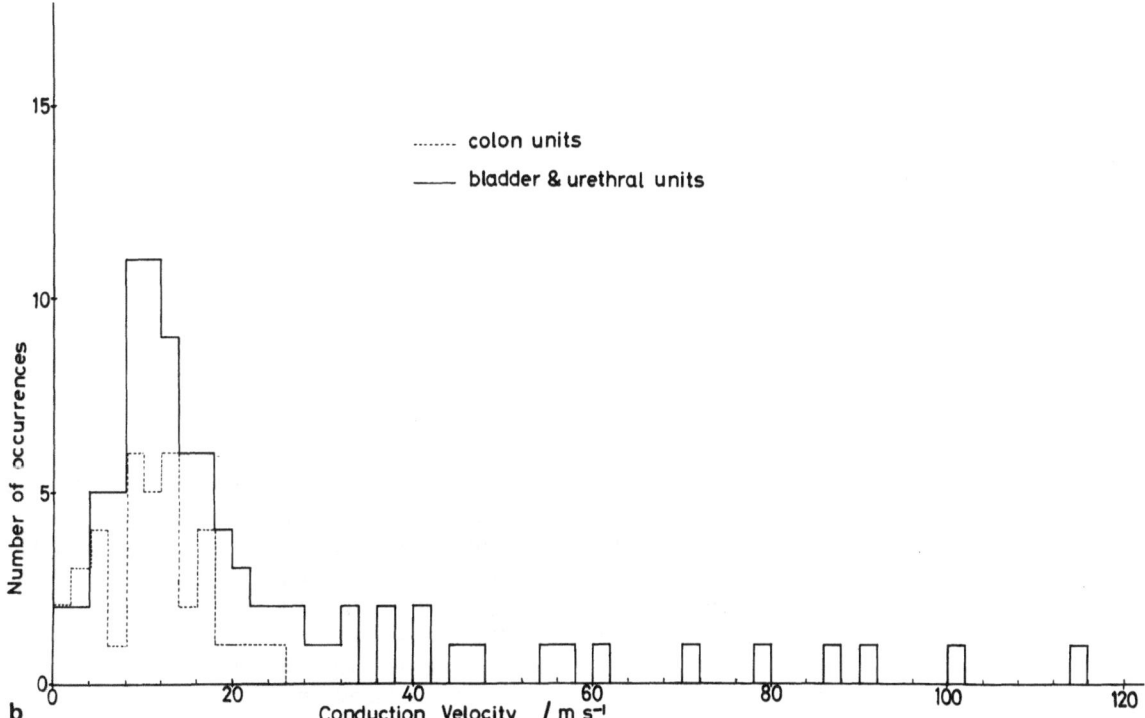

Fig. 4.10a,b. The conductions velocities of mechanosensitive afferents in (**a**) the hypogastric and (**b**) pelvic nerves of the cat. (Floyd et al. 1976; and courtesy of G. Lawrenson)

the peptide did not excite the afferent endings (Fig. 4.13). No specific chemoreceptors were found, and all endings which responded to the kinin were also mechanosensitive. There appear to be no recordings from pelvic nerve afferents during chemical stimulation.

Inflammation and Visceral Sensory Receptors

Inflammatory changes are often found in bladder mucosa and muscle in patients with sensory symptoms. Little physiological work has been performed on bladders that are inflamed, but there is no doubt that bladder inflammation can give rise to changes in the reflex activity of the viscus, and to an increased frequency and intensity of symptoms. One algesic agent that can be released by inflammatory states is bradykinin: intra-arterial injection of this peptide gives rise to large bladder contractions in cats that are dependent on the functional integrity of the micturition reflex pathway, i.e. connections between the cord and brain stem. The mechanosensitive sensory endings in the bladder discharge during the application of bradykinin (Floyd et al. 1977). Little is known of the action of capsaicin

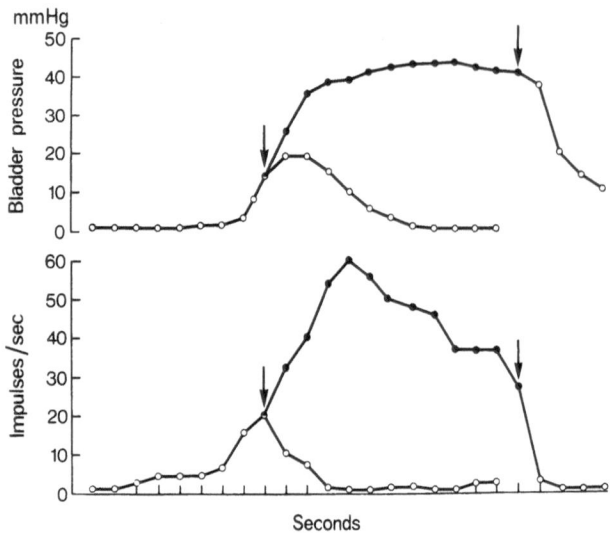

Fig. 4.11. The responses of pelvic nerve afferents during bladder contractions. (Iggo 1966)

on the functional properties of vesical nerve endings; the reports of Maggi et al. (1986), however, might suggest that depletion of substance P from afferent nerve endings in the bladder reduces their excitability to distension.

Responses of Receptors in the Urethra

The urethra of some species is surrounded by large numbers of pacinian corpuscles, which are rapidly adapting receptors. Todd (1964) reported that the urethra is innervated by pudendal nerve afferents that respond to the passage of urine. Bahns et al. (1986) have also recorded from afferents in the urethra that respond to distension, the threshold pressures, however, being rather higher than in the bladder. Brindley et al. (1974) suggested that the sensation that micturition was imminent was associated with opening of the bladder neck and detrusor

contraction; it is conceivable that endings in the proximal urethra with the properties described by Bahns et al. (1986) could mediate this sensation.

Summary

The bladder and urethra can contain pacinian corpuscles, which are rapidly adapting receptors with large myelinated fibres; these represent a small proportion of the afferent fibres that innervate these organs. The flow receptors described by Todd (1964) may have been pacinian corpuscles that responded to the vibrations associated with turbulent flow in the urethra. The majority of sensory endings have finely myelinated or unmyelinated axons and their conductions velocities are in the A-delta and C-fibre ranges, respectively. There appears to be little difference in function of these two groups of endings, which

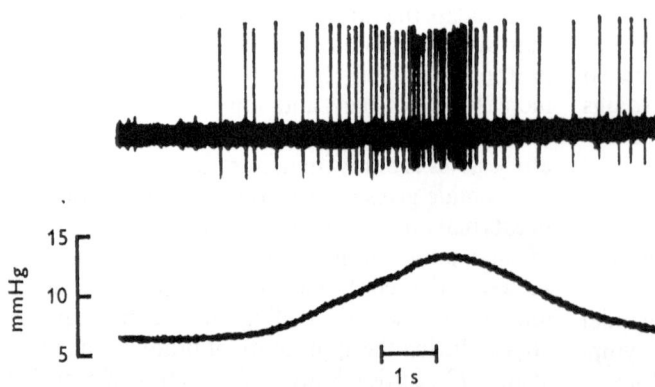

Fig. 4.12. The response of a hypogastric nerve afferent to a small contraction of the bladder. (Floyd et al. 1976)

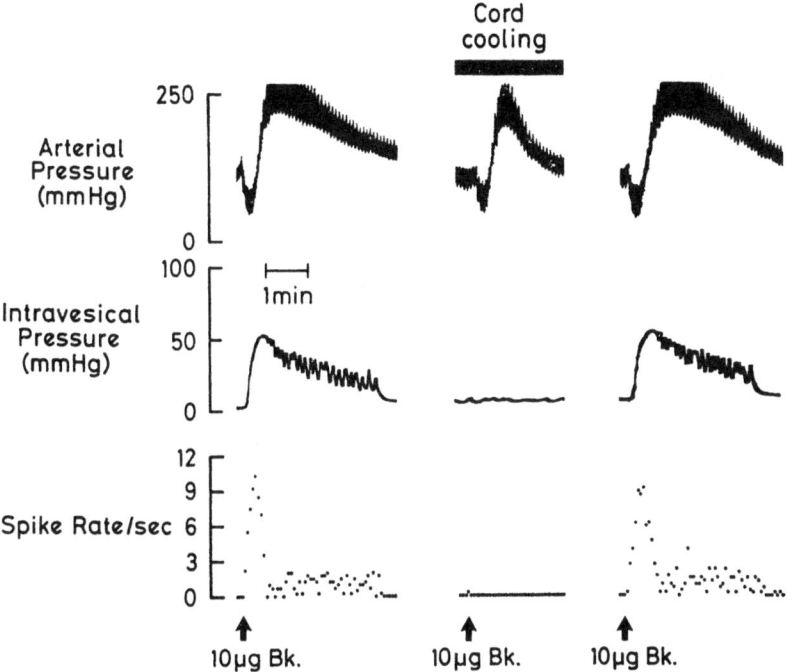

Fig. 4.13. The effects of intra-arterial injection of 10 μg bradykinin (*Bk.*) on a hypogastric nerve afferent from the bladder. The effects are shown before, during and after cold block of the spinal cord; the lack of contraction or response during cold block suggests that the afferent responds primarily to the mechanical event. (Floyd et al. 1977)

are slowly adapting mechanoreceptors; the sensory endings can be localized during single fibre recording experiments as spot-like endings in the bladder wall, serosa and peritoneal attachments. The pathways of the fibres are often perivascular, as are the sensory endings. They can signal the level of distension and the size and timing of bladder contractions, and probably do this because they can monitor the forces acting within the tissues in which they are embedded. During distensions, the lowest threshold fibres respond to distensions of 25%–75% of the volume at which the pressure–volume curve starts to become steep. All endings respond to contractions that produce bladder pressures akin to those occurring during micturition. The mechanoreceptors respond also to the intra-arterial injection of bradykinin, probably because they respond to the concomitant contraction of the viscus. There are few differences between the properties of the pelvic and hypogastric afferents from the bladder; a proportion of the hypogastric afferents have a resting discharge when the bladder is empty, but this is less common in the pelvic nerve endings, and the hypogastric endings appear to be mainly at the bladder base, whereas those of the pelvic nerve are more evenly distributed over the whole viscus. Little can be said of urethral receptors, except that "flow receptors" have been described, and that urethral nerve endings of A-delta and C-

fibres have higher thresholds of distending pressures than in the bladder.

Neurochemistry of Visceral Afferents

There is reasonable evidence that primary afferent endings, or endings of primary afferent collaterals participate in activities other than just the transmission of afferent or sensory information and the mediation of reflexes via the central nervous system. It has been known since Lewis's experiments on the triple response in skin that axon reflexes may be responsible for vasodilatation in response to afferent stimulation (Lewis 1942). A number of the peptides that can be released at the peripheral terminals of sensory axons have vasodilator activity, e.g. substance P, VIP and CGRP. About 80% of the substance P produced by the dorsal root ganglion cells of visceral afferents travels peripherally (see Dockray and Sharkey 1986). Antidromic activation of primary afferent axons induces vasodilatation, increased capillary permeability and plasma extravasation in skin, and possibly in some viscera. Rats treated with capsaicin during the neonatal period (i.e. rats without small afferent fibres, and depleted of substance P in their nerve terminals) do not show such activity, which suggests that the vasodilator neuropeptides may be involved in this response. In skin, peptides such as substance P may be released by the axon reflex which mediates the flare in the triple response: noxious stimulation of skin causes action potentials to travel centrally in small afferent fibres, and when they reach a branch point in the axon, antidromic impulses are also set up, and travel back towards a primary afferent terminal, where a vasodilator neurotransmitter is released.

There may be a similar action in some viscera; much of the work concerned with substance P release from visceral afferent fibres has concentrated on pathways from the gut, and it is now known that neurogenic oedema and extravasation of plasma can also occur in the bladder (McMahon and Abel 1987).

Relationship Between Afferent Activity and Sensation from the Lower Urinary Tract

The survey of sensations from the lower urinary tract suggested that there were a number of innocuous everyday sensations, of bladder filling and fullness, of desire to micturate and of imminence or presence of urine flow. These were distinguished from the sensations of urgency, and of pain, which are particularly marked if the tissues are inflamed. The lower urinary tract has been shown to contain a relatively homogeneous population of mechanoreceptors, impulses from which travel to the spinal cord in the pelvic and hypogastric nerves, and there are few functional differences between these two pathways, at least in the periphery. During distensions, the pressure thresholds of the sensory endings in the bladder form a unimodal population and cover the physiological range of pressures; the lowest threshold fibres became active at a point on the pressure–volume curve of the cat that approximates to the point on the human pressure–volume curve at which the first sensation of filling occurs. The pressure thresholds during contractions are similar to those during distensions. However, high intravesical pressures during micturition do not elicit pain, yet they do cause considerable activity in the mechanosensitive afferents. It would appear that sensations induced by pressure

are less noxious during detrusor contraction, even though the discharge from the endings may be greater, because of the dynamic activity of the endings. This cannot be a peripheral phenomenon, and the theoretical consideration of how it is mediated might be along the lines of modulation by "corollary discharges", possibly by means of recurrent inhibition (see Chap. 5, p. 133), or other connections between the autonomic and sensory systems, such as probably occurs in axon collaterals of the descending pathways from the raphe nuclei and the locus coeruleus.

The afferents from the urethra have higher thresholds than those in the bladder, but still within the range of pressures that can be applied to the urethra without inducing pain in humans. Jänig and Morrison (1986) argued that the relatively small range of pressure thresholds found in visceral receptors is attributable to the position of the endings with respect to the viscus, and physical factors, such as the dimensions of the organ, the amount of smooth muscle, and its tone or compliance. Therefore, the difference in thresholds of bladder and urethral endings may not be due to differences in the endings themselves, but to the physical state of the tissues in which they are located. The position of these endings is often outside the muscle layers, so the possibility exists that innocuous distending pressures when the urethral muscle is contracted (the normal state) may induce greater activity in these sensory endings and elicit sensation when the muscle is relaxed. A similar situation could occur at the bladder neck. Normally, the sphincteric muscle relaxes before the bladder contracts (see Chap. 9, p. 303). Thus, the state of the urethral muscle, its innervation and its ability to resist flow may be complicating factors in determining how the sensory endings respond.

Peripheral Basis of Visceral Sensation: Comparisons with Cutaneous and Deep Somatic Sensation

There have been many studies on how sensations are mediated by afferent nerve fibres during the last half century; relatively speaking, there are comparatively few studies of the visceral domain, and much of the quantitative information that is available has occurred as a result of studies in the skin or muscle. In order to relate the problems of visceral sensation to knowledge obtained from these systems, a few salient references to studies in nociception or proprioception will now be made, to draw attention to some of the similarities and differences between the different sensory pathways.

Mediation of Visceral Pain: Comparisons with Cutaneous Sensation

It is the view of some sensory physiologists that the neural apparatus for visceral sensation is likely to be less complex than that for somatic sensation; this opinion is based on the facts that visceral sensation is not localizable to the viscus of its origin, neither is it discriminative, and it is easily confused with somatic sensation (Willis 1986). Certainly, the numbers of afferent fibres from the viscera compose only a small proportion of the total afferent input to the spinal cord (see Chap. 3, p. 57), so if size is one factor that leads to complexity, this view would certainly be correct. These nerves, however, do mediate a number of sensations (innocuous sensations of filling and desire to void, as well as urgency and overt pain), and, in addition, these afferents have important reflex functions; these reflexes include parasympathetic, sympathetic and sphincteric reflexes concerned with the coordination of function in the lower urinary tract as well as reflexes influencing other structures, and when these are disordered the effects can be as disastrous for the individual as chronic pain. One of the major scientific questions in this topic is therefore how these multiple functions, innocuous and noxious sensations and a variety of reflex patterns elicited from the lower urinary tract, may be mediated.

In order to answer this complex question, experimenters have attempted to look at the properties of all the sensory fibres arising from the bladder, and to attempt a classification which will divide the whole population into different fibre types; separate groups of fibres with unique properties might then be described, and their properties related to different functions. It could be argued that the simplest situation is that in which one function is mediated by one (or more) fibre types, as appears to be the case in skin. However, current knowledge suggests that the viscera are not provided with a separate group of sensory endings for each of the functions detailed. This suggests either that some afferent fibres from the bladder must participate in multiple functions, and that some functions or sensations may be dependent on the activation of certain combinations of sensory endings; in the latter case, endings in different regions of the lower urinary

tract may interact and combine. Brindley et al. (1974), for instance, suggest that the sensation of imminent micturition depends on the combination of bladder contraction and opening of the bladder neck. This may be achieved by having a central pathway composed of neurons that combine inputs from these different locations. It is known that ascending spinal pathways do combine information arising from different sites (see later). That anatomical arrangement, however, would not solve the problem of sensations elicited only at high pressures, particularly as the afferent fibres from the bladder are a relatively homogenous low threshold population. (see Jänig and Morrison 1986). These workers, as well as others working on the testis and heart (Kumazawa 1986; Malliani et al. 1986) suggest that noxious sensations may be elicited by intense stimulation of low threshold receptors, rather than by stimulation of a separate group of high threshold endings.

In the bladder and urethra, the evidence for high threshold mechanoreceptors or specific nociceptors is poor. One might argue that there is no need for them because the stimulus–response function of the mechanoreceptors can cope with normal and supraphysiological pressures; however teleology can never provide a sound scientific basis for this or any other argument. The case for the intensity or summation theory of visceral sensation from the lower urinary tract has recently been argued by Jänig and Morrison (1986). The segregation of visceral impulses into pain pathways and into pathways concerned with other functions may then be a matter for the spinal cord (see later). The evidence presented in Chapter 3 suggests that visceral afferent information is distributed to a variety of regions of the grey matter of the spinal cord.

In inflammation, the sensitivity to bladder and urethral activity is increased, but we know nothing of the reasons for this phenomenon. It may be that the mechanoreceptors behave differently in inflammatory conditions, or there may be another group of primary afferents that have not previously been identified. If these receptors have more specific properties than the mechanoreceptors, then they may play a part in noxious visceral sensation associated with changes other than the purely mechanical ones.

Comparisons with Proprioceptive Sensations

A concept that has been useful in the analysis of sensations during movement of structures whose position is determined by motor activity initiated by the central nervous system was described by von Holst (1973). Similar ideas have been applied to proprioceptive sensations by Goodwin et al. (1972) and McCloskey (1981); the ideas contained in these papers may also have application in the study of sensation from the lower urinary tract and other visceral organs whose activity is regulated by the central nervous system. It is also appropriate to make some comparisons with proprioception, since visceral sensations are usually felt in the deep somatic domain rather than skin, and because the ascending pathways that carry information rostrally from the sacral cord often receive afferent inputs from the bladder and from the proprioceptive system (Milne et al. 1981; McMahon and Morrison, 1982a; see following discussion).

The muscle spindle is one example of a sensory ending which participates in more than one function; apart from the well-known reflexology induced by stretch of muscle spindles, these end-organs also play a major role in proprioception. A recent review of this subject by McCloskey (1981) states that the information passed to the spinal cord by the muscle spindle is influenced not only by the position of the limb, but also by the efferent discharge the spindle receives from the nervous system via gamma efferents; in order to work out the position, the brain must take into account these efferent discharges, and make comparisons between the signals it sends out to the muscle spindle along the gamma efferents and the afferent signals it receives from the primary afferents. Essentially, the signal from the muscle spindle (reafference in Fig. 4.14) is thought to be compared with a copy of the motor command (the "corollary discharge" or "efference copy") sent to the muscle spindle intrafusal fibres by the central nervous system via gamma efferents, and the difference is used in working out the position of the joint or the eye (Fig. 4.14). Goodwin et al. (1972), suggest that "corollary discharges" (copies of the command signals) "do exist and serve to modify the kinaesthetic actions of muscle afferent signals in the light of what the motor system is attempting". A comparable system may apply in the bladder and possibly other muscular viscera, where sensory endings are found to respond to stretch of the visceral wall by distensions, and to contractions induced by efferent discharges down autonomic nerves. In many viscera, motility is not controlled directly by efferent nervous activity because of a number of other factors, such as the intrinsic plexus and hormones; but in the case of the bladder, the brain could know which afferent discharges are due to distension and which are due to contractions, because the latter are initiated by the central nervous system. That might help to explain the vagueness of much visceral sensation, and why sensation is possibly a more consistent phenomenon in the lower urinary tract. In the case of the bladder, there is evidence for a recurrent inhibitory pathway from parasympathetic preganglionic neurons on to interneurons that process sensory information (de Groat and Ryall 1969); this is one potential source of "corollary discharges", but it need not be the only or the actual mechanism which allows comparisons of efferent and afferent activities to be compared.

The role of corollary discharges in visceral sensation is unknown, but the possibility cannot be denied. It seems not unlikely that some such mechanism is involved because ascending pathways that signal bladder contraction are either inhibited or excited by this event (Milne et al. 1981; McMahon and Morrison 1982a); the presence of pathways whose signals vary inversely with the contractile response is an essential component of the comparator that has been proposed by McCloskey (1981) and others (Fig. 4.14) Situations where such modulation or cancellation of sensory inputs by copies of efferent commands may occur are during sudden voluntary interruption of urinary flow, which produces sensations of fullness, and micturition against a high resistance, which is a common situation in men with prostatism, but not accompanied by noxious sensation. The sensations present in the former, relatively low pressure state, occur at a time when motor drive to the bladder is reduced; but in the latter, when much higher pressures are present, noxious sensations are absent, possibly as a result of motor drive being switched on. The ideas concerning corollary discharges are speculative, and it is possible to think up alternative explanations, but one cannot dismiss the corollary discharge hypothesis completely. Paintal (1986) has recently argued along similar lines for cardiac pain.

Ideas Based on the Regional Localization of Sensory Endings

One other concept currently in the literature on visceral sensation is the possibility that the sensory endings responsible for pain may be localized to particular regions of viscera. Thus, Malliani et al. (1986), who totally reject the concept of specific nociceptors in the viscera, believe that cardiac pain can be elicited more easily by spatially restricted mechanical abnormalities in, for instance, the coronary arteries, than from more widely distributed myocardial ischaemia; they suggest that the latter

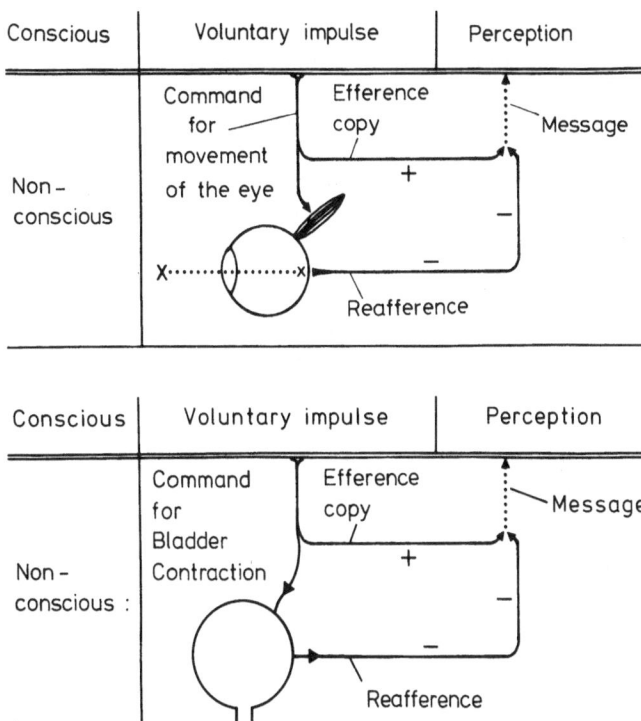

Fig. 4.14. Diagram to illustrate the role of corollary discharges in kinaesthesia, and possibly in bladder sensation. (Modified from von Holst 1973; McCloskey 1981)

is more likely to activate some central inhibitory modulation that would be more effective in preventing pain mechanisms. Cervero (1985), currently the main proponent of specificity theory, has suggested that specific nociceptors may occur in some regions and not others. It is known that the distending pressures necessary to elicit pain vary considerably in the gastrointestinal and urinary tracts; this could be explained by variations in the density of innervation of viscera, and there is no need to invoke the existence of specific nociceptors to explain this. The sensitivity of the trigone and urethra to light mechanical stimuli (inducing pain) is well described, whereas the detrusor is insensitive to this stimulus; thus if the sensations induced by distension of the lower urinary tract above the sphincter were to be attributed to a localized group of endings, it might be expected that these could be found in the bladder base and/or proximal urethra. Contractions of the bladder associated with voiding do not give rise to much sensation, but do if voiding is interrupted; similar symptoms occur in bladder instability, when the bladder also contracts against a closed sphincter. These are both situations in which sensory endings localized around the bladder base might be excited. This region is innervated by pelvic and hypogastric afferents. The absence of sensation in men with chronic prostatism who generate large bladder pressures may, in this case, be attributable to a change in the compliance of this tissue; if the tissue is less elastic, the high pressures may not give rise to the degree of distortion necessary to excite these endings. Thus there may be some evidence that mechanical stimulation of the bladder base elicits pain, whereas similar stimulation of the lower ureter does not (Risholm 1954), and the sensation could be explained either by excitation of nociceptors at this site, or by summation of impulses from low threshold mechanoreceptors that are present in high density. At the moment, there is little evidence for the former.

The preceding paragraphs have identified a number of mechanisms by which sensations from the lower urinary tract may be generated or modulated. The view that pain is mediated by specific nociceptors in these viscera has yet to be backed by evidence, and even if one considers the possibility that the nociceptive endings are strictly localized, the properties of endings in the most likely sites are not sufficiently different to justify a separate classification. The role of corollary discharges is hypothetical, but the existence of such a mechanism might explain some of the observed sensory phenomena.

It is not clear to what extent visceral sensation in the lower urinary tract occurs via mechanisms common throughout the viscera. The physiology of the spinal primary afferent input from different abdominal viscera (Morrison 1977; Blumberg et al. 1983; Jänig and Morrison 1986; Kumazawa 1986); and the anatomy of this visceral afferent input to the spinal cord appears to be similar at all levels (de Groat 1986). Nevertheless, there are major differences between the mechanisms of visceral sensation in the urinary bladder, the gallbladder and the heart. The ways in which the information is processed by some neurons in the sacral cord and in the upper lumbar cord may differ (Milne et al. 1981; McMahon and Morrison 1982a,b), and there are also discrepancies between studies of the thoracic and lumbosacral regions (Cervero 1985). This topic is beyond the scope of this book, but the reader is referred to a volume devoted to visceral sensation (Cervero and Morrison 1986).

Summary

Sensation occurs as a result of activity in afferent nerve fibres, and the existence of different modalities of sensation, such as touch and pain, or a sensation of bladder fullness and bladder pain, may be explained in different ways. Two main theories have to be considered. The first is the specificity theory, which was first enunciated in the early years of this century, and most physiologists agree that this theory can explain the variety of sensations that can be elicited from skin. In skin, painful sensation is subserved by a special group of slowly conducting fibres whose properties are quite specific: they respond only to stimuli that elicit tissue damage and painful sensation, and do not respond to stimuli that elicit innocuous sensations. Another theoretical possibility is that sensations of different character can be mediated by one type of nerve fibre, which can elicit an innocuous sensation when the discharge frequency is low, but can elicit noxious sensations at high frequencies of firing. This theory is the intensity theory or summation hypothesis, which was first formulated in 1920. It is doubtful whether painful sensations arising from the lower urinary tract are carried by a separate specialized group of fibres; rather it is likely that innocuous sensations and reflexes are subserved by the same nerve fibres that mediate noxious sensations, i.e. this system appears to work by the mechanisms implicit in the intensity theory.

One difference between skin and bladder receptors is that the latter can be excited by effector organ responses that are elicited via the autonomic nervous system. Thus, parasympathetic activity is a powerful stimulus to bladder afferent endings, which take part in reflex control as well as sensation. An analagous situation occurs in the control of the muscle spindle afferent firing by the gamma efferent pathway; these receptors, like those in the bladder, are involved in proprioceptive sensation and reflex control; comparisons have been made between the two sets of receptors because both are excited by stretch of the muscle in which they exist, and both can be excited by efferent nervous activity, viz. the gamma efferent supply to the muscle spindle and the parasympathetic efferent supply to the bladder.

The theoretical basis of kinaesthesia involves mechanisms which take account of the efferent activities of the gamma system, so that afferent firing due to motor activity initiated by the central nervous system can be recognized as such, and not attributed to stretch of the muscle. This theory suggests the existence of "corollary discharges", which occur at the same time as the efferent impulses, and are used to modulate the sensory experience induced by the afferent impulses from muscle spindles. In a sensory system such as that innervating the bladder, an analagous theoretical situation may apply; it might explain why the sensation of fullness is quickly relieved at the start of micturition, even though the bladder pressure is higher than in the passively distended condition. Furthermore, if noxious sensations are mediated by the same group of sensory endings that are responsible for reflex activities and innocuous sensations, as has been suggested, the possible existence of "corollary discharges" could explain why high intravesical pressure during micturition is painless, whereas similar pressures elicited by distension are painful.

This section has been speculative, in order to provoke thought; in the study of sensation, most hard information is available in the field of cutaneous sensation, but it might be unwise to follow that analogy without serious challenge, when a number of scientists have serious doubts as to whether the theoretical concepts involved fit the visceral situation. Reference has been made to the theoretical basis of kinaesthesia, where current knowledge is less complete, but where the functions of receptors fit better in a number of respects with the functions of receptors in the lower urinary tract.

Autonomic Nervous System and Visceral Pain

The visceral mechanoreceptors are situated in muscular tissues and are able to respond to contractions

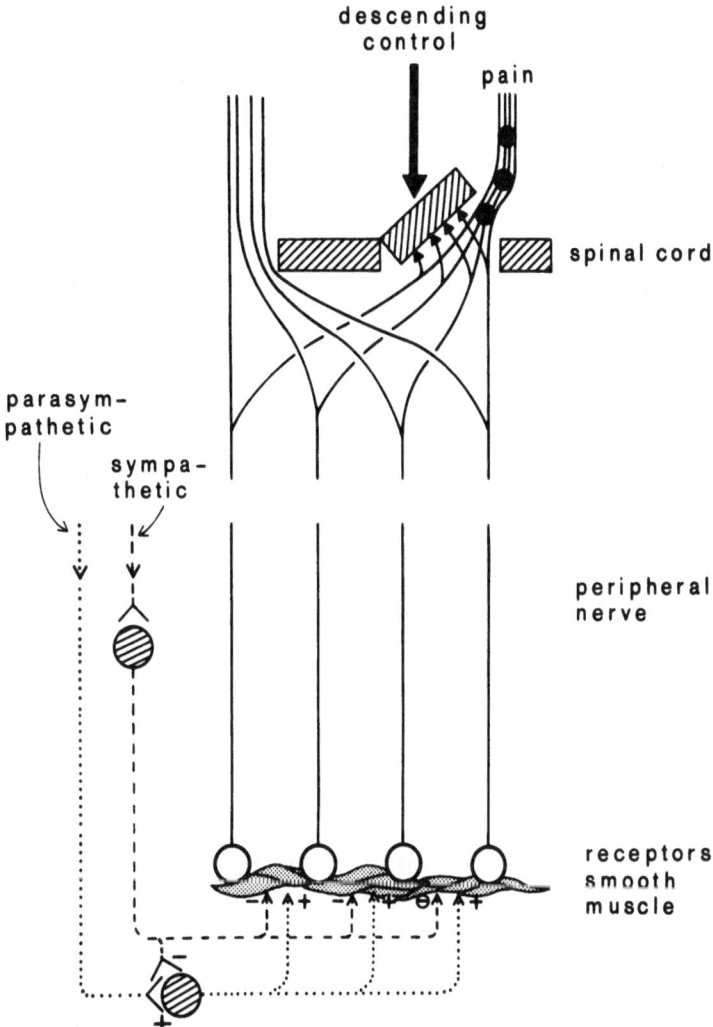

Fig. 4.15. Diagram to illustrate the possible site in the spinal cord in the periphery at which the central nervous system may modulate afferent information from the viscera. (Jänig and Morrison 1986)

of these viscera; in the case of the lower urinary tract, the motility is determined by the central nervous system. Sensation is influenced by the activity of the muscle, and the thresholds of sensation can be modified by drugs that act on adrenergic and other receptors in the smooth muscle (see earlier in this chapter). It has been suggested (Jänig and Morrison 1986) that the autonomic nervous system can mediate some antinociceptive functions in the periphery; this would be in addition to descending antinociceptive pathways that act in the spinal cord. Fig. 4.15 summarizes the pathways that might be responsible for central nervous influences on visceral sensation; in addition, this diagram suggests the ways in which visceral pain pathways might be influenced by visceral afferents that encode nociceptive events in the viscera by mechanisms implicit in the intensity theory.

Anatomical Basis of Referred Pain

One of the features of referred pain is that the pain is felt in somatic areas which receive innervation from the same spinal segments as the inflamed viscus that initiates the sensation (Head 1893; Lewis 1942). In recent years there has been particular interest in the possible role of visceral primary afferent collaterals in axon reflexes and in reflexes mediated by prevertebral ganglia (see Chap. 7, p. 217). The suggestion has also been made (Bahr et al. 1981; Pierau et al. 1982; Taylor and Pierau 1982) that visceral afferent neurons have collaterals in somatic nerves and that such viscerosomatic distribution of primary afferent endings might explain referred pain. This idea was originally proposed by Sinclair et al. (1948) (Fig. 4.16a), and the evidence

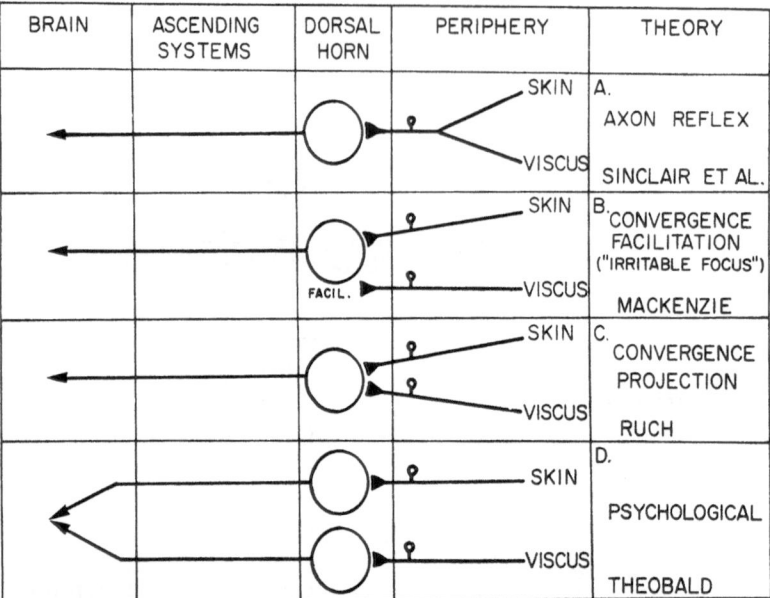

Fig. 4.16. Summary diagram illustrating various theoretical mechanisms of referred pain. *C* and *D* are both psychological and taken together bring out the point that convergence can occur at any synaptic level of the somatosensory system. (Selzer and Spencer 1969; Ruch 1979)

concerning the existence of such primary afferents is debatable. Jänig and Morrison (1986) concluded that any suggestions of the involvement of branching of primary afferent neurons in referred pain must be tentative until good evidence of function in this regard is available. In the lower urinary tract, Downie et al. (1984) found no evidence for sensory neurons that innervated both the bladder and the urethra. Referred pain can be explained better by the convergence–projection theory of Ruch (1946) (Fig. 4.16c, and later in this chapter). The facilitation suggested in Mackenzie's theory of convergence–facilitation (Fig. 4.16b; Mackenzie 1893, 1912) may be essential for hyperalgesia associated with visceral stimulation, but the presence of facilitation alone, without the ability of visceral impulses actually to discharge the postsynaptic neurons, is contrary to many of the experimental results. Viscerosomatic convergence is present in the spinal cord, and may well also exist at higher levels of the nervous system (Fig. 4.16d) as was suggested by Theobald (1941). Figure 4.16 contrasts the possible pathways and mechanisms that could give rise to referred pain.

Spinal Pathways that Transmit Sensory Information from the Bladder

Ascending pathways that carry information concerned with the state of the bladder can be found in the dorsal columns, the dorsolateral funiculus and the ventral columns. To some extent it is possible to separate anatomically the spinal pathways concerned with innocuous and noxious sensations, and reflex functions; however, the variety of function that can be recorded from each of these pathways makes it clear that anatomical separation of different functions is by no means precise.

Central Pathways: Ideas Based on Studies of the Effects of Lesions in Humans

Lateral Columns

The ascending pathway that subserves the desire to micturate is present in the superficial white matter of the lateral funiculi in humans (Nathan and Smith 1951), and is believed to be the spinothalamic pathway (Fig. 4.17). This tract is present in humans

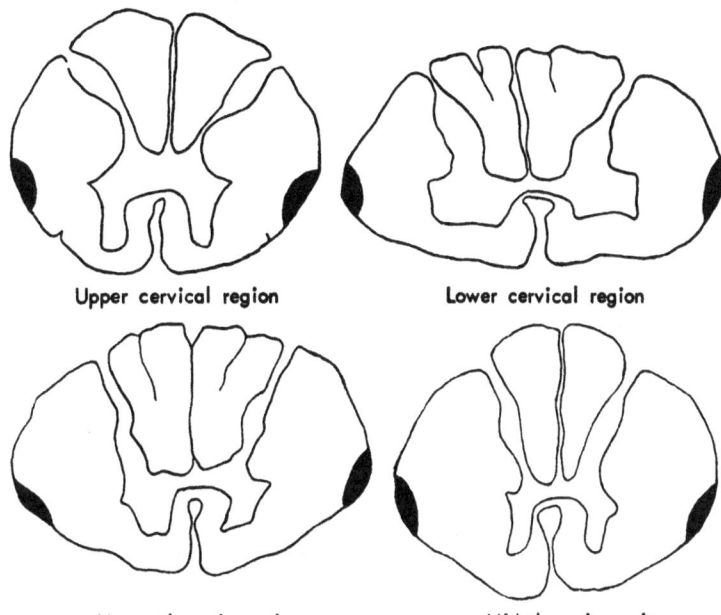

Upper cervical region   Lower cervical region

Fig. 4.17. The arrangement of ascending pathways for sensation from the lower urinary tract in humans (Nathan and Smith 1951; quoted in Bors and Comarr 1971)

Upper thoracic region   Mid-thoracic region

along the most lateral parts of the lateral columns, and either at the equator or slightly ventral to it. A bilateral lesion of the lateral part of the spinothalamic tracts causes a loss of somatic pain and thermal sensation, a loss of the desire to micturate or to defecate (Nathan 1956), a loss of pain sensation arising from the bladder, urethra and lower end of the ureter (Nathan and Smith 1958) and a loss of sexual sensations (Hyndman and Wolkin 1943; White and Sweet 1955). Anatomical and physiological studies on the ascending pathways transmitting information about the state of the bladder in animals also indicate that they travel up the dorsolateral funiculi from the lumbosacral cord to the thalamus, brain stem and other rostral locations.

Dorsal Columns

Sensations of touch and pressure arising from the urethra are believed to travel rostrally in the dorsal columns in humans, except for thermal sensation, which follows the spinothalamic pathway (Nathan 1952). White (1943) reports that the sensation of bladder distension survives cordotomy, and is presumed to be mediated by the dorsal columns. There are probably more than two ascending paths carrying visceral information, and exclusion of the

spinothalamic pathway should not necessarily focus attention on the dorsal columns; the dorsolateral funiculus of animals is known to convey sensory information concerned with bladder filling, and this possibility has not been excluded in humans. Nevertheless, there is good evidence for a dorsal column pathway from the bladder, urethra and abdominal or pelvic floor muscles. It is also known that a sensation of lower abdominal distension may be transmitted by the abdominal wall afferents, in which case the dorsal columns are likely to be involved.

Neurophysiological Studies on Animals

Dorsal Columns

The dorsal and lateral column pathways that transmit sensory information from the viscera have also been described in animals; the dorsal column pathway was called the "pelvic sensory vagus" by Kuru (1965), who traced a primary afferent pathway to the gracile nuclei in cats (Fig. 4.18). The existence of this pathway has recently been confirmed by de Groat and co-workers (see de Groat 1986). The dorsal columns are known also to contain a postsynaptic pathway (Brown 1981), i.e. axons whose cells of origin are located in the

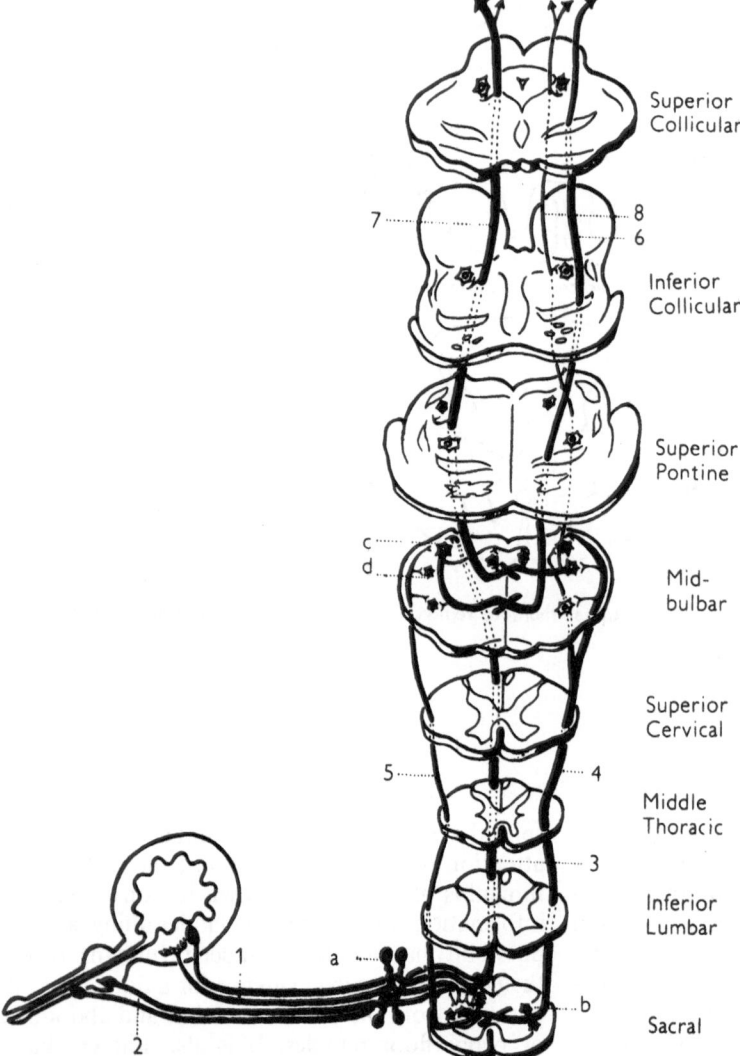

Fig. 4.18. Diagram to show the sensory nuclei and afferents related to micturition in the brain stem and spinal cord of the cat. *a*, Posterior root ganglion; *b*, dorsal group of intermediolateral column; *c*, para-alar nucleus; *d*, juxtasolitary nucleus; *1*, pelvic nerves; *2*, pudendal nerves; *3*, pelvic sensory vagus; *4*, sacrobulbar tract (contralateral ascending fibres); *5*, sacrobulbar tract (ipsilateral ascending fibres); *6*, para-alar thalamic tract; *7*, juxtasolitary thalamic tract (contralateral ascending fibres); *8*, juxtasolitary thalamic tract (ipsilateral ascending fibres). (Kuru 1965; quoted in Appenzeller 1982)

grey matter of the spinal cord; there is no evidence at present for a visceral component to this system.

Lateral Columns

The anatomical and immunocytochemical studies quoted previously show that visceral afferent fibres terminate in parts of the grey matter of the spinal cord remote from the normal termination sites of somatic primary afferents. Small-diameter somatic fibres tend to end in laminae I, V and X of the sacral cord (Light and Perl 1979), and unmyelinated

somatic afferents terminate in laminae I and II. Visceral afferents, however, terminate rarely in the substantia gelatinosa (lamina II); this separation represents a major difference between the ways in which cutaneous and visceral sensory information is handled.

In the cat, spinothalamic tract neurons are concentrated in laminae I, VII and VIII. In the monkey, which has a much larger spinothalamic tract, the neurons originate from all laminae, and are most concentrated in laminae I and V; neurons projecting to the lateral thalamus are mainly found in laminae I and V, whereas those projecting to the medial

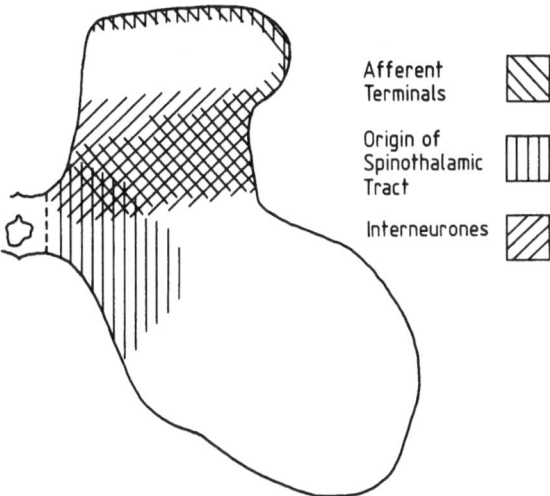

Fig. 4.19. The arrangement of primary afferent terminals with respect to the origin of ascending pathways in the cat. (Derived from de Groat 1986; Willis 1986; McMahon and Morrison 1982a)

Afferent Terminals

Origin of Spinothalamic Tract

Interneurones

thalamus are more common in laminae VII and VIII. The cat appears to lack the former projection, whereas both are present in the rat (Willis 1986).

Kuru (1965) reported that the lateral column pathway was in the superficial dorsolateral funiculus near the lumbosacral junction, and became more equatorial in its situation in cervical regions (Fig. 4.18); this has recently been confirmed electrophysiologically by McMahon and Morrison (1982a). It is well known that the cat spinothalamic tract follows a polysynaptic pathway through the brain stem, in contrast to the neospinothalamic tract of primates. Kuru (1965) concluded that the lateral pathway in cats and dogs was partially crossed in the cord and synapsed in the juxtasolitary nucleus of the medulla before proceeding to the thalamus. These pathways have their origin in the deeper laminae of the grey matter of the spinal cord, and visceral afferent fibres can be traced to these locations (Fig. 4.19; see also Fig. 3.3, p. 59, Fig. 3.4, p. 60). They respond to electrical stimulation of pelvic nerve afferents from the bladder, and the majority respond to bladder distension (Fig. 4.19b; see also Fig. 4.23).

Summary

The spinal pathways that transmit sensory messages from the visceral afferent terminations in the spinal cord to more rostral structures can be found in the dorsal, lateral and ventral columns. The dorsal column projections consist of primary afferent collaterals (i.e. there is no synapse in this pathway) which run ipsilaterally to synapse in the gracile (dorsal column) nuclei. This pathway transmits information about the sensation of touch and pressure in the urethra, and innocuous sensations from the pelvic floor muscles.

The lateral columns transmit information concerning temperature sensation in the urethra, and the sensations of bladder fullness and desire to micturate, and pain sensations from skin, bladder, urethra and the lower ureter, as well as sexual sensations. In humans, the pathway is situated superficially, just ventral to the equator of the cord, and is probably the spinothalamic tract. In animals, the ascending pathways can be found more widely, and particularly in the dorsolateral funiculi and the ventral column; there are marked species differences in the anatomy of the spinothalamic tract, and in some species the main pathway from the cord to the thalamus synapses, possibly more than once, in the brain stem.

In animals the visceral afferents end in laminae I and in deeper laminae of the dorsal horn, the intermediolateral column, the dorsal commissure and the area around the central canal. The spinothalamic tract originates from laminae I, VII and VIII in the cat and laminae I and V, with some participation from laminae VII and VIII, in the monkey. Most of the ascending neurons that carry information concerned with the state of the bladder have been found in the deeper layer of the dorsal horn and the area between the intermediolateral horn and the central canal. This corresponds to the laminae that receive visceral primary afferent terminals.

Properties of Spinothalamic Tract Neurons with Bladder Inputs at Different Segmental Levels of the Cord

Spinothalamic tract neurons carry information from somatic structures and viscera via neurons which can be excited by stimulation of viscera or of skin or muscle (Fig. 4.20); such neurons are said to show viscerosomatic convergence, and this arrangement may be responsible for the phenomenon of referred pain. It may seem surprising, but there is some doubt as to the existence of ascending neurons that respond only to visceral inputs; most if not all neurons which carry information from the viscera to the brain also carry somatic information, either from the skin or from muscles and joints. As there appear not to be pure visceral neurons, the term "viscerosomatic neurons" will be used in subsequent descriptions; these differ from somatic neurons in that the latter only transmit information about events in skin or deep somatic tissues. The following paragraphs show that viscerosomatic

neurons of different types are known to exist, and that there may be a predominance of one type over another in different regions of the lumbosacral cord.

Milne et al. (1981) studied the effects of bladder distension on spinothalamic tract neurons in monkeys (Figs. 4.20, 4.21). They reported that spinothalamic tract neurons at the thoracolumbar junction could be excited by bladder distension, by

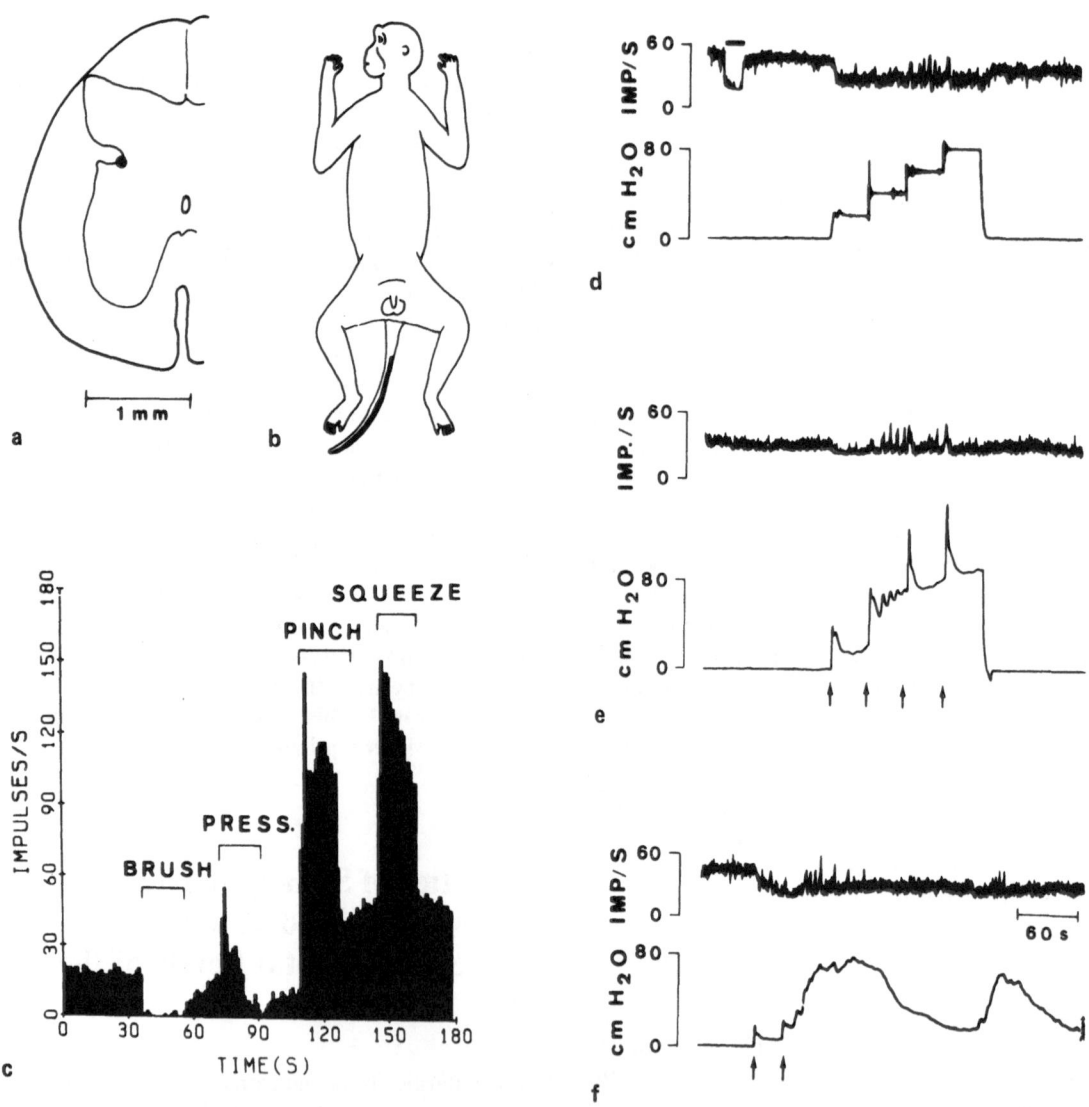

Fig. 4.20a–f. Somatovisceral convergence in spinothalamic tract neurons within the sacral cord of the monkey. Responses of a high threshold spinothalamic neuron to cutaneous and visceral stimuli. **a** Location of the cell in lamina V of the S-2 segment. **b** Location of the cutaneous receptive field on the tail. **c** Single pass peristimulus histogram of responses to innocuous and noxious stimuli applied to the cutaneous receptive field on the distal tail. **d** *Upper trace* is rate meter record of tonic inhibitory response to graded changes in intravesical pressure. *Lower trace* is bladder pressure (in cmH₂O). Note the transient discharges at high levels of pressure. The early downward deflection (*horizontal bar*) is an inhibitory response to brushing the hair within the cell's receptive field. **e** Phasic excitatory and tonic inhibitory responses to rapid injections of fluid into the urinary bladder. 10 ml fluid was injected at each *arrow*. Total accumulated volume was 40 ml. **f** Transient excitatory bursts of activity in phase with spontaneous isometric contractions of the urinary bladder. Note the accompanying tonic inhibition of background activity. 5 ml fluid was injected into the bladder at each *arrow* (Milne et al. 1981)

noxious heating of the skin and by squeezing the skin with forceps, but not usually by light mechanical stimuli; some of these cells were also excited by noxious compression of the testis. Some units were affected more by the visceral stimulus than by the cutaneous one, but in other cells the reverse was true. Some cells responded to distension with a phasic burst of activity, whereas in other cells the responses to bladder distension were maintained. In the sacral cord, spinothalamic tract neurons had mixed excitatory and inhibitory inputs in response to bladder distension, and had somatovisceral convergence from receptors in the deeper tissues of the tail and gluteal muscles. Neurons in the lower lumbar cord were inhibited by bladder distension. The differences between neurons at the tho-

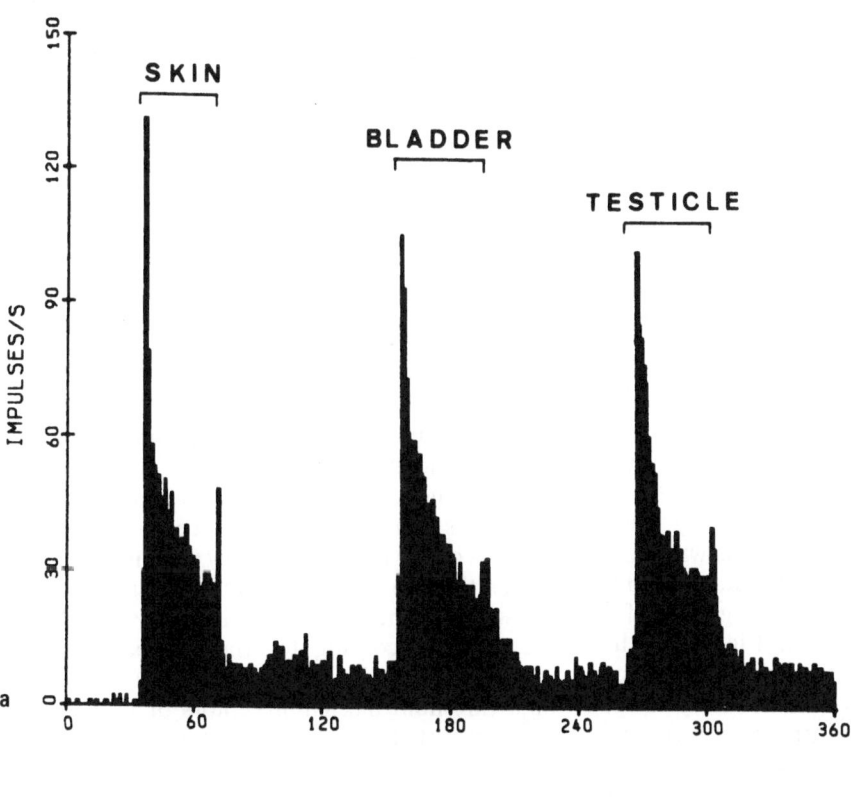

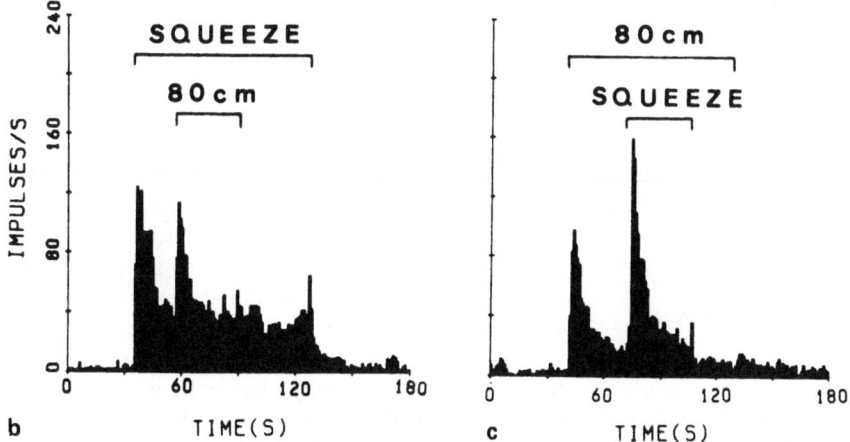

Fig. 4.21a–c. Effects of bladder distension and of cutaneous and testicular stimuli on a spinothalamic tract neuron in lamina V of the twelfth thoracic segment of the monkey spinal cord. **a** Responses to squeezing the skin of the flank, distension of the urinary bladder (60 ml volume injected rapidly) and manual compression of the exposed ipsilateral testicle. **b** Summation of responses to distension of the bladder (80 cmH$_2$O pressure) and squeezing the skin of the flank. **c** Reversed sequence of **b** (Milne et al. 1981)

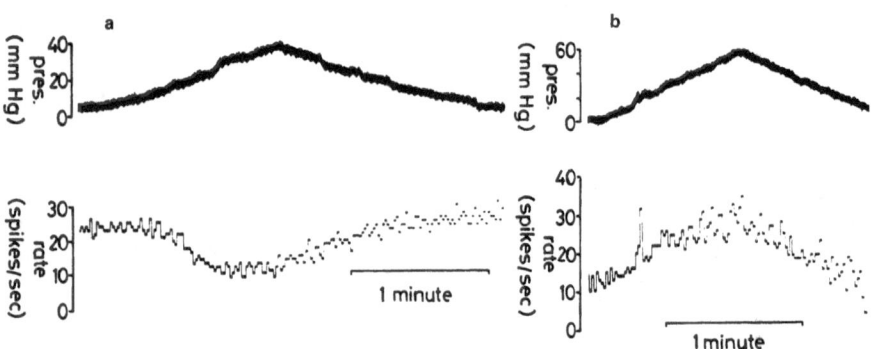

Fig. 4.22a,b. The effects of bladder distension on two different spinal neurons in the cat, showing (**a**) inhibition and (**b**) excitation. (Courtesy of S. B. McMahon)

racolumbar junction, the lower lumbar cord, and the sacral cord is interesting: all of these types of neurons, with cutaneous or deep somatic receptive fields, project to the thalamus and must presumably therefore be involved in sensation. McMahon and Morrison (1982a,b) also found neurons in the spinal cord that responded to bladder distension by being either excited or inhibited (Fig. 4.22); the inhibitory visceral receptive fields were found to be associated with a pelvic nerve input; neurons with a hypogastric, but without a pelvic nerve afferent pathway from the viscera were always excited by raising bladder pressure. In contrast, neurons with pelvic and/or hypogastric inputs could be either inhibited or excited by visceral distension. In segments of the cord that do not receive a visceral input via their dorsal roots (and only a relatively small visceral input via the tract of Lissauer), Cervero and Iggo (1978) and Cadden and Morrison (1984) found mainly inhibitory effects of raising bladder pressure in the cat and rat, respectively. It may be, therefore, that the segmental location of the cell body of a neuron is one factor in determining how it responds to visceral stimuli such as bladder distension. This view would also be supported by some of the results of McMahon and Morrison (1982a,b); the effects of visceral distension on neurons with only a lumbar afferent input were always excitatory, whereas the effects on neurons with only a pelvic nerve input were usually inhibitory.

Location of Spinal Neurons that Process Visceral Information

One major difference between the locations of spinal cells that process visceral information and those that do not is that the substantia gelatinosa (Rexed's lamina II) does not receive visceral inputs. Lamina I and the deeper layers of the dorsal horn

are the receiving sites for visceral information (McMahon and Morrison 1982a; Cervero 1985; de Groat 1986).

Classification of Viscerosomatic Neurons in the Spinal Cord

Neurons exhibiting viscerosomatic convergence can be subdivided in various ways, and physiologists who have worked predominantly on cutaneous sensation have tended to classify the neurons on the basis of how they respond to somatic rather than to visceral stimuli; for instance, one classification mentioned by Willis (1986) subdivides these cells according to whether the cutaneous inputs are from low threshold mechanoreceptors, high threshold mechanoreceptors or both, a generally accepted basis for classification of ascending neurons with cutaneous inputs. Spinothalamic neurons with these functions may be responsible for the patterns of referred pain seen in humans, and their function is consistent with the convergence–projection theory of Ruch (1946). However, this does not exclude the possibility that these can also be involved in reflex phenomena, as there is ample evidence that cells with cutaneous inputs may influence the motility of the bladder and other viscera (de Groat 1971; Sato et al. 1980).

An alternative method of classifying visceral neurons was used by de Groat et al. (1981) and McMahon and Morrison (1982a,b), who studied the types of viscerovisceral convergence seen in visceral afferent pathways in the spinal cord; these neurons could usually be activated also by cutaneous or muscular stimuli, but the somatic inputs were considered of secondary importance in the main study. These workers attempted a subdivision of cells with visceral inputs primarily according to the types of visceral information handled by these cells; this approach offered certain advantages in relating the activities of neurons to certain reflex events. Thus, different combinations of excitatory and inhibitory effects from the somatic and visceral areas were the basis for a classification which allows correlations to be made between neuronal and reflex properties (Table 4.5). In McMahon and Morrison's study, estimates of the length of any ascending axonal projections were also made, so that neurons with projections within the lumbosacral cord (their sacral interneurons), and neurons with axonal projections to the region between the thoracolumbar junction and the first cervical segment case, could be distinguished from neurons transmitting information to the brain. The exact site of projection of the latter, however, was not identified, and it is probable that these axons made contact with a number of sites in the brain stem and thalamus. This information is probably carried by the spinoreticular and spinothalamic tracts; the spinocervical tract and the dorsal column postsynaptic system are probably not involved. The neurons were identified by electrical stimulation of the pelvic or hypogastric nerves, unlike the neurons studied by Milne et al. (1981), who identified their sample by antidromic stimulation in the thalamus. Nevertheless, both groups of cells can be either excited or inhibited by bladder distension (Fig. 4.23); all exhibit viscerosomatic convergence and the differences observed between phasic and tonic responses in spinothalamic tract neurons were also found by McMahon and Morrison (1982a).

Table 4.5. The classification of visceral neurons in the spinal cord. (McMahon and Morrison 1982a,b)

Bladder pressure	Colonic pressure	Somatic receptive fields	Ascending projection	Possible functions
A. Neurons with long ascending axons				
(a) Neurons with receptive fields in one viscus only				
+	0	Yes	Supraspinal	Sensation,
−	0	Yes	Supraspinal	supraspinal
0	+	Yes	Supraspinal	reflexes or
0	−	Yes	Supraspinal	inhibitory phenomena [1]
(b) Neurons with receptive fields in more than one viscus				
+	+	Yes	Supraspinal	Sensory
−	−	Yes	Supraspinal	?reflex [1]
B. Neurons with axons that ascend to thoracic levels				
(a) Neurons with receptive fields in one viscus only				
+	0	Yes	T-13 to L-1	Sympathetic
−	0	Yes	T-13 to L-1	and
0	+	Yes	T-13 to L-1	somatic
0	−	Yes	T-13 to L-1	reflexes [2]
(b) Neurons with receptive fields in more than one viscus				
+	+	Yes	T-13 to L-1	Reflexes [2]
−	−	Yes	T-13 to L-1	
C. Neurons whose axons do not ascend to the thoracolumbar junction				
(a) Neurons with receptive fields in one viscus only				
+	0	Yes	Lumbosacral	Probably
−	0	Yes	Lumbosacral	reflexes [2]
0	+	Yes	Lumbosacral	?sensory
0	−	Yes	Lumbosacral	
(b) Neurons with receptive fields in more than one viscus				
+	+	Yes	Lumbosacral	?Sphincteric
−	−	Yes	Lumbosacral	reflexes [3]
+	−	Yes	Lumbosacral	Colonovesical
−	+	Yes	Lumbosacral	interactions [4]

Chapter 7 deals with the reflexes mentioned in this table. References are: [1] Evans and McPherson (1959), de Groat and Lalley (1972), McMahon and Morrison (1982a,c), McMahon (1986); [2] McMahon and Morrison (1982b), Morrison and McMahon (1982); [3] McMahon et al. (1982), Morrison and McMahon (1982); [4] de Groat (1971), de Groat et al. (1981), McMahon and Morrison (1982c), Floyd et at. (1982), Bartel et al. (1986), Jänig and McLachlan (1987).

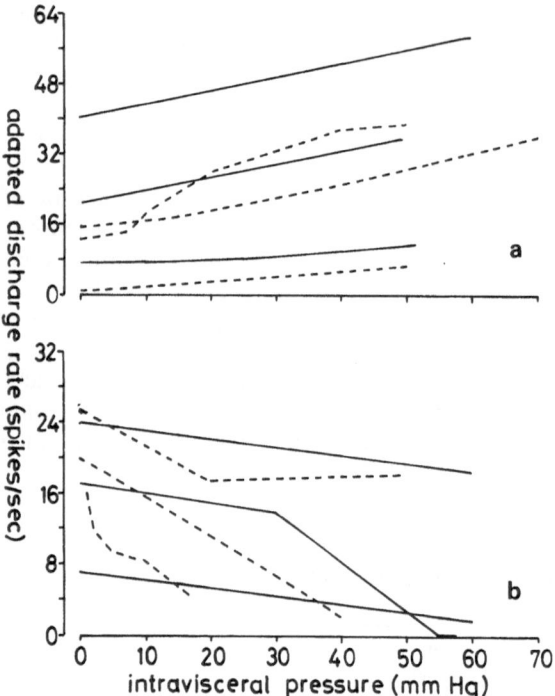

Fig. 4.23a,b. The stimulus–response curves of different spinal neurons for bladder and colon distension in the cat. The relationship between adapted discharge rate and intravesical (*dashed lines*) or intracolonic (*continuous lines*) pressure for 12 long ascending units which showed either a positive relationship (**a**) or an inverse relationship (**b**). For clarity error bars have not been shown, but in all cases there was a statistically significant difference between points in the lower and higher ranges of intravesical pressure. (McMahon and Morrison 1982a)

to activate these cutaneous inputs were either light or heavy touch or pressure (or both). Intra-arterial injections of bradykinin into the hind limb could also influence the activity of these cells (Fig. 4.24). There was therefore a difference between the results of Milne et al. (1981) and McMahon and Morrison (1982a,b) in that the viscerosomatic convergence observed in the sacral cord in the latter study was partly from skin and partly from deeper tissues. The difference may be related to the fact that the sacral neurons which projected to the brain in the latter study may have ended in a number of sites other than the thalamus. Modulation of neurons in the cerebellum, basal ganglia and brain stem have been observed during bladder activity (Bradley and Conway 1966; Porter 1967; Bradley and Teague 1968; Lumb and Morrison 1984).

The ascending pathways that carry information regarding the mechanical state of the bladder (McMahon and Morrison 1982a) transmit visceral information carried almost entirely by finely myelinated axons in the pelvic and hypogastric nerves. This conclusion comes from measurements of the latencies of the responses of central neurons following a stimulus to the peripheral nerve, and from measurements of the threshold current necessary to produce these responses. The mean latencies for the pelvic nerve responses were 21 and 17 ms (for excitatory and inhibitory effects) and the threshold currents were in the range that excited A-delta axons in the conditions that held in these experiments. Similarly the latencies following hypogastric stimulation averaged 15 ms; all of these latencies are consistent with conduction at A-delta velocities along the lengths of nerve trunk between the stimulating site and the recording site. Two-thirds of these neurons transmit visceral information from pelvic as well as hypogastric afferent

The types of somatic stimuli that were effective included cutaneous receptive fields (particularly in the skin of the perineum), and movements of muscles and joints (particularly at the base of the tail); the intensities of mechanical stimuli necessary

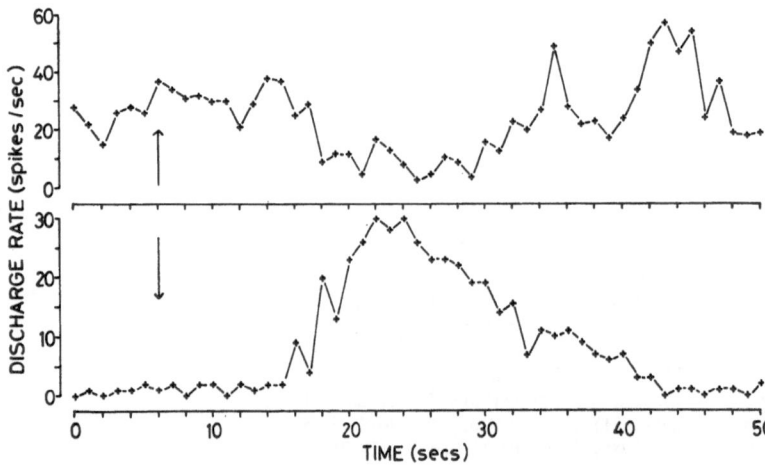

Fig. 4.24. The effects of intra-arterial bradykinin on two long ascending neurons that carry information from the bladder (10 μg in *top record*; 15 μg in *bottom record*). (McMahon and Morrison 1982a)

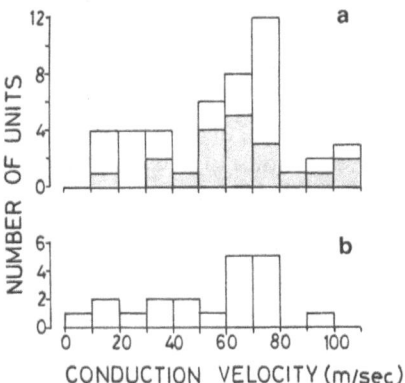

Fig. **4.25a,b**. Conduction velocities of **a** ascending and **b** descending spinal neurons that relay visceral information. *Stippled areas* refer to ascending units which possessed vesical mechanosensitive receptive fields.(McMahon and Morrison 1982a)

pathways, as shown by the patterns of convergence that could be seen when these nerves were stimulated electrically. About one-quarter of the axons respond to stimulation of pelvic nerve, but not of lumbar visceral afferents, and only 8% or so to respond to hypogastric stimulation alone. In each case in which the presence of a somatic input was investigated it could be found; thus these viscerosomatic neurons often had multiple inputs from viscera. The conduction velocities of the ascending axons showed a wide range, from about 10 m/s to over 100 m/s. Visceral information can therefore be transmitted rostrally quite rapidly (Fig. 4.25), as well as along the slower pathways that had been postulated by de Groat (1975). One explanation for the low conduction velocity estimated by de Groat (1975) is the presence of a considerable delay in the dorsal horn (McMahon and Morrison 1982a).

Summary

The ascending sensory pathways that transmit information from the bladder travel in the dorsal columns and in the lateral and ventral columns. Spinothalamic tract cells which relay messages from the bladder take their origin from the upper lumbar, lower lumbar and sacral segments of the spinal cord, and their properties appear to be somewhat different at these three levels: viscerosomatic convergence takes place at all sites, but the upper lumbar neurons, driven by hypogastric nerve inputs, are usually excited by bladder distension, whereas the sacral neurons, driven by pelvic nerve inputs, are often inhibited by this stimulus. Furthermore, the somatic input to viscerosomatic neurons in the sacral cord is often from muscle (and also from skin according to some studies) whereas the somatic input to this type of neuron in the upper lumbar cord is usually cutaneous. These properties may explain the occurrence and distribution of referred pain and deep muscle tenderness in disease of the lower urinary tract, and the explanation would be in keeping with the convergence–projection theory of Ruch (1946). However, the finding that viscerosomatic neurons may also have low thresholds during bladder distension suggest that some of these neurons could also contribute to sensations that are not generally regarded as noxious.

Mediation of Referred Visceral Pain and of Counterirritation Phenomena

The relationship between the properties of these different groups of visceral neurons in the spinal cord and the innocuous or noxious sensations or reflexes they may subserve is not entirely clear, but various attempts have been made to subdivide them, particularly in respect of reflex function. Thus, there exist cells in the sacral cord that respond to bladder distension, either by excitation or inhibition, and may also be either excited or inhibited by colonic distension. The influences of the colon and bladder may either be antagonistic or synergistic. These cells usually respond to touching the perineal skin, and often respond to raising and lowering of the tail. In addition some are affected by high threshold cutaneous, muscle or joint receptors. The responses of some of these cells correlate well with known reflex activities in intact or spinal animals (e.g. McMahon and Morrison 1982a,b,c; McMahon 1986; Bartel et al. 1986); surprisingly, the spinobulbospinal limb of the micturition reflex pathway described by de Groat and Ryall (1969) may be mediated by neurons that receive convergent information from several sources other than the bladder (McMahon 1986). The involvement of these neurons in reflexes will be discussed in the next chapter.

In other visceral sensory systems, such as that from the gallbladder, which do not give rise to sensations other than pain, Cervero (1985) has suggested that pain is mediated by neurons that have convergent inputs from the gallbladder and the skin, and that noxious levels of stimulation to the skin or the gallbladder are usually necessary to excite this type of viscerosomatic neurons. Cervero (1985) concludes that all the experimental findings support the convergence–projection theory of referred visceral pain (Ruch 1946). Not all the literature presents such a straightforward picture, however: Fields et al. (1970a,b) studied viscerosomatic neurons in the upper lumbar cord and found somatic inputs could be elicited by hair movement, light touch, joint movement, pressure on muscles and tendons, and noxious stimuli. A single unit sometimes possessed several widespread, discontinuous, polymodal receptive fields, within which excitatory and inhibitory components could be identified. The visceral receptive fields were sometimes present in a number of viscera. The results of Fields et al. (1970a,b) were obtained from unanaesthetized, decerebrated and spinalized cats. In the afferent pathway from the bladder, somatovisceral convergence is not only from high threshold receptors in the skin, but can be from high and low threshold receptors in muscle or skin, and these properties can be seen in animals with an intact neuraxis (McMahon and Morrison 1982a,b). These properties would fit in with the vagueness of visceral sensation generally, and the abnormal sensory experiences than can accompany micturition (see previous discussion); Nathan (1956) described referred sensations of an innocuous nature, associated with micturition. The deep rather than the superficial location of referred sensations may also fit with the existence of deep somatovisceral convergence. The association of these sensations with micturition rather than with overdistension is also interesting, and fits with the generally low pressure thresholds of afferents from the bladder. The range of bladder pressures necessary to excite the primary afferents and the central neurons on to which they project in the cat (Floyd et al. 1976; Floyd and Lawrenson 1979; Morrison 1981; McMahon and Morrison 1982a,b; Bahr et al. 1986; Jänig and Morrison 1986) is 4–30 mmHg, and 85% or so of these neurons respond to pressures of less than 15 mmHg (Fig. 4.26).

In the chapter on anatomical connections it was emphasized that visceral afferents enter a couple of sacral segments and some upper lumbar segments; in every species there are at least two segments whose dorsal root ganglia do not contain visceral afferents. These segments may receive some visceral input via the tract of Lissauer, but the density of the innervation is less than in the sacral cord or upper lumbar cord. Thus, the representation of the viscera in the sixth and seventh lumbar segments of the cat, and the fourth and fifth lumbar segments of the rat is relatively sparse. In this part of the cord, the effects of visceral stimuli on one category of neurons with cutaneous input (wide dynamic range, or convergent, neurons) is almost always inhibitory, whereas the effects on neurons with muscle or joint inputs may be either inhibitory or excitatory (Cadden and Morrison 1984). Inhibition of neurons with cutaneous inputs in the sixth lumbar segment of the cat during bladder distension has also been reported by Cervero and Iggo (1978).

Cadden and Morrison (1984, 1987) classified their cutaneous neurons according to a convention used in the study of somatic sensation (Table 4.6; see also p. 120). In the segments of cord between the autonomic outflows, the ascending neurons with low threshold cutaneous inputs were unaffected by visceral stimuli from the bladder or the colon; neurons with inputs from cutaneous nociceptors and low threshold mechanoreceptors (wide dynamic range neurons, which are believed to be an important part of the ascending pain pathway) were almost always inhibited by bladder distensions to pressure over 25 mmHg or by colonic distensions to rather higher pressures. However, the thresholds were not identified precisely, and in some neurons the thresholds were within the range found in the cat. The latter group of neurons are known to be inhibited by noxious cutaneous stimuli applied to any part of the body, and it seems probable that the inhibition originates from descending pathways with cell bodies in the brain stem. One of the

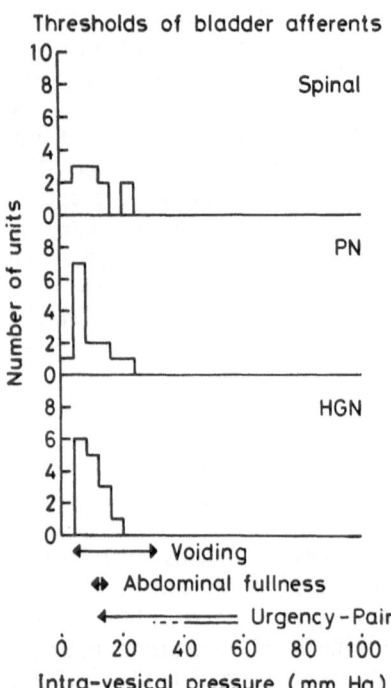

Fig. 4.26. The thresholds of primary afferents and spinal neurons that relay information from the bladder. The ranges of pressure that occur during voiding and that induce sensations are also shown. (Morrison 1981)

Table 4.6. Patterns of viscerosomatic convergence in the lower lumbar cord of the rat. (Cadden and Morrison 1984)

Somatic receptive fields in the foot	Bladder distension	Colonic distension	Noxious stimulation of skin distant to the receptive field
Cutaneous low threshold mechanoreceptors only	0	0	0
Low and high threshold mechanoreceptors in skin	− [a]	− [a]	− [a]
Receptors in muscles or joints	+ /0/ −	+ /0/ −	0

[a] Inhibition outlasts the stimulus and is probably mediated by a looped supraspinal pathway.

descending pathways which is known to receive information about the existence of noxious stimuli at any part of the body surface is the raphe–spinal tract that originates from the nucleus raphe magnus of the medulla. This pathway is also capable of regulating the transmission of somatic nociceptive information in the spinal dorsal horn; it is now well established that stimulation of the raphe–spinal pathway can induce analgesia, and that this can be elicited by noxious stimuli (Cadden et al 1984; Lumb 1984; Mokha et al. 1985). As such, the pathway may be a basis for pain relief by counterirritation (Fig. 4.27; Lumb 1986). A number of other structures have been implicated in stimulus-induced analgesia; these include the locus coeruleus, the parabrachial nuclei and the periventricular and periaqueductal grey matter (Katayama et al. 1984; Lumb 1984; Mokha et al. 1985).

Descending Control of Visceral Sensory Transmission

The preceding paragraph has already stated a possible involvement of the brain stem in modulating the transmission of nociceptive information from primary afferents into ascending pathways. High pressures in the bladder and the colon also inhibit the wide dynamic range neurons and it seems probable that at least part of this inhibition is due to activation of a looped pathway, beginning in the bladder or colon, and ascending the cord to synapse directly or indirectly on the raphe–spinal tract. The idea is supported by recordings from the raphe-spinal pathway by Lumb and Morrison (1984) and the importance of descending pathways in the regulation of visceral sensation has been reviewed by Lumb (1984, 1986); raphe–spinal neurons were excited by bladder pressures of 10–44 mmHg, and the threshold pressures were generally in excess of

25 mmHg (Fig. 4.28). These neurons were also excited by noxious mechanical and thermal stimuli to the skin of any part of the body. Thus, if the function of this pathway is to collect information about painful events, and to initiate a state of partial analgesia, then bladder pressures in excess of 25 mmHg induced by distension are generally required to activate this system. This range of pressures is generally considered to give rise to pain in humans.

These observations on the raphe–spinal pathway suggest that one of its functions is concerned with antinociception; another area of the brain stem that has an important role in descending control of noxious transmission through the dorsal horn is the grey matter surrounding the aqueduct in the midbrain (the periaqueductal grey, PAG). The PAG contains enkephalins and exerts its actions in the spinal cord by synapsing first in the nucleus raphe magnus, on the cells of origin of the raphe-spinal pathway. Thus the PAG and the raphe-spinal pathway are sites in the brain stem from which a state of anaesthesia can be induced in rats by electrical stimulation (Reynolds 1969); electrical stimulation of the PAG in rats appears also to abolish the somatic manifestations of visceral pain (Giesler and Liebeskind 1976). In humans, electrical stimulation of the PAG in patients with cancer pain arising from the viscera can produce pain relief (Hosobuchi et al. 1977; Richardson and Akil 1977a,b).

Spillane (1982) studied the effects of stimulation of the raphe nucleus on neurons in the sacral cord that had a pelvic nerve input. These studies in the cat showed that discharges in these neurons evoked by electrical stimulation of the vesical branches of the pelvic nerves could be inhibited (by approximately 90%) for about 100 ms starting 500–100 ms after a short train of stimuli had been given to the raphe nucleus. The total time course of the inhibition lasted some 400 ms and was followed by a small increase in the excitability of the neurons. This shows that raphe nucleus can exert an inhibi-

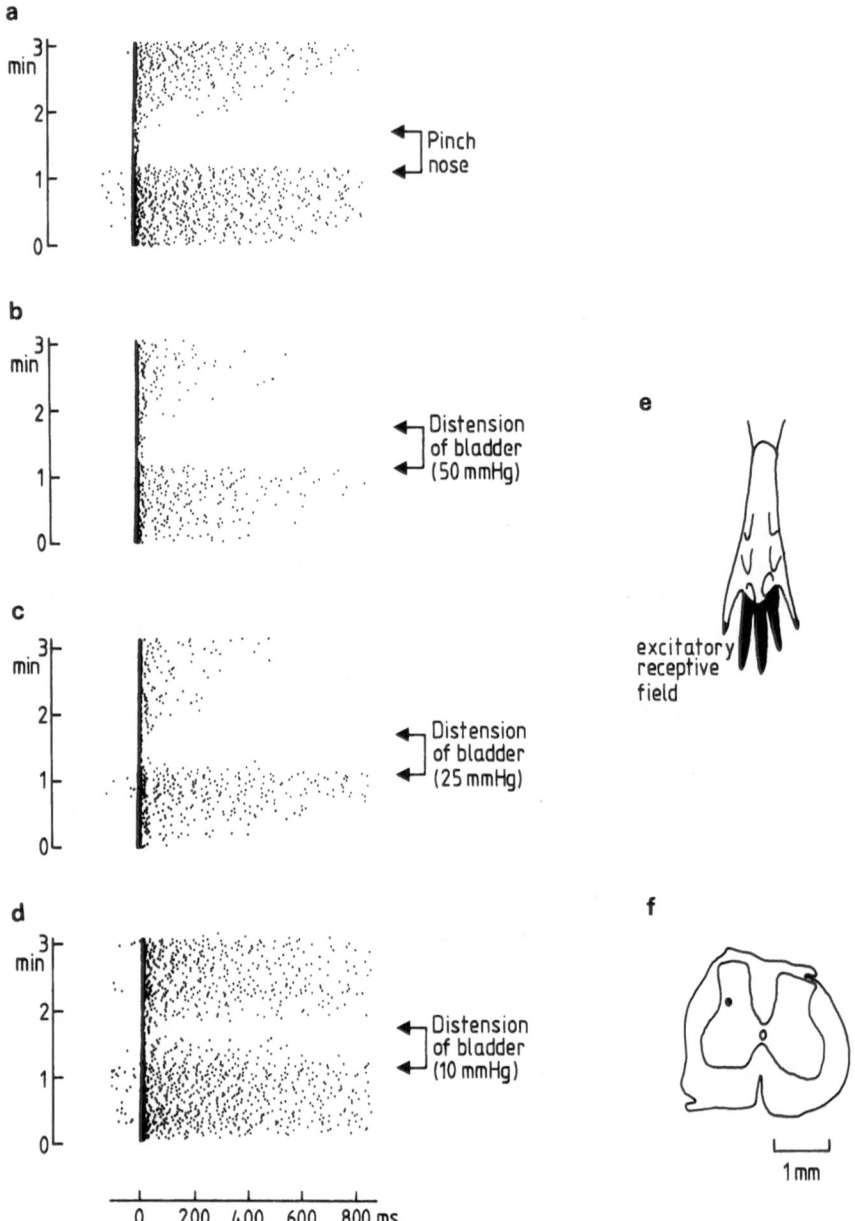

Fig. 4.27a–d. Responses of a convergent neuron in the dorsal horn of the rat spinal cord to stimulation of A- and C-fibres in the foot of the rat, and the effects of bladder distension on these responses. Results from experiment on a urethane-anaesthetized rat. **a–d** Dot display analyses showing successive responses (*bottom* to *top*) of a lumbar dorsal horn convergent neuron to repeated electrical stimulation of its receptive field (1.8 mA, 2ms; one stimulus every 1.5s). Note that although the presence of both short and long-latency responses is typical for this neuronal type, the asynchronous nature of the long-latency responses shown here is less commonly found. *Arrows* illustrate the periods of application of noxious somatic (**a**) and visceral (**b–d**) conditioning stimuli. **e** Excitatory receptive field on ipsilateral hind paw. **f** Recording site in dorsal horn. (S. W. Cadden and J. F. B. Morrison 1985, unpublished work; quoted in Lumb 1986)

tory descending control over the afferent input from the bladder.

The effects of raphe stimulation on the transmission of information from the bladder into sacral neurons are not the only way in which the raphe-spinal pathway can influence sensory messages from the lower urinary tract. One physiological change that could reduce the afferent input from the distended bladder in these circumstances would be an inhibition of micturition reflexes that are

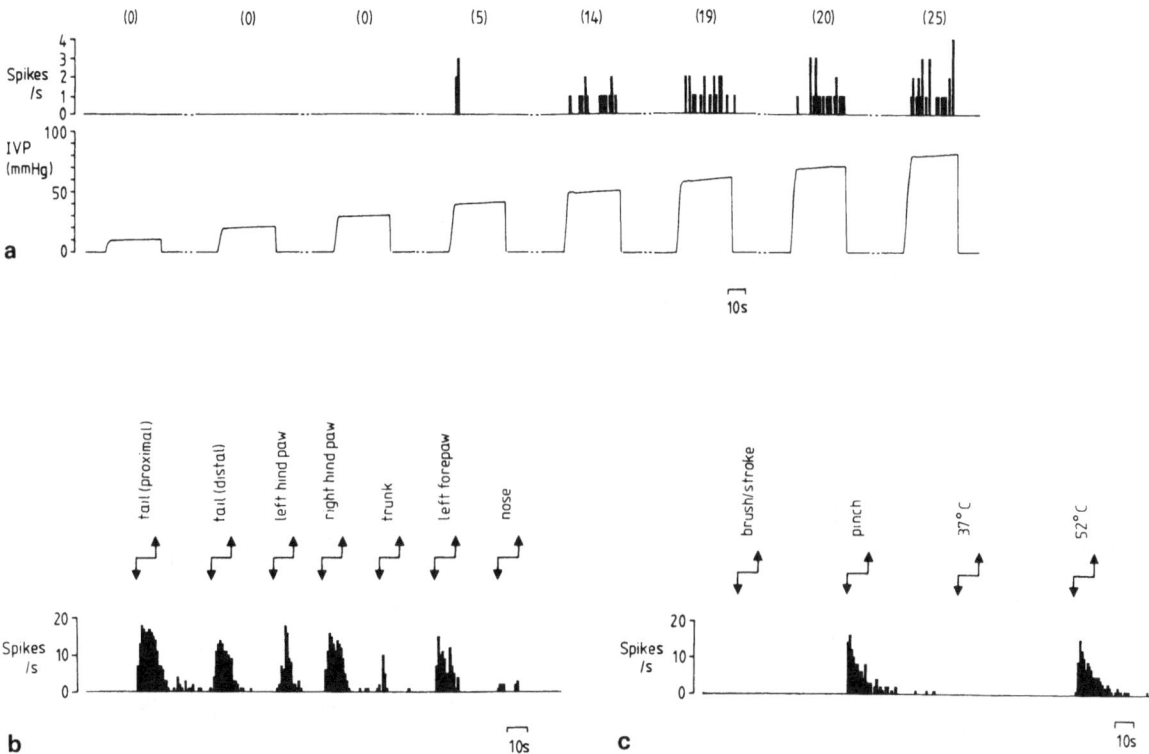

Fig. 4.28a–c. Results obtained from a recording made in the nucleus raphe magnus in a urethane-anaesthetized rat. Rate meter records showing the excitatory responses of an antidromically identified raphe–spinal neuron to visceral and noxious somatic stimuli. Stimuli applied: urinary bladder distensions of 10–80 mmHg (**a**), sustained pinches at the times indicated by the *arrows* (**b**), noxious and non-noxious mechanical and thermal stimuli applied to the tail (**c**). *Broken lines* indicate 30 s breaks in the record. Numbers in parentheses are total spike counts. *IVP*, intravesical pressure. (B. M. Lumb and J. F. B. Morrison 1986, unpublished work; quoted in Lumb 1986)

elicited by distension. In fact, the raphe–spinal pathway of the cat has a powerful inhibitory effect on the micturition reflex, and on certain somatic reflexes, in addition to its analgesic effect (Morrison and Spillane 1982). Similar effects can be induced by stimulation of the region around the nucleus reticularis gigantocellularis of the medulla (McMahon and Spillane 1982), an area which is also involved in the descending control of nociceptive transmission. The effects of the raphe–spinal pathway on the micturition reflex appear to be mediated, at least in part, by serotonin released in the lumbosacral cord, whereas the inhibitory effects of the reticulospinal pathway are not (Morrison and Spillane 1982, 1986). This is consistent with the known distribution of serotonin in descending pathways from the brain stem, and with the distribution of serotonin-containing terminals in the spinal cord. Light (1985) has shown that raphe–spinal descending axons terminate not only in the superficial dorsal horn, but also in the deeper laminae and the ventral horn. These sites overlap some of the terminals of visceral afferent fibres, and the autonomic nuclei.

Lumb and Morrison (1986) have found that an area of the periventricular grey matter in the anterior hypothalamus has an excitatory influence on the raphe–spinal pathway. This would fit in with behavioural and electrophysiological studies which suggest that this region of the cat's brain has a role to play in antinociception.

Influence of the Forebrain on Sensory Transmission

The effects of the cerebral cortex on sensory transmission are exerted through direct and indirect pathways that converge on the dorsal horn of the spinal cord. The corticospinal (pyramidal) tract is now recognized as having terminals within the dorsal horn as well as the ventral horn. In addition to this direct pathway, there are connections between the cortex and the bulbar reticular formation, which can also influence sensory processing at a spinal level (Brown 1981). The concept of a higher level regulating a lesser level has to be replaced by a more complicated system of controls over possibly all lower levels of the nervous system. We know little of the part played by the cortex in visceral sensation; almost all our knowledge is derived from studies of lesions produced by surgery, penetrating injuries or pathology. The main conclusion is that patients with lesions of the frontal lobe and anterior cingulate gyrus may suffer either from a lack of awareness of bladder sensation, or experience strong urinary tract symptoms at low bladder volumes (Andrew and Nathan 1964). The changes in sensory function, however, are usually not confined to the lower urinary tract, and the interpretation that the frontal lobe has a part to play in bladder sensation has to be seen in the light of the general state of these patients. The frontal lobes and anterior cingulate gyri are also involved in the regulation of bladder motility (see Chap. 8), and some of the sensory changes may be associated with contractions induced by autonomic pathways. Figure 8.20 (p. 266) shows the relationship between bladder filling and sensation in a patient before and after repair of an aneurysm of an anterior cerebral artery.

Summary

Descending control over sensory transmission can be demonstrated in electrophysiological and in behavioural experiments concerned with somatic sensation, but there is little available behavioural evidence on the role of descending pathways in the control of visceral sensation. A number of structures in the brain stem and forebrain are involved in inhibiting the onward passage of sensory messages (somatic and visceral) in the dorsal horn, and possibly other sites in sensory pathways. In respect of noxious information transfer, this descending control may explain phenomena such as the variation in pain thresholds in the same individual on different occasions, the use of counterirritation measures in the management of pain, the use of stimulation of structures within the brain, or of sensory nerves to induce analgesia, and the effects of acupuncture. Many of the descending pathways that are involved in antinociception originate in the brain stem, from structures including the nucleus raphe magnus, the nucleus reticularis gigantocellularis, the locus coeruleus, the parabrachial nuclei, the periaqueductal and periventricular grey matter.

References

Abrams P, Feneley R, Torrens MJ (1983) Urodynamics. Springer, Berlin Heidelberg New York, p 229

Andrew J, Nathan PW (1964) Lesions of the anterior frontal lobes and disturbances of micturition and defaecation. Brain 87:233–262

Appenzeller O (1982) The autonomic nervous system: an introduction to basic and clinical concepts, 3rd edn. Elsevier, New York, p 524

Arlhac A (1971) Données nouvelles sur la decharge des mecano-recepteurs vesicaux. Pflugers Arch 333:258–270

Bahns E, Jänig W (1985) Quantitative relations between distension and contraction of the urinary bladder, the blood pressure and the discharge rate of lumbar visceral afferents. Pflugers Arch 407:510–518

Bahns E, Ernsberger U, Jänig W, Nelke A (1986) Functional characteristics of lumbar visceral afferent fibres from the urinary bladder and the urethra. Pflugers Arch 407:510–518

Bahns E, Halsband U, Jänig W (1987) Functional characteristics of sacral visceral afferents from the lower urinary tract, colon and anus. Pflugers Arch (in press)

Bahr R, Blumberg H, Jänig W (1981) Do dichotomising afferent fibres exist which supply visceral organ as well as somatic structures? A contribution to the problem of referred pain. Neurosci Lett 24:25–28

Bahr R, Bartel B, Blumberg H, Jänig W (1986) Functional characterisation of preganglionic neurons projecting in the lumbar splanchnic nerves: neurons regulating motility. J Auton Nerv Syst 15:109–130

Bartel B, Blumberg H, Jänig W (1986) Discharge patterns of motility-regulating neurons projecting in the lumbar splanchnic nerves to visceral stimuli in spinal cats. J Auton Nerv Syst 15:153–163

Beck PW, Handwerker HO, Zimmermann M (1974) Nervous outflow from the cat's foot during noxious radiant heat stimulation. Brain Res 67:373–386

Behan RJ (1914) Pain. Appleton, London

Bessou P, Perl ER (1969) Response of cutaneous sensory units with unmyelinated fibres to noxious stimuli. J Neurophysiol 32:1025–1043

Blumberg H, Haupt P, Jänig W, Kohler W (1983) Encoding of visceral noxious stimuli in the discharge patterns of visceral afferent fibres from the colon. Pflugers Arch 398:33–40

Bors E, Comarr AE (1971) Neurological urology. Karger, New York

Bors E, Ma KT, Parker RB (1956) Observations on some modalities of bladder sensation. J Urol 76:566–575

Bradley WE, Conway CJ (1966) Bladder representation in the pontine-mesencephalic reticular formation. Exp Neurol 16:237–249

Bradley WE, Teague CT (1968) Spinal cord organisation of micturition reflex afferents. Exp Neurol 22:504–516

Brena SF, Sammons EE (1979) Phantom urinary bladder pain—case report. Pain 7:197–201

Brindley GS, Rushton DN, Craggs MD (1974) The pressure exerted by the external sphincter of the urethra when its motor nerve fibres are stimulated electrically. Br J Urol 46:453–462

Brown AG (1981) Organisation in the spinal cord: the anatomy and physiology of identified neurons. Springer, Berlin Heidelberg New York

Cadden SW (1985) A comparison of visceral and noxious somatic influences on neurons in the rat lumbar dorsal horn. In: Iggo A, Iverson LL, Cervero F (eds) Nociception and Pain. The Royal Society, London, p 414

Cadden SW, Morrison JFB (1984) The effects of visceral distension on the activities of lumbar dorsal horn neurons in the rat. J Physiol 350:71P

Cadden SW, Morrison JFB (1987) Effects of visceral distension on the activities of neurons receiving cutaneous inputs in the rat lumbar dorsal horn: comparison with effects of remote noxious somatic stimuli. (to be published)

Cadden SW, Villanueva L, le Bars D (1984) Modulation of dorsal horn neuronal activity by heterotopic noxious stimuli. In: Holden AV, Winlow W (eds) The neurobiology of pain. Manchester University Press, Manchester, pp 179–187

Cervero F (1985) Visceral nociception: peripheral and central aspects of visceral nociceptive systems. In Iggo A, Iverson LL, Cervero F (eds) Nociception and pain. The Royal Society, London, pp 197–120

Cervero F, Iggo A (1978) Natural stimulation of urinary bladder afferents does not affect transmission through lumbosacral spinocervical tract neurons in the cat. Brain Res 156:375–379

Cervero F, Morrison JFB (eds) (1986) Visceral sensation. Elsevier, Amsterdam (Progress in brain research, vol 67)

Cervero F, Morrison JFB (eds) (1986) Visceral sensation. Elsevier, Amsterdam (Progress in brain research, vol 67)

Clifton GL, Coggeshall RE, Vance WH, Willis WD (1976) Receptive fields of unmyelinated ventral root afferent fibres in the cat. J Physiol 256:573–600

de Groat WC (1971) Inhibition and excitation of sacral parasympathetic neurons by visceral and cutaneous stimuli in the cat. Brain Res 33:499–503

de Groat WC (1975) Nervous control of the urinary bladder of the cat. Brain Res 87:201–211

de Groat WC (1986) Spinal cord projections and neuropeptides in visceral afferent neurons. In: Cervero F, Morrison JFB (eds) Visceral sensation. Elsevier, Amsterdam, pp 165–187 (Progress in brain research, vol 67)

de Groat WC, Lalley PM (1972) Reflex firing in the lumbar sympathetic outflow to activation of vesical afferent fibres. J Physiol 226:289–309

de Groat WC, Ryall RW (1969) Reflexes to the sacral parasympathetic neurones concerned with micturition in the cat. J Physiol 200:87–108

de Groat WC, Nadelhaft I, Milne RJ, Booth AM, Morgan C, Thor K (1981) Organisaton of the sacral parasympathetic reflex pathways to the urinary bladder and large intestine. J Auton Nerv Syst 3:135–160

Denny-Brown D, Robertson EG (1933) On the physiology of micturition. Brain 56:149–190

Dockray GJ, Sharkey KA (1986) Neurochemistry of visceral afferent neurons. In: Cervero F, Morrison JFB (eds) Visceral sensation. Elsevier, Amsterdam, pp 133–148 (Progress in brain research, vol 67)

Downie JW, Champion JA, Nance DM (1984) A quantitative analysis of the afferent and extrinsic efferent innervation of specific regions of the bladder and urethra in the cat. Brain Res Bull 12:735–740

Evans JP (1936) Observations on the nerves of supply to the bladder and urethra of the cat, with a study of their action potentials. J Physiol 86:396–414

Evans MH, McPherson A (1959) The effects of distension of the bladder on somatic reflexes in the cat. J Physiol 146:438–458

Fields HL, Partridge LD Jr, Winter DL (1970a) Somatic and visceral receptive field properties of fibres in ventral quadrant white matter of the cat spinal cord. J Neurophysiol 33:827–837

Fields HL, Meyer GA, Partridge LD Jr (1970b) Convergence of visceral and somatic input on to spinal neurons. Exp Neurol 26:36–52

Floyd K, Lawrenson G (1979) Mechanosensitive afferent units in the cat pelvic nerve. J Physiol 290:51–52P

Floyd K, Hick VE, Morrison JFB (1976) Mechanosensitive afferent units in the hypogastric nerve of the cat. J Physiol 259:457–471

Floyd K, Hick VE, Koley J, Morrison JFB (1977) The effects of Bradykinin on afferent units in intra-abdominal sympathetic nerve trunks. Q Jl Exp Physiol 62:19–25

Floyd K, McMahon SB, Morrison JFB (1982) Inhibitory interactions between colonic and vesical afferents in the micturition reflex of the cat. J Physiol 322:45–52

Frimødt-Moller C (1972) A new method for quantitative evaluation of bladder sensibility. Scand J Urol Nephrol 6 (suppl 15): 135–142

Giesler GJ, Liebeskind JC (1976) Inhibition of visceral pain by electrical stimulation of the periaqueductal gray matter. Pain 2:43–48

Goodwin GM, McCloskey DI, Matthews PBC (1972) The contribution of muscle afferents to kinaesthesia shown by vibration induced illusions of movement and by the effects of paralysing joint afferents. Brain 95:705–748

Goldscheider A (1920) Das Schmerzproblem. Springer, Berlin

Gunterberg B, Norlén L, Stener B, Sundin T (1975) Neurologic evaluation after resection of the sacrum. Invest Neurol 13:183–188

Head H (1893) On disturbances of sensation, with special reference to the pain of visceral diseases. Brain 16:1–133

Holst E von (1973) Active functions of human visual perception. In: Selected papers of Erich von Holst: the behavioural physiology of animals and man, vol. 1. Methuen, London pp 192–219 (translated from German by Martin R)

Hosobuchi Y, Adams JE, Linchitz R (1977) Pain relief by electrical stimulation of the central gray matter in humans and its reversal by naloxone. Science 197:183–186

Hyndman OR, Wolkin J (1943) Anterior cordotomy. Further observations on physiological results and optimum manner of performance. Arch Neurol Psychiatry 50: 129–148

Iggo A (1955) Tension receptors in the stomach and the urinary bladder. J Physiol 128:593–607

Iggo A (1966) Physiology of visceral afferent systems. Acta Neuroveg 28:121–134

Jänig W (1986) Spinal cord integration of visceral sensory systems and sympathetic nervous system reflexes. In: Cervero F, Morrison JFB (eds) Visceral sensation. Elsevier, Amsterdam, pp 255–277 (Progress in brain research, vol 67)

Jänig W, McLachlan E (1987) Organisation of the lumbar sympathetic and afferent supply of colon and pelvic organs. Physiol Rev (in press)

Jänig W, Morrison JFB (1986) Functional properties of spinal visceral afferents supplying abdominal and pelvic organs, with special emphasis on visceral nociception. In: Cervero F, Morrison JFB (eds) Visceral sensation. Elsevier, Amsterdam, pp 87–114 (Progress in brain research, vol 67)

Kandel ER, Schwartz JH (1981) Principles of neural science. Arnold, London

Katayama Y, deWitt DS, Becker DP, Hayes RL (1984) Behavioural evidence for a cholinoceptive pontine inhibitory area: descending control of spinal motor output and sensory input. Brain Res 296:241–262

Klevmark B (1977) Motility of the urinary bladder in cats during filling at physiological rates. II. Effects of extrinsic bladder denervation on intraluminal tension and intravesical pressure patterns. Acta Physiol Scand 101:176–184

Klevmark B (1980) Motility of the urinary bladder in cats during filling at physiological rates. III. Spontaneous rhythmic bladder contractions in the conscious and anaesthetised animal. Influence of distension and innervation. Scand J Urol Nephrol 14:219–224

Knibestol M, Vallbo AB (1970) Single unit analysis of mechanoreceptor activity from the human glabrous skin. Acta Physiol Scand 80:178–195

Kumazawa T (1986) Sensory innervation of reproductive organs. In: Cervero F, Morrison JFB (eds) Visceral sensation. Elsevier, Amsterdam, pp 115–132 (Progress in brain research, vol 67)

Kuru M (1965) Nervous control of micturition. Physiol Rev 45:425–494

Lewis T (1942) Pain. MacMillan, New York

Light AR (1985) The spinal terminations of single, physiologically characterised axons originating from the pontomedullary raphe nuclei. J Comp Neurol 234:536–548

Light AR, Perl ER (1979) Spinal termination of functionally identified primary afferent neurons with slowly conducting myelinated fibres. J Comp Neurol 186:133–150

Lumb BM (1984) Neural basis of stimulation-produced analgesia. In: Holden AV, Winlow W (eds) The neurobiology of pain. Manchester University Press, Manchester, pp 188–196

Lumb BM (1986) Brainstem control of visceral afferent pathways in the spinal cord. In: Cervero F, Morrison JFB (eds) Visceral sensation. Elsevier, Amsterdam, pp 279–293 (Progress in brain research, vol 67)

Lumb BM, Morrison JFB (1984). Convergence of visceral and somatic information on to identified reticulo- and raphespinal neurons in the rat. J Physiol 357:33P

Lumb BM, Morrison JFB (1986) Electrophysiological evidence for an excitatory projection from ventromedial forebrain structures on to raphe- and reticulo-spinal neurons in the rat. Brain Res 380: 162–166

Maggi CA, Santicioli P, Borsini F, Giuliani S, Meli A (1986) The role of capsaicin-sensitive innervation of the rat urinary bladder in the activation of micturition reflex. Naunyn-Schmiedebergs Arch Pharmacol 332:276–283

Mackenzie J (1983) Some points bearing on the association of sensory disorders and visceral disease. Brain 16:321–354

Mackenzie J (1912) Symptoms and their interpretation, 2nd edn. Shaw, London

MacLennan AM, Goodell H (1943) Pain from the bladder, ureter and kidney pelvis. Proc Assoc Res Nerv Ment Dis 23:252–262

Malliani A, Lombardi F, Pagani M (1986) Sensory innervation of the heart. In: Cervero F, Morrison JFB (eds) Visceral sensation. Elsevier, Amsterdam, pp 39–48 (Progress in brain research, vol 67)

McCloskey DI (1981) Corollary discharges: motor commands and perception. In: Brookhart JM, Mountcastle VB (eds) Handbook of physiology. Section 1, The nervous system vol II, part 2. Waverly, Baltimore pp 1415–1447

McDonald DF, Murphy GP (1959) Quantitative studies of perception of thermal stimuli in the normal and neurogenic urinary bladder. J Appl Physiol 14:204–206

McMahon SB (1986) Sensory-motor integration in urinary bladder function. In: Cervero F, Morrison JFB (eds) Visceral sensation. Elsevier, Amsterdam, pp 245–253 (Progress in brain research, vol 67)

McMahon SB, Abel C (1987) A model for visceral pain states: chronic inflammation of rat urinary bladder by irritant chemicals. Pain 28: 109–127

McMahon SB, Morrison JFB (1982a) Spinal neurones with long projections activated from the abdominal viscera of the cat. J Physiol 332:1–20

McMahon SB, Morrison JFB (1982b) Two groups of spinal interneurons that respond to stimulation of the abdominal viscera of the cat. J Physiol 332:21–34

McMahon SB, Morrison JFB (1982c) Factors that determine the excitability of parasympathetic reflexes to the bladder. J Physiol 332:35–43

McMahon SB, Morrison JFB, Spillane Kathy (1982) An electrophysiological study of somatic and visceral convergence in the reflex control of the external sphincters. J Physiol 328:379–387

Milne RJ, Foreman RD, Giesler GJ, Willis WD (1981) Convergence of cutaneous and pelvic visceral nociceptive inputs onto primate spinothalamic neurons. Pain 11:163–183

Mokha SS, McMillan JA, Iggo A (1985) Descending control of spinal nociceptive transmission. Actions produced on spinal multireceptive neurons from the nuclei locus coeruleus and raphe magnus. Exp Brain Res 58:213–226

Morley J (1931) Abdominal pain. Livingstone, Edinburgh

Morrison JFB (1977) The afferent innervation of the gastrointestinal tract. In: Brooks FP, Evers PW (eds) Nerves and the gut. Slack, New York, pp 297–326

Morrison JFB (1981) Sensory processing in spinal pathways from the bladder. In Grastyan E, Molnar P (eds) Advances in physiological sciences, vol 16: Sensory functions. Pergamon, Oxford, pp 325–333

Morrison JFB, McMahon SB (1982) Colonic afferents and their reflexes. In: Weinbeck M (ed) Motility of the digestive tract. Raven, New York, pp 19–30

Morrison JFB, Spillane Kathy (1982) Characterisation of the raphe–spinal pathway which mediates inhibition of the micturition reflex in the cat. J Physiol 330:24–25P

Morrison JFB, Spillane Kathy (1986) Neuropharmacological studies on descending inhibitory controls over the micturition reflex. J Auton Nerv Syst (suppl) 393–397

Mountcastle VB (ed) (1980) Sensory receptors and neural encoding: introduction to sensory processes. In: Medical physiology 14th edn. Mosby, St Louis, pp 327–347

Murphy JJ, Schoenberg HW (1960) Observations on intravesical pressure changes during micturition. J Urol 84:106–110

Murray KHA, Feneley RCL (1982) Endorphins—a role in lower urinary tract function? The effect of opioid blockade on the detrusor and urethral sphincter mechanisms. Br J Urol 54: 638–640

Nathan PW (1952) Thermal sensation in the bladder. J Neurol Neurosurg Psychiatry 15:148–149

Nathan PW (1956) Sensations associated with micturition. Br J Urol 28:126–131

Nathan PW, Smith MC (1951) The centripetal pathway from the bladder and urethra within the spinal cord. J Neurol Neurosurg Psychiatry 14:262–280

Nathan PW, Smith MC (1958) Centrifugal pathway for micturition within the spinal cord. J Neurol Neurosurg Psychiatry 21:177–189

Paintal AS (1986) The visceral sensations—some basic mechanisms. In: Cervero F, Morrison JFB (eds) Visceral sensation. Elsevier, Amsterdam, pp 3–19 (Progress in brain research, vol 67)

Pedersen E (1983) Regulation of bladder and colon-rectum in patients with spinal lesions. J Auton Nerv Syst 7:329–338

Pierau FK, Taylor DCM, Abel W, Friedrich B (1982) Dichotomising peripheral afferent fibres revealed by intracellular recording of rat sensory neurones. Neurosci Lett 31:123–128

Porter RW (1967) A pallidal response to detrusor contraction. Brain Res 4:381–383

Powell PH, Feneley RCL (1980) The role of urethral sensation in clinical urology. Br J Urol 52:539–541

Procacci P (1969) A survey of modern concepts in pain. In Vinken PJ, Bruyn GW (eds) Handbook of clinical neurology, vol 1. North Holland, Amsterdam, pp 114–146

Procacci P, Francini F, Zoppi M, Maresca M (1975) Cutaneous pain threshold changes after sympathetic block in reflex dystrophies. Pain 1:167–175

Procacci P, Zoppi M, Maresca M (1986) Clinical approach to visceral sensation. In: Cervero F, Morrison JFB (eds) Visceral sensation. Elsevier, Amsterdam, pp 21–28 (Progress in brain research, vol 67)

Reynolds DV (1969) Surgery in the rat during electrical analgesia induced by focal brain stimulation. Science 164: 444–445

Richardson DE, Akil H (1977a) Pain reduction by electrical brain stimulation in man: Part 1: acute administration in periaqueductal and periventricular sites. J Neurosurg 47:178–183

Richardson DE, Akil H (1977b) Pain reduction by electrical brain stimulation in man: Part 2: chronic self-administration in the periventricular gray matter. J Neurosurg 47:184–194

Riddoch G (1921) Conduction of sensory impulses from the bladder by the inferior hypogastrics and the central afferent connections of these nerves. J Physiol 54:CXXXIV

Risholm L (1954) Studies on renal colic and its treatment by posterior splanchnic block. Acta Chir Scand [Suppl] 184:1–64

Ruch TC (1946) Visceral sensation and referred pain. In Fulton JF (ed) Howell's textbook of physiology, 15th edn. Saunders, Philadelphia, pp 385–401

Ruch TC (1979) The pathophysiology of pain. In: Ruch TC, Patton HD (eds) Physiology and biophysics, 20th edn, vol 1. Saunders, Philadelphia, pp 272–324

Sato A, Sato Y, Schmidt RF (1980) Reflex bladder activity induced by electrical stimulation of hind limb somatic afferents in the cat. J Auton Nerv Syst 1:229–241

Schmidt RF (1978) Fundamentals of sensory physiology. Springer, Berlin Heidelberg New York, p 286

Scott FB, Quesada EM, Cardus D (1964) Studies on the dynamics of micturition: observations on healthy men. J Urol 92:455–463

Selzer M, Spencer WA (1969) Convergence of visceral and cutaneous afferent pathways in the lumbar spinal cord. Brain Res 14:331–348

Shah PJR (1984) The assessment of patients with a view to urodynamics. In: Mundy AR, Stephenson TP, Wein AJ (eds) Urodynamics. Churchill Livingstone, Edinburgh, pp 53–61

Sinclair DC, Weddell G, Feindel WH (1948) Referred pain and associated phenomena. Brain 71:184–211

Spillane Kathy (1982) Brainstem influences on pelvic parasympathetic reflexes. PhD thesis, University of Leeds

Talaat M (1937) Afferent impulses in the nerve supplying the urinary bladder. J Physiol 89:1–13

Taylor DCM, Pierau FK (1982) Double fluorescent labelling supports electrophysiological evidence of dichotomising peripheral sensory nerve fibres in rats. Neurosci Lett 33:1–6

Theobald GW (1941) Referred pain; a new hypothesis. Times of Ceylon (Colombo)

Todd JK (1964) Afferent impulses in the pudendal nerves of the cat. Q J Exp Physiol 49:258–267

Torrens MJ, Abrams P (1979) Urine flow studies. Urol Clin North Am 6:71–88

Torrens MJ, Hald T (1979) Bladder denervation procedures. Urol Clin North Am 6: 283–294

Wall PD, Devor M, Inbal R, Scadding JW, Schonfield D, Seltzer Z, Tomkiewicz MM (1979) Autotomy following peripheral nerve lesions: experimental anaesthesia dolorosa. Pain 7:103–113

Wein AJ (1985) Drug therapy for detrusor hyperactivity: Where are we? Neurourol Urodyn 4:337–352

White JC (1943) Sensory innervation of the viscera: 'Studies on visceral afferent neurones in man based on neurosurgical procedures for the relief of intractable pain. Res Publ Assoc Nerv Mental Dis 23:373–390

White JC, Sweet WH (1955) Pain. Its mechanisms and neurosurgical control. Thomas, Springfield, Ill

Willis WD (1986) Visceral inputs to sensory pathways in the spinal cord. In: Cervero F, Morrison JFB (eds) Visceral sensation. Elsevier, Amsterdam, pp 207–225 (Progress in brain research, vol 67)

Willis WD, Coggeshall RE (1978) Sensory mechanisms of the spinal cord. Wiley, Chichester

Winter DL (1971) Receptor characteristics and conduction velocities in bladder afferents. J Psychiatr Res 8:225–235

Wolff HG (1963) Headache and other head pain. Oxford University Press, Oxford

Woolf CJ (1984) Long term alterations in the excitability of the flexion reflex produced by peripheral tissue injury in the chronic decerebrate rat. Pain 18:325–343

5 · The Functions of Efferent Nerves to the Lower Urinary Tract

John F. B. Morrison

The purpose of this chapter is to review the peripheral and central effects of excitation of efferent neurons innervating the lower urinary tract. The anatomy was reviewed in Section 1, and our main concern in this chapter will be with the physiology and pharmacology of ganglionic and/or neuro-effector transmission in sympathetic, parasympathetic and somatic efferent pathways. There are many excellent reviews on this subject and the reader may find monographs by Appenzeller (1982), Gabella (1976), Nilsson (1983) and Johnson and Spalding (1974), and reviews by Burnstock (1981, 1986), Morrison (1982), de Groat et al. (1981), de Groat (1975) and Andersson and Sjögren (1982) helpful for further study. The physiology of smooth muscle and of neuromuscular transmission in the bladder is dealt with in Chapter 6. However, the starting point of this chapter will deal with an effect which appears to be mediated by parasympathetic efferent neurons within the central part of their course, viz. the phenomenon of recurrent inhibition.

Preganglionic Parasympathetic Efferents

Recurrent Inhibition in the cord

In 1968, de Groat and Ryall observed that bladder contractions in the cat could be inhibited by stimulation of the central cut end of a sacral ventral root; the firing of contralateral preganglionic parasympathetic neurons whose discharges correlated with the occurrence of bladder contractions was also depressed. The hypothesis was that the depression of activity in the parasympathetic nerves to the bladder, and of bladder motility, was due to a recurrent inhibitory pathway, similar to that involved in the inhibition of somatic motoneurons; axon collaterals of parasympathetic efferent neurons, however, have not been detected in labelling experiments (Nadelhaft et al. 1980), and axons projecting into the cord via the ventral roots have been regarded as ventral root afferents rather than recurrent axon collaterals by Morgan et al. (1981). The anatomical basis for this phenomenon is therefore unclear. The recurrent inhibition of α-motoneurons is known to be mediated by recurrent axon collaterals synapsing on Renshaw cells (Eccles et al. 1954) and differs in a number of aspects from the inhibition observed by de Groat and Ryall (1968). Unlike the recurrent inhibition of somatic motoneurons, the effects of ventral root stimulation on bladder motility were prolonged, and often lasted for several minutes.

The inhibitory effects of ventral root stimulation were seen only when the central cut end of a ventral root which contained parasympathetic efferent fibres was stimulated at intensities which excited these or other fibres of similar diameter, and at rates greater than 10 s^{-1}. The effects were crossed, being relayed to the bladder via the parasympathetic efferents on the opposite side of the cord, and were seen in chronic spinal animals as well as chloralose-anaesthetized and decerebrate preparations. The phenomenon was still present when the bladder was deafferented, so the effects were not due to ventral root afferents (de Groat 1976). The recurrent inhibition was seen following section of the hypogastric nerves; the conclusion that it was not dependent on sympathetic efferent pathways may not be absolutely valid, because about 10% of the sympathetic fibres running to the pelvic viscera travel in the pelvic nerves (see Chap. 3, p. 79), and in addition, the ventral roots may contain sympathetic efferent fibres that innervate

pial blood vessels and possibly other structures (Risling et al. 1984).

Recurrent inhibitory effects (Fig. 5.1) were most effective at low bladder pressures, and were relatively weak when afferent inputs from the bladder were activated by higher levels of intravesical pressure; in contrast, high bladder pressures did not influence the firing of Renshaw cells. These recurrent inhibitory effects were explained by the presence of interneurons in the sacral autonomic cell columns that received recurrent axon collaterals from the parasympathetic efferents. de Groat and Ryall (1968) recorded from some cells which were synaptically activated by antidromic stimulation of the parasympathetic efferents in the ventral roots, at rather longer latencies than were observed in Renshaw cells, as one might expect from the low conduction velocities of the parasympathetic axons. They fired at high rates of discharge (600–800 spikes per second) for the first few spikes; the maximum number of spikes in a burst was seven, and the duration of the burst was 10–80 ms. This discharge is not as marked as seen in Renshaw cells in the same animals. These neurons followed frequencies of up to 40 per second. The interneurons that could be synaptically activated from a ventral root were not excited by stimulation of adjacent or contralateral ventral roots, but could be fired by stimulation of various ipsilateral leg nerves. Thus some form of somatovisceral convergence is seen even in this system. The recurrent inhibition was unaffected by intravenous injection of dihydo-β-erythrodine, which blocks the antidromic activation of Renshaw

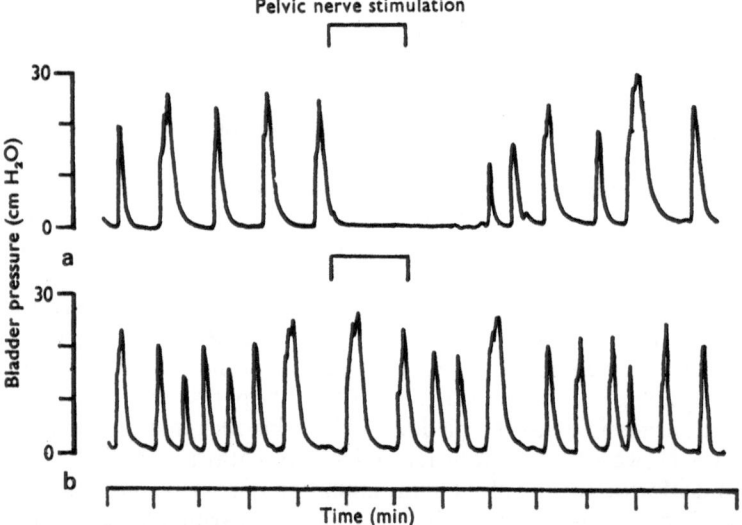

Fig. 5.1a,b. Recurrent inhibition of spontaneous bladder contractions by stimulation of the central end of a transected pelvic nerve. Stimulation applied at a frequency of 20 Hz during period indicated by *bar*. **a** Obtained with ipsilateral sacral dorsal roots transected, but the ventral roots intact; **b** after transection of the ipsilateral sacral ventral roots. (de Groat 1976)

cells. Atropine reduced the response of one cell to ventral root stimulation.

Thus, there is good evidence for recurrent inhibitory effects that can be elicited by electrical stimulation of small-diameter fibres in the ventral roots. Whether the ventral root fibres that elicit these responses in vesical parasympathetic efferents are in fact recurrent collaterals of parasympathetic efferents is as unclear as their peripheral connections, which may be with the bladder or with some other pelvic viscus. Inhibitory interactions between visceral nerves are known to occur, but the possibility that the efferent pathways might have reciprocal connections does not appear to have been explored experimentally; de Groat and Ryall (1968) concluded that if the recurrent collaterals came from the vesical efferents, this may be an additional reflex helping to maintain urinary continence and acting in coordination with guarding reflexes to the urethral sphincter. More recently, another function for the recurrent inhibitory effects that is in keeping with current ideas on the function of Renshaw cells has been proposed: de Groat et al. (1982) showed that the pelvic nerve efferents to the bladder show a bursting pattern of activity during reflex micturition, which continued to occur following spinalization. The irregular activity was thought to be due to recurrent inhibition rather than an intrinsic property of the neurons, as iontophoresis of excitatory amino acids on to the neurons elicited regular activity.

Physiology of Transmission in Pelvic Ganglia

There are relatively few studies of ganglionic transmission in the pelvic ganglia. Blackman et al. (1969) obtained intracellular records from neurons in the pelvic ganglia of guinea pigs, and most of these responded to electrical stimulation of the hypogastric (sympathetic) nerve; the effects of pelvic nerve stimulation were not investigated. They concluded that the function of the pelvic ganglion cells with hypogastric inputs differs from those in the sympathetic inferior mesenteric ganglion, but are like those in other parasympathetic ganglia, such as the avian ciliary ganglion (Martin and Pilar 1963). These pelvic ganglion neurons were thought to function as relays or distributing centres, passing information on from one preganglionic fibre to many postganglionic fibres, rather than as integrating centres, where there is scope for a considerable degree of interaction, such as occurs in the inferior mesenteric ganglion. The pelvic ganglion neurons that were excited by hypogastric nerve

stimulation appeared to be incapable of integrative ability, according to these authors. Thus, orthodromic stimulation of small myelinated or unmyelinated hypogastric nerve fibres gave rise to "all or nothing" responses, and these could not be graded on changing stimulus intensity. These ganglia were also said to resemble other parasympathetic relay stations in that they are located peripherlly, resistant to immunosympathectomy and to the action of certain specific ganglion stimulants. About 60% of the neurons responded to hypogastric nerve stimulation; there is a high safety factor in this process of neurotransmission.

In the cat, the effects of pelvic nerve stimulation were studied by Griffith et al. (1980), who observed three types of neurons, most of which were "tonic neurons", i.e. a prolonged suprathreshold depolarizing current gave rise to a maintained discharge of action potentials from the neurons; this is in keeping with other ganglion cells involved in the regulation of visceral motility (see Cassell et al. 1986). On orthodromic or antidromic stimulation, afterhyperpolarizations were absent, in contrast to neurons in the myenteric plexus, where these slower potentials are known to influence ganglionic transmission. The authors conclude, however, that the vesical parasympathetic ganglion of the cat is not a simple relay station, and that integration and modulation of ganglionic activity may involve up to three types of neurons (Fig. 5.2). Some of these cells are spontaneously active, and may modulate ganglionic transmission endogenously or through an afferent mechanism; these possibilities have also been recognized in sympathetic ganglia (Julé and Szurszewski 1983; Julé et al. 1983), where more work has been performed, particularly on afferent mechanisms (see Chap. 7, p. 217). There is reason therefore to doubt whether the ideas of Blackman et al. apply to all inputs to pelvic ganglion cells, particularly as other workers have shown a considerable potential for integration within the pelvic efferent pathway through these ganglia (e.g. de Groat and Saum 1972, 1976). Nevertheless, the observations of Blackman et al. (1969) have to be taken seriously, as they are the result of a careful study of an unusual aspect of these ganglia; it is of interest that Sundin (1972) found that hypogastric nerve stimulation could give rise to prolonged bladder *contractions* following preganglionic parasympathectomy. The possibility exists that there is more than one pathway through these ganglia, and that they may normally act in parallel, with one becoming dominant following a particular lesion. Furthermore, sexual dimorphism has been demonstrated in the neurons of the pelvic ganglia, and this may be one further source

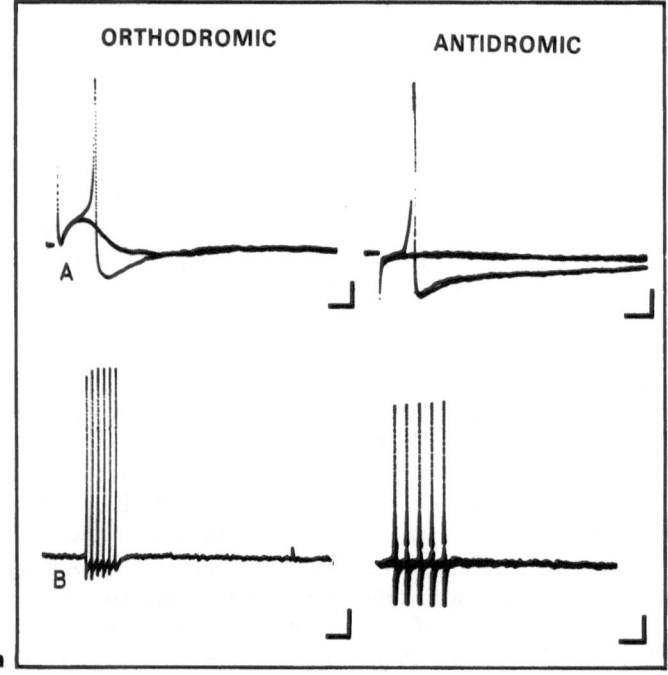

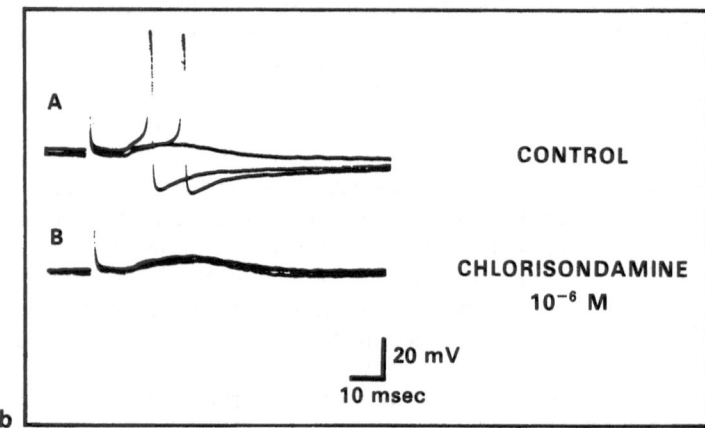

Fig. 5.2a. Orthodromic and antidromic response to electrical stimuli. *A*: Response of neurons to single stimuli. *B*: Response of neurons to orthodromic (6 Hz) and antidromic (5 Hz) train stimulation. Single-stimulus calibration: 10 mV × 10 ms. Train calibration: orthodromic 10 mV × 1 s: antidromic 10 mV × 0.5 s.
b Nicotinic blockade of the spike potential with chlorisondamine ($10^{-6} M$). *A*: Control response before the addition of chlorisondamine. Three superimposed traces are shown. *B*: Blockade of the spike potential after the addition of chlorisondamine. Calibration: 20 mV × 10 ms. (Griffiths et al. 1980)

of variation in the experimental results (see Chap. 3, p. 72).

Pharmacology of Neurotransmission in the Pelvic Ganglia

In recent years, it has become clear that neurons that contain a "classical" transmitter, such as acetylcholine, noradrenaline, adrenaline, dopamine, serotonin, γ-aminobutyric acid (GABA) or glycine, also contain other chemicals that could act as neurotransmitters. The list of substances that may coexist with these is long and is summarized in Table 5.1.

Table 5.1. Proven coexistence of neurotransmitters and neuromodulators (adapted from Hökfelt et al. 1986)

"Classical" transmitter	Coexisting petide
Acetylcholine	Enkephalin
	Substance P
	Vasoactive intestinal polypeptide (VIP)
	Calcitonin gene-related peptide (CGRP)
	Galanin
Noradrenaline	Enkephalin
	Neuropeptide Y
	Vasopressin
Serotonin	Substance P
	Thyrotrophin-releasing hormone (TRH)
	Cholecystokinin (CCK)
	Enkephalin

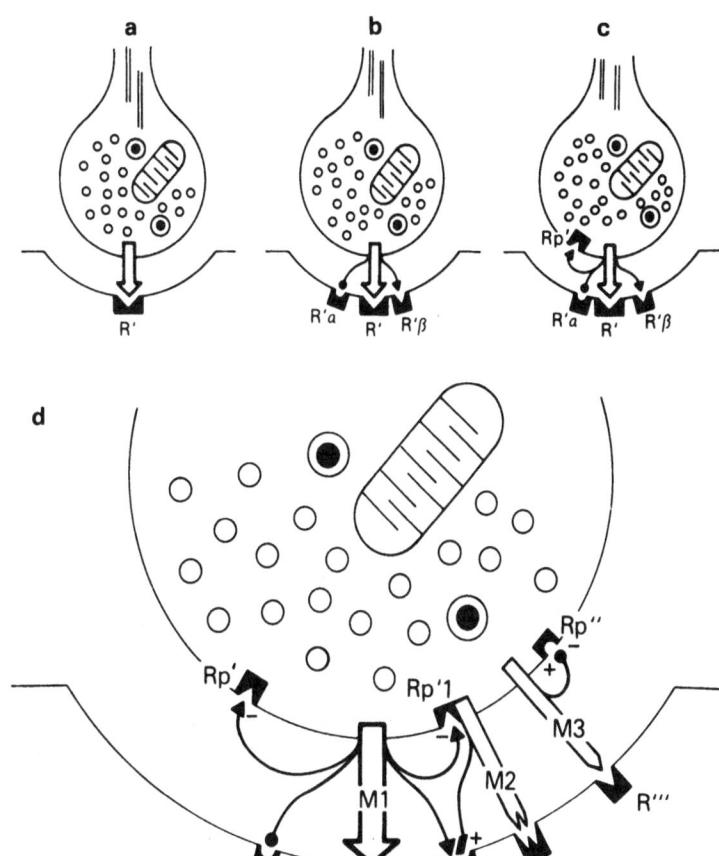

Fig. 5.3a–d. Schematic illustration of the development of the concept of chemical transmission. **a** One transmitter acts on one postsynaptic receptor (R'). **b** One transmitter acts on multiple types of postsynaptic receptors ($R'\alpha\ R'\beta$). **c** The transmitter acts in addition on a presynaptic receptor (Rp'). **d** Multiple compounds ($M1–3$), possibly differentially stored in small vesicles (classical transmitter) and in large dense-core vesicles (classical transmitter plus peptide), are released from the same nerve ending. The main possible interactions indicated are: (1) inhibition of release of the second messenger (peptide M2) by the classical transmitter (M1)[40] via presynaptic action[40] ($Rp'I$); (2) interaction at the postsynaptic receptor ($R'\beta$) level between M1 and M2[16,41]; (3) facilitation or inhibition of release of the classical transmitter by the peptide (M3) via action on a presynaptic receptor (Rb'')[45,46]; and (4) activation by the peptide (M3) of electrical activity in the presynaptic neuron via action on a presynaptic receptor (Rp'')[59]. (Lundberg and Hökfelt 1983)

The older view, that the classical transmitter is released and acts on one postsynaptic receptor, has been modified substantially during the last decade. Not only has more than one postsynaptic receptor site for the classical transmitter been demonstrated in some systems, the transmitter may also act pre-synaptically. The coexistence of two or more transmitters at a synapse has increased the combinations and permutations that can occur, and some idea of the complexity that might exist is provided in Fig. 5.3. This subject has recently been reviewed by Hökfelt et al. (1986).

The following summary deals with some of the actions and interactions of the classical transmitters and the putative neurotransmitters and neuromodulators. Table 5.2 summarizes some of the actions of the transmitters and putative transmitters in the pelvic ganglia.

Table 5.2. Pharmacological effects at pelvic ganglia

Drug	Effects	Reference
Acetylcholine	Depolarization Increased firing	Griffith et al. (1980)
VIP	Depolarization Increased firing Reduced K current	Akasu et al. (1986)
Leu-enkephalin	Reduced size of EPSPs Inhibition of high frequency discharges	Simmonds et al. (1983) Kawatani et al. (1983)
Catecholamines	Inhibition of transmission Facilitation } by α-receptors Excitation	de Groat and Saum (1972, 1976) Craggs and Stephenson (1981) de Sy et al. (1974)

Acetylcholine

The pelvic nerve ganglia cannot be considered as a simple relay station, with the straightforward function of faithfully transmitting preganglionic activity into postganglionic without any modification or fine tuning; indeed, there would be little point in having a ganglion in this situation if it were to function simply as an axon. A number of excitatory and inhibitory interactions between inputs to the ganglion are known to operate in the vesical ganglia. The pelvic parasympathetic, preganglionic input to these ganglia has cholinergic excitatory effects on the postganglionic neurons, mediated mainly by nicotinic, but also by muscarinic receptors (de Groat and Saum 1976). The process of ganglionic transmission was studied by these workers by recording the mass activity of postganglionic branches of the ganglion during stimulation of the preganglionic nerve trunk; recordings of this type are open to a variety of criticisms, which leave open to question the precise nature of the responses of individual postganglionic neurons or postulated subgroups of postganglionic fibres. In addition, this work was done before it was fully realized that the pelvic preganglionic trunk contains considerable numbers of sympathetic fibres (Kuo et al. 1984). Nevertheless, these authors showed that transmission in the vesical ganglia is facilitated upon repetitive stimulation of the preganglionic pelvic nerve trunk, such that the postganglionic response to a preganglionic stimulus increased 5- to 20-fold when the rate of stimulation was increased from $0.5-1 \, s^{-1}$ to $5-10 \, s^{-1}$ (Fig. 5.4). The facilitation is unaffected by atropine. It has been generally assumed, probably correctly, that the effects are due to synaptic activity of preganglionic parasympathetic efferents, and possible effects of the sympathetic efferents in the pelvic nerve have not been seriously considered. Thus, transmission is likely to be most effective when the preganglionic fibres fire at high rates, as during a reflex excitation of the bladder, and appears to be blocked if the preganglionic activity is low (de Groat and Saum 1976).

Vasoactive Intestinal Polypeptide

The presence of vasoactive intestinal polypeptide (VIP) in pelvic nerve afferents (Gibson et al. 1984a,

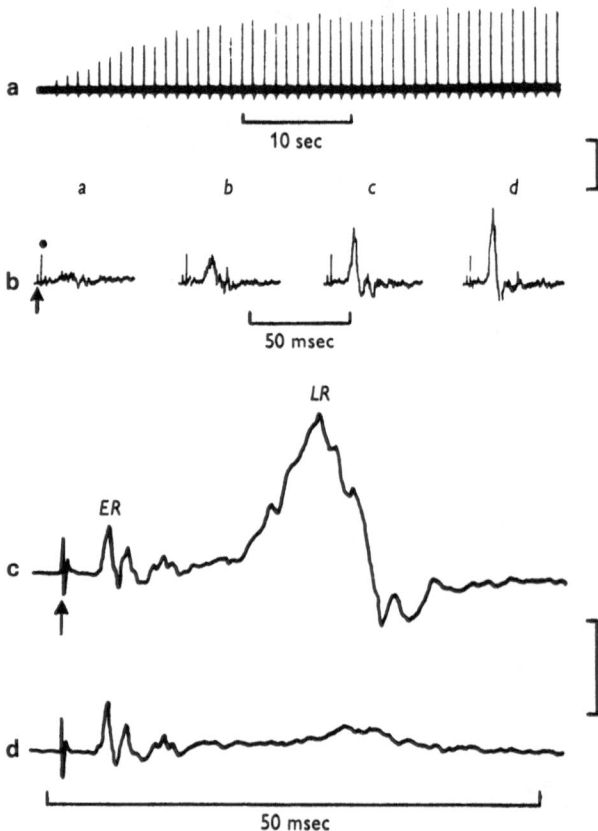

Fig. 5.4a–d. Recruitment of the postganglionic response elicited by repetitive stimulation of the pelvic nerves. **a** Action potentials recorded on a vesical postganglionic nerve filament in response to submaximal (5 V) stimulation of the preganglionic fibres at a frequency of 1 Hz. **b** Sample tracings from **a**, of the postganglionic responses obtained at varying times after the start of the stimulus train. In **b**, *a* was the first response elicited by preganglionic stimulation and *b*, *c* and *d* were obtained, respectively, 5, 10 and 20 s later. The *arrow* below **b** *a* denotes the stimulus artefact and the *dot* above denotes the early response. **c** Bimodal response to pelvic nerve stimulation (1.1 V, 1 Hz) recorded on nerve filaments arising from ganglia on the surface of the urinary bladder. *ER* and *LR,* respectively, are the early response and late response. The *arrow* below **c** denotes the stimulus artefact. **d** Effect of hexamethonium (100 μg, i.a.) on the bimodal response. Responses in **c** and **d** represent the average of 20 individual responses. Time calibration in **d** also applies to **c**; vertical calibration in **a** and **b** is 400 μV and in **c** and **d** is 200 μV, negativity upwards. (de Groat and Saum 1976)

1986), in ganglionic neurons of pelvic ganglia (Lundberg et al. 1979), and in axonal terminals around pelvic ganglion cells (Kawatani et al. 1985) makes the actions of VIP on parasympathetic neurons of particular interest. VIP excites central and peripheral neurons and facilitates the muscarinic responses in vesical parasympathetic ganglia (Kawatani et al. 1985). Akasu et al. (1986) have recently studied the effects of VIP on the membrane properties of neurons in the vesical ganglia of cats, and have found that it depolarized these cells, often sufficiently to evoke spontaneous action potentials in previously quiescent neurons (Fig. 5.5); the rate of firing of cells which had a resting discharge was increased. They concluded that the depolarization was due to a decrease in the potassium conductance of these membranes, and that the characteristics of the current was unlike previously described conductive mechanisms.

Enkephalins

Leu-enkephalin has been found in the sacral preganglionic efferent pathway to the urinary bladder as well as in terminals associated with the sacral

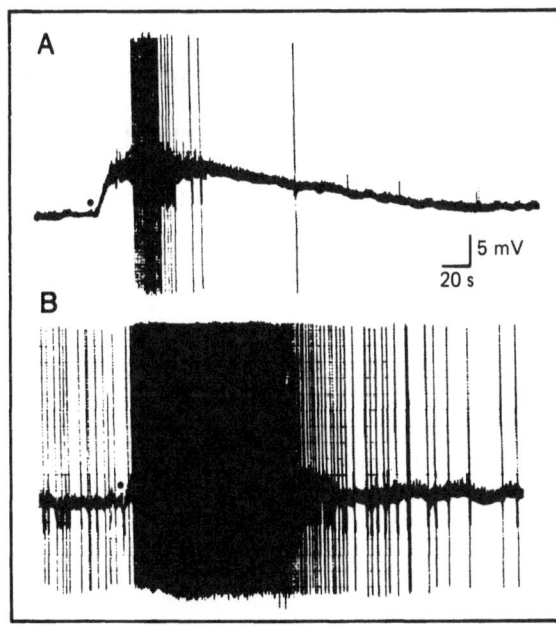

Fig. 5.5. a VIP depolarization is neuroexcitatory. VIP (5 μM) applied in a pressure pulse (50 ms, 20 lbf/in²) as indicated (*dot*) to a quiescent cell in *A* and to a spontaneously active neuron in *B*. Resting potential was − 60 mV in both cells. **b** Concentration–response relation for VIP. Micropipettes containing various concentrations (0.5–50 μM) of VIP were placed in the same position as close as possible to the recording electrode. VIP was applied at *arrow* to the same vesical parasympathetic neuron by pressure pulse (50 ms, 20 lbf/in²). *Right* and *left panels* show responses to VIP applied in concentrations as indicated; the amplitudes of the depolarizations obtained with application of various concentrations of VIP are illustrated in the *middle panel*. Resting potential was − 60 mV throughout. (Akasu et al. 1986)

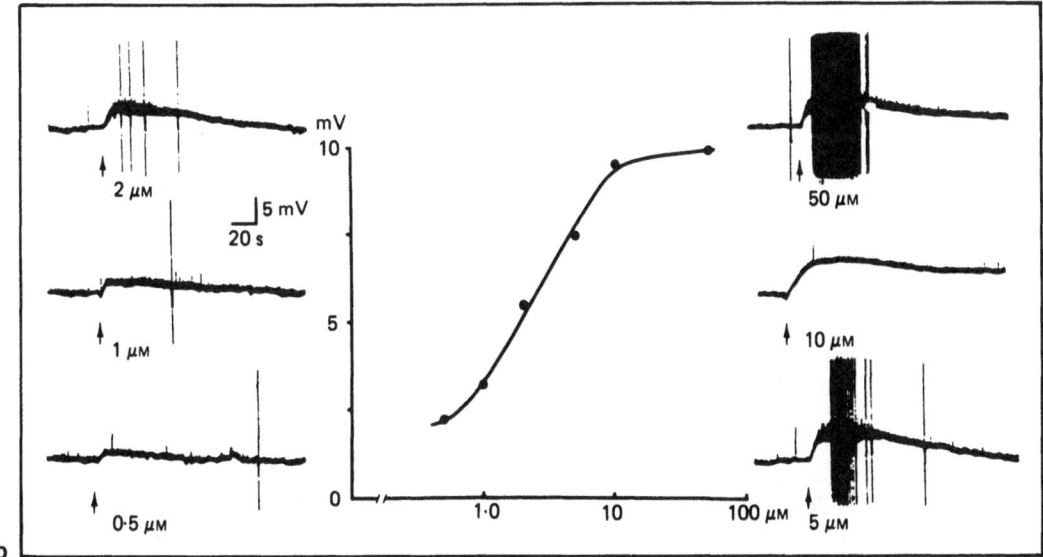

parasympathetic nucleus and Onuf's nucleus (Fig. 5.6), and has been found to inhibit transmission in bladder ganglia; these peripheral effects last a few minutes and are dose dependent. Intra-arterial morphine is 10–100 times less effective than Leu-enkephalin. Naloxone blocks the effects of exogenously applied Leu-enkephalin, and reduces inhibitory effects in the ganglia (Kawatani et al. 1983; de Groat et al. 1983; Simmonds et al. 1983). Leu-enkephalin depresses orthodromic transmission in vesical ganglia and reduces the size of excitatory postsynaptic potentials (EPSPs) recorded from ganglion cells, which suggests a presynaptic action on preganglionic parasympathetic neurons. Kawatani and de Groat (1982) found that the heterosynaptic inhibition elicited by repetitive stimulation of the pelvic nerves at high rates can be reduced by naloxone; however, there was no evidence of an involvement of Leu-enkephalin in ganglionic transmission when action potentials occur at low rates. The levels of activity necessary to produce an enkephalinergic depression of ganglionic transmission may be present during the micturition reflex.

Catecholamines

Transmission through the pelvic ganglia of the cat is modulated by sympathetic efferent pathways which influence the process via an α-adrenergic mechanism (Fig. 5.7; de Groat and Saum 1972). The anatomical basis for this mechanism could be seen in histological sections which showed adrenergic terminals close to vesical cholinergic ganglion cells (Hamberger and Norberg 1965). The influence of these sympathetic endings was found to be inhibitory, but is nowhere near as powerful as the direct inhibitory effect the sympathetic exerts on the bladder smooth muscle via β-adrenoceptors. Sympathetic inhibition of ganglionic transmission was not always present in de Groat and Saum's experiments, and was only effective at low levels of parasympathetic preganglionic input.

There is a suggestion that some sympathetic fibres may have the opposite effect to that described above: Craggs and Stephenson (1981) found that transmission in some proximal ganglia in the pelvic pathway to the bladder was facilitated during sym-

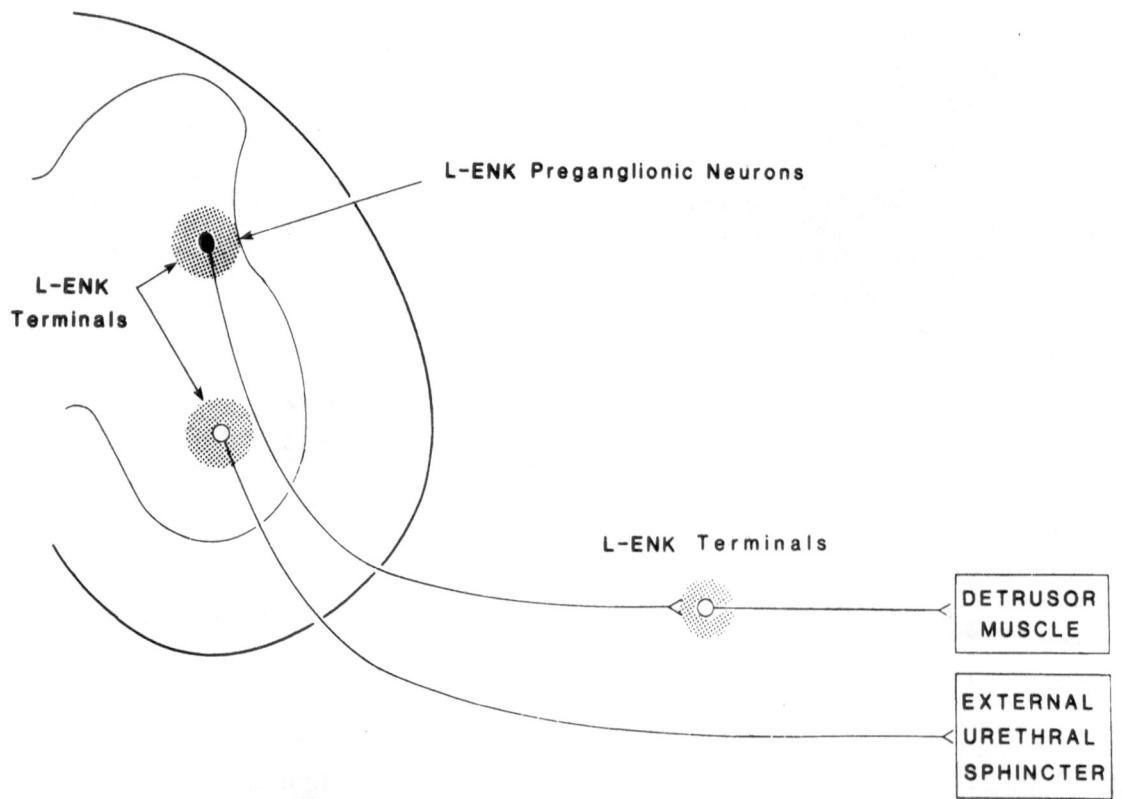

Fig. 5.6. The distribution of leucine-enkephalin (*L-Enk*) in the sacral efferent pathways to the lower urinary tract. Diagram shows the location of L-Enk terminals in the pudendal motor nucleus (Onuf's nucleus), which provides an innervation to the external urethral sphincter, and the sacral parasympathetic nucleus, which provides an innervation to the detrusor muscle. (de Groat et al. 1983)

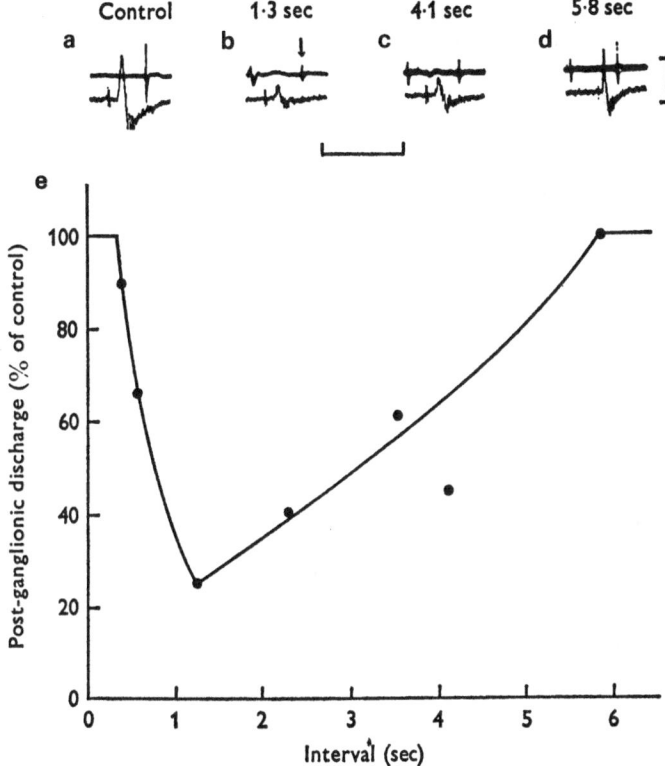

Fig. 5.7a–e. Time course of the inhibitory effect in pelvic ganglia to activation of the hypogastric nerve (HGN) with trains of stimuli (100 ms train duration, 20 V, 200 Hz intra-train frequency). **a** is a control response evoked by stimulation of the pelvic nerve (0.16 Hz) at submaximal stimulation (2 V). *Top trace* in this and remaining records is taken at slow sweep to show the time relation between the HGN-stimulus train (at the beginning of the sweep) and the pelvic-evoked response (at the *arrow*). *Bottom trace* is the pelvic-evoked response on an expanded sweep. Interval (in seconds) between the start of the HGN-train and the pelvic response is above each record. **e** is a plot of the depression of the amplitude of the pelvic discharge (*ordinate*) against the interval between the start of the train and the occurrence of the pelvic discharge. For the *upper traces* the horizontal calibration represents 2 s in **a** and **b**, 5 s in **c** and 10 s in **d**. For the *lower traces* the calibration represents 100 ms in **a–c** and 200 ms in **d**. Vertical calibration represents 240 μV for *top tracings* and 200 μV for *bottom tracings*. (de Groat and Saum 1976)

pathetic stimulation. Other work of a more pharmacological nature suggests that some sympathetic fibres excite cholinergic postganglionic neurons in the pelvic plexus and that these cholinergic neurons are responsible for the initial transient contraction of the bladder that can be seen upon stimulation of the hypogastric nerves in the cat (Sy et al. 1974).

The interpretation of these results is complicated by the fact that the transient rise followed by the fall in bladder pressure that can be seen during sympathetic stimulation could be due to differences in the responses of the trigone and the detrusor, differences between parasympathetic postganglionic neurons, their distribution, transmitters and actions (de Groat and Saum 1976; Burnstock 1986), differences between different populations of hypogastric nerve fibres, their transmitters and their connections (see Chap. 3, p. 77), or to differences in the techniques used by different workers.

Summary

The process of transmission of efferent information from pelvic nerve efferents to postsynaptic elements in the pelvic ganglia involves facilitatory and inhibitory processes. The classical transmitter at this site is acetylcholine, which acts on nicotinic receptors on postganglionic neurons, and gives rise to excitatory postsynaptic potentials, which cause the cell to fire once a threshold potential has been reached.

Some pelvic preganglionic fibres contain Leu-enkephalin, which inhibits transmission in these ganglia, the EPSPs are reduced in size by the action of Leu-enkephalin, and the probable site of action is presynaptic, i.e. to reduce the output of transmitter induced by preganglionic impulses. The effects of Leu-enkephalin are blocked by naloxone.

VIP is also present in pelvic ganglia, in afferent fibres, in axonal terminals and in some postsynaptic

neurons. Its action is excitatory, and VIP facilitates transmission mediated by muscarinic receptors in these ganglia. The exact role of VIP in these ganglia is unclear, but the presence of peptides such as VIP at this site should be viewed in the light of recent ideas such as those suggested in Fig. 5.3, viz. cotransmitters may act pre- or postsynaptically to modify the release or effects of a classical transmitter. Monoamines released from sympathetic efferent fibres appear to inhibit synaptic transmission in the pelvic efferent pathway, although some workers have suggested that catecholamines may also produce excitatory effects at this site.

Repetitive activity in pelvic preganglionic fibres facilitates parasympathetic transmission in these ganglia; the mechanism for this is not clear. Certain pelvic ganglion neurons are excited by hypogastric efferent fibres and appear to act as relay stations rather than integrative centres. There is additional electrophysiological evidence that at least three types of postsynaptic neurons are present in these ganglia. Some of these cells are spontaneously active and may modulate ganglionic transmission endogenously, or possibly through an afferent mechanism.

Thus, there may be a number of parallel pathways through the pelvic ganglia, and the properties of these different pathways may not be the same; there is sufficient variation in the results reported by different authors using different experimental approaches to support this tentative conclusion. If this were the case, some of the unexpected results obtained in unusual situations, e.g. those of Sundin (1972) following preganglionic parasympathectomy or those of de Sy et al. (1974) might be explained.

Physiology and Pharmacology of Transmission in the Bladder

The question of how autonomic nerves initiate contractions in the bladder is far from clear. Elbadawi and Schenk (1968, 1974) have concluded that the older ideas that suggested direct innervation of smooth muscle cells by sympathetic and parasympathetic postganglionic cells are rather simpler than the evidence would suggest. Instead they propose a system of intrinsic neurons in the bladder wall, some of which are adrenergic and others cholinergic, with physiological interactions between the two branches of the autonomic nervous system, as was demonstrated later by de Groat and Saum (1972, 1976). The intrinsic neurons are thought to innervate more than one structure at the vesicourethral junction, and to distribute activity

throughout the bladder and other muscles of the lower urinary tract. On anatomical grounds, Fehér et al. (1980) have argued that contractions of the bladder are initiated or controlled in the trigone, and that the other parts of the bladder wall are activated by local nerve processes and numerous nexal contacts (desmosome-like attachments) between smooth muscle cells. The trigone receives about three times as much innervation as the apex, and the contacts between the nerve and the smooth muscle can be as small as 150–250 nm. These authors concluded that there were four types of nerve terminal in the bladder musculature. There are a number of candidates for neurotransmitters at this site, and acetylcholine, ATP, noradrenaline, VIP and prostaglandins have been implicated in neurotransmission. Table 5.3 summarizes the actions of these agents.

Table 5.3. Actions of drugs and nerve stimulation on the bladder

Pelvic nerve	Atropine-resistant phasic contractions Atropine-sensitive tonic contractions
Hypogastric nerve	Initial contraction (α effect on ganglia) followed by relaxation (β effect on smooth muscle)
Acetylcholine	Contractions (tonic effects blocked by atropine)
ATP	Contraction; probably a NANC transmitter
Substance P	Contractions, but not likely to be a NANC transmitter
VIP	Relaxation in some species
Catecholamines	Relaxation via a β-action
Prostaglandins	May contribute to the contraction induced by ATP
Neurotensin Leu-enkephalin }	Excitation in some species

Acetylcholine

Bladder contractions are mediated by the parasympathetic nervous system. Brindley and Craggs (1976) found that bladder contractions in humans and in baboons are quite sensitive to atropine, and the idea that the predominant neurotransmitter in the parasympathetic supply to the human bladder appears to be acetylcholine is supported also by studies on isolated strips of human bladder (Sibley 1984; see also Chap. 6, p. 181). A species difference appears to exist, however, in that nonadrenergic noncholinergic mechanisms appear to be dominant in some animal species, but are weak in humans. The response of the feline bladder to pelvic nerve stimulation consists of phasic and tonic components; de Groat and Saum (1976) found that the former was resistant to atropine, but the latter was reduced by the drug. This conclusion is supported by older studies (Henderson and Roepke 1934) that also showed that the tonic component of the contractile response of the bladder during pelvic nerve stimulation could be blocked by atropine, as well as by more recent investigations that demonstrate nonadrenergic, noncholinergic as well as cholinergic neuroeffector transmission (Downie and Dean 1977; Mackenzie and Burnstock 1984; Burnstock 1986). Figure 5.8 shows the effects of atropine on the phasic and tonic responses of isolated bladder muscle, induced by nerve stimulation; it can be seen that atropine has a greater effect on

tonic than phasic responses, and that 40%–50% of the peak response can be blocked by atropine.

Sjögren et al. (1983) report that there is a small atropine-resistant contraction during transmural stimulation of the human bladder, but that the main mechanism is cholinergic (see also Sibley 1984, and Chap. 6, p. 181). In contrast, contractions of the detrusor in most nonprimate species show greater resistance to atropine (Ambache and Zar 1970; Taira 1972), and several groups of workers have been actively seeking to identify the neurotransmitter responsible. The putative neurotransmitters include a variety of peptides, and purines, such as adenosine triphosphate (ATP). The nonadrenergic, noncholinergic (NANC) transmission in the urinary bladder can be inhibited by drugs which block calcium entry into cells (Sjögren et al. 1982). Readers are referred to Chapter 6.

The muscarinic receptors in the guinea pig and rabbit bladders have been studied with ligand-binding techniques. In the guinea pig, the muscarinic receptors are present at a density of about 0.32 pmol per milligram protein (Nilvebrant and Sparf 1983a); in the rabbit, the density is 628 ± 80 fmol per milligram protein when estimated using the ligand quinuclidinyl benzilate (Johns 1983), whereas Lepor and Kuhar (1984) found a density of 21.5 fmol per milligram protein using tritiated N-methylscopolamine. There may be differences between different subgroups of muscarinic receptors (Nilvebrant and Sparf 1983b; Wamsley et al. 1984).

Substance P

Callahan and Creed (1986) concluded that the bladder receives excitatory noncholinergic innervation which mediates the large excitatory junction potential and the detrusor contraction in rabbits and guinea pigs. Substance P induces contractions, which were desensitized by prolonged contact with this peptide; the response to nerve stimulation in desensitized preparations was unaffected, however. In addition, the mechanical responses to substance P administration were reduced by chymotrypsin, whereas the responses to nerve stimulation were unaffected. Thus, the effects of substance P could be dissociated from the effects of nerve stimulation; the conclusion that substance P is probably not an NANC transmitter in the bladder was also made by Mackenzie and Burnstock (1984) (Fig. 5.9). They therefore thought it unlikely that the transmitter responsible for the NANC response is substance P, nor is it VIP (which induced relaxation), neurotensin, somatostatin, or Leu-enkephalin (none of

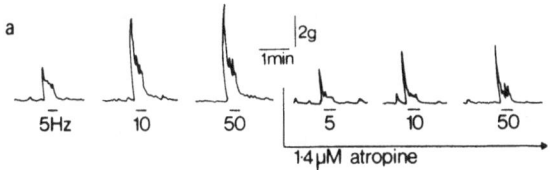

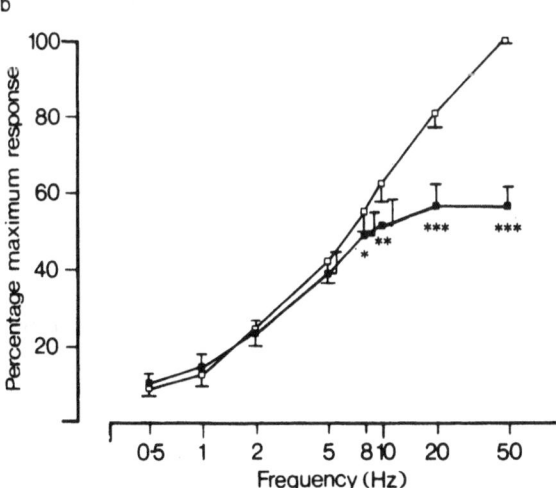

Fig. 5.8. a Field stimulation (20 s trains of 0.2 ms duration pulses) elicited frequency-dependent excitation of the muscle. This was characterized by an initial rapid contraction which relaxed to a relatively stable level within 15 s. Atropine (1.4 μM) reduced the maximum amplitude of the initial contraction and almost abolished the secondary tonic phase. **b** The relationship between stimulation frequency and the maximal amplitude of contractile responses is shown. The results are plotted as a percentage of the maximum control response against the frequency of the stimulation. □, mean (± SEM) control responses. ■, results in the presence of atropine (1.4 μM) (n = 12). Significant differences in the magnitude of the responses are indicated: *, $P < 0.05$; **, $P < 0.01$; ***, $P < 0.001$. Guanethidine (3.4 μM) was present throughout **a** and **b**. (Mackenzie and Burnstock 1984)

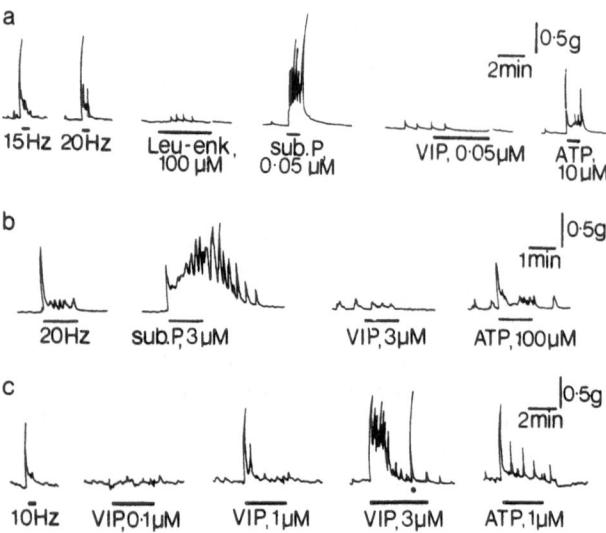

Fig. 5.9. a A comparison of the atropine-resistant responses of a guinea pig detrusor strip elicited by electrical field stimulation (NS: 15Hz or 20 Hz for 30 s), [Leu⁵]enkephalin (*Leu-enk*), substance P (*sub. P*), VIP and ATP at the given concentrations. **b** Another comparison of the atropine-resistant excitation (NS: 20 Hz for 75 s), substance P, VIP and ATP on a faster time scale. **c** A comparison of the atropine-resistant excitatory responses of bladder strips elicited by field stimulation (NS: 10 Hz for 30 s), VIP and ATP. *Horizontal bars* denote the duration of the stimuli. ●, NS: 10Hz for 3 s. Atropine (1.4 μM) and guanethidine (3.4 μM) were continuously present in all three preparations. (Mackenzie and Burnstock 1984)

which had observable actions on the bladder). Substance P also contracts the urethra in the rabbit and dog, but not the guinea pig.

Vasoactive Intestinal Polypeptide

The action of VIP on bladder muscle seems to be small: Johns (1979) and Mackenzie and Burnstock (1984) obtained small contractions with long latencies at high concentrations of VIP in the guinea pig whereas low concentrations are reported to induce relaxation in rabbit, pig, dog and human preparations (Levin and Wein 1981; Klarskov et al. 1984). Callahan and Creed (1986) also report a reduction in action potential frequency recorded from rabbit bladder strips; in addition, VIP did not potentiate the contraction induced by nerve

stimulation. VIP is reported to relax the bladder trigone in pigs, and the urethra in dogs, pigs and rabbits, but not in guinea pigs (Hill et al. 1984; Callahan and Creed 1986). It seems unlikely that VIP is a NANC transmitter in the bladder (Fig. 5.9; Mackenzie and Burnstock 1984). The possibility that VIP is involved in the vasodilatation that occurs during bladder filling has been studied by Andersson et al. (1985), who found no association between blood flow and VIP or substance P concentrations in the venous effluent.

Neurotensin and Enkephalins

Neurotensin and Leu-enkephalin are both reported to have excitatory effects on the canine bladder, but

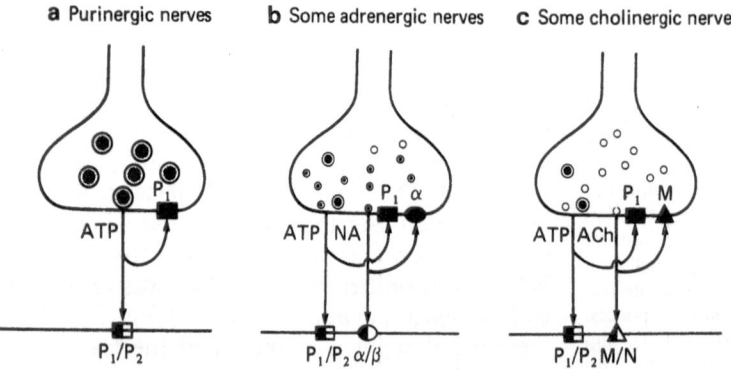

Fig. 5.10a–c. Schematic representation of presynaptic neuromodulation of transmitter release. **a** Reduction of ATP release by activation of presynaptic P₁ purinoceptors. **b** Reduction of noradrenaline release by activation of presynaptic alpha-adrenoceptors and P₁ purinoceptors. **c** Reduction of ACh release by activation of presynaptic muscarinic and P₁ purinoceptors. (Burnstock 1981)

have no effects on the lower urinary tract of rabbits and guinea pigs.

Adenosine Triphosphate

There is now a considerable body of evidence in favour of purinergic transmission in the bladder of a number of nonprimate species. Burnstock et al. (1978a) used quinacrine, which binds to ATP, in a fluorescence histochemistry technique to demonstrate the presence of putative purinergic nerves in the guinea pig bladder. ATP can be released by electrical stimulation of the bladder nerves (Burnstock et al. 1978b). Mackenzie and Burnstock (1984) compared the effects of different putative transmitters with the effects of field stimulation in the presence of atropine, and concluded that the response characteristics during nerve stimulation could best be mimicked by ATP (Fig. 5.9). Desensitization of the rabbit bladder with α,β-methylene-ATP also desensitizes the responses to nerve stimulation (Hoyle and Burnstock 1985). ATP contracts the rabbit and human bladder, the response consisting of phasic and tonic components; there is some evidence that ATP causes the release of prostaglandins from the bladder (Andersson et al. 1980). Figure 5.10 shows the possible sites of interaction between acetylcholine and ATP on the bladder, as well as some other interactions which involve cotransmitters acting pre- or post-synaptically to modify the response of a classical transmitter (Burnstock 1981).

Prostaglandins

Andersson et al. (1980) postulated that prostaglandins were released and played a part in the bladder contractions that can be induced by ATP. However, attempts to investigate the usefulness of prostaglandin synthetase inhibitors such as flurbiprofen have not yielded dramatic results in patients (Cardozo et al. 1980).

Monoamines

Sympathetic nerve terminals are about three times as numerous in the trigone and internal sphincter as in the apex of the bladder (Fehér et al. 1980). Each nerve fibre is thought to come into contact with only one or two smooth muscle cells at the apex, and six or seven muscle cells at the base.

Catecholamines act on bladder muscle to induce relaxation, and this is mediated by β-receptors.

There are more β-receptors (about 1000 fmol per milligram protein) than α-receptors (about 300 fmol/mg) in the bladder (Johns 1983), and there appears to be some differential distribution between the detrusor and trigone. Levin and Wein (1979) found that there were more β-adrenoceptors in the detrusor than the bladder base in the dog and rabbit; de Sy et al. (1974) have shown that catecholamines can excite pelvic ganglia and induce a transient contraction via this indirect route. Serotonin has been found to relax the urethra and trigone in the pig (Hills et al. 1984).

Physiology and Pharmacology of Transmission in the Urethral and Anal Sphincters

Autonomic Pathways to Smooth Muscle

The parasympathetic innervation of the human urethra appears to be less dense than that of the bladder (Ek et al. 1977), and there are fewer numbers of muscarinic receptors in the rabbit urethra than in the bladder (Johns 1983); the function of the parasympathetic supply to the urethra is controversial. Barrington believed that bladder distension could cause relaxation of the proximal urethra in cats, partly by a pelvic nerve efferent pathway (see Barrington's sixth reflex). Torrens (1978) showed that stimulation of sacral nerves in humans caused urethral relaxation (Fig. 5.11), and suggested a parasympathetic efferent pathway. Other workers have confirmed this effect and observed that it can be blocked by β-adrenoceptor antagonists. One possible explanation may be that the electrical stimuli induce antidromic action potentials in pelvic nerve afferents; in other species there is the possibility that some visceral afferent fibres have axon collaterals which synapse with postganglionic neurons in the prevertebral ganglia. The existence of adrenergic postganglionic neurons in the pelvic ganglia has been known for many years, and the observation that the urethral relaxation during sacral nerve stimulation can be blocked by β-blockers might be explained by a peripheral reflex pathway. α- and β-adrenoceptor densities in the dog and rabbit bladder and urethra have been investigated by Levin and Wein (1979), Larsson (1983) and by Johns (1983). In the rabbit urethra, α-receptors outnumber β-adrenoceptors; in the dog, the numbers of urethral β-receptors is intermediate between that in the bladder dome and the

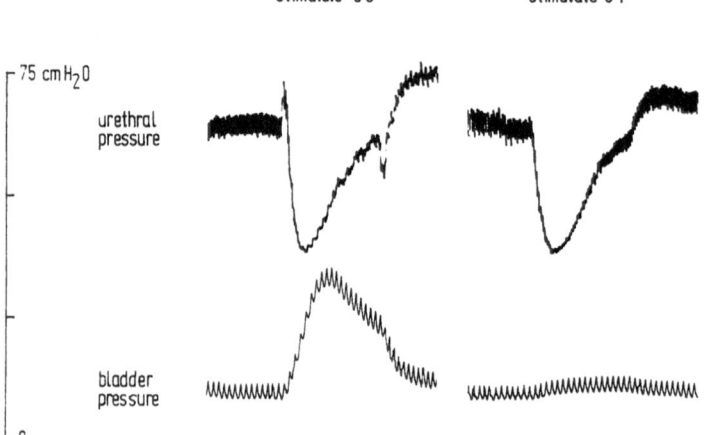

Fig. 5.11. The effects of stimulating the peripheral cut ends of the third and fourth sacral nerves on the urethral and bladder pressure in an anaesthetized, paralysed human. Urethral relaxation occurs independently of bladder contractions (Torrens 1978)

trigone, and the urethra and bladder base had more α-receptors than the bladder dome or ureter. Further evidence that the urethra can be relaxed by a pelvic nerve mechanism elicited by cortical stimulation was provided by Zuckerkandl (1893) and by Fränkl-Hochwart and Frölich (1904). In contrast to the relaxation induced by the pelvic nerves, acetylcholine has a weak contractile effect on the human urethra (Ek et al. 1978). There are considerable quantities of VIP in the urethra, and this peptide is known to relax this smooth muscle in rabbits and dogs. Hodson (1965) found that the sympathetic innervation, originating from cord segments rostral to L-2 caused contractions of the internal urethral sphincter during seminal emission in rabbits.

Classification of Skeletal Muscle Fibres

Striated muscle fibres can be divided into different types, depending on their speed of contraction and their susceptibility to fatigue. Since Ranvier's classical observations (1874) on slowly contracting "red" and rapidly contracting "white" muscle, it has often been assumed that there are two types of twitch fibres. It has only been relatively recently that the relationship between function and histochemical properties has been intensively investigated. Human muscles contain slow twitch and fast twitch fibres, but no tonic fibres. The contraction times during the twitch vary from 30 to 100 ms; the slow units tend to produce small forces, and are resistant to fatigue. The fast units often develop much larger forces, and those that produce the greatest force are fatiguable, whereas fast fibres that have a more moderate twitch tension are resistant to fatigue

(Fig. 5.12). The proportions of these vary between different muscles, but in the cat medial gastrocnemius (the site where we have the most extensive and reliable information) the proportions of fast fatiguable (FF), fast fatigue-resistant (FR), fast intermediate, and slow fibres are 45%, 25%, 5% and 25% respectively (Burke 1981).

Fast fibres are innervated by the faster α-motoneurons, and the slow twitch fibres are innervated by the slower α-motoneurons. The fast fatiguable fibres have larger diameters than the fast, fatigue-resistant fibres, which, in turn are larger than the slow fibres. There are also histochemical differences between the different fibre types, and the dominance of oxidative enzymes in slow twitch fibres and the lack of these in fast fatiguable fibres is one important histological correlate (Figs. 5.12, 5.13).

The presence of small-diameter fibres with oxidative enzymes in the human urethral sphincter suggests that these are slow fibres, which are adapted to produce a steady tension. Fibres with these histochemical characteristics are sometimes referred to as type I; types IIA and IIB have less ATPase reactivity, and have larger diameters. Type I muscle fibres have a rich blood supply, whereas the capillary supply in IIB fast-fatiguable fibres is sparse. There are correlations between the efficacy of the muscle spindle input and the Renshaw cell inhibitory input to motoneurons which innervate fast and slow muscle fibres (Burke 1981). Generally speaking, the efficacy of muscle spindle input is reduced and that from Renshaw cells is increased in the small motoneurons that innervate Type I fibres. It is of interhat pudendal motoneurons innervating the urethra do not receive muscle spindle inputs (there are none in the urethral sphincter) or Renshaw cell inhibitory inputs, and exhibit

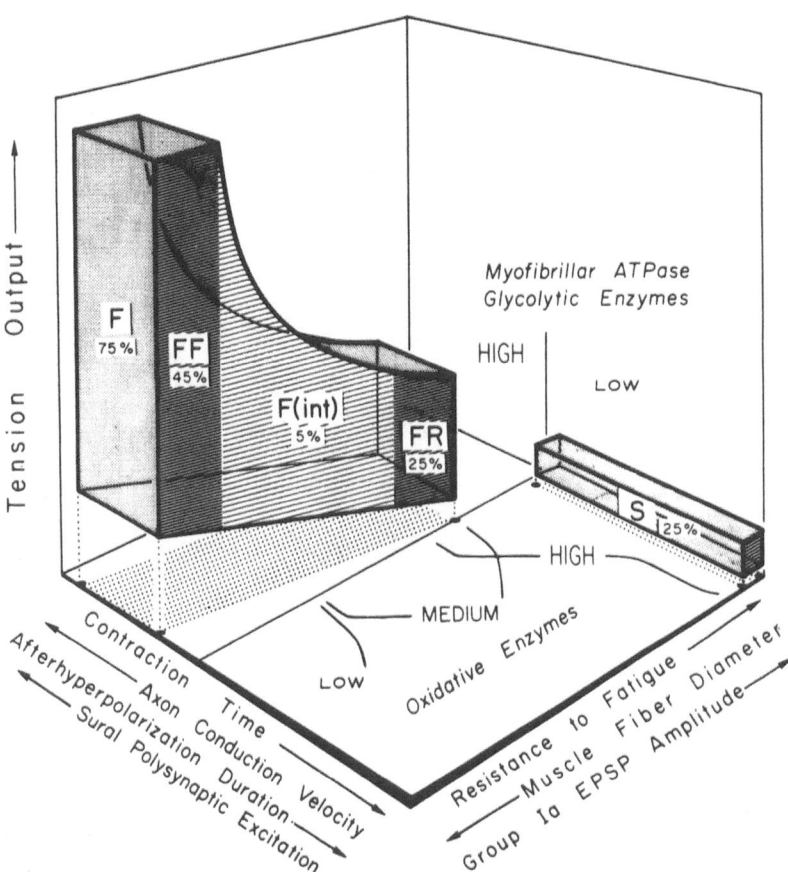

Fig. 5.12. Three-dimensional diagram summarizing interrelations between a variety of features of the physiological, morphological, and histochemical profiles of motor units found in the cat medial gastrocnemius (MG) population. Muscle unit properties include tetanic force output, twitch contraction time, resistance to fatigue, myofibrillar ATPase staining at alkaline pH, metabolic enzyme patterns (anaerobic and oxidative), and average muscle fibre diameter. Motoneuron properties included are axonal conduction velocity and duration of the afterhyperpolarization. Contrasting patterns of synaptic organization are represented by amplitude of monosynaptic group Ia EPSP (*right abscissa*) and by the apparent strength of polysynaptic excitatory input from distal hind limb skin (sural nerve: *left abscissa*). It should be clear that this summary somewhat distorts reality in that the various features specified are not all distributed in exactly the way displayed, but the general trends are as illustrated. The diagram is not intended to be all inclusive (other characteristics could be added to one or another axis) but rather emphasizes the fact that a great many properties of motoneurons and their muscle units display correlated variation, which can be communicated by the notion of motor unit "types". The types identified in the cat MG, along with their percentages in that population, are indicated on the *shaded boxes*, which denote the approximate loci of the clusters of data points seen in the actual data displays. (Burke 1981)

shorter afterhyperpolarizations (Fig. 5.14) than might be expected from moto-neurons innervating slow muscle (Jankowska et al. 1978; Mackel 1979); the resting discharge in these units may be related to these properties.

Pudendal Innervation of the Urethral and Anal Sphincters

Brindley et al. (1974) found that stimulation of the pudendal motoneurons in humans give rise to intraurethral pressures of up to about 120 mmHg. When repetitive stimuli were used, the pressure fell, presumably because the skeletal muscle of the rhabdosphincter fatigued during the first 2 min stimulation to leave a steady level of contraction which persisted for 10 min or more.

Electrical activity in the anal sphincter persists after spinal transection, and following transection of the autonomic nerves carrying afferents from the gut and pelvic viscera. This activity appears to be generated within the spinal cord, and afferent activity from pelvic skin, and from muscle spindles

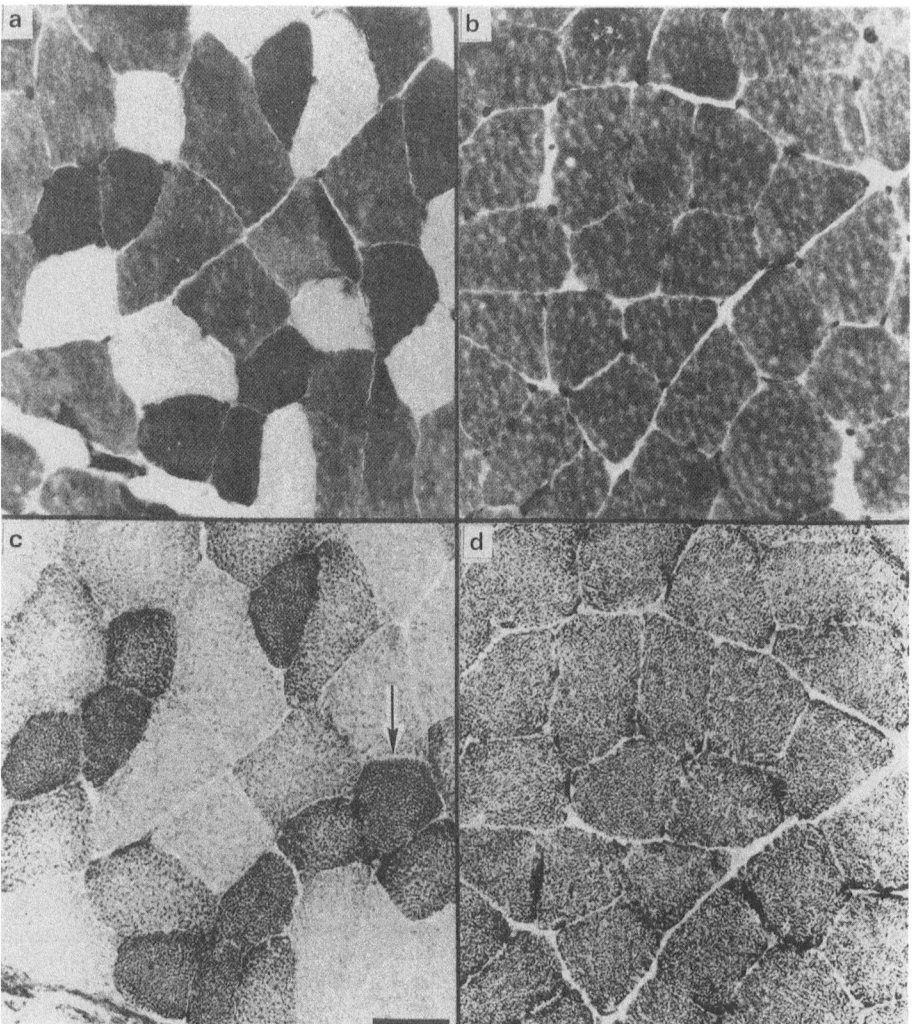

Fig. 5.13a–d. Histochemical profiles of fibres in heterogeneous (**a, c**) and homogeneous (**b, d**) cat muscles. Photomicrographs of representative serial sections of heterogeneous lateral gastrocnemius (LG; **a, c**) and homogeneous soleus (SOL; **b, d**) muscles in the same cat processed as a single block of tissue and photographed under identical conditions. **a** and **b** are from a section stained for myofibrillar ATPase activity after incubation in an acidic buffer at pH 4.65. **c** and **d** are from a serial section stained for oxidative enzyme NADH-dehydrogenase. When comparing AC-ATPase and NADH staining patterns for individual fibres (the histochemical profile), at least 3 fibre types can be recognized in LG but only 1 in SOL. Note that appearance of an LG fibre with the profile characteristic of type S muscle units (**c**, *arrow*) is not identical to appearance of SOL fibres. *Calibration bar* in **c** is 100 μm. (Burke 1981)

in the external anal sphincter have excitatory effects on the pudendal motoneurons; however, less than one-third of all pudendal motoneurons receive segmental monosynaptic inputs, and these neurons exhibit spikes which are followed by an after-hyperpolarization that is shorter than would be expected of tonic motoneurons (Fig. 5.14; Mackel 1979). These motoneurons may innervate muscle fibres on both sides of the anal sphincter.

Electrical activity is present in the external urethral and anal sphincter at rest (Sundin and Petersén 1975). Denny-Brown and Robertson (1933) claimed that the external urethral sphincter can be contracted voluntarily, but cannot be relaxed at will; Sundin and Petersén (1975), however, have found the electromyographic activity of this muscle can be extinguished at low bladder volumes or high bladder volumes, and in the latter case micturition is absent. This is in keeping with studies on pudendal nerve block or section, which show that continence is maintained in this situation (Bors and Comarr 1971).

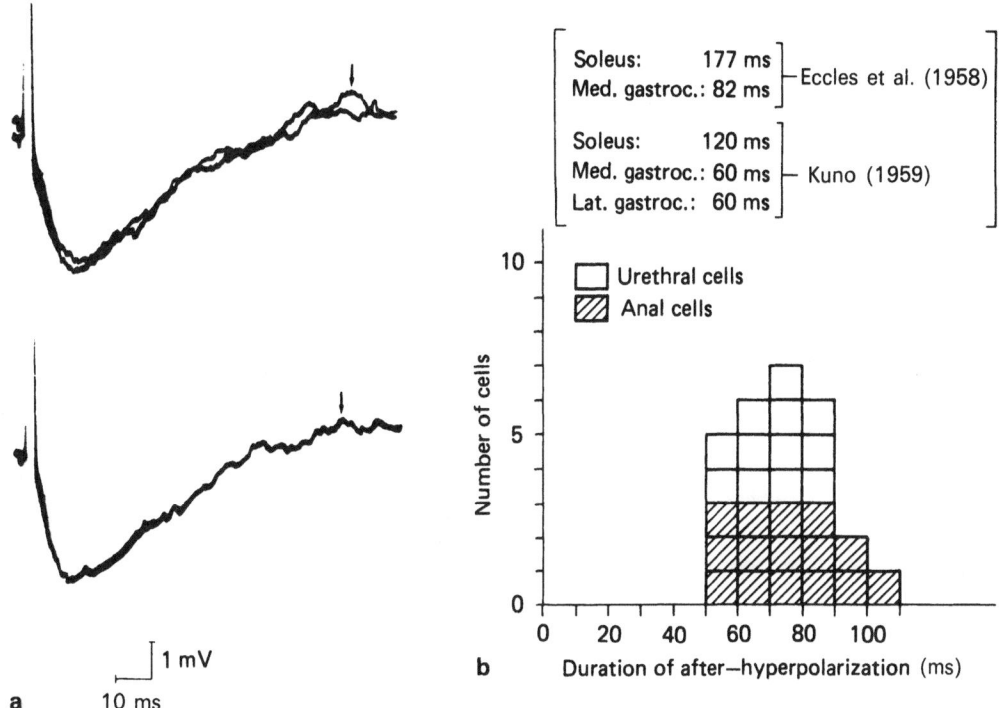

Fig. 5.14a, b. Duration of the spike afterhyperpolarization in motoneurons innervating the external urethral and external anal sphincters. **a** Two examples of spike afterhyperpolarizations. The *arrows* indicate where the negativity reaches the base line again. **b** A summarizing histogram of the spike afterhyperpolarizations for 27 sphincter motoneurons. The *inset above* shows the average values obtained in hind limb motoneurons. (Mackel 1979)

Electromyographic techniques have been used to demonstrate changes in the innervation of the external urethral and anal sphincters in humans (Blaivas et al. 1977; Parks et al. 1977; Fowler et al. 1984; Womack et al. 1986b). Patients with cauda equina lesions have changes in the shape of individual motor units, and the fibre densities are increased; this is one of a number of tests that have been used recently to assess the extent of peripheral denervation and reinnervation in the sphincters.

The reflexology of the external anal sphincter has been reviewed recently by Swash (1982) and by Gonella et al. (1987). Parks et al. (1977) and Swash (1982) found that the response of the sphincter to electrical stimulation of the perianal skin consisted of a number of components of differing latencies. The shortest latency responses were too fast to be reflex in origin, and Wright et al. (1985) came to the conclusion that they were attributable to direct excitation of motoneurons within the muscle, because they persisted after spinal anaesthesia had abolished all resting and reflex activity in the sphincter (Fig. 5.15). Garry et al. (1959) suggest that a guarding reflex is present in the skeletal muscle components of the urethral sphincter of the cat, and that the electrical activity in this muscle increases as the bladder is distended. A similar guarding reflex is certainly present in the human external anal sphincter and recruitment of activity can be induced by raising intra-abdominal pressure and by the Valsalva manoeuvre in normal and spinal men (Melzak and Porter 1964; Womack et al. 1986a).

There is a suggestion that the pudendal nerves in some species contain parasympathetic efferent fibres (Hulsebosch and Coggeshall 1982), but functional studies on these have not yet been performed.

Large neurons such as α-motoneurons do not usually stain for peptides by immunocytochemical techniques. One of the newer peptides, calcitonin gene-related peptide (CGRP), however, is known to occur in motoneurons, and is present in the sacral cord of a number of mammalian species (Gibson et al. 1984b), and appears to be present in pudendal as well as other motoneurons.

Summary

Skeletal muscle fibres can be classified into different types using physiological and anatomical information. The largest diameter fibres are innervated by the fastest α-motoneurons into the largest motor

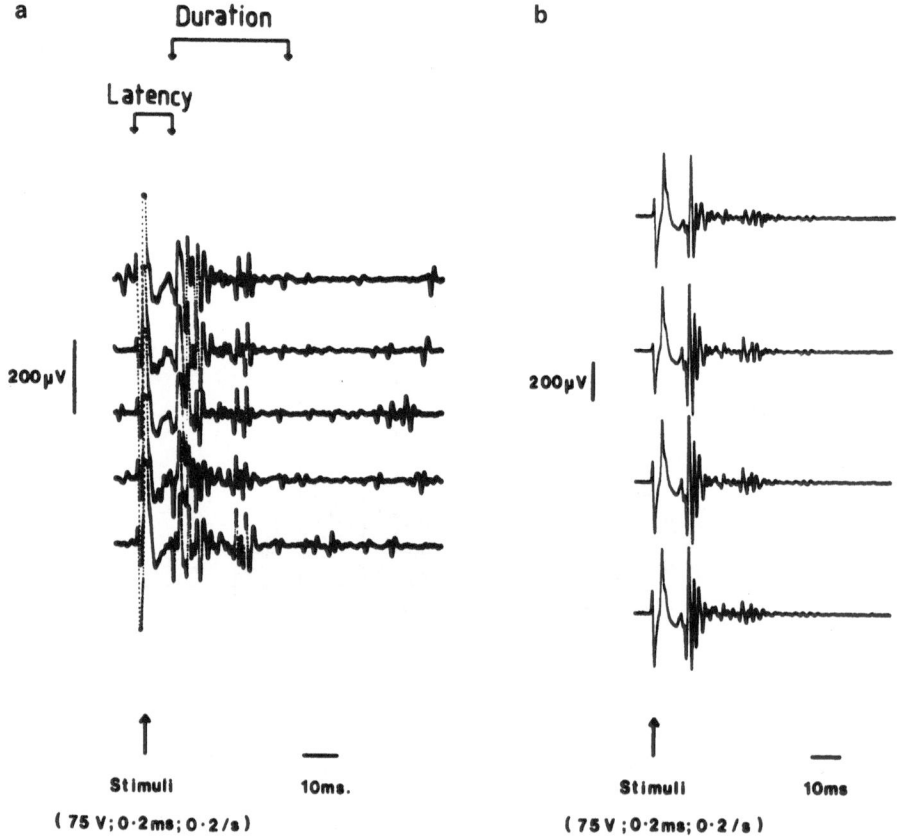

Fig. 5.15a,b. Electromyographic responses of the external anal sphincter in a normal man immediately following electrical stimulation of the perianal skin. **a** Before (latency 9.4 ms) and **b** during (latency 9.2 ms) spinal anaesthesia, showing the early response is still evoked during spinal anaesthesia. (Wright et al. 1985)

units, and are specialized to produce high levels of force over a short period of time, in that they are relatively lacking in oxidative enzymes. The smallest diameter muscle fibres are organized into smaller motor units, innervated by smaller diameter α-motoneurons, and are specialized to produce lower levels of force over a long period; they have a rich blood supply, and are resistant to fatigue. The latter are sometimes termed type I and the former, type II. An intermediate group can also be defined. The urethral sphincter appears to consist almost entirely of type I fibres, and are innervated by the lower range of diameters of α-motoneurons, which is in keeping with the morphology of Onuf's nucleus (see Chap. 3, p. 73). These motoneurons have a resting discharge in normal and in spinal subjects, and differ from motoneurons innervating the limb muscles in that they have little monosynaptic input, no Renshaw cell inhibition, no crossed disynaptic inhibition, and shorter afterhyperpolarization than might be expected. Pudendal motoneurons may contain CGRP, but act on the rhabdosphincter by the release of acetylcholine, which acts on nicotinic synapses.

The parasympathetic innervation causes relaxation of urethral smooth muscle, and the sympathetic supply can contract it.

Function of Sympathetic Nerves to the Lower Urinary Tract

Elliot (1907) described the excitatory pelvic and inhibitory hypogastric innervation of the cat bladder. Stimulation of the hypogastric nerves in the cat gives rise to a transient increase in intravesical pressure, followed by a maintained relaxation of the viscus (de Groat and Saum 1972). Bladder contractions are diminished during sympathetic stimulation, and the compliance of the bladder is increased; the steep part of the pressure–volume curve is also shifted to the right by sympathetic stimulation. The sympathetic efferents give rise to these effects partly by direct innervation of smooth muscle, but also via interactions with the para-sympathetic supply to the bladder in the pelvic ganglia; the effects of the sympathetic on transmission in the pelvic ganglia has been described previously. Nilsson (1983) presents a useful review of the methodology and the species differences that are found in the autonomic nervous system. Jänig and McLachlan (1987) have reviewed the sympathetic innervation of the pelvic viscera.

Ganglionic Transmission in the Inferior Mesenteric Ganglia

Noncholinergic excitatory postsynaptic potentials have been described in bullfrog and in mammalian sympathetic ganglia, where their time course is slow, possibly indicating that the receptor for the transmitter is distant from the ionic channel it activates, There are various candidates for the neurotransmitter in this slow EPSP, and luteinizing hormone-releasing hormone, substance P, VIP, cholecystokinin, serotonin and vasopressin may be involved (Jan and Jan 1982; Otsuka and Konishi

1983; Peters and Kreulen 1984; Love and Szurszewski 1985; Kreulen and Peters 1986).

Cassell et al. (1986) compared the properties of two groups of sympathetic neurons in the guinea pig concerned with vasoconstriction and with visceral motility. Two distinct patterns of response of sympathetic postganglionic neurons in a variety of ganglia to the intracellular injection of current have been described (Fig. 5.16): A prolonged suprathreshold depolarizing current passed through the intracellular electrode leads to either a brief burst of activity or to a maintained firing of the cell, and neurons with these different characteristics have been designated "phasic" and "tonic" neurons, respectively. The work of Cassell et al. (1986) addressed the possibility that neurons with these different characteristics might subserve different physiological functions. The vascular innervation of pelvic viscera appears to originate from ganglion cells along the hypogastric nerves and in the pelvic plexuses (Langley and Anderson 1895d; Sjöstrand 1965), whereas that of the hind limb originates from the lumbar sympathetic chain. Hypogastric axons are mainly concerned with the regulation of visceral activity (Langley and Anderson 1895a–c), and orig-

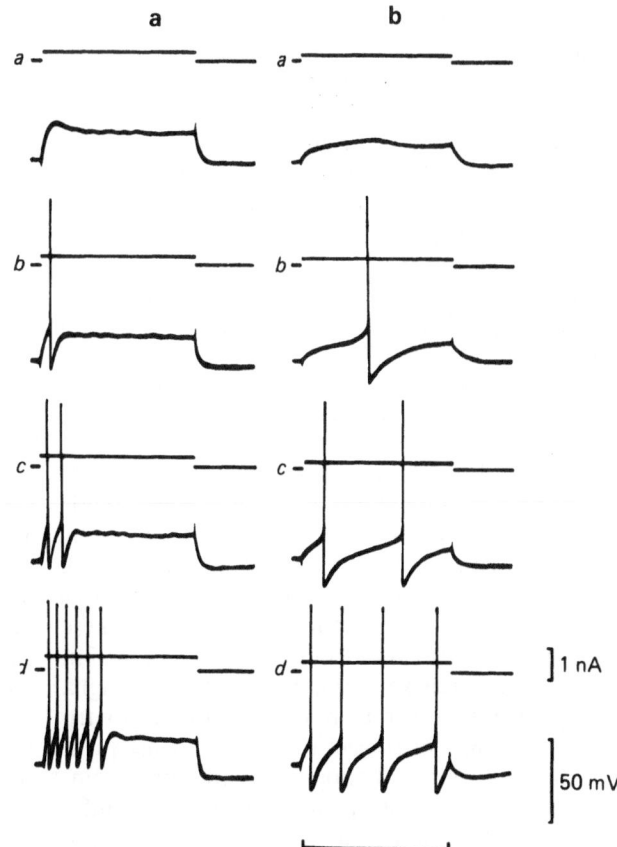

Fig. 5.16a,b. Responses of different guinea pig sympathetic neurons to depolarizing current passed through the recording microelectrode. **a** Cell in lumbar sympathetic ganglion L-5; **b** cell in the distal lobe of the inferior mesenteric ganglia. Magnitude of current increased progressively from *a* (just subthreshold) to *d*. The neuron in **a** discharges phasically at the onset of the current pulse, while the neuron in **b** fires tonically throughout the passage of current. (Cassell et al. 1986)

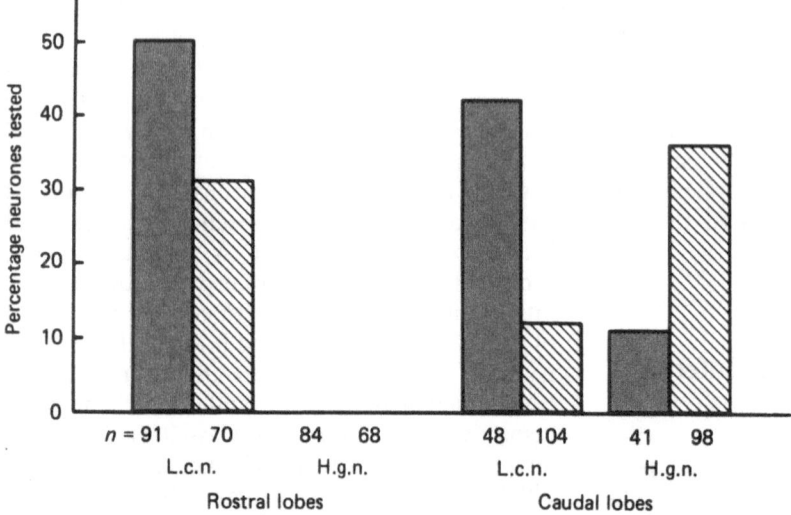

Fig. 5.17. Recordings were made from cat inferior mesenteric ganglion cells in the rostral and caudal poles of the ganglion using intracellular microelectrodes. The graph shows the percentage of nonspontaneous ▨ and irregular discharging ◩ neurons in the rostral and caudal lobes which responded with antidromic responses to stimulation of the indicated peripheral nerves. Height of columns represents percentage of cells which exhibited responses during stimulation of the different peripheral nerves. Although not shown, none of the regular discharging neurons in the rostral lobes responded with antidromic responses. *L.c.n.*, lumbar colonic nerve; *H.g.n.*, left and right hypogastric nerves. (Julé et al. 1983)

inate from the distal lobe of the inferior mesenteric ganglion (Fig. 5.17; Julé et al. 1983; Baron et al. 1985). They compared the membrane properties of inferior mesenteric ganglion cells from the distal lobe of the ganglion and neurons from the caudal lumbar sympathetic ganglia, and found that the former were usually tonic neurons, whereas the latter vasoconstrictor neurons usually exhibited phasic properties. The differences in properties can be attributed to differences in the occurrence of voltage-dependent potassium channels present in their cell membranes.

The classification of postganglionic sympathetic neurons in the inferior mesenteric ganglia of the guinea pig by Cassell et al. (1986) was based on the responses of these neurons to current injection; an alternative method of classification of sympathetic postganglionic neurons in the cat was described by Julé and Szurszewski (1983) and Julé et al. (1983), and utilized the spontaneous patterns of discharge as well as the effects of current injection in these neurons. These results are described in Chapter 7 in the section on reflexes through prevertebral ganglia (see p. 217). It should be remembered that in vitro techniques may be subject to artefacts consequent on the lack of blood supply and the ionic composition of bathing media (which may differ between laboratories), and sometimes of hypothermia, all of which may affect the responses of neurons. This is not intended as criticism, but a note of caution is appropriate in interpreting results of this nature, particularly when making comparisons between different laboratories or between in vivo and in vitro preparations.

Sympathetic Neuroeffector Transmission

Bladder

The influence of sympathetic stimulation is a transient rise followed by a prolonged fall in intravesical pressure and a suppression of motility. The main effect on the smooth muscle is inhibitory and is more powerful than the indirect inhibitory effect exerted on the pelvic ganglia (de Groat and Saum 1972); the direct effect on bladder smooth muscle is mediated by β-adrenoceptors. The initial rise in bladder pressure has been attributed to excitation of pelvic parasympathetic ganglion cells by catecholamines, as the effect is blocked by phentolamine. Ligand-binding studies show that β-receptors are more numerous than α-receptors in the rabbit bladder; in the rat approximately 20% of the β_2-adrenoceptors are in the mucosa (Morrison et al. 1986).

It is interesting that stimulation of the hypogastric nerve following transection and degeneration of the pelvic nerves results in a prolonged bladder contraction. It has been suggested that this is because of changes in the adrenoceptor innervation and of the populations of adrenoceptors present in the vesical ganglia (Sundin 1972). Normally α-adrenoceptors are present at this site and there is some evidence that, at least in some pelvic ganglia, the hypogastric innervation can excite parasympathetic postganglionic neurons. Some paraplegic patients are known to have damage to the lower segments of the spinal cord, even though the level of the lesion is much higher (Thomas 1984; see Chap. 7, p. 224). In some of these patients anomalous activity of the bladder can be seen, and can be diminished by administration of α-adrenergic antagonists. In this situation, there can be a decrease in residual volume and urethral resistance, and an increase in the bladder volume (Norlén and Sundin 1978).

Urethra

The human urethra has a sparse innervation with adrenergic nerve terminals (Ek et al. 1977), except in the preprostatic part, which is densely innervated by noradrenaline-containing fibres (Gosling et al. 1977). The guinea pig appears to have a visible internal urethral sphincter, innervated by a dense plexus of adrenergic fibres (Ulmsten et al. 1977); other laboratory animals are known to have a rich supply of sympathetic postganglionic terminals, and in the cat it is known that a proportion of these arise from the sacral sympathetic ganglia and travel in the pelvic nerves (Downie et al. 1984). The sympathetic nerves are thought to contract the urethra, and contraction of the preprostatic part of the urethra is an essential component of ejaculation. There is some doubt whether this part of the urethra has any function in maintaining continence as surgical resection of this area does not usually give rise to incontinence.

Summary

The lower urinary tract receives sympathetic efferent innervation from the pre- and postganglionic fibres in the hypogastric nerve, and the postganglionic sympathetic efferents of the pelvic nerve (see Chap. 3, p. 77). Postganglionic neurons in the inferior mesenteric ganglia can be divided into different types on the basis of their responses to injected currents, or on the degree of regularity or irregularity of their discharges. Some neurons have synaptic inputs from preganglionic fibres and from afferent collaterals, and the excitatory effects are mediated by acetylcholine, which acts on nicotinic receptors. In addition, there is evidence for facilitation of ganglionic transmission by means of a slow EPSP, that appears to be mediated by substance P, possibly released from afferent collaterals (see Chap. 7, p. 217). The postganglionic noradrenergic fibres have direct effects on bladder smooth muscle mediated by β-adrenoceptors, and less powerful α-adrenoceptor actions on pelvic ganglia. The effect of hypogastric nerve stimulation is to produce a transient contraction of the bladder, followed by a prolonged relaxation. The latter is mediated by β-adrenoceptors on the smooth muscle, and the former may be due to excitation of cholinergic postsynaptic elements in the pelvic ganglia.

Neurotoxins that Influence Bladder Function

Kreiss et al. (1980) have reported that a catalyst that has been used in the production of polyurethane foam has toxic effects on the nerves of the bladder. Contact with diethylaminopropionitrile (DMAPN) gives rise to sensory and motor abnormalities which affect particularly the nerves to the bladder, rather than somatic peripheral neuropathies. The workers who were in contact with this material developed hesitancy and a need to strain.

Capsaicin is known to modify the reflex activity of the bladder when acutely administered to animals. The immediate effects include hyperactive reflexes, and the toxin leaves diminished reflexes as a residual effect; these changes may be due to the loss of substance P from these nerve endings (Santicioli et al. 1985; Maggi et al. 1986).

Clinical Pharmacology of the Lower Urinary Tract

There have been three recent reviews on this subject by Andersson and Sjögren (1982), Caine (1984) and Wein (1984). This is an area in which empirical approaches have dominated the scene, and there are relatively few high quality, controlled, clinical trials. Both reviews deal with specific questions about what might be done for patients with different groups of symptoms and/or urodynamic findings, and stress the need for proper investigation of the patients. This topic is dealt with in Section 3 (see p. 275).

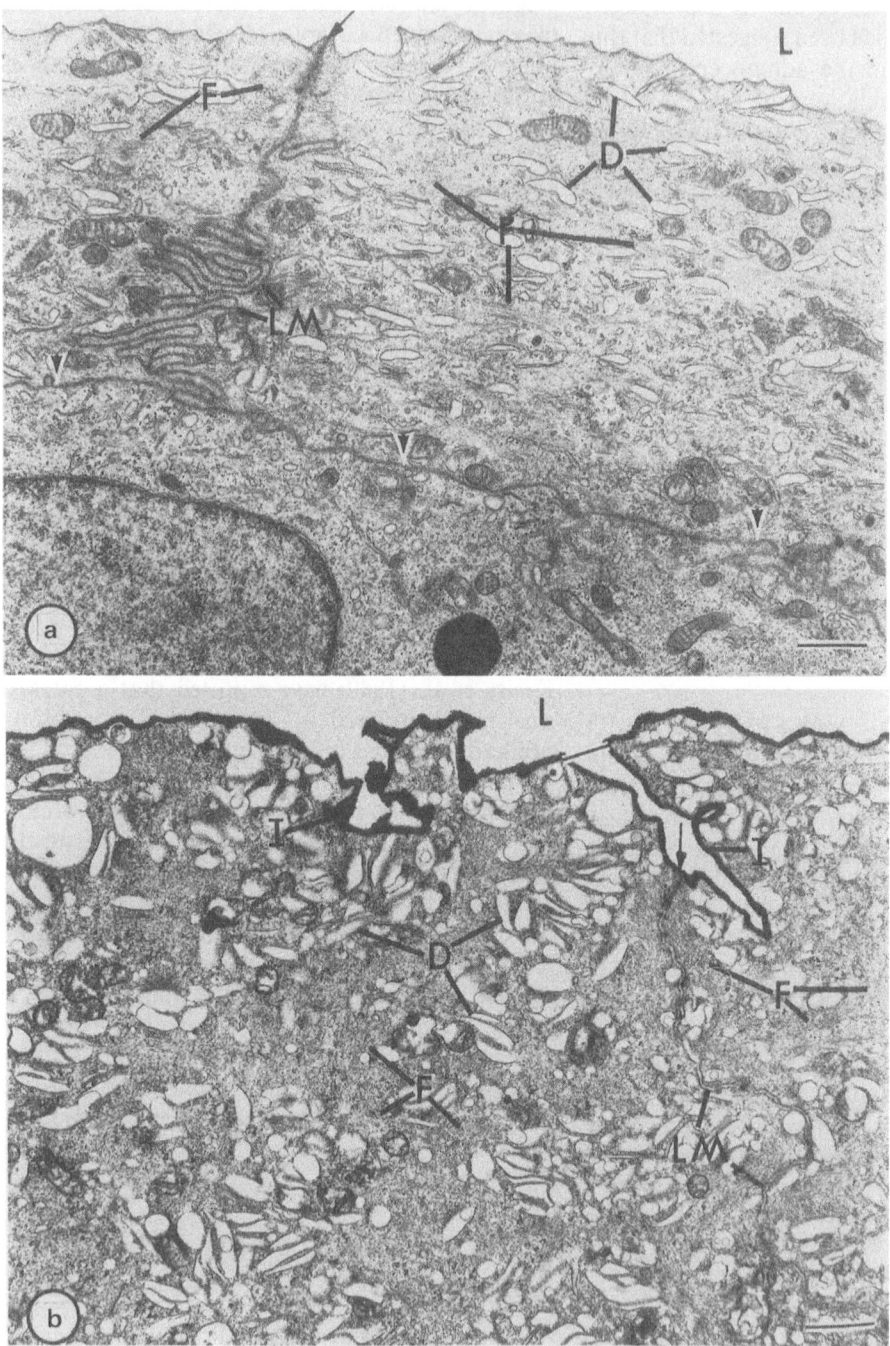

Fig. 5.18. a An electron micrograph of transitional epithelial cells of an expanded gerbil urinary bladder. Apical transitional epithelial cells from expanded gerbil bladder demonstrate characteristics similar to those from rat bladder. These properties include (a) an orientation of groups of cytoplasmic filaments (*F*) that are approximately parallel to the lumenal surface; (b) an association of the filaments with the junctional complex (*arrow*); (c) highly folded lateral membranes (*LM*); and (d) an unfolded border composed of the basal plasma membrane of the apical cells and the apical plasma membrane of the intermediate cells (*arrowheads*). Note the discoidal vesicles (*D*) composing the residual pool. *L*, lumen (× 11 275). **b** An electron micrograph of apical transitional epithelial cells of contracted gerbil urinary bladder stained en bloc with ruthenium red. Apical cells from contracted gerbil bladders appear similar to those from rat bladders. Ruthenium red-stained infoldings of lumenal membrane (*L*) are observed. An increase in the number of discoidal vesicles (*D*) seems apparent relative to the expanded state. The filamentous network (*F*) is reorientated to a seemingly random pattern and can be seen throughout the cytoplasm of the field shown. Note the unfolded appearance of the lateral membranes (*LM*) between adjacent apical cells. *I*, infolded membrane; *arrow*, junctional complex (× 11 275). (Minsky and Chlapowski 1978)

Physiology of Epithelia in the Lower Urinary Tract: Possible Influences of Hormones and Neurotransmitters

Structure of Uroepithelium

In a recent review, Lewis (1986) challenged the idea that the epithelium of the mammalian bladder is an inert barrier, and focused attention on some of the dynamic, living, characteristics of this tissue. The transitional epithelium is composed of three layers: the deepest layer is composed of small basal cells (5–10 μm diameter) attached to the basement membrane by structures similar to hemidesmosomes; the intermediate layer is composed of larger cells (20–30 μm diameter) that contain lysosomes and cytoplasmic vesicles resembling those of the superficial cell layer; and the superficial layer itself, which is composed of large hexagonal cells (50–100 μm diameter, derived by fusion of intermediate cells) with a dense packing of cytoplasmic vesicles and large lysosomes. Sometimes the superficial cells are binucleated. The apical membrane of these superficial cells appears to be unique. There appears to be a system of cytoplasmic vesicles in the apical portions of the surface epithelium; about 78% of the surface of the apical epithelium is made up of rigid plaques which are joined together to form vesicles, and the remainder of this surface is a normal flexible lipid bilayer (Fig. 5.18).

The vesicles are connected by cytoplasmic filaments that coalesce at the tight junctions between neighbouring cells. These vesicles appear to be incorporated into the apical membrane of the epithelial cells during bladder distension (Fig. 5.19). Microfilaments within the cell appear to be respon-

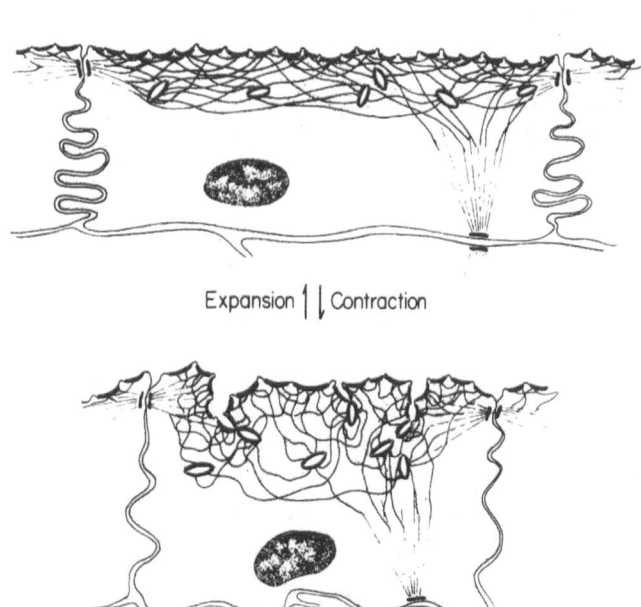

Expansion ↑↓ Contraction

Expansion ↑↓ Contraction

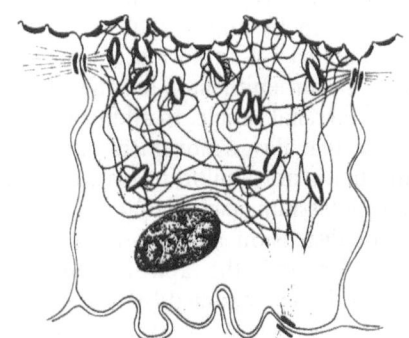

Fig. 5.19. A highly simplified diagram of the changes in morphology of an apical transitional epithelial cell during the expansion–contraction cycle. The hypothesized lumenal membrane-filament-discoidal vesicle interaction is illustrated. (Minsky and Chlapowski 1978)

sible for the translocation of the vesicles into apical membrane, as the process can be blocked by cytochalasin B or D, but not by colchicine, which would interfere with the microtubular system. The microtubular system of the cell does have a part to play in the maintenance of cellular geometry (Minsky and Chlapowski 1978). Lewis (1986) considers all uroepithelia to be similar; however, Morrison et al. (1986) found there to be differences in the density of β_1-adrenergic binding sites between the base and the dome of the bladder. β_1-Adrenoceptors were plentiful at the bladder neck, but absent from the dome of the moderately distended bladder; β_2-adrenoceptors are present in both sites, but in lower density than the β_1-subtype. It is known that sympathetic reflexes are active during bladder distension (de Groat and Lalley 1972; Floyd et al. 1982; Jänig 1986), and whether adrenergic mechanisms, acting on epithelial β-adrenoceptors, contribute to the changes observed in the epithelial distribution of vesicles is unknown.

Barrier Functions and Transport Mechanisms in Bladder Mucosa

Hicks (1976) concluded that the mammalian bladder, in vivo, is relatively impermeable to electrolytes and other small molecules, and lacks active transport systems. Clearly, the bladder can maintain an osmotic concentration of about four times the plasma osmolarity in humans, and can function as an even more effective barrier in animals that concentrate their urine to higher osmolarities. It is therefore interesting to find that in vitro studies of the mammalian uroepithelium demonstrate the presence of active transport processes in these cells. In these studies, the epithelium was dissected from the muscle and mounted in a chamber which allows the measurement of surface area, resistance, transmembrane voltages, and active transport of sodium in the presence of different environmental electrolyte concentrations. The average membrane transport of sodium is around 21 pequiv. $cm^2 \ s^{-1}$. The value is increased by 100%–200% by aldosterone and is inhibited by amiloride or ouabain. In these respects, the uroepithelium has the characteristics of the distal renal tubule. In addition, there is an ionic channel that allows movement of sodium and potassium, but is not amiloride sensitive, and another sodium/potassium channel that is unstable. During cycles of distension and contraction, the cytoplasmic vesicles and the apical membrane exchange, and this is accompanied by an eightfold increase in amiloride-sensitive sodium transport, and a 2.5-fold increase in the other leak pathways.

The amiloride-sensitive pathway is sensitive to the presence of endogenous enzymes, and can be degraded by serine proteases such as urokinase, kallikrein, and plasmin. These can be released by the distal nephron, and kallikrein can probably be released in the bladder itself during inflammation. These enzymes are seen as able to convert a normal sodium-selective channel into a nonselective one, and ultimately into a subunit that can partition between a hydrophobic and hydrophilic phase. The explanation provided by Lewis (1986) for the discrepancy between in vivo and in vitro results is in the probable presence of these deactivating enzymes in the in vivo studies. This suggests a role for these enzymes in a physiological situation; whether or not this is the case has yet to be proven.

Hormonal Influences

It is clear from this discussion that aldosterone can increase the transport of sodium across the uroepithelium by a factor of 2–3. Little is known about the actions of other drugs or neurotransmitters, but the catecholamine receptors are present in these epithelia; the role of these receptors in the modulation of transport functions, or in the production of a mucopolysaccharide layer on the mucosal surface (Mooreville et al. 1983) is unknown. One other hormone that might be mentioned in this context is oestrogen, that has long been known to be of use in the treatment of urinary frequency in older women; this hormone is said to help because of its action on atrophic epithelia in the lower urogenital tract.

Summary

Uroepithelium may not be as inert as has been thought in the past. There is evidence of specialized membranes in the surface cells that become incorporated into the surface as the bladder is distended, and the properties of the surface cells have some similarities with those of the distal renal tubule, e.g. in their responses to aldosterone.

References

Akasu T, Gallagher JP, Hirai K, Shinnick-Gallagher P (1986) Vasoactive intestinal polypeptide depolarisations at cat bladder parasympathetic ganglia. J Physiol 374:457–474

Ambache N, Zar MA (1970) Non-cholinergic transmission by postganglionic motor neurones in the mammalian bladder. J Physiol 210:761–784

Andersson KE, Sjögren C (1982) Aspects of the physiology and pharmacology of the bladder and urethra. Prog Neurobiol 19:71–89

Andersson KE, Husted S, Sjögren C (1980) Contribution of prostaglandins to the adenosine triphosphate induced contraction of the rabbit urinary bladder. Br J Pharmacol 70:443–452

Andersson PO, Bloom SR, Mattiasson A, Uvelius B (1985) Changes in the vascular resistance in the feline urinary bladder in response to bladder filling. J Urol 134:1041–1046

Appenzeller O (1982) The autonomic nervous system: an introduction to basic and clinical concepts, 3rd edn. Elsevier, New York, p 524

Baron R, Jänig W, McLachlan EM (1985) The afferent and sympathetic components of the lumbar outflow to the colon and pelvic organs in the cat. II. The lumbar splanchnic nerves. J Comp Neurol 238:147–157

Blackman JG, Crowcroft PJ, Devine EE, Holman ME, Yonemura K (1969) Transmission from preganglionic fibres in the hypogastric nerve to peripheral ganglia of male guinea pigs. J Physiol 201:723–743

Blaivas JG, Labin KB, Bauer SB, Retik AB (1977) A new approach to electromyography of the external urethral sphincter. J Urol 117:773–777

Bors E, Comarr AE (1971) Neurological urology. Karger, New York

Brindley GS, Craggs MD (1976). The effect of atropine on the urinary bladder of the baboon and of man. J Physiol 256:55P

Brindley GS, Rushton DN, Craggs MD (1974) The pressure exerted by the external sphincter of the urethra when its motor nerve fibres are stimulated electrically. Br J Urol 46:453–462

Burke RE (1981) Motor units: anatomy, physiology and functional organisation. In: Brookhart JM, Mountcastle VB (eds) Handbook of Physiology. Section 1, The nervous system, vol II, part 1. Waverley, Baltimore, pp 345–422

Burnstock G (1981) Neurotransmitters and trophic factors in the autonomic nervous system. J Physiol 313:1–35

Burnstock G (1986) The changing face of autonomic neurotransmission. Acta Physiol Scand 126:67–91

Burnstock G, Cocks T, Crowe R, Kasakov L (1978a) Purinergic innervation of the guinea pig urinary bladder. Br J Pharmacol 63:125–138

Burnstock G, Cocks T, Kasakov L, Wong HK (1978b) Direct evidence for ATP release from non-adrenergic, non-cholinergic ("purinergic") nerves in the guinea pig taenia coli and bladder. Eur J Pharmacol 49:145–149

Caine M (1984) The pharmacology of the lower urinary tract musculature. In: Chisholm GD, Williams DI (eds) Scientific foundations of urology, 2nd edn. Heinemann, London, pp 433–442

Callahan SM, Creed KE (1986) Non-cholinergic neurotransmission and the effects of peptides on the urinary bladder of guinea pigs and rabbits. J Physiol 374:103–115

Cardozo L, Stanton S, Robinson H, Hole D (1980) Evaluation of flurbiprofen in detrusor instability. Br Med J 280:281–282

Cassell JF, Clark AL, McLachlan EM (1986) Characteristics of phasic and tonic sympathetic ganglion cells in the guinea pig. J Physiol 372:457–484

Craggs MD, Stephenson JD (1981) Sympathetic facilitation of parasympathetic pathways to the urinary bladder. J Physiol 319:90–91P

de Groat WC (1975) Nervous control of the urinary bladder of the cat. Brain Res 87:201–211

de Groat WC (1976) Mechanisms underlying the recurrent inhibition of the sacral parasympathetic outflow to the urinary bladder. J Physiol 257:503–514

de Groat WC, Lalley PM (1972) Reflex firing in the lumbar sympathetic outflow to activation of vesical afferent fibres. J Physiol 226:289–309

de Groat WC, Ryall RW (1968) Recurrent inhibition in sacral parasympathetic pathways to the bladder. J Physiol 196:579–591

de Groat WC, Saum WR (1972) Sympathetic inhibition of the urinary bladder and of pelvic ganglionic transmission in the cat. J Physiol 220:297–314

de Groat WC, Saum WR (1976) Synaptic transmission in parasympathetic ganglia in the urinary bladder in the cat. J Physiol 256:137–158

de Groat WC, Nadelhaft I, Milne RJ, Booth AM, Morgan C, Thor K (1981) Organisation of the sacral parasympathetic reflex pathways to the urinary bladder and large intestine. J Auton Nerv Syst 3:135–160

de Groat WC, Booth AM, Milne RJ, Roppolo JR (1982) Parasympathetic preganglionic neurons in the sacral spinal cord. J Auton Nerv Syst 5:23–43

de Groat WC, Kawatani M, Hisamitsu T, Lowe I, Morgan C, Roppolo J, Booth AM, Nadelhaft I, Kuo D, Thor K (1983) The role of neuropeptides in the sacral autonomic reflex pathways of the cat. J Auton Nerv Syst 7:339–350

Denny-Brown D, Robertson EG (1933) On the physiology of micturition. Brain 56:149–190

de Sy W, Lacroix E, Leusen I (1974) An analysis of the urinary bladder response to hypogastric nerve stimulation in the cat. Invest Urol 11:508–516

Downie JW, Dean DM (1977) The contribution of cholinergic postganglionic neurotransmission to contractions of the rabbit detrusor. J Pharmacol 203:417–425

Downie JW, Champion JA, Nance DM (1984) A quantitative analysis of the afferent and extrinsic efferent innervation of specific regions of the bladder and urethra in the cat. Brain Res Bull 12:735–740

Eccles JC, Fatt P, Koketsu K (1954) Cholinergic and inhibitory synapses in a pathway from motor axon collaterals to motoneurones. J Physiol 126:524–562

Elbadawi A, Schenk EA (1968) A new theory of the innervation of bladder musculature. Part 1. Morphology of the intrinsic vesical innervation apparatus. J Urol 99:585–587

Elbadawi A, Schenk EA (1974) A new theory of the innervation of bladder musculature. part 4. Innervation of the vesicourethral junction and external urethral sphincters. J Urol. 111:613–615

Ek, A, Alm P, Andersson K-E, and Persson CGA (1977) Adrenergic and cholinergic nerves of the human urethra and urinary bladder. Acta Physiol Scand 99:345–352

Ek A, Andersson K-E, Ulmsten U (1978) The effects of norephedine and bethanecol on the human urethral closure pressure profile. Scand J Urol Nephrol 12:97–104

Elliot TR (1907) The innervation of the bladder and the urethra. J Physiol 35:367–445

Fehér E, Vajda J, Csanyi K (1980) Quantitative analysis for innervation of the smooth muscle cells of the wall of the urinary bladder. J Auton Nerv Syst 2:71–80

Floyd K, Hick VE, Morrison JFB (1982) The influence of visceral mechanoreceptors on sympathetic efferent discharge in the cat. J Physiol 323:65–75

Fowler CJ, Kirby RS, Harrison MJG, Milroy EJG, Turner-Warwick R (1984) Individual motor unit analysis in the diagnosis of disorders of urethral sphincter innervation. J Neurol Neurosurg Psychiatry 47:637–641

Fränkl-Hochwart LV, Frölich A (1904) Über die cortikale Innervation der Harnblase. Neurol Zentralbl (Leipzig) 23:646 655

Gabella G (1976) Structure of the autonomic nervous system. Chapman and Hall, London, p 214

Garry RC, Roberts TDM, Todd JK (1959) Reflexes involving

the external urethral sphincter in the cat. J Physiol 149:653–665

Gibson SJ, Polak JM, Anand P, Blank MA, Morrison JFB, Kelly JS, Bloom SR (1984a) The distribution and origin of VIP in the spinal cord of six mammalian species. Peptides 5:201–207

Gibson SJ, Polak JM, Bloom SR, Sabate IM, Mulderry PM, Ghatei MA, McGregor GP, Morrison, JFB, Kelly JS, Evans RM, Rosenfield MG (1984b) Calcitonin gene-related peptide immunoreactivity in the spinal cord of man and eight other species. J Neurosci 4:3101–3111

Gibson SJ, Polak JM, Anand P, Blank MA, Yiangou Y, Su HC, Terenghi G, Katagiri T, Morrison JFB, Lumb BM, Inyama C, Bloom SR (1986) A VIP/PHI-containing pathway links urinary bladder and sacral spinal cord. Peptides 7 (Suppl 1): 205–219

Gonella J, Bouvier M, Blanquet F (1987) The extrinsic innervation of small and large intestines, and of related sphincters. Physiol Rev (in press)

Gosling JA, Dixon JS, Lendon KG (1977) The autonomic innervation of the human male and female bladder neck and proximal urethra. J Urol 118:302–305

Griffith WH, Gallagher JP, Shinnick-Gallagher P (1980) An intracellular investigation of the cat vesical pelvic ganglia. J Neurophysiol 43:343–354

Hamberger B, Norberg KA (1965) Adrenergic synaptic terminals and nerve cells in bladder ganglia of the cat. Int J Neuropharmacol 4:41–45

Henderson VE, Roepke MH (1934) The role of acetylcholine in bladder contractile mechanisms and in parasympathetic ganglia. J Pharmacol Exp Ther 51:97–111

Hicks RM (1976) The mammalian urinary bladder: an accommodating organ. Biol Rev 50:215–246

Hills J, Meldrum L, Klarskov P, Burnstock G (1984) A novel non-adrenergic, non-cholinergic nerve-mediated relaxation of the pig bladder neck: an examination of possible neurotransmitter candidates. Eur J Pharmacol 90:287–293

Hodson N (1965) Sympathetic nerves and reproductive organs in the male rabbit. J Reprod Fertil 10:209–220

Hökfelt T, Everett B, Meister B, Melander T, Schalling M, Johansson O, Lundberg JM, Hulting A-L, Werner S, Cuello C, Hemmings H, Ouimei C, Walaas I, Greengard P, Goldstein M (1986) Neurons with multiple messengers, with special reference to neuroendocrine systems. Recent Prog Horm Res 42:1–70

Hoyle CHV, Burnstock G (1985) Atropine resistant excitatory junction potentials in rabbit bladder are blocked by a,b-methylene ATP. Eur J Pharmacol 114:239–240

Hulsebosch CE, Coggeshall RE (1982) An analysis of the axon populations in the nerves to the pelvic viscera in the rat. J Comp Neurol 211:1–10

Jan LY, Jan YN (1982) Peptidergic transmission in sympathetic ganglia of the frog. J Physiol 327:219–246

Jänig W (1986) Spinal cord integration of visceral sensory systems and sympathetic nervous system reflexes. In: Cervero F, Morrison JFB (eds) Visceral sensation. Elsevier, Amsterdam, pp 255–277 (Progress in brain research, vol 67)

Jänig W, McLachlan E (1987) Organisation of the lumbar sympathetic and afferent supply of colon and pelvic organs. Physiol Rev (in press)

Jankowska E, Padel Y, Zarzecki P (1978) Crossed disynaptic inhibition of sacral motoneurones. J Physiol 285:425–444

Johns A (1979) The effect of vasoactive intestinal polypeptide on the urinary bladder and taenia coli of the guinea pig. J Physiol Pharmacol 57:106–108

Johns A (1983) Alpha- and beta-adrenergic and muscarinic cholinergic binding sites in the bladder and urethra of the rabbit. Can J Physiol Pharmacol 61:61–66

Johnson RH, Spalding JMK (1974) Disorders of the autonomic nervous system. Blackwell, Oxford, p 300

Julé Y, Szurszewski JH (1983) Electrophysiology of neurones of the inferior mesenteric ganglion of the cat. J Physiol 344:277–292

Julé Y, Krier J, Szurszewski JH (1983) Patterns of innervation of neurones in the inferior mesenteric ganglion of the cat. J Physiol 344:293–304

Kawatani M, de Groat WC (1982) Enephalinergic inhibition in parasympathetic ganglia of the cat urinary bladder. Soc Neurosci Abstr 8:552

Kawatani M, Lowe IP, Booth AM, Backes MG, Erdman SL, de Groat WC (1983) The presence of leucine-enkephalin in the sacral preganglionic pathway to the urinary bladder of the cat. Neurosci Lett 39:143–148

Kawatani M, Rutigliano M, de Groat WC (1985) Selective facilitatory effect of vasoactive intestinal polypeptide (VIP) on muscarinic firing in vesical ganglia of the cat. Brain Res 336:223–234

Klarskov P, Gerstenberg T, Hald T (1984) Vasoactive intestinal polypeptide influence on the lower urinary tract smooth muscle from human and pig. J Urol 131:1000–1004

Kreiss K, Wegman DH, Niles CA, Siroky MB, Krane RJ, Feldman RG (1980) Neurological dysfunction of the bladder in workers exposed to dimethylaminopropionitrile. J Am Med Assoc 243:741–745

Kreulen DL, Peters S (1986) Non-cholinergic transmission in a sympathetic ganglion of the guinea pig elicited by colon distension. J Physiol 374:315–334

Kuo DC, Hisamitsu T, de Groat WC (1984) A sympathetic projection from sacral paravertebral ganglia to the pelvic nerve and to postganglionic nerves on the surface of the urinary bladder and large intestine of the cat. J Comp Neurol 226:76–86

Langley JN, Anderson HK (1895a) On the innervation of the pelvic and adjoining viscera. Part I. The lower portion of the intestine. J Physiol 18:67–105

Langley JN, Anderson HK (1895b) On the innervation of the pelvic and adjoining viscera. Part II. The bladder. J Physiol 19:71–84

Langley JN, Anderson HK (1895c) On the innervation of the pelvic and adjoining viscera. Part IV. The internal generative organs. J Physiol 19:122–130

Langley JN, Anderson HK (1895d) On the innervation of the pelvic and adjoining viscera. Part V. Position of the nerve cells on the course of the efferent nerve fibres. J Physiol 19:131–139

Larsson B (1983) Demonstration of alpha-adrenoceptors in the rabbit bladder base and urethra with H3-dihydroergocryptine ligand binding. Acta Pharmacol Toxicol 52:188–194

Lepor H, Kuhar MJ (1984) Characterisation of muscarinic cholinergic receptor binding in the vas deferens, bladder, prostate and penis of the rabbit. J Urol 132:392–396

Levin RM, Wein AJ (1979) Quantitative analysis of alpha and beta adrenergic receptor densities in the lower urinary tract of the dog and rabbit. Invest Urol 17:75–77

Levin RM, Wein AJ (1981) Effect of vasoactive intestinal polypeptide on the contractility of the rabbit urinary bladder. Urol Res 9:217–218

Lewis SA (1986) The mammalian urinary bladder: it's more than accommodating. News Physiol Sci 1:61–65

Love JA, Szurszewski JH (1985) Effects of vasoactive intestinal polypeptide on neurons of the guinea pig inferior mesenteric ganglion. Fed Proc 44:1718

Lundberg JM, Hökfelt T (1983) Coexistence of peptides and classical neurotransmitters. Trends Neurosci 6:325–333

Lundberg JM, Hökfelt T, Schultzberg M, Uvnas-Wallensten K, Kohler L, Said S (1979) Occurrence of VIP-like immunoreactivity in cholinergic neurones of the cat: evidence from combined immuno-histochemistry and acetylcholinesterase staining. Neuroscience 4:1539–1559

Mackel R (1979) Segmental and descending control of the external urethral and anal sphincters in the cat. J Physiol 294:105–122

Mackenzie I, Burnstock G (1984) Neuropeptide action on the guinea pig bladder: a comparison with the effects of field stimulation and ATP Eur J Pharmacol 105:85–94

Maggi CA, Santicioli P, Borsini F, Giuliani S, Meli A (1986) The role of capsaicin-sensitive innervation of the rat urinary bladder in the activation of micturition reflex. Naunyn-Schmiedebergs Arch Pharmacol 332:276–283

Martin AR, Pilar G (1963) Dual mode of synaptic transmission in the avian ciliary ganglion. J Physiol 168:443–463

Melzak J, Porter NH (1964) Studies of the reflex activity of the external sphincter ani in spinal man. Paraplegia 1:277–296

Minsky BD, Chlapowski FJ (1978) Morphometric analysis of the translocation of lumenal membrane between cytoplasm and cell surface of transitional epithelial cells during the expansion-contraction cycles of mammalian urinary bladder. J Cell Biol 77:685–697

Mooreville M, Fritz RW, Mulholland SG (1983) Enhancement of the bladder defense mechanism by an exogenous agent. J Urol 130:607–609

Morgan C, Nadelhaft I, de Groat WC (1981) The distribution of visceral primary afferents from the pelvic nerve within Lissauer's tract and the spinal gray matter and its relationship to the sacral parasympathetic nucleus. J Comp Neurol 201:415–440

Morrison JFB (1982) The neural control of the bladder. In: Bloom SR, Polak JM, Lindenlaub E (eds) Systemic role of regulatory peptides. Schattauer, Stuttgart, pp 381–396

Morrison JFB, Nimmo AJ, Whitaker EM (1986) The localization of beta-adrenoceptor subtypes in the rat urinary bladder. J Physiol 381:29P

Nadelhaft I, de Groat WC, Morgan C (1980) Location and morphology of parasympathetic preganglionic neurons in the sacral spinal cord of the cat revealed by retrograde axonal transport of horseradish peroxidase. J Comp Neurol 193:265–281

Nilsson S (1983) Autonomic nerve function in the vertebrates. Springer, Berlin Heidelberg New York

Nilvebant L, Sparf B (1983a) Muscarinic receptor binding in the guinea-pig urinary bladder. Acta Pharmacol Toxicol 52:30–38

Nilvebrant L, Sparf B (1983b) Differences between binding affinities of some antimuscarinic drugs in the parotid gland and those in the urinary bladder and ileum. Acta Pharmacol Toxicol 53: 304–313

Norlén L, Sundin T (1978) Alpha-adrenolytic treatment in patients with autonomous bladders. Acta Pharmacol Toxicol 43:31–34

Otsuka M, Konishi S (1983) Substance P—the first peptide neurotransmitter? Trends Neurosci 6:317–320

Parks AG, Swash M, Urich H (1977) Sphincter denervation in anorectal incontinence and rectal prolapse. Gut 18:656–665

Peters S, Kruelen DL (1984) A slow EPSP in mammalian inferior mesenteric ganglion persists after in vivo capsaicin. Brain Res 303:186–189

Ranvier L (1874) De quelques faits relatifs à l'histologie et à la physiologie des muscles striés. Arch Physiol Norm Pathol 1:5–18

Risling M, Dahlsgaard C-J, Cukierman A, Cuello AC (1984) Electronmicroscopic and immunohistochemical evidence that unmyelinated ventral root axons make U-turns or enter the spinal pia mater. J Comp Neurol 225:53–63

Santicioli P, Maggi CA, Meli A (1985) The effect of capsaicin pretreatment on the cystometrograms of urethane anaesthetised rats. J Urol 133:700–703

Sibley GNA (1984) A comparison of spontaneoous and nerve mediated activity in bladder muscle from man, pig and rabbit. J Physiol 354:431–444

Simmonds WF, Booth AM, Thor KB, Ostrowski NL, Nagel JR, de Groat WC (1983) Parasympathetic ganglia: naloxone antagonises inhibition by leucine-enkephalin and GABA. Brain Res 271:365–370

Sjögren C, Andersson K-E, Husted S, Mattiasson A, Moller-Madsen B (1982) Atropine-resistance of the transmurally stimulated, isolated human bladder. J Urol 128:1368–1371

Sjöstrand NO (1965) The adrenergic innervation of the vas deferens and the necessary male genital glands. Acta Physiol Scand 65: (suppl 257) 1–82

Sundin T (1972) Reinnervation of the urinary bladder. Scand J Urol Nephrol [Suppl] 17:1–25

Sundin T, Petersén I (1975) Cystometry and simultaneous electromyography from the striated urethral and anal sphincters and from levator ani. Invest Urol 13:40–46

Swash M (1982) Early and late components in the human anal reflex. J Neurol Neurosurg Psychiatry 45:767–769

Taira N (1972) The autonomic pharmacology of the bladder. Annu Rev Pharmacol 12:197–208

Thomas DG (1984) The urinary tract following spinal cord injury. In: Chisholm GD, Williams DI (eds) Scientific foundations of urology. Heineman, London, pp 431–433

Torrens MJ (1978) Urethral sphincteric responses to stimulation of the sacral nerves in the human female. Urol Int 33:22–26

Ulmsten U, Sjöberg NO, Alm P, Andersson K-E, Owman C, Walles B (1977) Functional role of an adrenergic sphincter in the female urethra of the guinea pig. Acta Obstet Gynecol Scand 56:387–390

Wamsley JK, Gehlert DR, Roeske WR, Yamamura HI (1984) Muscarinic antagonist binding site heterogenicity as evidenced by autoradiography after direct labelling with 3H-QNB and 3H-pirenzepine. Life Sci 34:1395–1402

Wein AJ (1984) Pharmacology of the bladder and urethra. In: Mundy AR, Stephenson TP, Wein AJ (eds) Urodynamics. Churchill Livingstone, Edinburgh, pp 26–41

Womack NR, Morrison JFB, Williams NS (1986a) Impaired recruitment of the pelvic floor musculature by intra-abdominal pressure in faecal incontinence. Gut 27:A624

Womack NR, Morrison JFB, Williams NS (1986b) The role of pelvic floor denervation in the aetiology of idiopathic faecal incontinence. Br J Surg 404–407

Wright AL, Williams NS, Gibson JS, Neal DE, Morrison JFB (1985) Electrically evoked activity in the human external anal sphincter. Br J Surg 72:38–41

Zuckerkandl E (1893) In: Eulenberg's Realenzyhlopädie der gesammten Heilkunde, 3rd edn, vol III, p 331

6 · Physiology of Bladder Smooth Muscle

Alison Brading

Structure and Organization of Bladder Smooth Muscle
Mechanical Properties of Bladder Smooth Muscle
Electrical Properties of Bladder Smooth Muscle
Contractile Mechanisms of Smooth Muscle
Intracellular Calcium in Smooth Muscle
Contractile Behaviour and Excitation–Contraction Coupling in Bladder Smooth Muscle
Neuromuscular Transmission in Smooth Muscle

An understanding of the basic physiological properties and innervation of the human bladder is likely to be the keystone for the design of rational pharmacological treatments of bladder dysfunction. Unfortunately studies on smooth muscles in general have lagged behind those on striated muscles, although the growth in our understanding has blossomed in the last two decades. It has become apparent that smooth muscles are extremely varied in their properties, and show many very interesting features which make them exciting to study in their own right. As a result of our increased knowledge, unexpected avenues have opened up which may lead to development of new clinical treatments.

Smooth muscles of the urinary tract have been less well studied than many other types, and although the contractile behaviour of strips of bladder smooth muscle has been used for some time in pharmacological studies, it is only recently that electrophysiological studies have been carried out on these tissues. To my knowledge there are no published values even of the membrane potential in human bladder smooth muscle. The problem is that there are too few collaborations between basic scientists and practising urologists, and therefore those with the experience and interest to make the desirable physiological studies often have little access to suitable human material. Clearly small mammals are the most convenient for laboratories to handle, and much work has been carried out on the rat, guinea pig and rabbit. Some of the earlier studies were on cat and dog bladder, but these animals are becoming very expensive to use. None of the bladders of these smaller mammals is very similar to human bladder, and therefore we have recently been studying pigs, since the pig bladder shares many features with the human. It also responds to outflow obstruction with hypertrophy and the development of instability, in a manner analogous to the human bladder (Sibley 1985).

This chapter will describe what is known of the properties of human bladder; however, because so much of the literature is concerned with animal work, this will also be included where appropriate. Much of our understanding of the behaviour of bladder smooth muscles is at present extrapolated from experiments that have been performed on other smooth muscles. Some sections of this chapter will therefore review the general properties of smooth muscle, so that what is described of the detailed behaviour of bladder can be seen in context.

Structure and Organization of Bladder Smooth Muscle

Gross Structure

The human bladder is conventionally divided into the dome (detrusor muscle), the trigone and the bladder neck. The trigone is delineated by the ureteric orifices and the internal urethral meatus. The smooth muscles from these three areas show differences both in their anatomical organization and in their behaviour. The bladder of the Landrace pig is about the same size as the human bladder, and the organization of its smooth muscle is also very similar. In smaller mammals, however, the trigone is not anatomically well defined, although the behaviour of the smooth muscle from the bladder base differs from that of the dome. All bladders are lined by a specialized mucosa or urothelium supported by a layer of loose connective tissue. The smooth muscle is separated by connective tissue from the parietal peritoneum in the dome and pelvic structures in the immobile bladder base. More experimental work has been published about the detrusor muscle than the trigone or bladder neck regions, and the detailed anatomy of the bladder neck and the proximal urethra is yet to be described in many species.

Arrangement of Smooth Muscle

Dome

The arrangement of the smooth muscle cells appears fairly uniform throughout the dome of the bladder. The individual muscle cells are grouped into well-defined bundles which can be seen under the dissecting microscope. In the human, and to a lesser extent in the pig, the muscle bundles are arranged in a complicated pattern, running and interdigitating with each other through most of the wall, so that it is difficult to dissect out long bundles. However, longitudinally arranged muscle bundles occur on the inner and outer aspects of the wall. In smaller mammals such as guinea pig and rabbit, the muscle bundles follow a simpler path, running parallel to each other from the base over the dome, so that it is relatively easy to dissect long strips of interconnected fibres. Most physiological and pharmacological studies have been carried out on such strips.

Electron microscopy reveals a clearer picture of the muscle bundles. Although most smooth muscles have their cells arranged in bundles, in bladder these are more discrete than in other tissues, and each bundle is surrounded by connective tissue containing much collagen. Electron micrographs of detrusor smooth muscle from pig and human bladder show bundles of up to a few millimetres in diameter, each composed of groups of smaller bundles with thin connective tissue between them, as shown in Fig. 6.1. The orientation of the larger bundles is, as expected, rather random (Sibley 1984a). In smaller animals the muscle bundles are much less complex, and are similar in size to the small "sub-bundles" seen in human and pig muscle. In the guinea pig, for example, Uvelius and Gabella (1980) describe bundles which are slightly elliptical

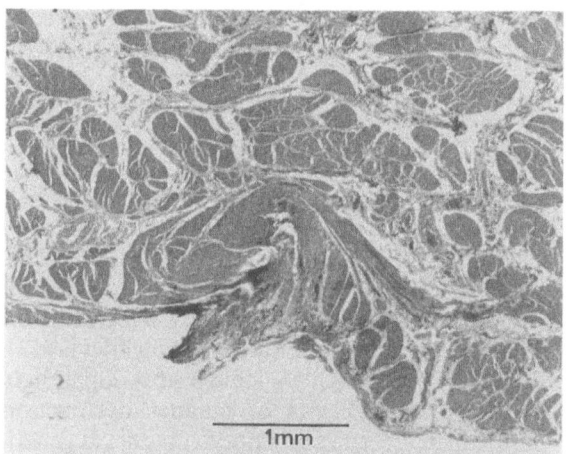

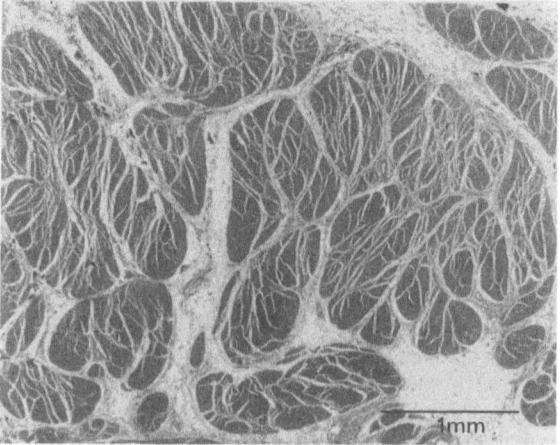

Fig. 6.1a, b, Low magnification cross section of muscle bundles; **a** from normal pig detrusor, **b** from normal human detrusor. Note similar arrangement of muscle cells into small bundles grouped together into larger bundles; these are divided from each other by connective tissue containing collagen.

in cross section, and with a diameter in relaxed tissue of about $100\,\mu$m at the widest point. One bundle may contain some 200–300 cells at any point along its length. Figure 6.2 shows a muscle bundle from the guinea pig bladder cut in cross section, and in near longitudinal section. Electron microscopy supports the impression that in these smaller animals most bundles run meridionally, but also reveals that there is variation in their orientation and that some of the bundles nearer the mucosa run equatorially.

With reference to the collagen fibres surrounding the muscle bundles, most of the fibres are arranged in parallel to the long axis of the cells at resting length. In many smooth muscles, however, when the muscle shortens, the collagen fibres become wound in helices around the cell. Smooth muscles are known to have synthetic activity, and to be able to lay down collagen in some circumstances. In hypertrophied smooth muscles, for instance, it has been suggested that the increased collagen present is formed by the smooth muscle cells themselves (for references see Burnstock 1981b), and this could account for the increased collagen seen in hypertrophied detrusor from patients with outflow obstruction.

Trigone

In the human bladder, the trigone is a triangular structure about the size of a large postage stamp. It consists of a superficial layer of smooth muscle

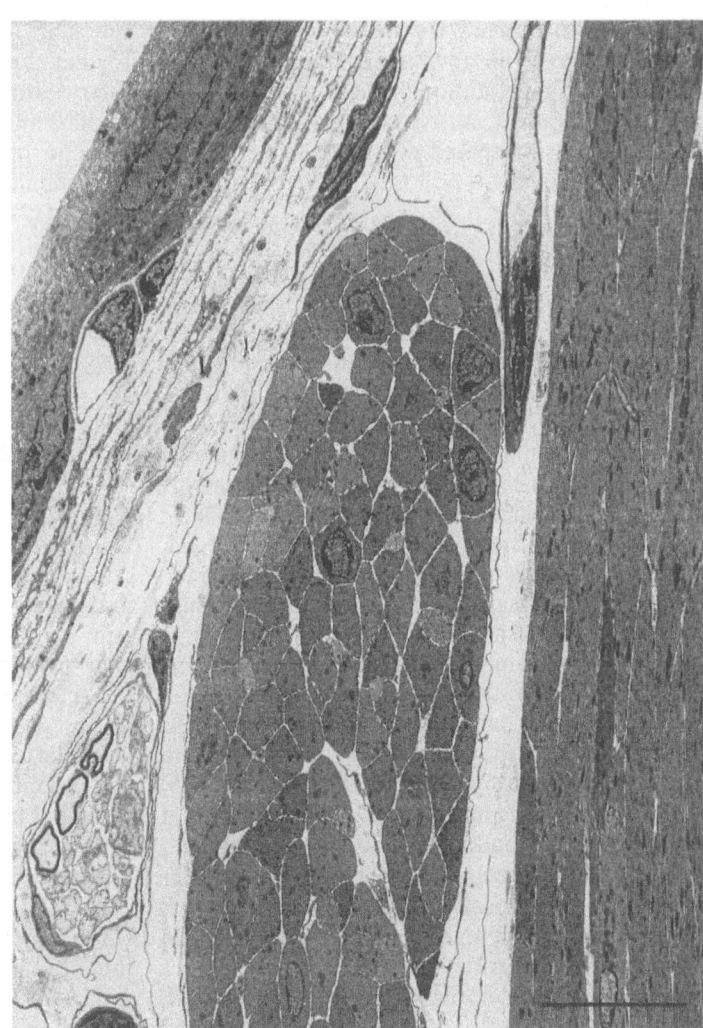

Fig. 6.2. Electron micrograph of guinea pig detrusor; the bladder was distended with 2 ml fluid before fixation. *Top left:* urinary epithelium and a capillary running close to it. *Bottom left:* nerve bundle containing several unmyelinated and three myelinated axons. *Centre:* muscle bundle in tranverse section and to the *right* another bundle in oblique section. *Calibration bar* = 10 μm. (Adapted from Uvelius and Gabella 1980)

under the mucosa, and a deep layer continuous with the detrusor. The upper edge of the trigone is particularly well seen through the mucosa, because of the presence of a thickening of the superficial trigone called the interureteric bar. When the mucosa is removed, the lateral thickenings of the superficial trigone can be more clearly seen. In these thickened areas, the muscle bundles run parallel to the thickenings, whereas towards the centre of the superficial trigone the bundles are more random in their orientation. In the trigone, the bundles are much smaller than in the detrusor, and discrete, like those in the dome of the guinea pig bladder. The deep trigone appears indistinguishable from the detrusor muscle under the dissecting microscope, although, as will be shown later, the behaviour of the tissues is somewhat different.

In the pig the trigone is smaller in size than in the human. The superficial trigone is much less well developed and contains more connective tissue. The deep trigone again microscopically resembles the detrusor muscle in structure, but shows different behavioural properties. In many of the smaller mammals there is no obvious structure resembling the human trigone, although there are differences in the behaviour of muscle strips taken from the bladder base as compared with the dome.

Bladder Neck and Urethra

The smooth muscle of the bladder neck is histologically, histochemically and pharmacologically distinct from the detrusor proper (Gosling et al. 1983; Klück 1980; Nergardh and Boreus 1972). In both sexes the muscle bundles are much smaller than in the detrusor. In the male these are mainly circularly orientated, so that there is a collar of smooth muscle around the bladder neck, contraction of which probably prevents retrograde ejaculation and may at times assist in the maintenance of continence. Distally these fibres merge with the musculature in the stroma and capsule of the prostate. In the female, the muscle bundles extend obliquely and longitudinally from the bladder neck into the urethral wall, and there is no true internal sphincter. Gosling and Dixon (1975) have also published an account of the structure of the bladder neck and urethra in cats, rabbits and guinea pigs. In these animals, again the bladder neck is morphologically distinct from the dome. The muscle bundles break up so that the cells run singly or in small groups separated by connective tissue with elastic fibrils. In males, the smooth muscle forms a well-developed circularly orientated

layer which extends along the urethra as far as the ejaculatory ducts. In females this layer is less well developed, but still can be identified along the length of the urethra.

Ultrastructure

Smooth Muscle Cells

Higher magnification reveals that bladder smooth muscle cells are very similar to other visceral smooth muscles. When relaxed, the cells have a smooth surface, and are polygonal in cross section. When the muscle is contracted, the cells become scalloped in appearance with an irregular surface.

The cell membranes contain many of the features characteristic of all smooth muscle cells. There are cavaeoli (flask-shaped invaginations of the membrane) and elements of subsurface sarcoplasmic reticulum often associated with them. There are regular thickenings apparent on the membrane at intervals around the cells which are known as dense bodies, and are thought to be sites for insertion of actin filaments. The cytoplasm contains many mitochondria and, with appropriate fixation, three types of filament can be seen: thick (15 nm diameter, myosin), thin (6–8 nm diameter, actin) and intermediate (10 nm diameter, desmin or vimentin). The cells are uninucleate with prominent central nuclei. For further details readers are referred to Gabella (1981).

Junctions Between Cells

Many cell-to-cell junctions link the smooth muscle cells. Processes from cells can regularly be seen contacting neighbouring cells, often forming peg-and-socket arrangements, with intermediate and gap junctions (Gabella 1981). Although Daniel et al. (1983) have not seen gap junctions between the cells in human detrusor muscle, these junctions are present in the guinea-pig and rabbit detrusor (Gabella 1981), and in rat and cat internal sphincter smooth muscle (Elbadawi 1982). Gap junctions are thought to consist of symmetrically arranged protein channels in each membrane, joined to provide a continuous pathway from one cell to the other, of dimensions such that ions and small water-soluble molecules can pass between the cells. Gap junctions interconnect the cells electrically, making groups of cells into a functional syncytium, so that current injected into one cell will spread throughout the bundle, and action potentials can propagate from cell to cell. Smooth muscle cells in which no gap junctions have been seen can also be shown to be electrically interconnected, and it is not yet known whether in life transient gap junctions are formed which are not preserved during fixing for electron microscopy, or whether the other types of cell-to-cell junction seen, such as the intermediate junctions, also allow current to flow between the cells.

Neuromuscular Junctions

In all species, isolated nerve profiles, and occasional small groups of fibres penetrate into the bundles, and can be seen between the cells. The number of nerve profiles seen is fairly high—for example Daniel et al. (1983) suggest that in human detrusor every muscle cell comes within 200 nm of a nerve, although there is disagreememt as to whether this constitutes a close connection. In most cross sections of bladder muscle from any species, there are far fewer nerve profiles than smooth muscle cells. It appears that only a small percentage of smooth muscle cells come into really close synaptic contact with a nerve (with a gap of about 20 nm), although careful quantitative measurements are necessary to prove this. Certainly, the density of innervation is less than in the classical densely innervated smooth muscles, e.g. the sphincter pupillae and vas deferens, where quantitative analysis has shown each smooth muscle to be closely contacted by more than one nerve ending (Gabella 1981). The possibility of there being many bladder smooth muscle cells that do not receive direct input from a nerve is supported, in the guinea pig, by the relative infrequency with which junction potentials can be recorded from single smooth muscle cells by microelectrodes (J.L. Mostwin 1985, personal communication).

Summary

The smooth muscle cells in the bladder have the typical features of all smooth muscles. They are long and thin, and contain the usual intracellular elements. The cells are arranged in bundles which interdigitate and surround the lumen of the bladder. The individual cells within a bundle are connected together to form a functional syncytium. Nerve terminals run within the bundles, making occasional close connections with the cells, although it is unlikely that every cell receives direct synaptic input.

Mechanical Properties of Bladder Smooth Muscle

Measurement of Mechanical Properties

There have been many studies on the mechanical properties of intact bladders and of bladder strips in humans and in many different species of animals. Attempts have been made to define some of the passive and active properties of the bladder wall in the hope of being able to distinguish changes that may occur in the diseased state. It would clearly be of value to be able to assess changes systematically, and correlate them with the ability of various treatments to improve symptoms. Passive properties such as elasticity, viscoelasticity and compliance have been studied, and attempts made to define active properties such as "contractility". Such studies are unfortunately extremely difficult to interpret, because of the complexity of the bladder wall, the involvement of tissues other than smooth muscle, and the fact that the smooth muscle cells are spontaneously active, even during passive filling. A brief description of some of the results will be given here, however, as they point to some interesting specializations of bladder smooth muscle. Readers who are interested in this approach are referred to recent reviews by Coolsaet (1985) and van Duyl (1985a).

Mechanical Behaviour During Filling

During normal bladder filling, intravesical pressure rises remarkably little as the bladder distends. In humans, for example, at filling rates of up to about 100 ml/min, the pressure in a normal bladder rises less than 5 cm/100 ml water. The same is true for the pig (Sibley 1985), and a similar volume–pressure relationship is found in the rabbit bladder in vitro (Levin and Wein 1982). This high compliance is an important feature of the bladder wall, and as yet inadequately understood. One implication is that the bladder wall is unusually elastic, capable of stretching to a wide range of lengths while in the unstimulated state. Passive length–tension curves of pig detrusor strips (Griffiths et al. 1979) would support this, although it is hard to define the resting length of each strip, and above a certain elongation passive tension rises quite steeply, as also seen in strips from the rabbit detrusor (Uvelius 1976).

If the bladder is artificially filled at faster rates, the pressure rises, but will fall after filling has ceased, and at a given volume the pressure reached will be transiently higher with faster filling rates. This behaviour is also seen in isolated strips, and is not consistent with the detrusor showing only elasticity, but is reminiscent of the stress relaxation shown by viscoelastic solids. However, the behaviour is more complicated, because the unstretched length of the tissue is not constant. Models of the passive behaviour of the bladder can take account of this by including an element of plastic flow (Griffiths 1980).

Structural Changes During Filling

In order to interpret the mechanical behaviour of the bladder during filling, it is necessary to investigate the structural changes of the bladder wall that accompany passive bladder distension. These have been carefully investigated in rabbit and guinea pig bladder by Uvelius and Gabella (1980). As the bladder is filled, the thickness of the wall decreases as does the number of its component bundles. The shape of the bundles changes so that they become rather flattened in cross-sectional area. The length of the individual muscle cells increases linearly with bladder radius over normal functional volumes, the mean length of the cells in guinea pig bladder increasing nearly fourfold (from 160 to $580\,\mu m$) during filling. There is a proportional decrease in the diameter of both the cells and the bundles, so that the cells occupy a smaller cross-sectional area. The packing density of the cells thus changes from 36 000 (cells)/mm^2 in nearly empty bladder to 170 000/mm^2 in distended bladder. These results show that the muscle bundles and individual smooth muscle cells are simply stretched as the bladder fills, without the cells slipping longitudinally past each other in a bundle. The cells within the individual bundles can therefore presumably still function as a syncytial unit whatever their length, although the bundles themselves clearly do undergo considerable rearrangement.

Spontaneous Mechanical Activity

Another factor that will be of importance in determining the mechanical behaviour is the spontaneous activity of the smooth muscle cells that occurs in the absence of any nervous activity. This will be discussed in more detail later, but at this point it is worth pointing out that muscle strips of all species so far studied show spontaneous contractions.

It seems likely that in the intact bladder the individual muscle bundles are also contracting spontaneously, although the organization of the bundles is such that, at least in the larger animals, there is probably insufficient synchrony to produce significant intravesical pressure fluctuations during filling of the bladder. The role of the activity may be in keeping the bladder in an optimal shape within the body. However, this activity does make it difficult to interpret the length–tension relationships measured in whole bladder and in bladder strips.

Length–tension Relationship During Activation

One of the most fascinating aspects of the bladder is its ability to contract efficiently over a wide range of volumes. In the intact guinea pig bladder the active force produced by the "longitudinal" musculature in response to supramaximal pelvic nerve stimulation has been calculated from the volume–pressure relationship and muscle cross-sectional area (Uvelius and Gabella 1980). This active force remains surprisingly constant. It is nearly maximal over a wide range of cell lengths, and is achieved at filling volumes of roughly 25%–75% of bladder capacity. At maximum capacity the active force is 75% of its highest value and about 25% when the bladder is empty. The maximum intravesical pressure on activation, however, is obtained at much lower volumes as seen in the volume–pressure relationship (Fig. 6.3), and declines to about 30% of maximum when the bladder is fully distended (Uvelius and Gabella 1980).

It has already been shown that bladder distension is accompanied by a remarkable change in the length of each smooth muscle cell. The properties quoted imply that the ability of individual cells to generate force is independent of length over a range far greater than in striated muscles. In addition, the maximum force produced by each cell is 5–6 mN which would be equivalent to some 60 N/cm^2 cross-sectional area, higher than figures quoted for force generated in skeletal muscles (Gabella 1976).

The complex events of micturition are initiated and controlled through the nervous pathways to the smooth muscles. In order to understand these events and to learn more rational ways of modifying them in pathological conditions, we need to discuss some of the basic physiological properties of smooth muscle cells in general, and bladder smooth muscle in particular. In the next sections the electrical properties of the cells will be discussed, followed by an account of the contractile machinery and how it can be activated. Finally, we shall con-

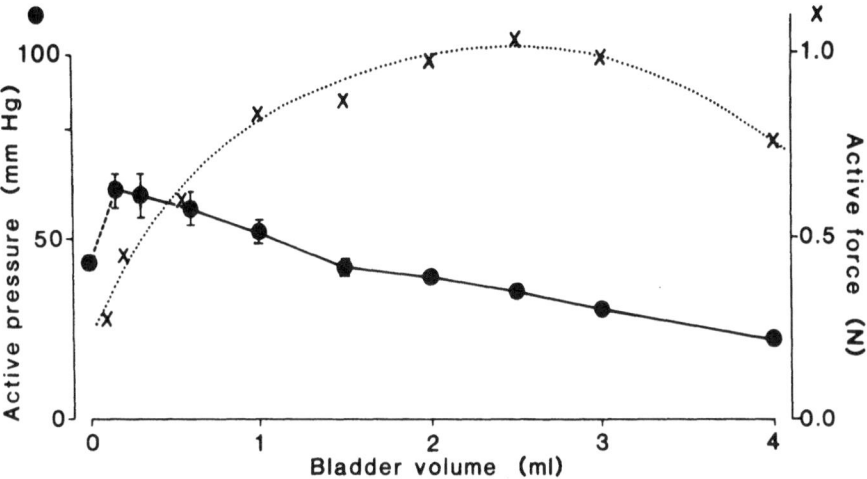

Fig. 6.3. Relationship between bladder volume, active pressure (in response to supramaximal stimulation of the pelvic nerves), and active force generated in the longitudinally orientated musculature of the guinea pig bladder. Note how the active force remains nearly constant and maximal over a wide range of volumes, and therefore cell lengths. (Adapted from Uvelius and Gabella 1980)

sider how the nervous pathways can exert their effects on the smooth muscle cells.

Summary

During filling, the intravesical pressure does not normally rise significantly: the passive behaviour of the bladder can be described in terms of viscoelasticity and plasticity, but the interpretation is complicated because individual cells almost certainly exhibit spontaneous activity during filling. As the bladder fills, the muscle bundles undergo reorganization, but the muscle cells within a bundle are simply stretched, and can extend to about four times their length in the empty bladder. Individual cells are capable of exerting nearly maximum force over a wide range of lengths.

Electrical Properties of Bladder Smooth Muscle

Measurement of Electrical Properties

Microelectrodes have been used to record membrane potentials and membrane properties in the bladder smooth muscles of relatively few species. Guinea pig and rabbit bladder have been the most extensively studied (Ursillo 1961; Creed 1971a, b; Callahan and Creed 1981, 1985; Creed et al. 1983; Brading et al. 1986) but recordings have also been made from the bladders of wallabies and dogs (Callahan and Creed 1985) and from pigs (J. Heaton and A.F. Brading 1986, unpublished work). Additional information has been obtained with the "sucrose gap" method of extracellular recording. This method relies on the fact that the cells form a functional syncytium, and a thin strip of tissue can be treated as an electrical cable. A small central section of muscle cells in such a strip (the node) is superfused with warmed Krebs solution, and neighbouring areas are separately perfused by isotonic sucrose solution, which prevents extracellular current flow. Current can still flow intracellularly between the cells in the sucrose. The summed activity of the nodal cells can then be measured by recording differences in extracellular potential between the node and one end of the strip (often depolarized with high K solution) across a sucrose gap. A double sucrose gap apparatus may be used, in which current can be injected into the nodal cells across one sucrose gap, and the potential changes recorded across the other (details of the method can be found in Coburn et al. 1975). Tension responses of the nodal cells can also be measured simultaneously, which is extremely difficult to do when recording with microelectrodes.

Passive Properties

The size of the resting membrane potential (conventionally measured inside the membrane relative to earth, and thus negative in the resting state) depends on the relative permeability of the membrane to ions, and the concentration gradients of ions across the membrane. For each ion one can specify an equilibrium potential, which is the membrane potential which would exactly balance the tendency of that ion to move down its concentration gradient. At any time, the membrane potential is usually nearest to the equilibrium potential for the most permeant ion, which in the resting state is normally K. This means that an increase in the membrane permeability to one ion will cause a movement of the membrane potential towards the equilibrium potential for that ion; for example, increasing K permeability will hyperpolarize the cell (make the membrane potential more negative). A change in the extracellular concentration of an ion will alter its equilibrium potential, and if the membrane is permeable to this ion, this change will result in a movement of the membrane potential towards the new equilibrium potential. For example, increasing extracellular K will reduce the transmembrane K gradient and its equilibrium potential, and depolarize the cell (Fig. 6.4).

The ionic gradients which exist across the membrane owe their existence to the presence of nonpermeant anions inside the cell, and to the actions of various active mechanisms which can move ions against their electrochemical gradients. The most important active mechanism for the resting potential is the Na–K pump, which establishes the high intracellular K concentration and its negative equilibrium potential. Smooth muscles also seem actively to accumulate Cl ions, and the Cl equilibrium potential is usually positive to the resting membrane potential. This arrangement is illustrated in Fig. 6.4.

Published values of the membrane potentials of the smooth muscle of the guinea pig bladder are -36.3 ± 4.3 mV in the dome, and -42.2 ± 4.1 mV in the urethra (Callahan and Creed 1981). For comparison, the ureter smooth muscles have a membrane potential of -49.7 ± 4.9 mV (Brading and Aickin 1985) and the vas deferens -75 mV (Aickin and Brading 1983). Published values for the rabbit are -37.5 ± 3.9 mV for the dome and -42.2 ± 4.1 mV for the urethra (Callahan and Creed 1985).

There is little information available about the resting permeabilities of the membrane of bladder smooth muscle to ions, and no published figures of their intracellular ionic activities. However, measurements of total ion content of guinea pig bladder smooth muscle made in this laboratory suggest that bladder smooth muscle has an ionic distribution similar to that of other smooth muscles (see Brading 1981). The relatively low membrane

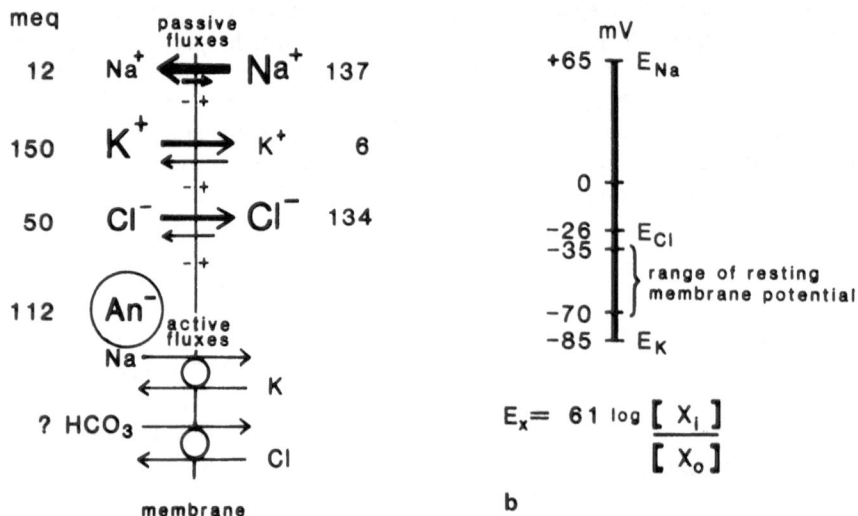

Fig. 6.4a, b. Diagram to show the distribution of ions across the smooth muscle cell membrane, and the equilibrium potentials. **a** The passive fluxes are represented above, and the thickness of the *arrow* indicates the size of the flux. Note that none of the major ions is at equilibrium, and there is a net inward driving force for Na ions and a net outward driving force for K and Cl. Below are represented the active transport processes. *An⁻* represents nonpermeant anions. Concentrations in milliequivalents (mEq). **b** Voltage scale in millivolts (mV). The equilibrium potentials (E_x) are calculated from concentration gradients (X_i and X_o are intra- and extracellular ionic concentrations) using the Nernst potential (formula given). The membrane potential varies in different smooth muscles, but is usually between E_K and E_{Cl}.

potential in bladder suggests either that the K permeability is lower than usual, or that the Cl or Na permeabilities are comparatively high. We have used [86]Rb (a radioactive isotope of Rb which behaves in a manner similar to K) to estimate K movements in guinea pig bladder, and the rate constant of the efflux suggests that the permeability is probably low—the steady rate of loss of Rb from the cells was some 1.5 times faster in the guinea pig taenia (resting potential about −51 mV) than in the bladder dome.

Kurihara and Creed (1972) have looked at the effects of changes in the ionic environment on the resting potential of cells in the dome of the guinea pig bladder. Increase in extracellular K concentration depolarizes the cells. Theoretically, a tenfold change in extracellular K causes a 61 mV change in the equilibrium potential, and if the membrane were permeable only to K, this would also be the change in membrane potential. In practice, the maximum slope in guinea pig bladder cells was 31 mV per tenfold change in extracellular K. This compares with a maximum slope of 42 mV in the taenia, and is consistent with the K permeability being rather low. Lowering the extracellular K concentration surprisingly depolarized the cells, a result that is not explicable in terms of the change in equilibrium potential, since this should lead to hyperpolarization. However, the addition of ouabain and the removal of extracellular Na also caused depolarization. These treatments would all tend to reduce the activity of the Na pump, and the results suggest that the pump is electrogenic, pumping more Na out than K into the cell, and thus generating an outward current which will contribute a small amount to the membrane potential. Removing Cl from the medium (using a nonpermeant substitute) caused a transient depolarization, consistent with the Cl equilibrium potential's being positive to the resting potential in the normal solution.

The individual muscle cells within a bundle are electrically coupled so that current injected into one cell will spread in three dimensions. The electrical properties of the detrusor muscle bundles can be determined from intracellular recordings of the membrane potential in response to extracellular current injection (using the "Tomita bath", a method for uniform and synchronous current application: Abe and Tomita 1968). The time constant of the membrane (the time taken for the voltage to change by 63% of the steady state value in response to a step change in current) is the product of the membrane capacity and resistance, and was found to be about 127 ms in the guinea pig (Creed 1971a). This value is similar to that found in other smooth

muscles (Tomita 1970), and is about an order of magnitude greater than found in skeletal muscle—a reflection of the nearly tenfold greater membrane resistance of smooth muscles. Also in the guinea pig, the length constant (the distance over which the voltage change initiated by the injected current declines by 63% of its size at the injection site) varied in different preparations from 1.23 to 1.95 mm (mean 1.7 mm), and in the rabbit dome the length constant varied from 1.85 to 2.61 mm (Creed et al. 1983). These values are again similar to those found in other smooth muscles. Since the length of the individual cells in the guinea pig and rabbit bladder is probably only some 200–300 μm, the value of the length constant is about the length of 6–7 cells, and again illustrates the fact that the cells must be coupled together electrically.

Active Properties

Action Potential Mechanisms

The mechanisms underlying action potential generation in all excitable tissues involve a sequence of voltage-dependent changes in membrane permeability, leading to ion flow down pre-existing electrochemical gradients. There is an initial ion flow leading to an inwardly directed current which is responsible for the depolarizing phase of the action potential. This is normally followed by a switching off of the inward current and the activation of an outward repolarizing current. There are two ionic species which are distributed so that there is a large electrochemical gradient which could generate inward current: Na and Ca. The equilibrium potentials for these ions is considerably *positive* to the membrane potential, so that increases in their permeability will depolarize the cells. K ions are usually involved in generating outward repolarizing currents.

In fast conducting tissues such as nerve and striated muscle, the inward current is carried by Na ions, and since the concentration of Na in the outside medium is large, the current flow per unit area of the membrane can be high. This leads to a rapid upstroke (up to 100 V/s), and fast propagation of the signal. The Na channels are blocked by tetrodotoxin (TTX), a highly specific blocker of these voltage-sensitive Na channels.

Smooth muscles do not need such rapid propagation of action potentials, and the upstroke of the action potential is caused by inflow of Ca ions (phylogenetically a much earlier mechanism). The Ca-based spikes have a much slower rising phase (less than 10 V/s) because of the smaller current density flowing through the Ca channels, are insensitive to lowering extracellular Na, but reduced by lowering extracellular Ca, and are blocked by the "Ca antagonist" drugs such as D600, nifedipine, etc., but not by TTX.

Bladder Action Potentials

The first person to publish intracellular records of spikes from mammalian bladder was Ursillo (1961), in the rabbit dome. He recorded membrane potentials of -20 to -70 mV, but his action potentials were small and irregular, and never showed an overshoot. More recently, Creed et al. (1983) have demonstrated that spikes with overshoots of up to 18 mV can be recorded from all parts of the rabbit bladder, although there is considerable variation in shape, and often both afterhyperpolarizations and depolarizations are present. In the guinea pig dome and urethra, the action potentials are simple in shape, the rising phase and falling phase being similar in duration with a pronounced afterhyperpolarization. The parameters measured by Kurihara (1975) in the guinea pig are a peak spike potential of 9.9 ± 4.7 mV from a resting potential of -38.9 mV (total height 48.7 mV), a maximum rate of rise of 4.2 V/s and rate of fall of 5.4 V/s in isotonic Krebs solution, Hypertonic solution (the tonicity being doubled by the addition of sucrose) is frequently used for microelectrode recordings, since tissues do not move in this solution. Under these conditions, the peak potential gets larger $(15.2 \pm 5.9$ mV) and the rates of rise and fall get slower (3.6 and 3.5 V/s, respectively), though with little change in resting potential. The duration of the spike at half amplitude in the guinea pig was some 10–15 ms, with an afterhyperpolarization of about 500 ms (Mostwin 1986).

The effects of drugs and of changes in the ionic composition of the medium on the spike parameters have been studied in the guinea pig to give some idea of the underlying ionic mechanisms. The spikes are not affected by TTX, but the rate of rise and the peak potential are progressively diminished on lowering the extracellular Ca or on application of Ca antagonists (Fig. 6.5; Mostwin 1986). A preliminary report has appeared of a study on single cells isolated from the guinea pig dome. Isenberg and Klöckner (1985) have applied a whole cell voltage-clamp to such cells, and have been able to record the inward currents responsible for the action potential using patch electrodes. With this technique, a microelectrode with a flame-polished tip of a few micrometres in diameter is placed against the enzymatically cleaned surface of an isolated cell,

and gentle suction is applied, which results in the formation of a high resistance (GΩ) seal between the electrode and the membrane. A short pulse of pressure then disrupts the patch of membrane, allowing the interior of the cell to become electrically connected to the electrode, so that currents can be injected to clamp the potential at a desired value. Isenberg and Klöckner abolished outward K currents by blocking them with Cs. In a medium containing 3.6 mM Ca, depolarizing the cell to -5 mV evoked a transient inward current that peaked within 4 ms to -12 μA/cm^2 and declined with complex kinetics. The current was a Ca current; it was abolished when Ca was omitted from the medium, and was blocked by Ni and by D600. This demonstrates that bladder smooth muscle is similar to other smooth muscles in that Ca ions carry the current during the rising phase of the action potential. However, in the absence of extracellular divalent cations, and the presence of a Ca chelator (EGTA), inward current can be carried through the same pathway by Na ions.

The falling phase of the spike is probably due to the switching on of a voltage-sensitive outward K current as well as the inactivation of the inward Ca current. Tetraethylammonium ions block K channels, and prolong the falling phase of the action potential (Fig. 6.6). The afterhyperpolarization seen in the guinea pig may be due to the opening of another set of K channels. It has been proposed that smooth muscles may possess K channels that are opened when intracellular Ca rises (Ca-activated K channels). The afterhyperpolarization is diminished as extracellular Ca is lowered, suggesting that Ca entering during the spike is necessary to activate K channels, but the hyperpolarization is also blocked by small depolarizations of the membrane, and is not blocked by concentrations of Ca antagonists that reduce the inward Ca current, and therefore some type of voltage-sensitive channels may be involved. The mechanisms underlying the complex afterpotentials seen in rabbit bladder smooth muscles have not been investigated.

Summary

Bladder smooth muscle in small mammals has a relatively low resting potential (about -40 mV), and can generate propagated action potentials. Electrical connections between the smooth muscle cells in the muscle bundles allow the bundles to behave as cables along which the action potentials can propagate. Intracellularly recorded action potentials depolarize the cells by about 50 mV, and have a duration at half amplitude of 10–15 ms. They show similar rates

Ca Removal

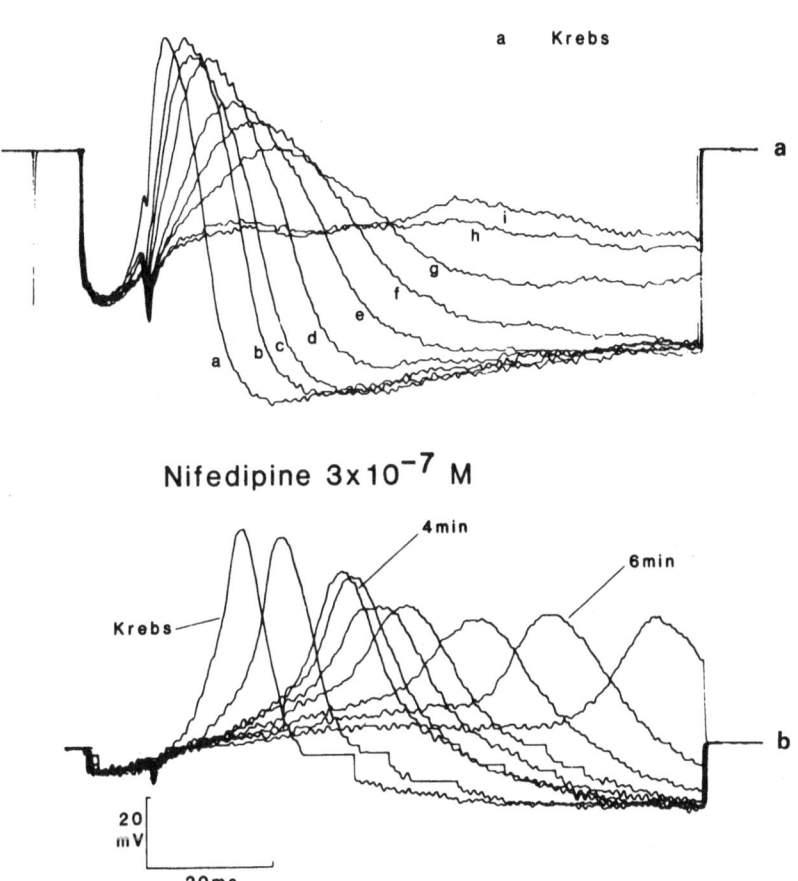

Nifedipine 3x10⁻⁷ M

Fig. 6.5a, b. The effects of Ca removal (**a**) and addition of the Ca antagonist, nifedipine (**b**) on the action potential of the guinea pig detrusor. In **a**, trace *a* is control action potential, with typical pronounced afterhyperpolarization. Traces *b–e* show progressive change in action potential shape after reducing Ca to 0.25 mM (spikes recorded at about 5-min intervals). Traces *f–i* show Ca reduced to 0.1 mM, and 1 mM EGTA added (effective Ca concentration very low). Note reduction in rate of rise and amplitude of the spike, and progressive loss of afterhyperpolarization, **b** the effects of $3 \times 10^{-7} M$ nifedipine. Again, the rate of rise and amplitude of the spike are progressively diminished, but there is no obvious change in the afterhyperpolarization. Direct muscle stimulation. The "squaring" of the waveform in **b** is due to digitization of the signal with inadequate resolution. (Mostwin 1986)

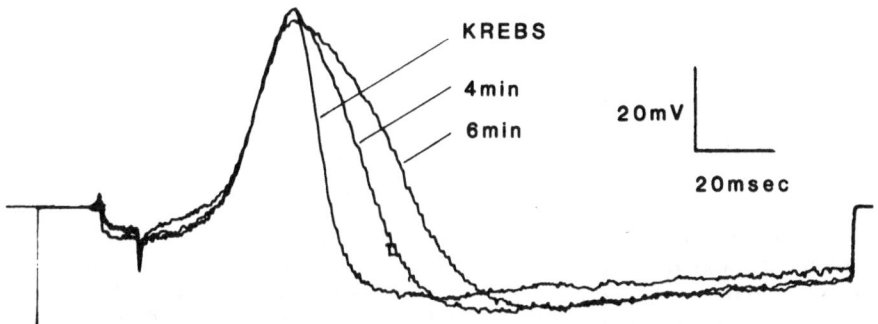

Fig. 6.6. Effect of 4 mM tetraethylammonium chloride (TEA) on the action potential of guinea pig detrusor. TEA has little effect on the rising phase of the spike or the afterhyperpolarization, but prolongs the falling phase. (Mostwin 1986)

of rise and fall, and a pronounced afterhyper-polarization which may last 500 ms. The upstroke is generated by inward Ca current, and the downstroke probably involves inactivation of the Ca current and activation of an outward K current.

Contractile Mechanisms of Smooth Muscle

Sliding Filament Theory

In all muscles, shortening occurs through sliding of thick and thin filaments past each other. The filaments are arranged in parallel, and sliding is achieved by the formation of crossbridges between the filaments. Thick filaments are composed of an array of myosin molecules, each of which has a double projecting head (the crossbridge) which can attach to sites on the thin actin filaments. The myosin heads have enzymic activity (ATPase) which is triggered by interaction with the actin molecules, and is able to break down ATP to liberate energy which is used to power a change in the crossbridge conformation leading to pulling of the two filaments past each other. In striated muscles there is an orderly arrangement of filaments into sarcomeres separated by Z-bands onto which the actin filaments insert. Myosin filaments are in the centre of the sarcomeres, and the thin filaments interdigitate regularly with two thin filaments for each thick one. The crossbridges are polarized so that they point in one direction at one end of the thick filament and the other direction at the opposite end, and there is an area devoid of crossbridges in the centre. Thin filaments are thus drawn together towards the centre of each thick filament, and crossbridge turnover results in tension development, or shortening of the individual sarcomeres.

Structure of Smooth Muscle Contractile Machinery

Smooth muscles do not have their thick and thin filaments arranged into sarcomeres, and they do not have Z-bands. Instead actin filaments insert either onto the plasma membranes through the "dense bands" seen at intervals along the membranes, or onto "dense bodies" in the cyto-plasm. The insertions of the actin onto the cell membrane results in the scalloped appearance of the cells when they are contracted. The thick filaments lie in the cytoplasm in parallel with the length of the cell, and are somewhat different from those found in striated muscles. The thick filaments in smooth muscle do not show the end polarization of the crossbridges, and there is no central area free of crossbridges. It has been suggested (Craig and Megerman 1977) that they may have crossbridges arranged in alternate rows of each orientation (side polarization). This appears to be an adaptation to allow the thin filaments to slide the whole length of the thick filaments, and might account for the remarkable ability of these cells to produce active force over a great range of lengths (Fig. 6.7). There is a marked difference between striated and smooth muscles in the proportion of thick to thin filaments. In smooth muscles there are many more thin filaments, and between 5 and 27 thin filaments for each

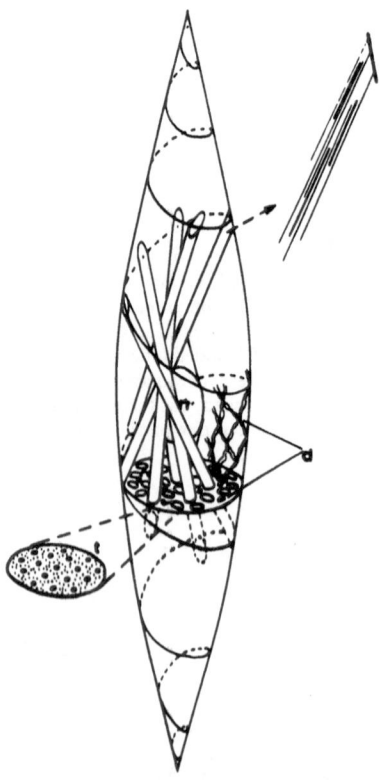

Fig. 6.7. Diagram of possible arrangement of myofilaments in a smooth muscle cell: *t* is a transverse section of a bundle of filaments, showing thick and thin filaments in relatively orderly array (the *arrow* shows insertion of thin filaments onto the cell membrane), and longitudinal arrangement of the thick and thin filaments; *d* is the "skeletal" structure of the dense bodies and bands, and the intermediate filaments. (Small 1977). Recent work by Small does not support the spiral arrangement of the filament bundle insertions onto the cell membrane.

thick filament have been reported (Gabella 1981). There is a third type of filament present in smooth muscle, the 100 nm or intermediate filament. These are thought to be noncontractile, and to be cytoskeletal in function, possibly aiding re-extension of the smooth muscle after contraction.

Activation of the Contractile Machinery

An increase in the intracellular free Ca concentration is the main trigger of the contractile machinery. The means by which this activation is brought about again differs between striated and smooth muscle. In relaxed striated muscle interaction between the crossbridges and the actin filaments is prevented by the presence of tropomyosin, a filamentous protein that stretches along the actin filaments, and prevents the heads of the myosin crossbridges binding to the reactive sites of the actin molecules. Tropomyosin is held in this position by troponin, a complex globular protein which occurs at intervals along the thin filament. Ca binds to the troponin, causing it to change shape. This allows the tropomyosin to slide away from the active sites on the actin molecules, so that crossbridge cycling can begin; thus Ca is acting to *disinhibit* or release the system.

In smooth muscles this complex control system is not fully developed, and the main role of Ca is to cause *activation* of the myosin crossbridges. In these tissues the crossbridges are only active when two small protein components (light chains) on the myosin heads are phosphorylated. The number of crossbridges that are phosphorylated at any time depends on the relative activities of two enzymes, the myosin light chain kinase which is responsible for the phosphorylation, and the myosin light chain phosphatase which can dephosphorylate the crossbridges. The role of Ca is in activating the kinase, whereas the phosphatase is independent of the free Ca concentration. Ca binds to calmodulin, a specific calcium-binding protein, which then binds to and activates the myosin light chain kinase, resulting in phosphorylation of the myosin light chains and crossbridge turnover.

Although an increase in intracellular Ca leading to phosphorylation of the myosin light chains is the initial step for smooth muscle contraction, there may be several other modulating pathways. This is again in contrast with skeletal muscle, where the contractile apparatus has evolved into an extremely efficient, rapid, but very stereotyped system. One form of regulation which operates in some smooth muscles during maintained tension is the conversion of the crossbridges into a "latch" state, in which crossbridge turnover and ATP utilization is considerably reduced, without significant loss of force (Askoy et al. 1982). This may be an energy conserving mechanism in smooth muscles, such as in blood vessels, that are required to maintain force for long periods of time, and consequently may turn out to be relatively unimportant in bladder. Another mechanism that has recently been established involves an inhibitory protein, caldesmon, which plays a role similar in some respects to the inhibitory unit of troponin in striated muscle (Marston and Smith 1985). It is associated with the thin filaments, and prevents interaction of actin and myosin molecules except when it is bound to a Ca–calmodulin complex, which causes it to release its inhibition. Thirdly, recent work suggests that the absolute sensitivity of the contractile machinery to calcium ions can be altered. For example, the stimulation of β-adrenergic receptors can trigger pathways reducing the affinity of the myosin light chain kinase to the Ca–calmodulin complex, thus resulting in a reduction in the degree of phosphorylation of the myosin light chains. This could result in a drop in tension without a fall in the intracellular Ca concentration (see for example Rüegg and Pfitzer 1985). It has also been shown by using intracellular Ca-sensitive probes that different excitatory stimuli may cause the same tension development whilst elevating the free calcium to different extents (Morgan and Morgan 1984).

This question, far from being of purely academic interest, may be illustrating the theoretical basis for future therapeutics. It is becoming clear that many of these regulatory mechanisms are tissue specific, and their metabolic pathways are potentially susceptible to modulation by drugs. We may now be entering an era in which the new possibilities of directly regulating the contractile machinery will be exploited.

Summary

Smooth muscles possess thick and thin filaments, composed of myosin and actin, respectively. The thick filaments are found in the cytoplasm, and the thin filaments mainly attached to the cell membranes. Contraction occurs by crossbridge formation sliding the filaments past each other. Activation of the contractile machinery involves a Ca-dependent phosphorylation of the myosin heads, and there are several other regulatory processes which may be able to alter the relationship between free Ca and the amount of tension generated.

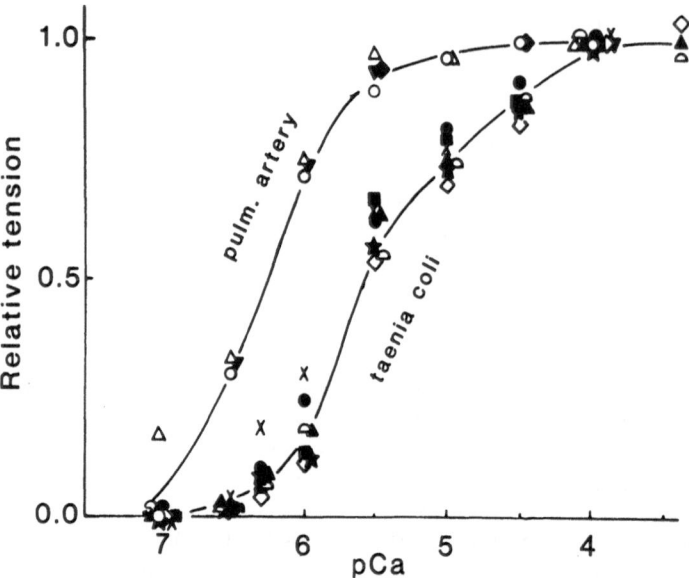

Fig. 6.8. Relationship between tension and intracellular Ca concentration in saponin-skinned smooth muscles from the guinea pig. *Abscissa:* negative logarithm of the free Ca concentration (*pCa*). (Endo et al. 1977)

Intracellular Calcium in Smooth Muscle

Relationship Between Free Ca and Tension

Studies have been carried out with glycerol-extracted smooth muscles, in which the membranes have been dissolved by prolonged exposure to glycerol, and with saponin-skinned smooth muscles, in which exposure to saponin produces large holes in the plasma membrane (for references see Kuriyama 1981). In these systems the contractile machinery is in contact with the bathing fluid, and the relationship between tension and the ionized Ca concentration can be measured. There is a continuous relationship between the two, which is similar to that found in striated muscles. Figure 6.8 illustrates the relationship in skinned smooth muscle from guinea pig pulmonary artery and taenia. Both tissues are relaxed below $10^{-7} M$ Ca, have a threshold just above $10^{-7} M$, and are fully contracted at $10^{-4} M$ Ca. These results are typical of what has been found in other smooth muscles. Arterial muscle appears to be somewhat more sensitive to Ca than visceral muscles, but the patterns are similar. No studies are yet in print about this relationship in the bladder.

Intracellular Ca in Relaxed Smooth Muscles

To maintain the relaxed state, the intracellular free Ca ion concentration has to be kept at less than $10^{-7} M$. This is achieved by extrusion of Ca across the cell membrane, binding of Ca to intracellular sites and uptake into intracellular organelles.

Ca extrusion out of the cell occurs against both potential and concentration gradients, and requires energy. There are at least two known mechanisms, an ATP-dependent Ca pump, and a Na-Ca exchange mechanism. The latter exchange uses the potential energy of the inwardly directed Na gradient to power extrusion of Ca against its electrical and chemical gradients; its importance and even its existence in smooth muscles is disputed by some.

The total intracellular Ca is probably of the order of 5×10^{-4} mol/kg wet weight (Brading and Widdicombe 1977), but the majority of this Ca is either bound to cellular proteins or membranes, or sequestered inside organelles. The sarcoplasmic reticulum of smooth muscle consists of vesicles often found in association with the cavaeoli at the surface of the cell, and also deeper in the cytoplasm. It has been shown that Ca can be pumped into these vesicles by an ATP-dependent Ca pump that differs somewhat in its properties from that found in the plasma membrane (Daniel 1985a; Wuytack et al). The sarcoplasmic reticulum can accumulate Ca to quite high values (at least 30 mmol/kg dry weight; Somlyo and Franzini–Armstrong 1985)—enough when released into the cytoplasm for maximal activation of the contractile machinery. If the cell becomes overloaded with Ca, then the mitochondria will also accumulate this ion.

Mechanisms for Increasing Free Intracellular Ca

There is a background passive leak of Ca into the cells down an electrochemical gradient, and an increase in intracellular Ca leading to activation of the contractile machinery could theoretically occur if the mechanisms for extruding Ca were inhibited. More commonly, however, a rise is achieved either by activating release of Ca from internal stores, or by increasing the permeability of the membrane to Ca. The changes in transmembrane Ca permeability are brought about by regulation of Ca channels in the membrane. Two classes of channel are postulated: voltage-sensitive channels, and receptor-operated channels. Activation of either will permit entry of Ca down its electrochemical gradient.

Voltage-Sensitive Channels

There are probably two types of Ca channel activated by membrane depolarization: those discussed earlier, that are involved in the production of action potentials (spikes), which first open and then close (inactivate) in reponse to maintained depolarization, and are often called "fast" Ca channels, and those that remain open during maintained depolarization, and are often called "slow" Ca channels.

In striated muscle Ca does not appear to enter in significant amounts from the outside during the action potential, but even if it did, the surface area:volume ratio is such that it would be impossible for Ca to reach the deeper contractile machinery in quantities sufficient to trigger contraction in the time taken for activation. In smooth muscle, the small diameter of the cells with their large surface area:volume ratio makes this less of a problem. Calculations of the amount of Ca entering during the upstroke indicate that a single spike could elevate the free intracellular Ca by about $10^{-6}M$, which would exceed the threshold for contraction (Kuriyama 1981). However, it is very unlikely that the Ca does remain free, because of the extensive buffering of Ca that occurs in smooth muscle (as

discussed earlier), and it is now thought that the role of the Ca entering may be to trigger release of more Ca from internal stores. If Ca release from the stores is blocked by procaine, Ca entering during the spike does not trigger contraction, although the sensitivity of the contractile machinery to Ca is not altered (Itoh et al. 1981).

Any excitatory agent that causes depolarization of the smooth muscle membrane will also trigger opening of the slow Ca channels, Ca entry and contraction. Application of high K solutions for example is a good way of initiating depolarization. The resulting contractions are easily blocked by low concentrations of Ca antagonist drugs, suggesting that the continuing entry of Ca through the membrane is necessary to keep the contractile machinery activated.

Receptor-operated Channels

Receptor-operated channels are channels that are opened in response to interaction of transmitters or agonist drugs with specific receptors on the cell surface. Many agonists lead to the opening of channels with high Na permeability, resulting in depolarization of the membranes, and subsequent opening of the voltage-sensitive Ca channels and contraction. It has also been suggested that some drugs may activate receptor-operated Ca channels, to allow Ca entry without depolarization (Bolton 1979).

Release of Calcium from Internal Stores

There is excellent evidence that Ca released from internal stores does play a major role in excitation–contraction coupling in many smooth muscles (for references see Itoh et al. 1985). It has been suggested that the release can be triggered by a Ca-activated Ca release process similar to that known to exist in heart muscle. This would presumably be the route through which Ca entering during the spikes triggers Ca release. It has also long been known that many smooth muscles will contract to some agonist drugs without membrane depolarization—a process sometimes called pharmacomechanical coupling (Somlyo and Somlyo 1968). Indeed, many tissues will exhibit short-lived contractions in Ca-free solutions. This suggests that the process also involves Ca release from internal stores, and that receptor activation must lead to the production of some second messenger other than Ca, which can cause the Ca release. Recent studies suggest that in some cases, drug–receptor interaction results in the breakdown of a membrane lipid (phos-phatidylinositol-4,5-bisphosphate), leading to the production of inositol trisphosphate (IP_3) and 1,2-diacylglycerol as second messengers. IP_3 has recently been shown to be able to release Ca from the Ca stores in skinned smooth muscles, and may therefore play a role in pharmacomechanical coupling (Hashimoto et al. 1986).

Summary

An increase in the free intracellular Ca levels triggers activation of the contractile machinery. In relaxed smooth muscle, intracellular Ca is kept low by extrusion of Ca across the membrane, intracellular Ca buffering and Ca uptake into intracellular stores. The rise in intracellular free Ca can occur by entry through fast or slow voltage-sensitive channels, through receptor-operated Ca channels, and by release of Ca from internal stores.

Contractile Behaviour and Excitation–Contraction Coupling in Bladder Smooth Muscle

Spontaneous Activity in the Detrusor and Trigone

Mechanical Activity

Spontaneous mechanical activity has been recorded from strip preparations of most bladder, trigone and urethral smooth muscles that have been studied. Sibley (1984a, b), for example, compared muscle strips from bladders of human, pig and rabbit. He found spontaneous activity in all strips from rabbit bladder (mean frequency 7.3 contractions per minute), and in about 20% of strips from pig (1.1 contractions per minute) and human bladder (2.2 contractions per minute). More recent work from this laboratory suggests that the great majority of pig and human detrusor strips will contract spontaneously if left to equilibrate for a sufficient period.

van Duyl (1985b) has also studied spontaneous activity simultaneously from different regions of a large strip of pig detrusor muscle, and has shown that different rhythms occur in different areas. We have found that both trigonal and detrusor muscle from humans and pigs can contract spontaneously and that the activity is not diminished by the

addition of atropine, tetrodotoxin or the ganglion-blocking drug hexamethonium. We have not been able to find any clear relationship yet between the length of the bladder strips and the frequency of the contractions—certainly spontaneous activity occurs over a wide range of lengths. Several drugs will, however, modify the frequency of contraction and it has been suggested that the activity may be under the control of prostaglandins produced by the tissues (Downie and Slack 1983; Satake et al. 1984).

Electrical Activity

Spontaneous electrical activity has also been recorded in isolated strips of the bladder dome from all species so far studied. It normally takes the form of regular spikes which occur with a frequency varying from about 6 to 30 spikes per minute (see for example Callahan and Creed 1985). The frequency can be modulated by injections of depolarizing and hyperpolarizing current (Fig. 6.9), and is also increased by application of excitatory agents such as acetylcholine and substance P (Callahan and Creed, 1986; Mostwin 1986), and by agents which decrease K permeability, such as procaine (Kurihara 1975) and tetraethylammonium (TEA) ions (Mostwin 1986).

Spontaneous Activity in the Urethra

Spontaneous mechanical activity has also been seen in urethral smooth muscle strips from wallabies, dogs, rabbits and, more rarely, from guinea pigs (Callahan and Creed 1981, 1985), and spontaneous electrical activity has been recorded in isolated strips from the urethra. In the female rabbit, the activity is similar to that seen in the dome, although the spike shapes are somewhat different. In the female guinea pig, however, the urethral smooth muscle shows bursts of spontaneous spiking, occurring once every few minutes, each burst containing some 15–30 spikes (Callahan and Creed 1985).

Link Between Mechanical and Electrical Activity

Spikes

Simultaneous recordings of electrical and mechanical activity using the sucrose gap method show clearly that contractile activity is related to spike

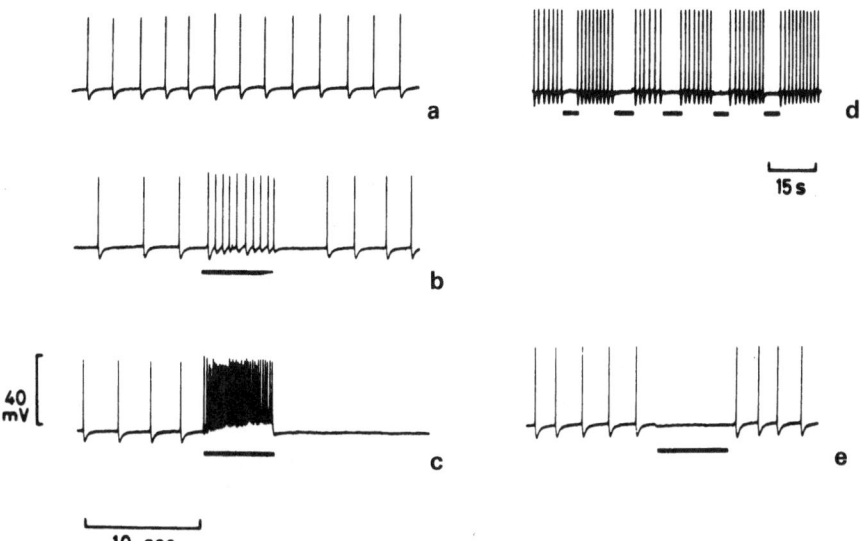

Fig. 6.9a–e. Modulation of spontaneous electrical activity in the guinea pig detrusor by depolarizing and hyperpolarizing current pulses. **a** Control activity at 36°C. **b,c** Spike frequency increases in response to depolarizing current pulse. **d,e** Cessation of spontaneous activity on application of a hyperpolarizing current pulse, on two time scales. (Mostwin 1986)

production by the tissue. Mostwin (1986) has studied the factors underlying the generation of tone in guinea pig detrusor. Figure 6.10 shows an example from the guinea pig bladder, where spikes are elicited by depolarizing current pulses of increasing strength. Each spike clearly leads to a phasic contraction of the tissue, and if the frequency is fast enough, some fusion occurs to give an increase in the basal tone (Mostwin 1986). Most excitatory agents which cause contraction of the tissue do so by increasing the frequency of the spikes.

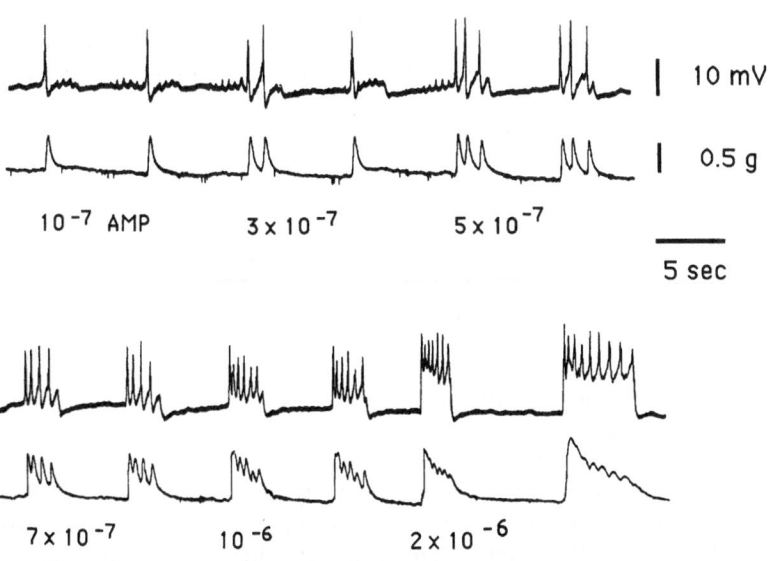

Fig. 6.10. Electrical and mechanical responses of guinea pig bladder to outward (depolarizing) current injection (double sucrose gap). Pulses of increasing intensity (3-s duration) were applied, inducing action potentials, associated with phasic tension development. With the larger current pulses, repetitive activity can be seen which results in progressive fusion of the contractile responses, until the phasic activity is superimposed on a more tonic contraction. (Mostwin 1986)

Depolarization

There is also evidence that tension can be initiated by depolarization per se without spike activity. Depolarizing the membrane with high concentrations of K results in a smooth increase in tone. Electrophysiological evidence shows that the depolarization in this case is considerable and the spikes absent, suggesting that the spike mechanism has been inactivated. The fact that tone is maintained suggests that bladder smooth muscle does possess the slow type of Ca channel. In contrast to high K, carbachol, even at concentrations producing maximum contractions, does not depolarize enough to inactivate the spike mechanisms, and the contractile response always shows evidence of the phasic contractions superimposed on the increase in tone. Spontaneous activity like the responses to high K concentrations, and low concentrations of carbachol, is abolished by Ca antagonist drugs such as nifedipine, suggesting that these mechanisms all require entry of Ca from the outside.

Store Release

There is evidence that there may be an intracellular Ca store in detrusor smooth muscle. At present there is no evidence about whether the contractile response to each spike involves release of this store, but it seems likely that it may. Evidence for the presence of a store has been provided by Mostwin (1985), and involves the use of Ca-free solutions, in which free Ca is kept very low by the use of the Ca chelator ethylene glycol bisaminoethylether tetra-cetic acid (EGTA). Tissues were exposed to Ca-free solution, and tested at intervals for their ability to contract to application of high K concentrations or maximal doses of carbachol. The contractile response of the tissue to K is lost far more rapidly than to carbachol. The carbachol response is reduced to 50% of the initial value in 10 min, whereas after 2 min exposure the response to K is only 10% of the initial value, and after 10 min the response to high K is abolished. The response to high K concentration is also lost immediately in normal Ca if the Ca antagonist D600 is added, suggesting that the response is entirely dependent on entry of extracellular Ca. In contrast, the tissue will produce one large response to high carbachol however long the exposure to the Ca antagonist: subsequent responses are reduced in size, but not abolished. These results suggest that high concentrations of carbachol can release Ca from an internal store.

After the ability of the tissue to contract has been lost in Ca-free solution, a brief application of high Ca (5 s, 71.5 mM) reprimes the system, and allows the tissue to respond once to high concentrations of carbachol (but not to K depolarization). It seems that low (and probably more physiological) concentrations of muscarinic agonists cannot release the store. A very similar situation has been found in the taenia of the guinea pig caecum (Brading and Sneddon 1980).

Phasic Nature of Bladder Smooth Muscle

One notable feature of bladder smooth muscle which contrasts with several other smooth muscles, is the extremely phasic nature of its contractile behaviour. It is very difficult to find any condition in which the tissue strips can maintain significant continuous tone for more than a few minutes. Continuous exposure to high doses of any agonist result in an initial fast contraction that rapidly fades to a low level. The contractile response fades whether the tone is induced by low concentrations of agonists which are clearly associated with increased frequency of spike production, or whether high concentrations of K are used which depolarize the tissue sufficiently to block spike production.

The mechanisms involved in this rapid fading are not yet understood. Presumably it reflects either a reduction in the level of free intracellular Ca, or a reduction in the sensitivity of the contractile machinery to Ca. The former could involve reduction of Ca entry by a progressive inactivation of the Ca channels, some enhancement of the Ca translocating systems preventing a sustained increase in intracellular Ca, or loss of Ca release from the store. Whatever mechanism is involved, phasic behaviour is clearly appropriate to a muscle which is required to contract only for long enough to empty the bladder. For those interested in the phasic and tonic behaviour of smooth muscles in general, more detailed discussion is given by Golenhofen (1976, 1981).

Summary

Bladder smooth muscle shows spontaneous electrical and mechanical activity. Each action potential gives rise to a phasic tension increase, and normal control of the tension is through alterations in the frequency of the action potentials. If the tissue is depolarized sufficiently, the spike mechanisms become inactivated, and there is probably Ca entry through slow voltage-sensitive Ca channels. There is also evidence

for an internal store of Ca that can be released by high concentrations of muscarinic agonists, but not by depolarization with high K. The detrusor is remarkably phasic in nature, and is unable to maintain a significant tone for more than a few minutes despite the continuing presence of agonists.

Neuromuscular Transmission in Smooth Muscle

Modulation of Tone Through Innervation

Some smooth muscles show no spontaneous tone in the absence of nervous activity, and in these cases tension is initiated by release of transmitter from excitatory nerves, leading to excitatory junction potentials, propagated spikes and phasic contraction. Most smooth muscles, including the detrusor, show some form of spontaneous tone, even in the presence of TTX (which ensures that there is no activity in nerves innervating the tissue). If, as in the detrusor, the tone is associated with the production of spontaneous spikes, then activity in the nerves may modify this spontaneous spike generation. Nerve impulses may initiate or inhibit action potential production transiently through excitatory or inhibitory junction potentials, or cause longer term changes in the frequency of the action potentials, correspondingly increasing or decreasing the tone. The transmitters may also affect tension through modulation of the voltage-sensitive Ca channels or through pharmacomechanical coupling mechanisms as described.

Innervation of Smooth Muscle

All efferent nerve fibres penetrating smooth muscles are unmyelinated and postganglionic. The intramuscular nerves contain bundles of fibres supported by Schwann cells. Towards the nerve endings, the axons become varicose, with expansions of about $1-2 \mu$m, containing synaptic vesicles and mitochondria, alternating with narrow lengths containing predominantly microtubules. The varicosities appear to be the release sites for the transmitters, and at these sites the nerve terminals are naked of Schwann cells. Beautiful electron micrographs of these sites can be found in the work of Gabella (e.g. Gabella 1978, 1981).

There is enormous variation in the pattern of innervation between different smooth muscles, and this has been extensively reviewed (e.g. Burnstock

1970; Bennett 1972). In general, the smooth muscles which show no spontaneous tone, but rely on nerve activity to initiate contractions, are the most densely innervated; the varicose nerve terminals penetrate between the cells within the muscle bundles, so that each cell may have close contact (20Å) with several varicosities. Activation of these nerves is likely to lead to discrete excitatory or inhibitory junction potentials. Smooth muscles that have spontaneous myogenic activity controlled by nerves are usually less densely innervated, and not all the cells have direct contact with a varicosity; release sites also occur at varicosities which may be some distance (up to 10 000Å) away. In these muscles activation of the nerves may evoke discrete potentials only in those cells directly innervated, while release from more distant sites may result in slower less discrete membrane potential changes.

The classical cholinergic and adrenergic autonomic postganglionic nerve fibres innervate smooth muscles, and have characteristic structure and physiology. Burnstock (1972) has produced useful diagrams showing the main features of their varicosities (Fig. 6.11). Over the last few decades it has become apparent, however, that the autonomic innervation to smooth muscle is much more complicated and subtle than was originally thought. There is increasing evidence for the presence and release of other active substances from axons (Burnstock 1972, 1981a): these include amines, such as histamine, serotonin, and dopamine; amino acids such as GABA; purines such as ATP, and a large number of peptides. Such substances may possibly be released on their own or as cotransmitters with the amines or acetylcholine. They may have pre- and postsynaptic effects, and may act either as primary transmitters, or as modulators of the transmission caused by the more classical transmitters. The nonadrenergic, noncholinergic nerve fibres may account for the presence of varicosities with features thought uncharacteristic of adrenergic and cholinergic nerve terminals. An illustration of the possible characteristics of a "purinergic" varicosity is included in Fig. 6.11.

Neuromuscular Transmission in the Detrusor

Activation of Intrinsic Nerves

It is possible to activate the intrinsic nerves in strips of smooth muscles from the bladder dome, trigone and urethra, without directly activating the smooth muscle cells. To do this, very short duration trans-

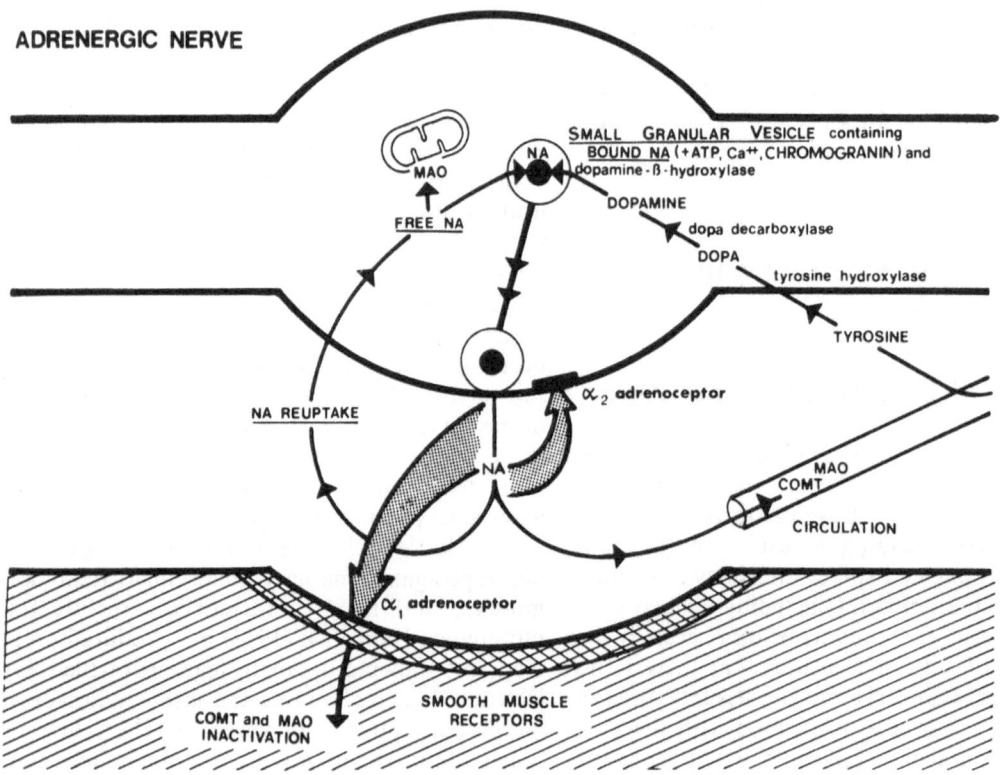

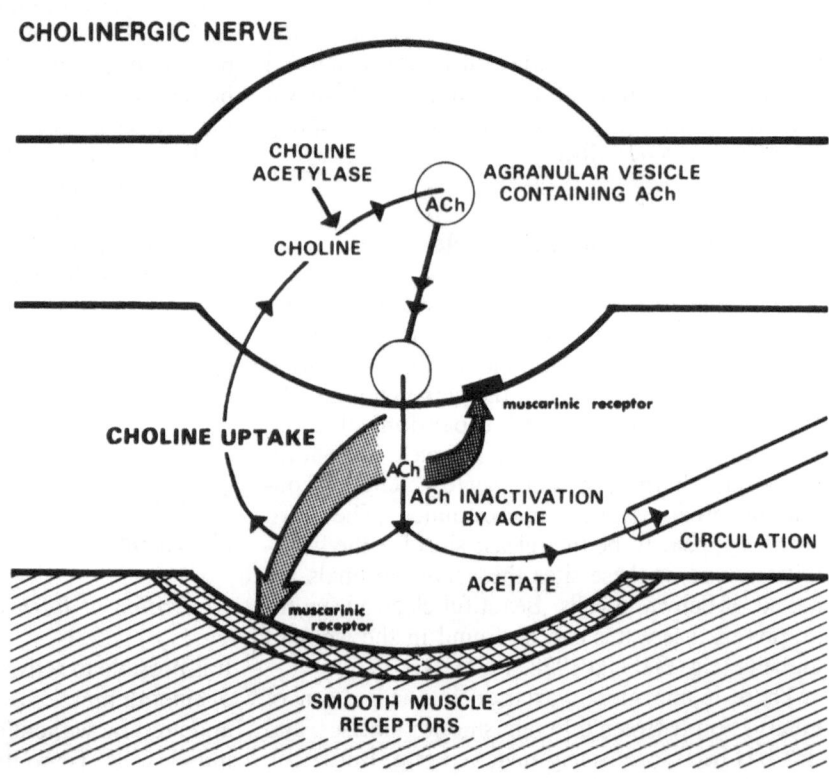

Fig. 6.11

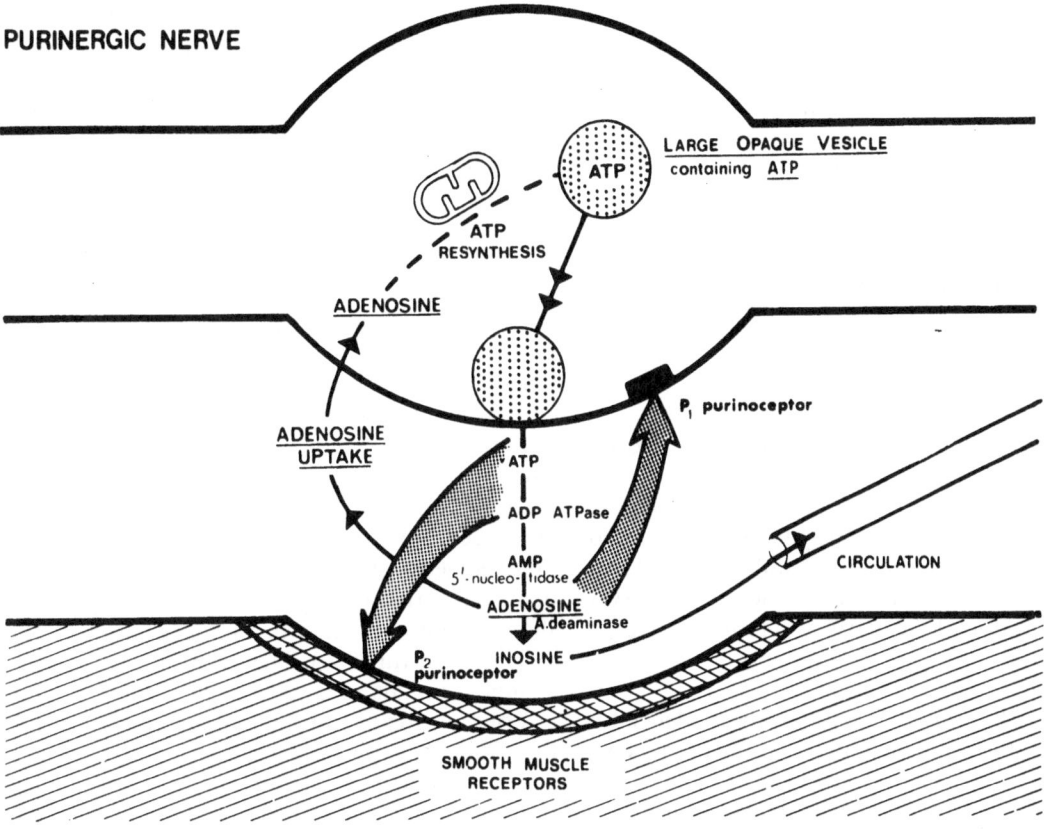

Fig. 6.11. Diagrammatic representation of adrenergic, cholinergic and "purinergic" varicosities. (Modified from Burnstock 1972)

mural current pulses are given. Because the time constant of nerves is very short (a few milliseconds) in comparison with the time constant of smooth muscle (more than 100 ms), currents can depolarize the nerves to threshold at a time when very little change in the smooth muscle membrane potential will have occurred. Any response of the muscle that is triggered by these pulses is thus likely to have been induced indirectly through release of transmitters from the nerves. To be certain that responses are nerve mediated, most workers test that they are abolished by TTX, since this stops action potential propagation in nerves, without affecting activity in the smooth muscles. TTX-resistant activity is then thought to be due to direct effects on the smooth muscle, although it is possible that current could directly release some transmitter from nerve terminals. (For known transmitters this possibility can be tested using blocking drugs.)

In the apparatus used in this laboratory, muscle responses to transmural pulses of 0.05 ms and 50 V were abolished by TTX at all frequencies used in the guinea pig and rabbit. In the pig they were abolished below 15 Hz, and were only 3% of the control at 50 Hz. In humans the TTX-resistant com-

ponent was 0.5% at 10 Hz, rising to 7% at 50 Hz (Sibley 1984a, b).

Contractile Responses

The response to intramural nerve stimulation of the detrusor from all mammals studied is a contraction. There is no evidence that I am aware of for any direct inhibitory innervation of this smooth muscle. Studies we have carried out in Oxford show that it is usually possible to see a contractile response to a single nerve pulse in most nonhuman bladders, especially when there is not too much spontaneous activity. Human bladder requires repetitive nerve stimulation before a significant response can be seen. Contractile responses are not antagonized by blockade of intramural ganglia with hexamethonium, suggesting that the response is evoked by postganglionic nerve stimulation.

Cholinergic Innervation The excitatory innervation to the detrusor muscle is via the parasympathetic pathway. This is classically regarded as a cholinergic system, with both pre- and postganglionic neurons releasing acetylcholine, and the receptors on the smooth muscle being muscarinic receptors. In bladder, acetylcholine has long been known to be released on stimulation of the parasympathetic nerves (e.g. Henderson and Roepke 1934), and it can be shown to come from the postganglionic neurons (e.g. Carpenter and Rand 1965). Anatomically, the nerves within the muscle bundles stain positively for acetylcholinesterase, and the varicosities of the terminals contain small agranular vesicles that are thought to be characteristic of cholinergic nerves. Functionally, acetylcholine applied in the bathing solution contracts the smooth muscle, an effect abolished by atropine and enhanced by cholinesterase inhibitors. Nerve-mediated responses of muscle strips are at least partially inhibited by atropine, particularly at higher frequencies, and the responses are enhanced by cholinesterase inhibitors (e.g. Sibley 1984a). All this is consistent with a role for cholinergic nerves in the excitation of the detrusor smooth muscle.

Atropine Resistance In spite of the evidence for cholinergic innervation, it has also long been recognized that in most mammalian bladders, contractile responses to parasympathetic nerve stimulation cannot be totally abolished by the acetylcholine muscarinic receptor blockers. Atropine resistance was reported by Langley and Anderson in 1895. These acute observers noted the contractile response of dog, rabbit and cat bladders by eye, and showed that atropine appeared to reduce only slightly the effectiveness of nerve stimulation. This partial resistance to atropine was further examined by Henderson and Roepke (1934) recording intraluminal pressure in the perfused dog bladder, and Ursillo and Clark (1956) using transmural stimulation of isolated strips of rabbit bladder. In both cases only a partial block by atropine could be shown. Since then many people have examined this phenomenon, and there has been much controversy about it. It has been suggested that acetylcholine is the transmitter, but that resistance to atropine is due to the large amounts of acetylcholine released during repetitive stimulation, overcoming the competitive antagonism. Alternatively, atropine may not reach the postjunctional receptors, or the receptors themselves may have an unusually low affinity for it. It has also been suggested that acetylcholine is not the transmitter, or not the only transmitter. A careful study by Ambache and Zar (1970) using guinea pigs, rabbits and cats, and single short-duration supramaximal nerve stimuli (totally blocked by TTX) showed that such responses were only very slightly affected by atropine at concentrations that

required an increase of 1000- to 2500-fold in bath-applied acetylcholine to produce a matching response to that obtained in the absence of atropine. The size of the nerve-evoked responses in the presence of atropine were not enhanced by cholinesterase inhibition, even though the postjunctional receptors were clearly not saturated with the transmitter, since two pulses produced a much larger response. These authors concluded that it was illogical to retain the concept that the motor innervation was cholinergic. At present, most people believe that there is either a dual excitatory innervation, or that cholinergic nerves release a cotransmitter along with acetylcholine.

Comparative studies of frequency–response curves to transmural nerve stimulation of detrusor muscle strips from different species in the presence and absence of atropine, show that the magnitude of atropine-resistant responses is frequency dependent, and varies from species to species. Figure 6.12 shows results obtained in this laboratory (Sibley 1984b; Mostwin 1986). The size of the response to a single stimulus as a percentage of the maximum response is greatest in the guinea pig ($\sim 30\%$); in the rabbit and pig it was about 20%, and there was no response in the human. In the nonhuman bladders, the atropine resistance was greatest at low frequencies of stimulation, atropine having almost no effect on the response to a single stimulus. At high frequencies, the guinea pig showed the most atropine resistance; about 76% of the contractile response remained in the presence of atropine ($10^{-6}M$), whereas in the rabbit about 56% and in the pig only some 15% remained in $5 \times 10^{-7}M$ atropine. The atropine-resistant component was removed by TTX, leaving only a very small TTX-resistant contraction in the pig. In human bladder, the small contractions remaining in the presence of atropine were not TTX sensitive, and therefore were unlikely to have been caused by nerve activity.

Some studies have been carried out on primates. In Old World monkeys (such as the rhesus monkey and baboon) bladder contraction in response to sacral ventral root stimulation was almost completely abolished by atropine, whereas the response of the more primitive New World primates (such as the marmoset and Cebus monkey) still showed some atropine resistance, particularly at low frequencies of stimulation (Craggs and Stephenson 1982, 1986). These authors suggest that the noncholinergic mechanism present in the New World monkeys may have evolved for the purposes of urinary marking, a behaviour also common in nonprimate mammals.

There is still argument with respect to human material. It is not always easy to get specimens of

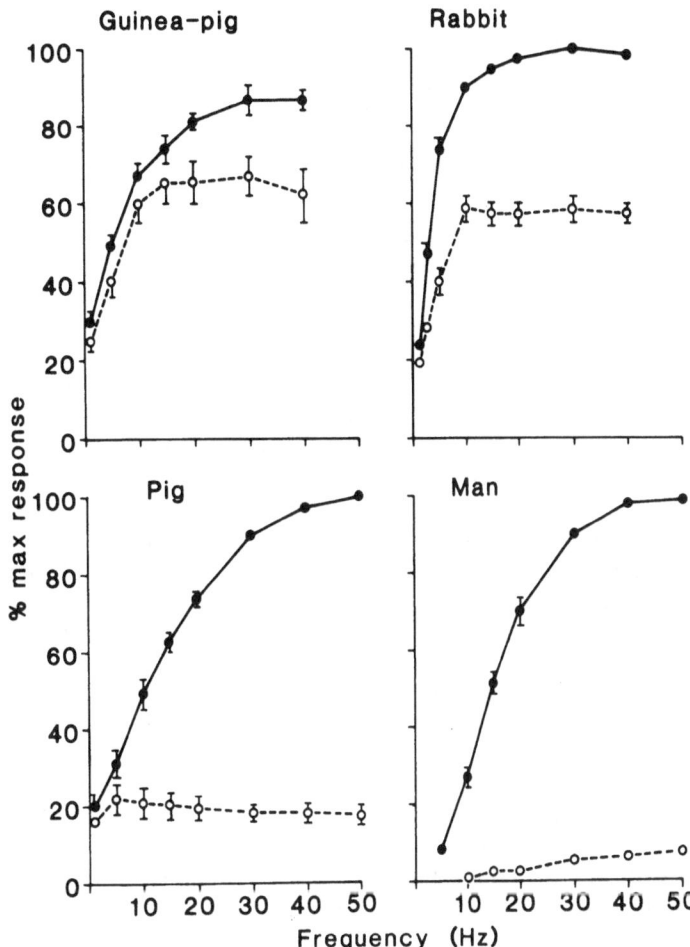

Fig. 6.12. The effect of atropine ($10^{-6} M$ in guinea pig, and $5 \times 10^{-7} M$ in other species) on the frequency–response curves of the detrusor from different species to transmural nerve stimulation. *Solid symbols*, control; *open symbols*, atropine-resistant response.

normal human bladder, but in those cases where we have studied material from cadavers, we have no evidence for any noncholinergic excitatory innervation to the detrusor. The fact that these strips require trains of stimuli before a response is elicited, also fails to provide support for a noncholinergic innervation since it is the response to a single stimulus that is the most atropine resistant in nonhuman mammals. However, large atropine-resistant contractions to transmural nerve stimulation have been seen in some specimens of abnormal human material by Sjögren et al. (1982b) and Nergardh and Kinn (1983), for instance in tissues obtained from patients with prostatic hypertrophy. In these studies the TTX sensitivity of the atropine-resistant contractions does not seem to have been tested. We have also seen a large amount of atropine resistance of the contractions to trains of 0.05 ms transmural stimuli in strips of muscle from unstable human bladders, but these contractions remain in the presence of TTX. Work in our laboratory strongly suggests that unstable bladder muscle

becomes hypersensitive to excitatory agents (Sibley 1984a; Speakman et al. 1986a), and we believe that in this state the membrane potential of the muscle cells may be so near threshold that direct muscle activation can be achieved even with the very short pulses used. Cowan and Daniel (1983), however, believe that these contractions are still nerve evoked, even if TTX resistant.

Nature of Noncholinergic Excitation The noncholinergic excitatory innervation is also nonadrenergic. Activation of β-adrenoceptors causes relaxation of detrusor smooth muscle, and although α-adrenoceptor agonists are excitatory in many species (e.g. Taira 1972), the frequency–response curves of muscle strips to transmural nerve stimulation are unaltered in the presence of doses of phentolamine that block contractions to bath-applied α-adrenoceptor agonists (e.g. Sibley 1984a). The present jargon is thus to call this innervation nonadrenergic, noncholinergic, or NANC.

The most likely candidate for the NANC transmitter seems to be ATP. Largely following the work of Burnstock and his collaborators, the idea has grown that ATP may play a role as primary transmitter or a cotransmitter in many postganglionic neurons in the autonomic nervous system, and may be either excitatory as in the bladder, vas deferens and some blood vessels, or inhibitory as in many gut muscles (for references see Burnstock 1972, 1981a, 1985; Daniel 1985b). It can be associated with other excitatory transmitters, such as noradrenaline in the vas deferens, or acetylcholine, as in the bladder. In the vas deferens, it is thought to be a cotransmitter with noradrenaline from adrenergic nerve fibres, as suggested by the use of neurotoxins specific for adrenergic nerves, which also block the NANC response. In the bladder it is not known whether ATP is a cotransmitter with acetylcholine from the postganglionic cholinergic nerves, or whether there are specific "purinergic" nerves releasing ATP.

The specific evidence suggesting the involvement of ATP in the NANC innervation of the bladder will now be considered. In 1972 Burnstock et al. tested the effects of several purine nucleotides on bladders of rats and guinea pigs. They showed that most of these substances can contract the bladders, but that ATP was the most potent, and at the correct dose, its application could match the electrically stimulated atropine-resistant contractions. Both were blocked by quinidine, although the response to acetylcholine in the absence of atropine persisted. Later work from this group using the guinea pig bladder has shown that ATP is released when the bladder is stimulated, and that the release is blocked by TTX and is Ca dependent (Burnstock et al. 1978a, b). The effects of various analogues suggested that the receptors involved were of the subclass P_2 (Burnstock et al. 1983).

A fairly specific antagonist (unfortunately irreversible) for ATP receptors has been found (arylazidoaminopropionyl-ATP, or ANAPP$_3$; Hogaboom et al. 1980). This blocks the effects of applied ATP, and partially blocks the pelvic nerve-evoked contractions of the bladder of the cat, without antagonizing the effects of acetylcholine (Theobald 1982). Another analogue, α,β-methylene-ATP (Kasakov and Burnstock 1983) has been found which produces specific desensitization of the P_2 purinoceptor. In the presence of this drug, after the initial excitation has diminished, the guinea pig bladder loses its ability to respond to the NANC innervation, but still responds to the cholinergic innervation and cholinergic agonists; the same is true in the pig. Similar results with adenylyl-5-(β,γ-methylene)-diphosphonate were obtained in the guinea pig by Hourani (1984).

Many other groups of workers have studied NANC excitatory innervation and the role of purines using the bladders of several different animals, for instance rabbit bladder has been studied by Dean and Downie (1978a, b), by Levin et al. (1981, 1982) and by Husted et al. (1980); guinea pig bladder by Krell et al. (1981) and by Westfall et al. (1983). The effects of purines on human bladder have also been studied by Husted et al. (1983). These authors found that ATP and ADP produced contractile responses, whereas AMP and adenosine produced no response. Hypertrophic bladders were more responsive to ATP than normal bladders, and indomethacin reduced any tonic component of the response that was present, suggesting that ATP might also induce the release of prostaglandins.

We have studied the effects of α,β-methylene-ATP on the pig and guinea pig bladder. Figure 6.13 shows the effects of atropine and α,β-methlyene-ATP alone and together on the frequency–response curve of detrusor strips from the guinea pig to transmural nerve stimulation (Mostwin 1986). As can be seen, α,β-methylene-ATP has a more pronounced affect on the low frequency response, and significantly reduces the response to a single stimulus, whereas atropine preferentially reduces the high frequency response. Together at the concentrations used, they are almost as effective as TTX in abolishing the effects of transmural nerve stimulation. Similar results were obtained in the pig.

Responses to Other Smooth Muscle Modulators
Bladder smooth muscle of humans and animal species may contract or relax in response to a variety of naturally occurring agents, and frequently the response may be mediated by specific receptors on the smooth muscle cells. In some instances there is no evidence that the substances are released by intravesical nerves, or that they are likely to reach the bladder smooth muscle under normal circumstances. Nevertheless, the activation or inhibition of such receptors may be of therapeutic value.

Human detrusor contracts in response to prostaglandins, and it is possible that prostaglandin release may contribute to the effects of other agents and even to nerve stimulation (Husted et al. 1983). It also contracts to histamine and serotonin, but does not normally respond to α-adrenoceptor stimulation. Relaxation is produced by β-adrenoceptor stimulation. Details of these responses may be found in Caine (1984).

Recent experiments on the effects of peptides have shown that substance P will cause animal bladders to contract, although it is not thought a likely candidate for the excitatory NANC transmitter

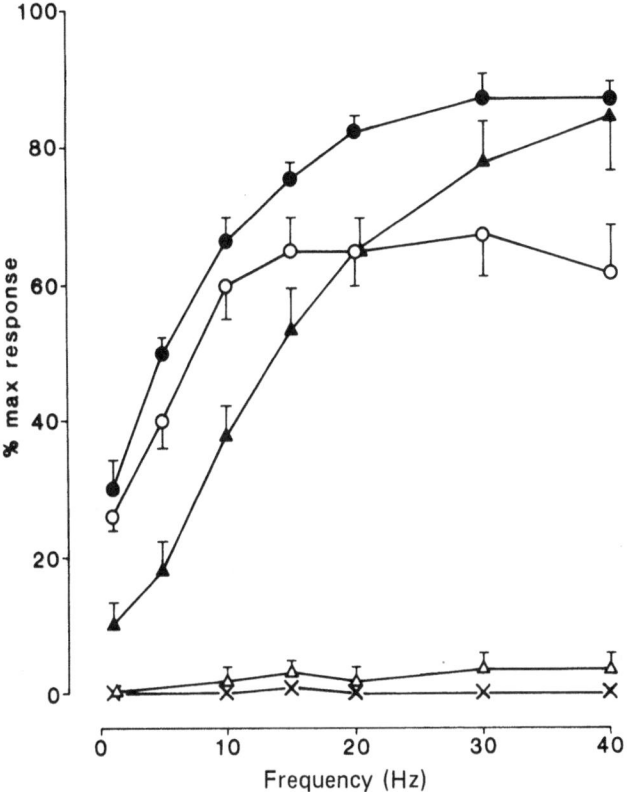

Fig. 6.13. Frequency–response curves of guinea pig detrusor to transmural nerve stimulation. *Solid circles*, control; *open circles*, in the presence of atropine ($10^{-6} M$); *solid triangles*, in the presence of α, β-methylene-ATP ($3 \times 10^{-5} M$); *open triangles*, in the presence of both drugs; *crosses*, in the presence of TTX ($2 \times 10^{-7} M$).

(Sjögren et al. 1982a; Mackenzie and Burnstock 1984; Callahan and Creedy 1986). VIP normally relaxes bladder smooth muscle (Levin and Wein 1981; Klarskov et al. 1984; Callahan and Creed 1986). Since there is some evidence for a sparse peptidergic innervation of the bladder (Alm et al. 1978; Hökfelt et al. 1978), peptides could in some species be playing a minor role in physiological control.

Electrical Responses

The effects of transmural nerve stimulation on the membrane potential of detrusor strips have been studied using rabbit (Creed et al. 1983) and guinea pig (Mostwin 1986). In the rabbit, double sucrose gap records show that a single supramaximal stimulus of the transmural nerves (responses blocked by TTX) produced an action potential and twitch from detrusor muscle strips. When the current amplitude was reduced, an early depolarization without a spike was seen, which was an excitatory junction potential (e.j.p.). This potential lasted about 500 ms.

It was often followed by a late depolarization, which had a latency of 250 ms and a duration of 600–1200 ms. The late depolarization was small or absent after single stimuli, but got larger if several stimuli at 20 Hz were applied. It was also markedly enhanced in the presence of the cholinesterase inhibitor, neostigmine, and it was abolished by atropine. This suggests that the late depolarization is a muscarinic response, resulting from stimulation of cholinergic nerves. In contrast, the early e.j.p. was not affected by these drugs. It was also not affected by adrenergic receptor blockers or guanethidine, suggesting that noradrenaline was not involved, or by blockers of 5-hydroxytryptamine or histamine receptors. These results confirm that the rabbit detrusor receives both cholinergic and NANC innervation, and additionally demonstrate that the NANC innervation is responsible for the e.j.p.s, and the contractile response to single stimuli. Creed et al. (1983) tried out the effects of some drugs that had been shown to interfere with purinergic transmission (quinidine, theophylline; Burnstock 1978), and found that likewise they did not interfere with e.j.p. production. At this time, however, the more specific drugs interacting with ATP receptors were not available, and the authors concluded that the NANC innervation was not purinergic, although ATP itself was shown to depolarize the tissues and cause a reduction in size of the e.j.p.s. Repetitive stimulation (0.5–2 Hz) caused a rundown in the e.j.p. amplitude. More recently, Callahan and Creed (1986) have investigated the effects of various peptides on the rabbit bladder. Although substance P could depolarize and contract the detrusor (and has been a candidate for the NANC transmitter), desensitization to substance P did not abolish the e.j.p.s and thus it is an unlikely candidate.

In the guinea pig, microelectrodes were used to look at the effects of transmural stimuli. As in the rabbit, spikes could be evoked by single transmural nerve stimuli. If conditioning hyperpolarizing current pulses were applied, the spike potential could be abolished, but in about one in four or five cells, an e.j.p. was seen. The amplitude ranged from 8 to 12 mV, and the duration from 400 to 600 ms. TTX or elevation of the concentration of Mg in the bathing solution could selectively abolish the e.j.p.s without altering the ability to elicit action potentials with direct stimulation of the muscle cells with long pulses (TTX resistant). Nerve-evoked e.j.p.s or action potentials were not abolished by atropine. In the presence of neostigmine, a slowly developing and prolonged late depolarization could sometimes be elicited by transmural nerve stimulation, and this was abolished by atropine. These experiments again suggest that both cholinergic and NANC innervation is present, the NANC innervation being responsible for the e.j.p.s. The fact that e.j.p.s could be recorded in only a small percentage of the cells, although nerve-evoked action potentials could be seen in all cells, suggests that only a proportion of the cells in this species receives close innervation from a nerve, and that in cells that do not have close innervation, the action potential is propagated from adjacent cells. In the guinea pig, the effect of α,β-methylene-ATP was tried. The tissue was progressively desensitized with the drug, and after desensitization, single transmural nerve stimuli were no longer able to initiate spikes, although direct muscle stimulation (TTX resistant) could still elicit spikes (Mostwin 1986). This is again consistent with ATP being the transmitter responsible for initiation of the e.j.p.s.

Neuromuscular Transmission in the Trigone, Bladder Base and Urethra

The trigone has, with the exception of a few noteworthy papers (Tanagho et al. 1968; Tanagho 1976), received less attention in the literature. The smooth muscle of the trigone consists of two distinct layers, and classically the innervation of the superficial trigone is described as entirely adrenergic, while the deep trigone is thought to be indistinguishable both morphologically and in its innervation from the detrusor.

We have studied the contractile responses from smooth muscles of the human and pig deep and superficial trigone and compared them with the responses of the detrusor. Human material was obtained from patients undergoing total cystectomy. It proved comparatively easy unambiguously to dissect viable strips from both layers. In the pig, the superficial trigone was very thin, and contained much connective tissue, so that preparations tended to respond rather poorly to electrical and chemical stimuli, and we did not study them extensively. The main trigonal structure seemed to be the deep trigone in the pig, and it was easy to dissect strips from this layer.

Superficial Trigone

We have evidence for transmural nerve stimulation activating four different mechanisms in the human superficial trigone, three excitatory and one inhibitory. The major excitatory innervation seems to be noradrenergic, as we expected from the literature, and consistent with the vigorous contractile response of the strips to phenylephrine, an α-adrenoceptor agonist. However, there is also a considerable cholinergic response to nerve stimuli, and the tissue can also produce about 40% of its maximum response to applied acetylcholine. Figure 6.14 shows frequency–response curves of the contraction to transmural nerve stimulation in the presence and absence of phentolamine (an α-adrenoceptor blocker), and atropine. Atropine ($10^{-6}M$) significantly reduced the contractile responses, particularly at higher frequencies, whereas phentolamine ($10^{-6}M$) reduced the responses more evenly at all frequencies. A combination of the two drugs did not, however, completely abolish the TTX-sensitive responses to transmural stimulation, particularly at lower frequencies, suggesting that there may be a small NANC component.

Electrical stimulation regularly produced TTX-sensitive relaxations of the superficial trigone at low stimulation frequencies. At higher frequencies the contractile responses appeared, and gradually replaced the relaxations. Both muscarinic and α-receptor blockade increased the size of the relax-

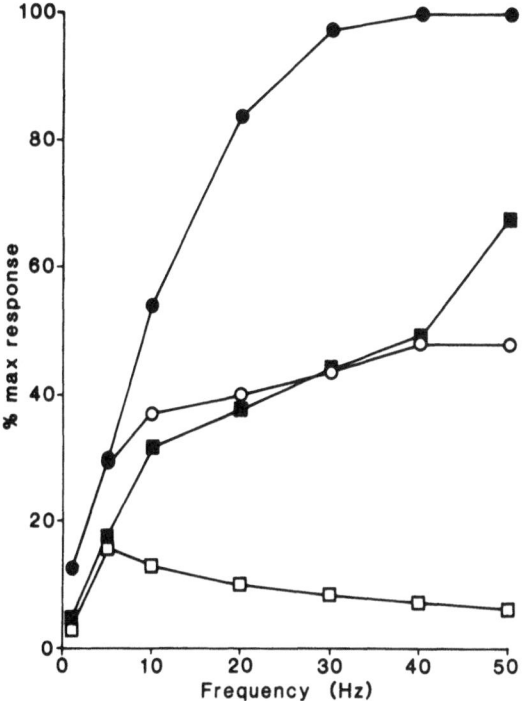

Fig. 6.14. Frequency–response curves of the human superficial trigone to transmural nerve stimulation. *Solid circles*, control; *open circles*, in the presence of $10^{-6}M$ atropine; *solid squares*, in the presence of $10^{-6}M$ phentolamine; *open squares*, in the presence of both drugs. Residual response blocked by TTX, indicating that there is a small NANC component to the innervation.

ations by approximately 50%. Propranolol had no effect on the relaxations at β-receptor blocking concentrations ($\leqslant 10^{-6}M$), but at higher concentrations ($> 10^{-5}M$) reduced both the relaxations and the contractile responses (probably a local anaesthetic effect). These results suggest that the relaxation is also mediated by NANC nerves (Speakman et al. 1986b).

Deep Trigone

In human tissue, the deep trigone responded vigorously to carbachol, but also showed a small contractile response to α-adrenergic stimulation (18%). In comparison, the detrusor as expected showed a strong muscarinic contraction, but did not contract (and in most cases actually showed a small relaxation) to α-adrenergic activation with either phenylephrine or noradrenaline. (This muscle relaxation may be mediated through an inhibitory action of α-adrenoceptors on intrinsic parasympathetic ganglia.) In the pig, the deep trigone was much less

similar to the detrusor, and more like the human superficial trigone. It had a predominantly adrenergic innervation, and produced its most powerful contraction to applied noradrenaline. The cholinergic component was smaller, and applied acetylcholine could produce a contraction of only about 40% of the maximum. There was again some evidence of an NANC excitatory component. As discussed already, nerve-mediated relaxations of the trigonal muscle were also seen.

Bladder Base and Proximal Urethra

In humans it can be seen anatomically that there is an extensive adrenergic innervation of the bladder neck in the male, and only a sparse cholinergic one (Gosling et al. 1983). Recent work (Walker and Bates 1985) on biopsy specimens of the bladder neck (taken during open and transurethral prostatectomy) suggests that there may also be an excitatory NANC innervation associated with the adrenergic innervation. The tissues contracted to transmural stimulation of the nerves, and also to applied noradrenaline. Doses of α-blockers sufficient to abolish the responses to noradrenaline had no effect on the nerve-mediated response. This situation is very reminiscent of the behaviour of the vas deferens in most species studied, in which there is now gathering evidence that ATP may be coreleased with noradrenaline from the sympathetic nerves (Sneddon and Westfall 1984).

More recently some work has been published on the mechanical and electrical responses of the bladder base and proximal urethra of rabbits to transmural nerve stimulation (Ito and Kimoto 1985). The behaviour of the bladder base was very similar to the dome. In the proximal urethra, however, stimulation evoked a phasic contraction followed by relaxation, and electrical recording showed excitatory and inhibitory junction potentials and a late depolarization. The size of the fast e.j.p. was reduced to about half by guanethidine or phentolamine (suggesting that there is an adrenergic component) and was further reduced, but not abolished, by atropine, leaving an NANC component. The late depolarization was abolished by atropine, as in the detrusor. The inhibitory junction potential (i.j.p.) was also mediated by NANC innervation.

Klarskov et al. (1983) have also demonstrated an NANC inhibitory innervation to the trigone, bladder neck and urethra of humans and pigs (see also Torrens 1978), and further studies as to the possible identity of the transmitter involved have been carried out on the pig bladder neck by Hills

Table 6.1. Innervation of bladder muscles

	Excitatory			Inhibitory
	Cholinergic	Adrenergic	NANC	NANC
Human				
Detrusor	***	0	0	0
Deep trigone	***	**	−	−
Superficial trigone	**	***	√	√
Urethra	*	***	√	−
Animal				
Detrusor	***	0	√	0
Trigone	**	***	√	√
Urethra	**	**	√	√

Symbols: *** present and of major importance; ** present and of medium importance; * present and of minor importance; 0 absent; √ present, but relative importance unknown or variable with species; − not tested.

et al. (1984). The only substances out of a wide range studied that were potential candidates for the inhibitory transmitter were vasoactive intestinal polypeptide (VIP), and 5-hydroxytryptamine (5-HT). These substances were able to relax the tissue at reasonable concentrations, but substances which were able to block the relaxant effects of the applied drugs (methysergide and chymotrypsin) had no effect on the nerve-mediated relaxations.

These studies on the base of the bladder and urethra suggest that the innervation here may be rather complex. The fact that there is almost certainly more than one NANC innervation is of considerable interest, and it seems probable that more than one transmitter is involved. When the identity of the transmitters has been unequivocally established, the development of specific agonists and antagonists to the receptors mediating the responses may play an important role in the future treatment of abnormal micturition. Table 6.1 summarizes the innervation of the smooth muscles of the lower urinary tract.

Summary

Detrusor

Activation of intrinsic nerves causes only excitation of the detrusor. In nonhuman bladders there is evidence for two types of excitatory innervation: one cholinergic, and the other nonadrenergic, noncholinergic (NANC). The cholinergic innervation occurs in human and animal bladders; it does not elicit e.j.p.s, and is most effective with repetitive stimulation. In the presence of cholinesterase inhibitors its effectiveness is enhanced, and slowly devel-

oping depolarizations can be seen in response to transmural nerve stimulation. Stimulation of the NANC nerves elicits e.j.p.s and contractions will result from a single stimulus. These nerves probably have ATP as a transmitter. Human detrusor contracts poorly to single stimuli of the intrinsic nerves, and does not appear to possess the NANC innervation.

Trigone, Bladder Base and Urethra

The innervation of these structures is more complicated than that of the bladder dome. The superficial trigone appears to be innervated by three excitatory pathways (adrenergic, cholinergic and NANC), and an NANC inhibitory pathway. The deep trigone is not just an extension of the detrusor, and contracts in response to both cholinergic and α-adrenergic agonists. The bladder base and proximal urethra have an extensive excitatory adrenergic innervation (but there is probably an excitatory NANC cotransmitter with noradrenaline), a small excitatory cholinergic innervation and an NANC inhibitory pathway, with an unknown transmitter.

Acknowledgements. I should like to thank all my colleagues for their help and encouragement in the writing of this chapter, and for allowing me to use their results. In particular, I thank Jo Barber, Paul Dennis and Roneen Hobbs for their help in preparing the manuscript, and Jeremy Heaton, Jacek Mostwin, Gary Sibley, Mark Speakman and Doug Walmsley for their results.

References

Abe Y, Tomita T (1968) Cable properties of smooth muscle. J Physiol (Lond) 196:87–100

Aickin CC, Brading AF (1983) Towards an estimate of chloride permeability in the smooth muscle of guinea-pig vas deferens. J Physiol (Lond) 336:179–197

Alm P, Alumets J, Brodin E et al. (1978) Peptidergic (substance P) nerves in the genito-urinary tract. Neuroscience 3:419–425

Ambache N, Zar MA (1970) Non-cholinergic transmission by post-ganglionic motor neurons in the mammalian bladder. J Physiol (Lond) 210:761–783

Askoy MO, Murphy RA, Kamm KE (1982) Role of Ca^{2+} and myosin light chain phosphorylation in regulation of smooth muscle. Am J Physiol 242:C109–C116

Bennett MR (1972) Autonomic neuromuscular transmission. Cambridge University Press, Cambridge

Bolton TB (1979) Mechanisms of action of transmitters and other substances on smooth muscle. Physiol Rev 59:606–718

Brading AF (1981) Ion distribution and mechanisms of transmembrane ion movement. In: Bülbring E, Brading AF, Jones AW, Tomita T (eds) Smooth muscle: an assessment of current knowledge. Arnold, London, pp 66–92

Brading AF, Aickin CC (1985) Sodium, calcium and contraction in smooth muscle. In: Mironneau J (ed) Calcium regulations in smooth muscle: biochemical and physiological aspects. Inserm, Paris

Brading AF, Sneddon P (1980) Evidence for multiple sources of Ca for activation of the contractile machinery of guinea-pig taenia coli on stimulation with carbachol. Br J Pharmacol 70:229–240

Brading AF, Widdicombe JH (1977) The use of lanthanum to estimate the numbers of extracellular cation-exchanging sites in the guinea-pig taenia coli, and its effects on transmembrane monovalent ion movements. J Physiol (Lond) 266:255–273

Brading AF, Mostwin JL, Sibley GNA, Speakman MJ (1986) The role of smooth muscle and its possible involvement of diseases of the lower urinary tract. Clin Sci 70 (suppl 14):7S–13S

Burnstock G (1970) Structure of smooth muscle and its innervation. In: Bülbring E, Brading AF, Jones AW, Tomita T (eds) Smooth muscle. Arnold, London, pp 1–69

Burnstock G (1972) Purinergic nerves. Pharmacol Rev 24:509–581

Burnstock G (1978) A basis of distinguishing 2 types of purinergic receptor. In: Straub RW, Bolis L (eds) Cell membrane receptors for drugs and hormones. Raven, New York

Burnstock G (1981a) Neurotransmitters and trophic factors in the autonomic nervous system. J Physiol (London) 313:1–35

Burnstock G (1981b) Development of smooth muscle and its innervation. In: Bülbring E, Brading AF, Jones AW, Tomita T (eds) Smooth muscle: an assessment of current knowledge. Arnold, London, pp 431–457

Burnstock G (1985) Nervous control of smooth muscle by transmitters, cotransmitters and modulators. Experientia 41:869–874

Burnstock G, Dumsday B, Smythe A (1972) Atropine resistant excitation of the urinary bladder: the possibility of transmission via nerves releasing a purine nucleotide. Br J Pharmacol 44:457–461

Burnstock G, Cocks T, Crowe R, Kasakov L (1978a) Purinergic innervation of the guinea-pig urinary bladder. Br J Pharmacol 63:125–138

Burnstock G, Cocks T, Kasakov L, Wong A (1978b) Direct evidence for ATP release from non-adrenergic, non-cholinergic ("purinergic") nerves in the guinea-pig taenia coli and bladder. Eur J Pharmacol 49:145–149

Burnstock G, Cusack NJ, Hills JM, Mackenzie I, Meghji P (1983) Studies on the stereoselectivity of the P_2-purinoceptor. Br J Pharmacol 79:907–913

Caine M (1984) The pharmacology of the urinary tract. Springer, Berlin Heidelberg New York

Callahan SM, Creed KE (1981) Electrical and mechanical activity of the isolated lower urinary tract of the guinea-pig. Br J Pharmacol 74:353–358

Callahan SM, Creed KE (1985) The effects of oestrogens on spontaneous activity and responses to phenylephrine of the mammalian urethra. J Physiol (Lond) 358:35–46

Callahan SM, Creed KE (1986) Non-cholinergic neurotransmission and the effects of peptides on the urinary bladder of guinea-pigs and rabbits. J Physiol (Lond) 374:103–115

Carpenter FG, Rand SA (1965) Relation of acetylcholine release to responses of the rat urinary bladder. J Physiol (Lond) 180:371–382

Coburn RF, Ohba M, Tomita T (1975) Recordings of intracellular electrical activity with the sucrose gap method. In: Daniels EE, Paton DM (eds) Methods in pharmacology, vol 3. Plenum, New York, pp 231–245

Coolsaet B (1985) Bladder compliance and detrusor activity during the collection phase. Neurourol Urodyn 4:263–273

Cowan WD, Daniel EE (1983) Human female bladder and its noncholinergic contractile function. J Physiol Pharmacol 61:1236–1246

Craggs MD, Stephenson JD (1982) The effects of parasympathetic blocking agents on bladder E.M.G.'s and function in conscious and anaesthetised cats. Neuropharmacology 21:695–703

Craggs MD, Stephenson JD (1986) A non-cholinergic urinary bladder mechanism in New World primates. J Physiol (Lond) 377:78p

Craig R, Megerman J (1977) Assembly of smooth muscle myosin into side polar filaments. J Cell Biol 75:990–996

Creed KE (1971a) Membrane properties of smooth muscle membrane of the guinea-pig urinary bladder. Pflügers Arch 326:115–126

Creed KE (1971b) Effects of ions and drugs on the smooth muscle cell membrane of the guinea-pig urinary bladder. Pflügers Arch 326:127–141

Creed KE, Ishikawa S, Ito Y (1983) Electrical and mechanical activity recorded from rabbit urinary bladder in response to nerve stimulation. J Physiol (Lond) 338:149–164

Daniel EE (1985a) The use of subcellular membrane fractions in analysis of control of smooth muscle function. Experientia 41:905–913

Daniel EE (1985b) Nonadrenergic, noncholinergic (NANC) neuronal excitatory interactions with smooth muscle. In: Grover AK, Daniel EE (eds) Calcium and contractility. Humana, Clifton, NJ

Daniel EE Cowan W, Daniel VP (1983) Structural basis for neural and myogenic control of human detrusor. Can J Physiol Pharmacol 61:1247–1273

Dean DM, Downie JW (1978a) Contribution of adrenergic and purinergic neurotransmission to contraction in rabbit detrusor. J. Pharmacol Exp Ther 207:431–445

Dean DM, Downie JW (1978b) Interaction of prostaglandins and A.T.P. in the non-cholinergic neurotransmission in rabbit detrusor. Prostaglandins 16:245–251

Downie JW, Slack BE (1983) Sensitivity to indomethacin of tetrodotoxin-resistant contractions of smooth muscle from the base of rabbit bladder. Br J Pharmacol 79: 334:336

Elbadawi A (1982) Ultrastructure of vesicourethral innervation: 1. Neuroeffector and cell junctions in male internal sphincter. J Urol 128:180–188

Endo M, Kitazawa T, Yagi S et al. (1977) Some properties

of chemically skinned smooth muscle fibres. In: Casteels R, Godfraind T, Rüegg JC (eds) Excitation-contraction coupling in smooth muscle. Elsevier/North Holland, Amsterdam, pp 199–209

Gabella G (1976) The force generated by a visceral smooth muscle. J Physiol (Lond) 263: 199–213

Gabella G (1978) The structure of the autonomic nervous system. Chapman Hall, London

Gabella G (1981) Structure of smooth muscles. In: Bülbring E, Brading AF, Jones AW, Tomita T (eds). Smooth muscle: an assessment of current knowledge. Arnold, London, pp 1–46

Golenhofen K (1976) Theory of P and T systems for calcium activation in smooth muscle. In: Bülbring E, Shuba MF (eds) Physiology of smooth muscle. Raven, New York, pp 197–202

Golenhofen K (1981) Differentiation of calcium activation processes in smooth muscle using selective agonists. In: Bülbring E, Brading AF, Jones AW, Tomita T (eds) Smooth muscles: an assessment of current knowledge. Arnold, London, pp 157–170

Gosling JA, Dixon JS (1975) The structure and innervation of smooth muscle in the wall of the bladder neck and proximal urethra. Br J Urol 47:549–558

Gosling JA, Dixon JS, Humpherson JR (1983) Functional anatomy of the urinary tract. Churchill Livingstone, Edinburgh

Griffiths DJ (1980) Urodynamics. Medical physics handbooks 4, Adam Hilgar, Bristol

Griffiths DJ, Van Mastrigt R, van Duyl WA et al. (1979) Active mechanical properties of the smooth muscle of the urinary bladder. Med Biol Eng Comput 17:281–290

Hashimoto T, Hirata M, Itoh T et al. (1986) Inositol 1,4,5-trisphosphate activates pharmacomechanical coupling in smooth muscle of the rabbit mesenteric artery. J Physiol (Lond) 370:605–618

Henderson VE, Roepke MH (1934) The role of acetylcholine in bladder contractile mechanisms and in parasympathetic ganglia. J Pharmacol Exp Ther 51:97–111

Hills J, Meldrum L, Klarskov P et al. (1984) A novel non-adrenergic, non-cholinergic nerve mediated relaxation of the pig bladder neck; an examination of possible neuro-transmitter candidates. Eur J Pharmacol 99:287–293

Hogaboom GK, O'Donnell JP, Fedan JS (1980) Purinergic receptors: arylazide photoaffinity analogue of ATP is a specific pharmacological antagonist of ATP. Science 208:1273

Hökfelt T, Schultzberg M, Elde R et al. (1978) Peptide neurones in peripheral tissues including the urinary tract: immunohistochemical studies. Acta Pharmacol Toxicol 43:79–89

Hourani SMO (1984) Desensitization of the guinea-pig urinary bladder by the enantiomers of adenylyl s-(β,γ-methylene) diphosphonate and by Substance P. Br J Pharmacol 82:161–164

Husted S, Sjögren C, Andersson K-E (1980) Role of prostaglandins in the responses of rabbit detrusor to non-cholinergic, non-adrenergic nerve stimulation and to ATP. Arch Int Pharmocodyn Ther 246:84–97

Husted S, Sjögren C, Andersson K-E (1983) Direct effects of adenosine and adenine nucleotides on isolated human urinary bladder and their influence on electrically induced contractions. J Urol 130:392–398

Isenberg G, Klöckner V (1985) Calcium currents of smooth muscle cells isolated from the urinary bladder of guinea pig: inactivation, conductance and selectivity is controlled by micromolar amounts of [Ca]°. J Physiol (Lond) 358:60p

Ito Y, Kimoto Y (1985) The neural and non-neural mechanisms involved in urethral activity in rabbits. J Physiol (Lond) 367:57–72

Itoh T, Kuriyama H, Suzuki H (1981) Excitation-contraction coupling in smooth muscle cells of the guinea-pig mesenteric artery. J Physiol (Lond) 321:513–535

Itoh T, Ueno H, Kuriyama H (1985) Calcium-induced calcium release mechanism in vascular smooth muscles—assessments based on contractions evoked in intact and saponin treated skinned muscles. Experientia 41:989–996

Kasakov L, Burnstock G (1983) The use of the slowly degradable analog, α,β-methylene ATP, to produce desensitisation of the P_2-purinoceptor: effect on non-adrenergic, non-cholinergic responses of the guinea-pig urinary bladder. Eur J Pharmacol 86:291–294

Klarskov P, Gerstenberg T, Ramirez O et al. (1983) Non-cholinergic, non-adrenergic nerve mediated relaxation of trigone, bladder neck, and urethral smooth muscle in vitro. J Urol 129:848–850

Klarskov P, Gerstenberg T, Hald T (1984) Vasoactive intestinal polypeptide influence on lower urinary tract smooth muscle from human and pig. J Urol 131:1000–1004

Klück P (1980) The autonomic innervation of the human urinary bladder, bladder neck and urethra: a histochemical study. Anat Rec 198:439–447

Krell RD, McCoy JL, Ridley PT (1981) Pharmacological characterization of the excitatory innervation to the guinea-pig urinary bladder in vitro: evidence for both cholinergic and non-adrenergic, non-cholinergic neurotransmission. Br J Pharmacol 74:15–22

Kurihara S (1975) The effect of procaine on the mechanical and electrical activities of the smooth muscle cells of the guinea-pig urinary bladder. Jpn J Physiol 25:775–788

Kurihara S, Creed KE (1972) Changes in membrane potential of the smooth muscle cells of the guinea-pig urinary bladder in the various environments. Jpn J Physiol 22:667–683

Kuriyama H (1981) Excitation-contraction coupling in various visceral smooth muscles. In: Bülbring E, Brading AF, Jones AW, Tomita T (eds) Smooth muscle: an assessment of current knowledge. Arnold, London, pp 171–197

Langley JN, Anderson HS (1895) The innervation of the pelvic and adjoining viscera. Part II. The bladder. J Physiol (Lond) 19:71–84

Levin RM, Wein AJ (1981) Effect of vasoactive intestinal peptide on the contractility of the rabbit urinary bladder. Urol Res 9:217–218

Levin RM, Wein AJ (1982) Response of the in vitro whole bladder (rabbit) preparation to autonomic agonists. J Urol 128: 1087–1090

Levin RM, Jacoby R, Wein AJ (1981) Effect of adenosine triphosphate on contractility and adenosine triphosphatase activity of the rabbit urinary bladder. Mol Pharmacol 19:525–528

Levin RM, Jacoby R, Wein AJ (1982) High affinity, divalent, ion-specific binding of tritiated A.T.P. to homogenate derived from rabbit urinary bladder. Mol Pharmacol 23:1–7

Mackenzie I, Burnstock G (1984) Neuropeptide action on the guinea-pig bladder; a comparison with the effects of field stimulation and ATP. Eur J Pharmacol 105:85–94

Marston SB, Smith CWJ (1985) The thin filaments of smooth muscles. J Muscle Res Cell Motil 6:669–708

Morgan JP, Morgan KG (1984) Stimulus-specific patterns of intracellular calcium levels in smooth muscle of ferret portal vein. J Physiol (Lond) 351:155–167

Mostwin JL (1985) Receptor operated intracellular stores in the smooth muscle of the guinea-pig bladder. J Urol 133:900–905

Mostwin JL (1986) Electrical and mechanical aspects of bladder contractility. DPhil Thesis, Oxford

Nergardh A, Boreus LO (1972) Autonomic receptor function in the lower urinary tract of man and cat. Scand J Urol Nephrol 6:32–36

Nergardh A, Kinn A-C (1983) Neurotransmission in activation

of the contractile response in the human urinary bladder. Scand J Urol Nephrol 17:153–157

Rüegg JC, Pfitzer G (1985) Modulation of calcium sensitivity in guinea-pig taenia coli: skinned fiber studies. Experientia 41:997–1001

Satake N, Shibata S, Veda S (1984) Phentolamine-induced rhythmic contractions in bladder detrusor muscle of guinea-pig. Br J Pharmacol 83:965–971

Sibley GNA (1984a) The response of the bladder to lower urinary tract obstruction. DM Thesis, Oxford

Sibley GNA (1984b) A comparison of spontaneous and nerve-mediated activity in bladder muscle from man, pig and rabbit. J Physiol (Lond) 354:431–443

Sibley GNA (1985) An experimental model of detrusor insta-bility in the obstructed pig. Br J Urol 57:292–298

Sjögren C, Andersson K–H, Husted S (1982a) Contractile effects of some peptides on the isolated urinary bladder of guinea-pig, rabbit and rat. Acta Pharmacol Toxicol 500:175–184

Sjögren C, Andersson K–H, Husted S et al. (1982b) Atropine resistance of transmurally stimulated isolated human bladder muscle. J Urol 128:1368–1371

Small JV (1977) Studies on isolated smooth muscle cells: the contractile apparatus. J Cell Sci 24:327–349

Sneddon P, Westfall DP (1984) Pharmacological evidence that ATP and noradrenaline are co-transmitters in the guinea-pig vas deferens. J Physiol (Lond) 347:561–580

Somlyo AV, Frazini-Armstrong C (1985) New views of smooth muscle structure using freezing, deep-etching and rotary shadowing. Experientia 41:841–856

Somlyo AP, Somlyo AV (1968) Vascular smooth muscle. 1. Normal structure, pathology, biochemistry and biophysics. Pharmacol Rev 20:197–272

Speakman MJ, Brading AF, Dixon JS (1986a) The patho-physiology of detrusor instability in obstructed micturition. Proceedings of the 16th Annual Meeting of the International Continence Society, Boston

Speakman MJ, Walmsley D, Brading AF (1986b) An in vitro pharmacological study of the human trigone—a site of non-adrenergic, non-cholinergic transmission. Proceedings of the British Association of Urological Surgeons, London meeting

Taira N (1972) The autonomic pharmacology of the bladder. Annu Rev Pharmacol 12:197–208

Tanagho EA (1976) The ureterovesical junction: anatomy and physiology. In: Williams DI, Chisholm GD (eds) Scientific foundations of urology, vol 2. Year Book Medical Publishers, Chicago, p 32

Tanagho EA, Smith DR, Meyers FH (1968) The trigone: ana-tomical and physiological considerations. 2. In relation to the bladder neck. J Urol 100:633-639

Theobald R (1982) Arylazido aminopropionyl A.T.P. (ANAPP3) antagonism of cat urinary bladder contractions. J Auton Pharmacol 3:175–179

Tomita T (1970) Electrical properties of mammalian smooth muscle. In: Bülbring E, Brading AF, Jones AW, Tomita T (eds) Smooth muscle. Arnold, London, pp 197–243

Torrens MJ (1978) Urethral sphincteric responses to stimulation of the sacral nerves in the human female. Urol Int 33:22–26

Ursillo RC (1961) Electrical activity of the isolated nerve-urinary bladder strip preparation of the rabbit. Am J Physiol 201:408–412

Ursillo RC, Clark B (1956) The action of atropine on the urinary bladder of the dog and on an isolated nerve-bladder strip preparation of the rabbit. J Pharmacol Exp Ther 118:338–347

Uvelius B (1976) Isometric and isotonic length-tension relations and variations in cell length in longitudinal smooth muscle from rabbit urinary bladder. Acta Physiol Scand 97:1–12

Uvelius B, Gabella G (1980) Relation between cell length and force production in urinary bladder smooth muscle. Acta Physiol Scand 110:357–365

van Duyl WA (1985a) A model for both the passive and the active properties of urinary bladder tissue related to bladder function. Neurourol Urodyn 4:275–283

van Duyl WA (1985b) Spontaneous contractions in urinary bladder smooth muscle: preliminary results. Neurourol Urodyn 4:301–307

Walker J, Bates CP (1985) Responses of isolated strips of smooth muscle from the human bladder neck. Proceedings of the 15th Annual Meeting of the International Continence Society, London, pp 168–169

Westfall DP, Fedan JS, Colby J et al. (1983) Evidence for a contribution by purines to the neurogenic response of the guinea-pig urinary bladder. Eur J Pharmacol 87:415–422

Wuytack F, Raemaekers L, Casteels T (1985) Ca-exchange, Ca-channels and Ca-antagonists. Experientia 41:900–905

7 · Reflex Control of the Lower Urinary Tract

John F. B. Morrison

Descending Spinal Pathways Concerned with Micturition
Reflex Integration of Bladder and Sphincteric Function
Reflexes Regulating the Force of Bladder Contractions
Reflex Control of Sphincters: Connections Within Onuf's Nucleus
Sympathetic Reflexes
Effects of Spinal Transection on the Function of the Lower Urinary Tract

Descending Spinal Pathways Concerned with Micturition

Reflexes that regulate the lower urinary tract may have pathways entirely within the spinal cord or may involve looped pathways coursing between the cord and the brainstem. The ascending pathways were described in Chapter 4 (see p. 115). Sensory pathways and reflex pathways are not so easily distinguished, and the pathways described for sensation may also be involved, via axon collaterals in reflex control, at any level between the spinal cord and the thalamus. Before we deal with the reflexes that regulate the activities of the lower urinary tract, we need to consider the main descending pathways within the central nervous system that influence the activities of the autonomic and Onuf's nucleus.

Holstege and Kuypers (1982) have described in some detail the descending pathways to the intermediolateral column, using the retrograde HRP method, and microinjections of labelled amino acids into different brain stem nuclei; the labelled amino acid can be traced to axonal endings in the spinal cord by autoradiography. These techniques show the origin and termination as well as the paths of the descending fibres, and, when combined with neurochemistry, different groups of neurons can be identified within these pathways. In 1964, Dahlström and Fuxe, using histochemical techniques, described the locations of neurons that contained monoamines. The A group cells contain either noradrenaline or adrenaline, and the B group neurons contain 5-hydroxytryptamine (serotonin); certain groups of cells in both categories are known to project to the spinal cord, and the A1, A5, B1 and B3 groups as well as the raphe nucleus (a site of serotonergic neurons project to the spinal cord, where they are distributed to the autonomic outflows as well as the ventral and superficial dorsal horns. Light (1985) has shown that the spinal terminals of single raphe–spinal axons are distributed to more than one of these sites (see Fig. 7.5). The descending pathways from the raphe nuclei and the

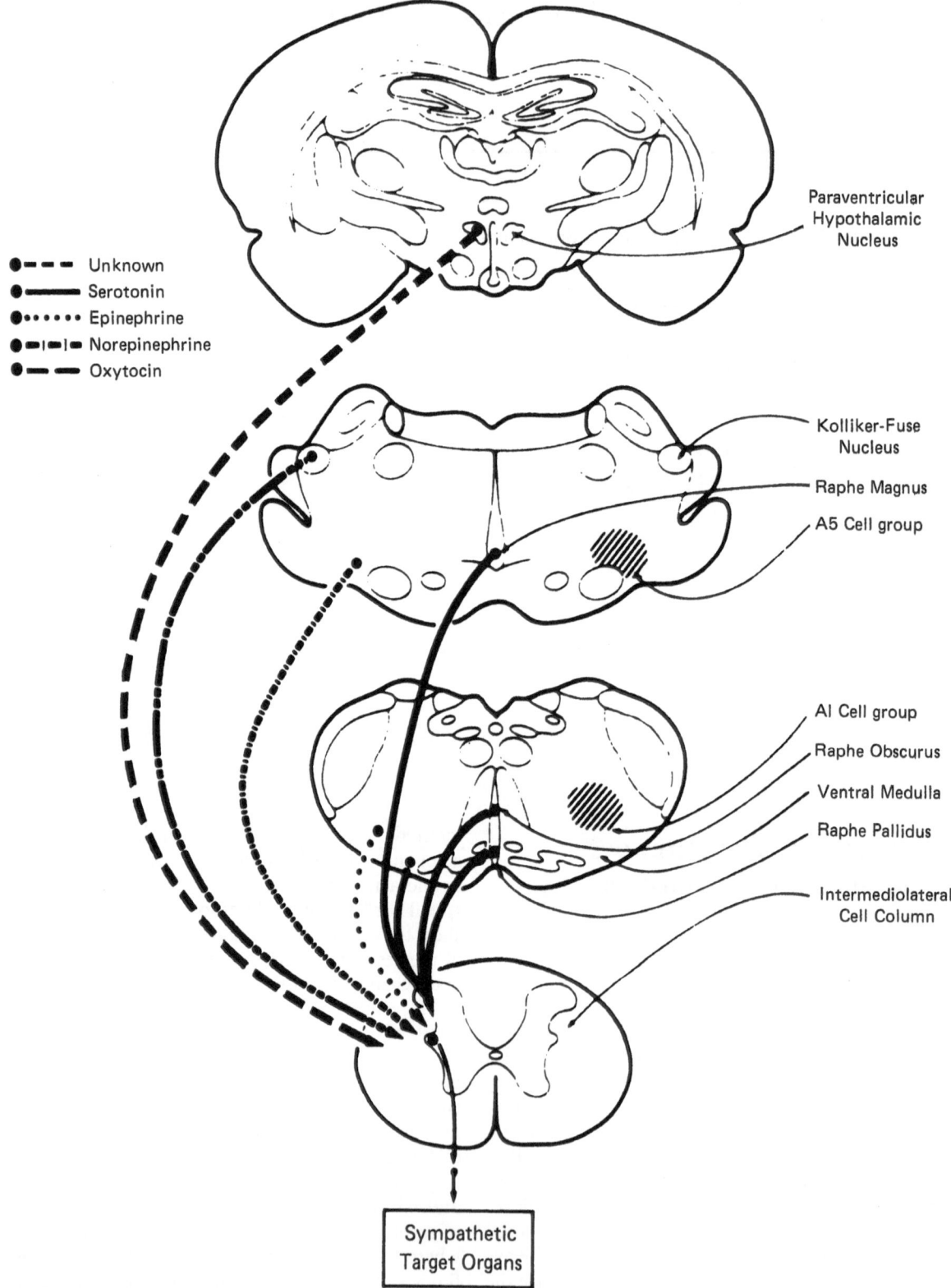

Fig. 7.1. The origins of some of the main descending pathways that converge on the intermediolateral column. (Reproduced from Jänig 1985; based on Loewy and Neil 1981)

locus coeruleus are the main descending pathways containing serotonin and noradrenaline, respectively.

More recently, other histochemical maps have been produced showing the distribution of peptidergic neurons in the central nervous system; Cuello et al. (1983) for instance have outlined the distribution of substance P neurons, and Hökfelt et al. (1979) have described the enkephalinergic neurons that project from the brain stem to the spinal cord. These techniques have also shown that at least one of the classical peptide hormones, oxytocin, is present in some of the fibres that run from the hypothalamus to the spinal cord (Swanson and Sawchenko 1983), where they terminate in the intermediolateral cell columns (Fig. 7.1), possibly producing inhibition of the sympathetic efferents (Coote et al. 1982).

Location of Descending Pathways in the Spinal Cord

In humans, Nathan and Smith (1958) state that the descending pathways relevant to micturition are located in the equatorial plane of the cord within the lateral columns. The efferent fibres which control the bladder are medial, next to the grey matter, whereas those innervating the skeletal muscle of the sphincters and pelvic floor lie more laterally, in the middle of the lateral columns. Section of the medial part of the lateral columns bilaterally interrupts this descending autonomic pathway, and gives rise to a change in bladder function to a state similar to that present after complete cord transection (Nathan 1976). In fact, the anatomy of the pathways that pass through this region of the spinal cord is quite complex, and the following sections outline some of the main types of neurons that may contribute to control of the bladder.

In cats, Kerr and Alexander (1964) located the descending pathway to the bladder in the dorsolateral funiculus, whereas McMahon and Morrison (1982c) found it in the lateral funiculus, in an equatorial region, and ventral to it in the cervical segments of the cord. In the latter study, small stimulating currents were applied through a bipolar electrode in the lateral funiculus of the cervical cord, and the descending pathway was found to be quite discretely localized (Fig. 7.2). The position of the descending pathway in the ventrolateral columns in the cat is supported by the studies of Wang and Ranson (1939a,b) who found that the excitatory effects elicited by hypothalamic stimulation were transmitted through this region of the cord.

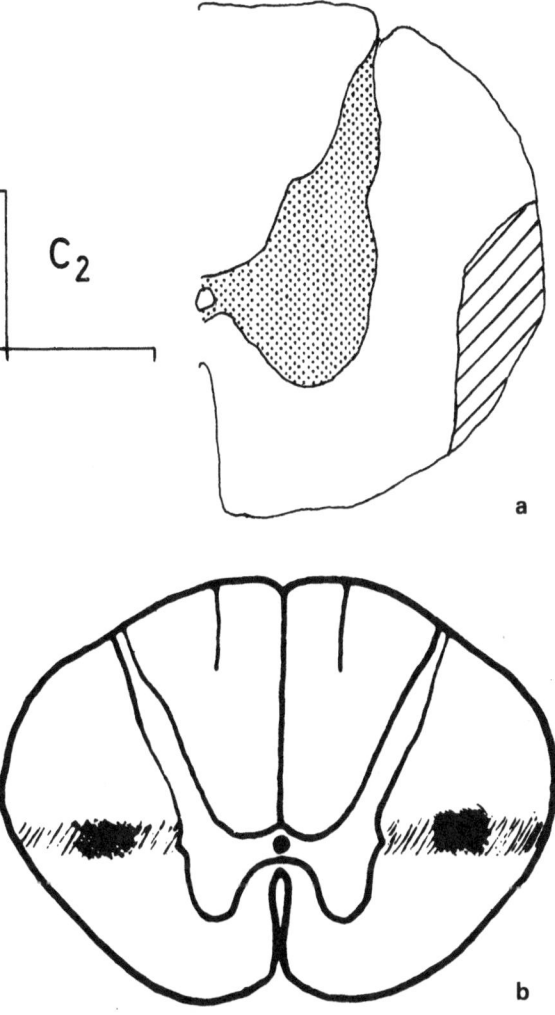

Fig. 7.2a,b. The locations of descending pathways that influence the pelvic efferents to the bladder in (a) cat and (b) human. (a from McMahon and Morrison 1982c; b from Nathan and Smith 1958)

Origin of Descending Pathways in the Brain Stem

In the cat, Barrington (1925) found that destruction of a medial area of the rostral pons (medial to the pontine micturition centre) caused a loss of the desire to micturate and urinary retention; stimulation of this area causes inhibition of micturition (Appenzeller 1982; Kuru 1965). Lesions of a more lateral region associated with the locus coeruleus caused an inability to contract the bladder, and this region was described as the pontine micturition centre or the pontine detrusor centre. Kuru (1965) studied the locations of degenerating myelin associated with lesioned descending tracts; lesions were made using electric currents delivered to sites where

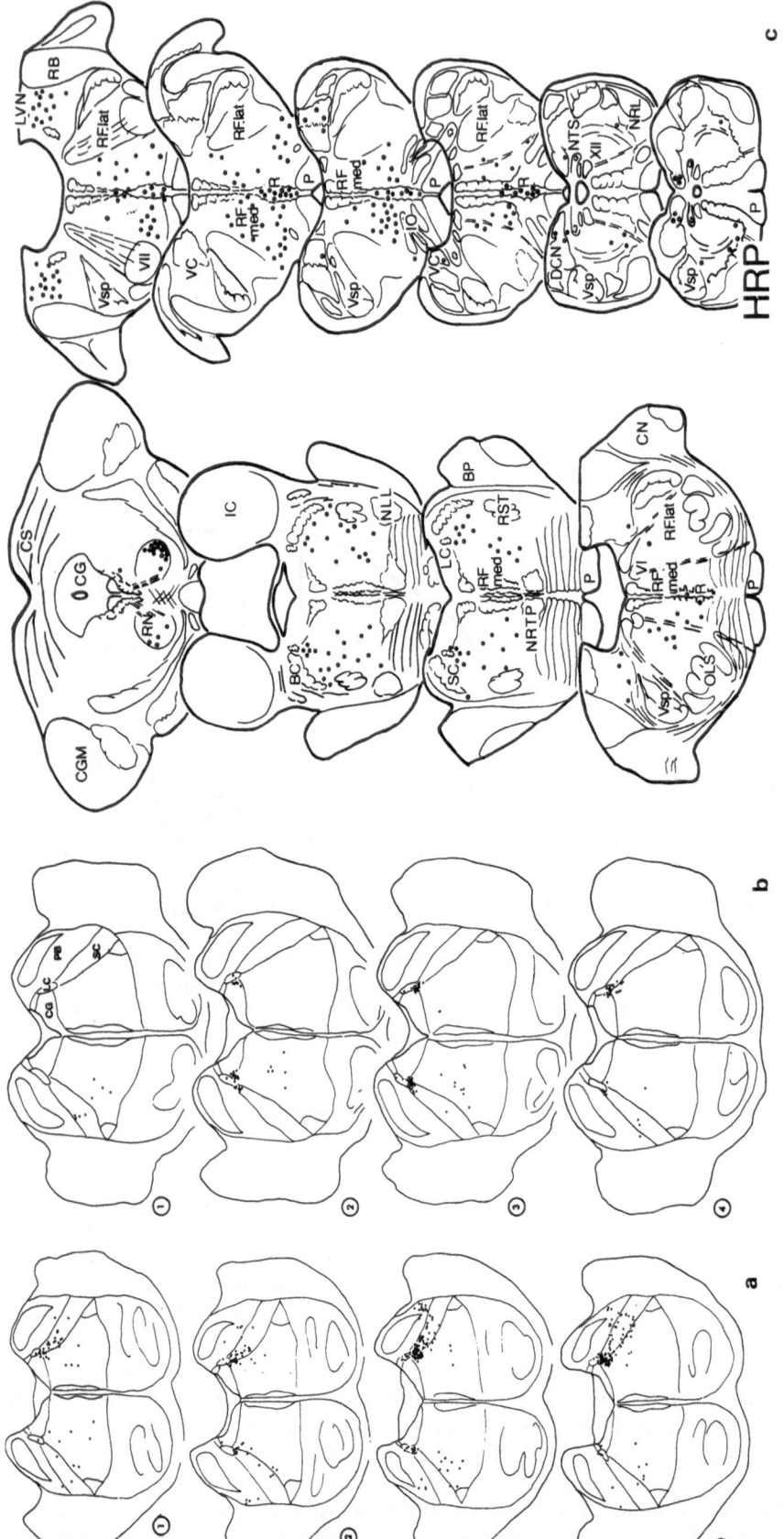

Fig. 7.3. a,b Origins of spinal projections from the locus coeruleus and subcoeruleus/parabrachial regions in the monkey. The locations of neurons in the locus coeruleus, subcoeruleus, parabrachial nuclei and pontine tegmentum labelled retrogradely with HRP from the left upper cervical cord (**a**) and sacral spinal cord (**b**). Each *dot* corresponds to a single labelled neuron. Drawings show all labelled neurons in a single 50 μm section. Note, in **b**, the relatively few labelled neurons in the subcoeruleus/medial parabrachial nuclei following an injection of HRP restricted to the sacral spinal cord. **c** Distribution of the retrogradely labelled neurons in the cat brain stem after left S-1 HRP injection. (**a,b** from Westlund and Coulter 1980; **c** from Holstege and Kuypers 1982)

stimuli elicited changes in bladder motility, and it was assumed that these descending pathways may all have had some function in micturition. Kuru described three descending tracts, the lateral, ventral and medial reticulospinal tracts; Nathan (1976) comments that bilateral lesions of the medial and ventral reticulospinal tracts in humans are not associated with disturbances of micturition. Spillane (1982), and McMahon and Spillane (1982) investigated the sites in the brain stem that give rise to bladder contractions and efferent activity in the pelvic nerve, and found only one, rather diffuse, pathway running caudally through the brain stem concerned with bladder contractions; this pathway probably originates within the locus coeruleus or lateral tegmental nucleus (see p. 199 and Fig. 7.3). In addition, two descending inhibitory pathways ran from: (a) the raphe nucleus in the caudal medulla; and (b) a more lateral site around the nucleus gigantocellularis reticularis of the pons and medulla (see Fig. 8.10, p. 250).

Descending Pathways from the Medulla: Raphe–Spinal and Reticulospinal Pathways

Figure 7.3b shows the distribution of HRP-containing cell bodies in the cat brain stem following injection of HRP into the first sacral segment of the cat (Holstege and Kuypers 1982); cell bodies in the raphe nuclei, the medial reticular formation, the locus coeruleus/medial parabrachial region, and the nucleus retroambiguus contain the tracer. Microinjections of tritiated acids into the midline of the medulla and medial tegmental fields resulted in labelling of axonal terminals at different sites in the spinal cord, depending on the level of the injection. The raphe nuclei and medial reticular formation project to the thoracolumbar and sacral autonomic motoneuronal cell columns (Fig. 7.4). The intermediate zones of the spinal grey also receive projections from all levels of the medial medulla. In addition, injections into the midline of the brain stem rostral to the facial nucleus label terminals in the dorsal horn, while injections into more caudal sites show up axonal terminations in motoneuronal cell groups.

The effects of raphe stimulation on bladder motility appear to be mediated by bilateral pathways in the cord, and include not only inhibition of bladder motility and reflexes, but also of vesicosphincteric reflexes (McMahon et al. 1982), and transmission through the dorsal horn. Morrison and Spillane (1986) found that the inhibitory effects of the raphe–spinal pathway on the micturition reflex were reversed to a large extent by lysergic acid diethylamide, an antagonist of serotonin; this is consistent with other studies on the raphe–spinal pathway that suggest that some of its actions are mediated by serotonin. Gilbey et al. (1981) also studied the effects of the raphe–spinal pathway, but on sympathetic reflexes recorded from the reflex responses of preganglionic neurons; they too found inhibition, which could be partially reversed by lysergic acid diethylamide or cyproheptadine, suggesting that the pathway was serotonergic. The distribution of terminals of single raphe–spinal axons has recently been investigated by Light (1985), using intra-axonal injection of HRP near the terminals; Fig. 7.5 shows the distribution in the lower lumbar cord of the cat (a region where the autonomic outflow is absent in this species).

The intermediolateral cell column is also innervated by descending pathways from the ventrolateral medulla. Two groups of serotonergic neurons, the B1 group with cell bodies in the nucleus interfasciculus hypoglossi, and the B3 group originating from the nucleus reticularis gigantocellularis, pars *alpha* directly innervate the autonomic nuclei and the ventral horn (Loewy et al. 1981). These descending fibres reach lumbar and sacral levels via the dorsolateral columns in the thoracolumbar cord, but their functions with respect to micturition are not known. They also project to more rostral sites that send descending pathways to the spinal cord, e.g. the dorsolateral tegmental nucleus (which may be the location of the pontine micturition centre in the rat; see p. 250), the locus coeruleus and the lateral parabrachial nuclei in the pons, the central grey matter of the midbrain and the paraventricular nucleus of the hypothalamus. Following microinjections of tritiated amino acids into the locus coeruleus/medial parabrachial area of the dorsolateral pons, Holstege and Kuypers (1982) found evidence of axon terminals at all levels of the grey matter of the cord, but preferentially apposed to the autonomic cell groups of the sacral cord, while neurons in the centrolateral pons could be traced to the upper thoracic sympathetic nuclei, the phrenic nucleus and Onuf's nucleus.

Other catecholamine-containing neurons in the brain stem also project to the intermediolateral column, e.g. the A1 group appears to be the main adrenaline-containing input, and the A5 group the main noradrenaline containing pathway. Sympathetic preganglionic neurons are inhibited by noradrenaline and excited by serotonin, but the effects of stimulating neuronal pathways that contain these appear to have the opposite effects. Stimulation of the raphe nuclei and the norad-

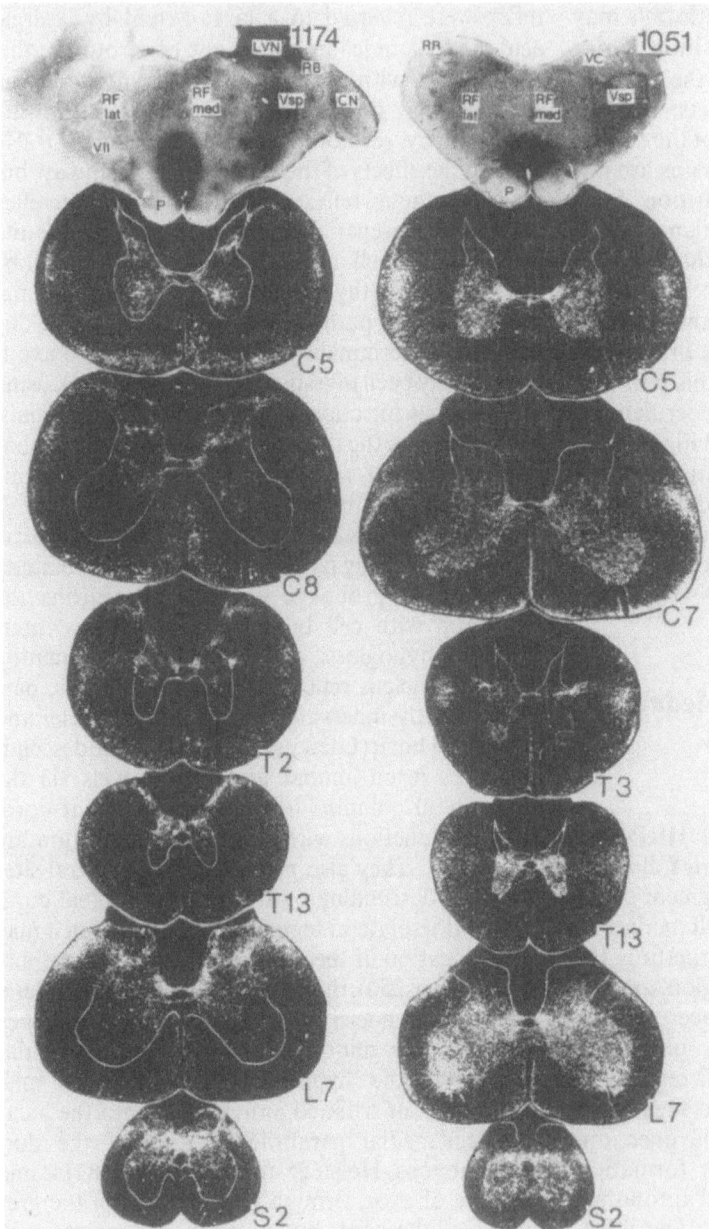

Fig. 7.4. Bright field photomicrographs of the autoradiographs showing the tritiated leucine injection areas and dark field photomicrographs showing the spinal distributions of labelled fibres in case 1174 (*left*), with an injection in the nucleus raphe magnus and the adjoining reticular formation, and in case 1051 (*right*), with an injection in the raphe nucleus at the rostral pole of the inferior olive. In case 1174 (*left*), the labelled fibres are distributed mainly to the dorsal horn, the intermediate zone and the autonomic motoneuronal cell groups. The latter projection is most striking in the T-2 photomicrographs. In contrast (*right*), the labelled fibres are distributed mainly to the intermediate zone and the autonomic and somatic motoneuronal groups, but not the dorsal horn. (Holstege and Kuypers 1982)

renergic pathway respectively inhibit and excite sympathetic efferents; the discrepancy is unexplained (Loewy and Neil 1981).

Intracerebroventricular administration of opioid peptides or morphine inhibit bladder contractions in cats (Hisamitsu and de Groat 1984). The site or sites of action of these peptides in producing these effects are still to be determined, but it is possible that the enkephalins may either inhibit excitatory descending pathways or excite inhibitory descend-

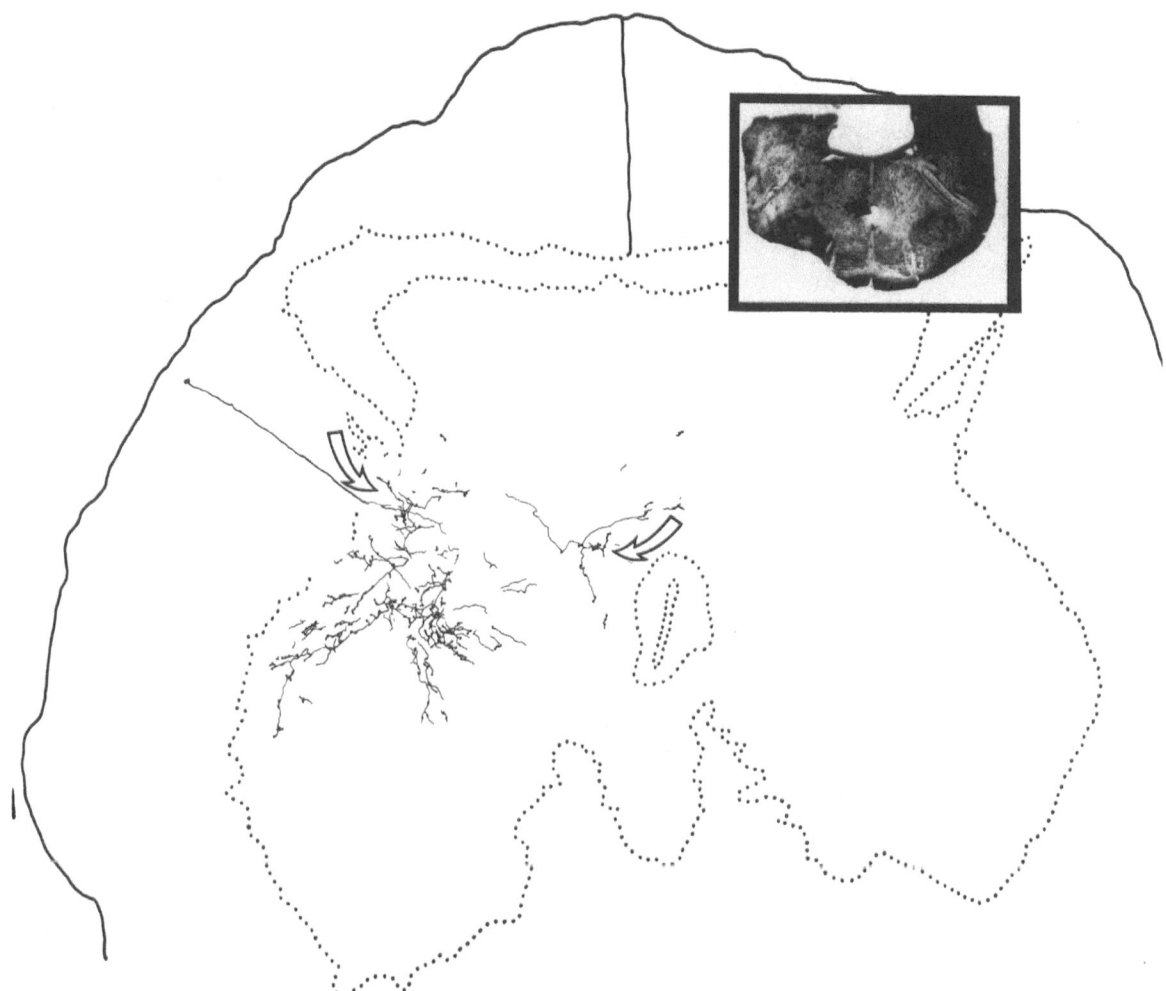

Fig. 7.5. Camera lucida reconstruction from five 50-μm thick transverse sections from an axon which was injected with HRP in the dorsolateral funiculus (DLF) and which had its cell soma in the brain stem, probably near the midline of the medulla (see the *insert*). A small portion of the axon is shown in the DLF, and a single axon collateral and its terminations are reconstructed. (Light 1985)

ing pathways. In this regard, neurons of the raphe nuclei in the medulla can be excited by enkephalins (Lumb 1984, 1986).

Pathways from the Dorsolateral Pons: The Locus Coeruleus and Lateral Tegmental Nucleus

A number of workers have suggested that the origin of the descending excitatory pathway in the micturition reflex was situated in the region of the locus coeruleus or the dorsal tegmental nucleus, just rostral to the locus coeruleus (Barrington 1925;

Tang and Ruch 1956; de Groat 1975; Satoh et al. 1979a,b; Westlund and Coulter 1980). The descending pathway from the locus coeruleus projects to all levels of the cord in the monkey (Westlund and Coulter 1980); the adjacent, but more lateral medial parabrachial nucleus has a much less dense projection to the sacral cord. Noradrenaline is present in the locus coeruleus and the lateral tegmental nucleus, but noradrenaline is probably not excitatory to the parasympathetic efferent neurons (Ryall and de Groat 1972); it is known, however, that noradrenaline coexists with enkephalin, neuropeptide Y and vasopressin in the neurons of locus coeruleus (Hökfelt et al. 1986). The role of the nordrenaline-containing axons remains to be estab-

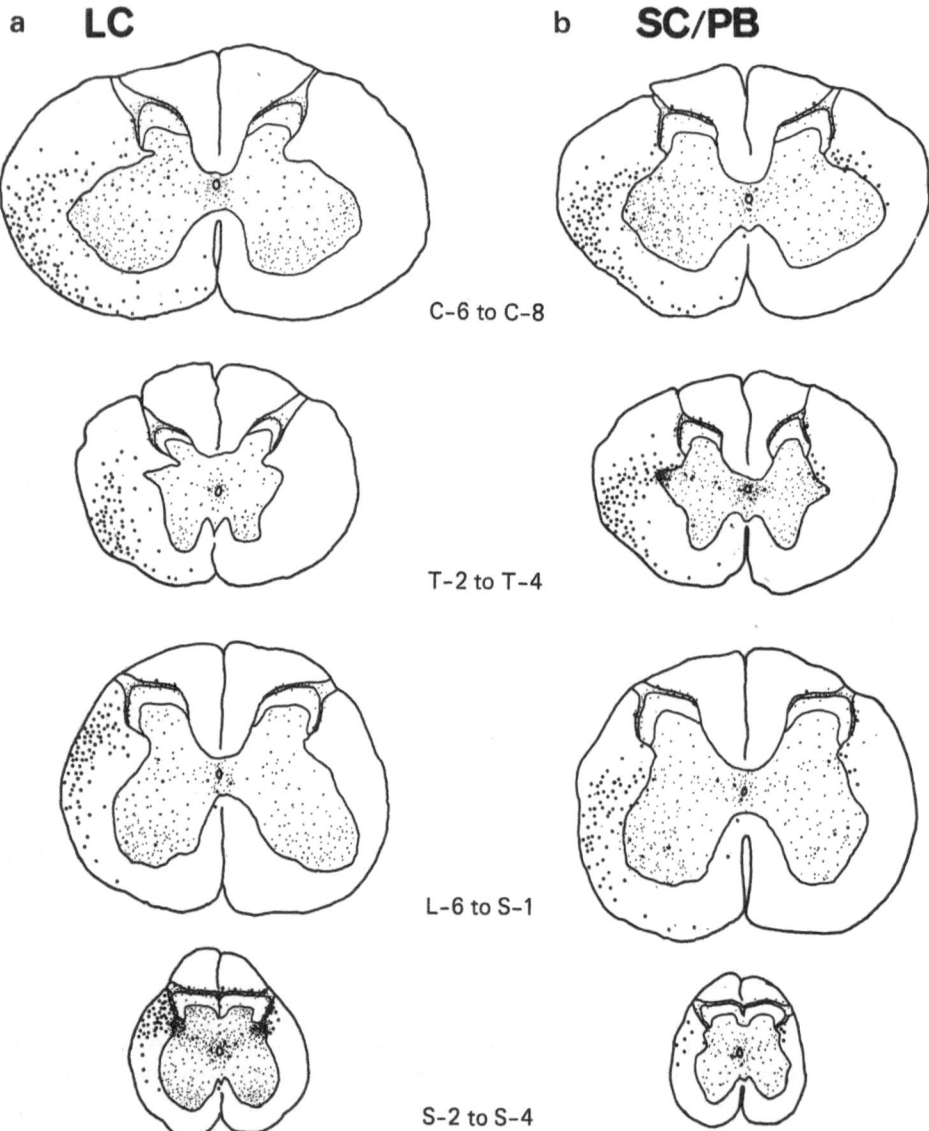

a LC

b SC/PB

C-6 to C-8

T-2 to T-4

L-6 to S-1

S-2 to S-4

Fig. 7.6a,b. Spinal cord terminations of descending projections from the locus coeruleus and subcoeruleus/medial parabrachial nuclei. Distribution of anterograde axonal (*large dots*) and terminal labelling (*fine stipple*) are plotted from autoradiographs on representative sections of the spinal cord taken from within the indicated spinal segmental levels. **a** Labelling pattern from injection of isotope into the left locus coeruleus (*LC*) shows heaviest concentrations of silver grains bilaterally over the sacral cord segments (*S-2 to S-4*) in the intermediolateral cell column and dorsal horn. Labelling is also seen at all spinal cord levels in the superficial dorsal horn, intermediate zone, the ventral horn and around the central canal. **b** Labelling pattern from injections of isotope into the left subcoeruleus/medial parabrachial nuclei (*SC/PB*) shows heaviest concentrations of silver grains bilaterally in the thoracic cord segments (*T-2 to T-4*), as well as in the superficial dorsal horn, the ventral horn and around the central canal. Note the presence of labelled fibres in the dorsal part of the lateral funiculus contralateral to injections of the subcoeruleus/medial parabrachial nuceli. In the drawings the borders of the substantia gelatinosa (laminae II and III) of the dorsal horn are indicated. (Westlund and Coulter 1980)

lished, and they are thought to be involved in anti-nociception (Mokha et al. 1985), in micturition and defecation (de Groat 1975; Elam et al. 1986) or in linking autonomic or visceral functions and behaviour (Elam et al. 1986).

Satoh et al. (1978a) have studied another more medially situated nordrenaline-containing nucleus, the lateral tegmental nucleus, in the rat. This nucleus does not have an equivalent in the monkey, but in the rat it projects to the thoracolumbar cord.

(Unfortunately possible projections to the sacral cord were not examined.) Satoh et al. (1978b) have suggested the involvement of this nucleus in the micturition reflex, because rats with bilateral lesions of this nucleus developed urinary retention. This was interpreted as localizing the pontine micturition centre to the lateral tegmental nucleus in the rat. Lumb and Morrison (1987) have found that electrical stimulation with currents as low as $5–10\,\mu A$ in the parabrachial nucleus of the rat can cause bladder contractions; furthermore, microinjections of DL-homocysteic acid, an excitatory amino acid that selectively excites cell bodies, but not axons, also causes bladder contractions when injected into this region of the dorsolateral rostral pons. This suggests that some cell bodies in the parabrachial region can cause bladder contractions in the rat. In the cat, cholinoceptive stimulation of this region can produce antinociceptive effects and muscular atonia (Katayama et al. 1984); in addition some visceromotor functions were suppressed, but no mention was made of micturition or sphincteric function.

The precise origin of the descending pathway of the micturition reflex has therefore still to be clarified, although the general region of its location is not in doubt. The results are complicated by the presence of inhibitory pathways, which are generally present medial to the detrusor regions. The distribution of axonal terminals of descending spinal axons from the locus coeruleus is shown in Fig. 7.6; tritiated amino acid has been injected into: (a) the locus coeruleus; and (b) the locus sub-coeruleus and medial parabrachial region, and the protein containing the radioactive label has been transported towards the axon terminals, where it has been demonstrated using autoradiography. Axons descend in the lateral columns (mainly), and adopt a more dorsal position as the pathway reaches more caudal segments of the spinal cord; there is a marked distribution to the superficial laminae of the dorsal horn (Westlund and Coulter 1980).

Pathways from the Paraventricular Nucleus of the Hypothalamus

The paraventricular nucleus of the hypothalamus is well known for its oxytocin content and its projections to the posterior pituitary. The nucleus contains about 10 000 neurons in the rat, and is unusually complex. Recent evidence has shown that this nucleus also projects to the spinal cord as well as other areas of the brain, and at least five different cell types within the nucleus are known to project

to the medulla and spinal cord (Swanson and Sawchenko 1983). An oxytocinergic pathway is known to pass from the paraventricular nuclei to the parabrachial nucleus and locus coeruleus of the pons, the dorsal vagal motor nucleus and tractus solitarius of the medulla. Fibres also descend through the ventrolateral reticular formation and the dorsolateral funiculus of the cord to proceed caudally to the filum terminale. They end in the intermediolateral column, the central grey matter and the superficial dorsal horn. Some contain vasopressin as well as oxytocin. In addition there are about four times as many cells that project to the spinal cord from the paraventricular nucleus that do not contain either vasopressin or oxytocin. It is possible therefore that the paraventricular nucleus can influence the autonomic outflow via a direct spinal projection and through one or more other descending systems, such as the noradrenergic descending pathway from the locus coeruleus.

Pathways from the Cortex

Sherrington (1892) studied the control of the anal and vaginal sphincters in the monkey, and concluded that the peripheral innervation was essentially unilateral. However, the anal contractions that occurred when the motor cortex was stimulated was not entirely unilateral, although it was more marked on the contralateral side. The reticulospinal and raphe–spinal pathways can also influence the activity of pudendal nerve efferents (Mackel 1979; McMahon et al. 1982). Gjone and Setekliev (1963) found that electrical stimulation of the frontal lobes of the cerebral cortex could cause bladder contraction or relaxation, but it is not clear whether these events are mediated by direct corticospinal pathways or via brain stem nuclei.

Functional Interaction of Descending Pathways in the Control of Bladder and Sphincters

The α-motoneurons that innervate the rhabdosphincter will be dealt with first, as the descending pathways that regulate their activity appear to be similar to those controlling skeletal muscle in general. There is a direct pathway from the motor cortex, as well as indirect pathways that influence the sphincters via reticulospinal tracts (Sherrington 1892; Mackel 1979). The influence of descending pathways on the tone of skeletal muscle is thought of as a balance between inhibitory and excitatory influences; the excitatory pathways concerned with postural muscle include the medullary reticulospinal pathway (inhibitory) and the vestibulospinal and pontine reticulospinal pathways (excitatory); higher centres can act on these descending bulbospinal paths. Thus, the cerebral cortex, basal ganglia and the majority of the cerebellum exert an overall inhibitory influence via their connections with these descending paths; in addition, the flocculonodular lobe of the cerebellum exerts a facilitatory influence via the vestibulospinal tracts. Spasticity may result from a loss of inhibitory influence or excessive facilitation (see Fig. 7.9; Kandel and Schwartz 1981). Mackel (1979) has shown that descending pathways from the medulla can cause EPSPs or IPSPs in sphincteric motoneurons, depending on the site of the stimulating electrode in the medulla (Fig. 7.7; see also Fig. 8.10, p. 250). The influence of the vestibulospinal pathway on the urethral sphincter is unclear, but there is evidence that the medullary reticulospinal pathway has inhibitory effects on urethral motoneurons (Fig.

7.7; Mackel 1979). Yamamoto and Araki (1962) and Yamamoto et al. (1962) found that the lateral medullary reticular formation was necessary for the "pelvicoperineal–abdominal" reflex, in which there is concomitant contraction of the diaphragm and abdominal wall muscles and relaxation of the perineal muscles following an electrical stimulus to the pelvic nerve. This may be the pathway by which somatic musculature has synergistic activity in voiding.

In the control of the bladder, Kuru (1965) described descending vesicoconstrictor and vesicorelaxer influences (Fig. 7.8), that were thought to act reciprocally; the balance of inhibitory and excitatory effects was thought to determine the contractile state of the bladder. Tang and Ruch (1956) had similar ideas about the organization of descending pathways from the hypothalamus and midbrain in the regulation of micturition (Fig. 7.9). The functional aspects of these influences on the excitability of the micturition reflex will be considered in Chapter 8 (see p. 253).

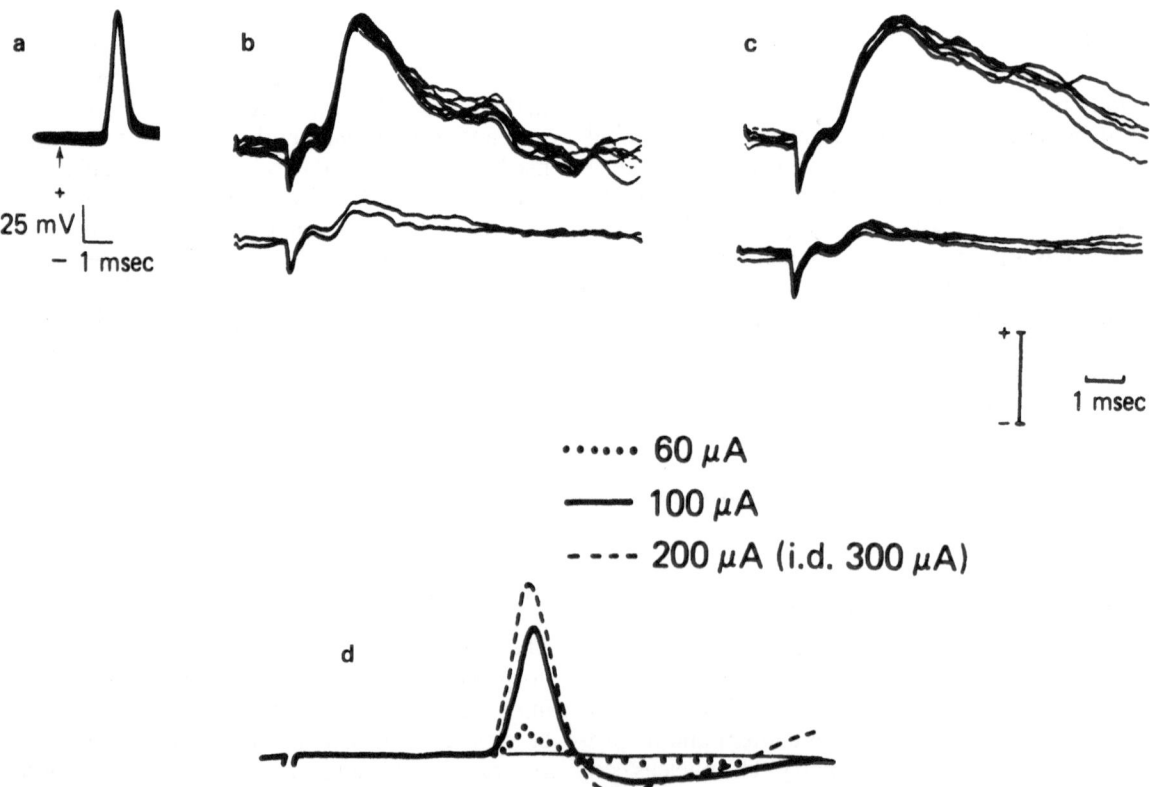

Fig. 7.7a–d. The effects of exciting afferent and descending pathways from the brain stem on the membrane potential of sphincteric motoneurons. **a–c** Intracellularly recorded responses following stimulation of ipsilateral dorsal roots and peripheral nerve branches and brain stem. **a** The antidromic spike for the cell in **b.** *Arrow* indicates stimulus onset. Calibration underneath. **b,c** Oscillographic recordings in two sphincter motoneurons following ipsilateral S-2 dorsal root stimulation. The *lower traces* in **b** and **c** show the extracellular fields. **d** Computer-averged responses recorded in an external urethral sphincter motoneuron following brain stem stimulation, showing gradation of the responses upon increasing stimulus strength. Calibration: **b,c** 1 mV; **d** 250 μV. (Mackel 1979)

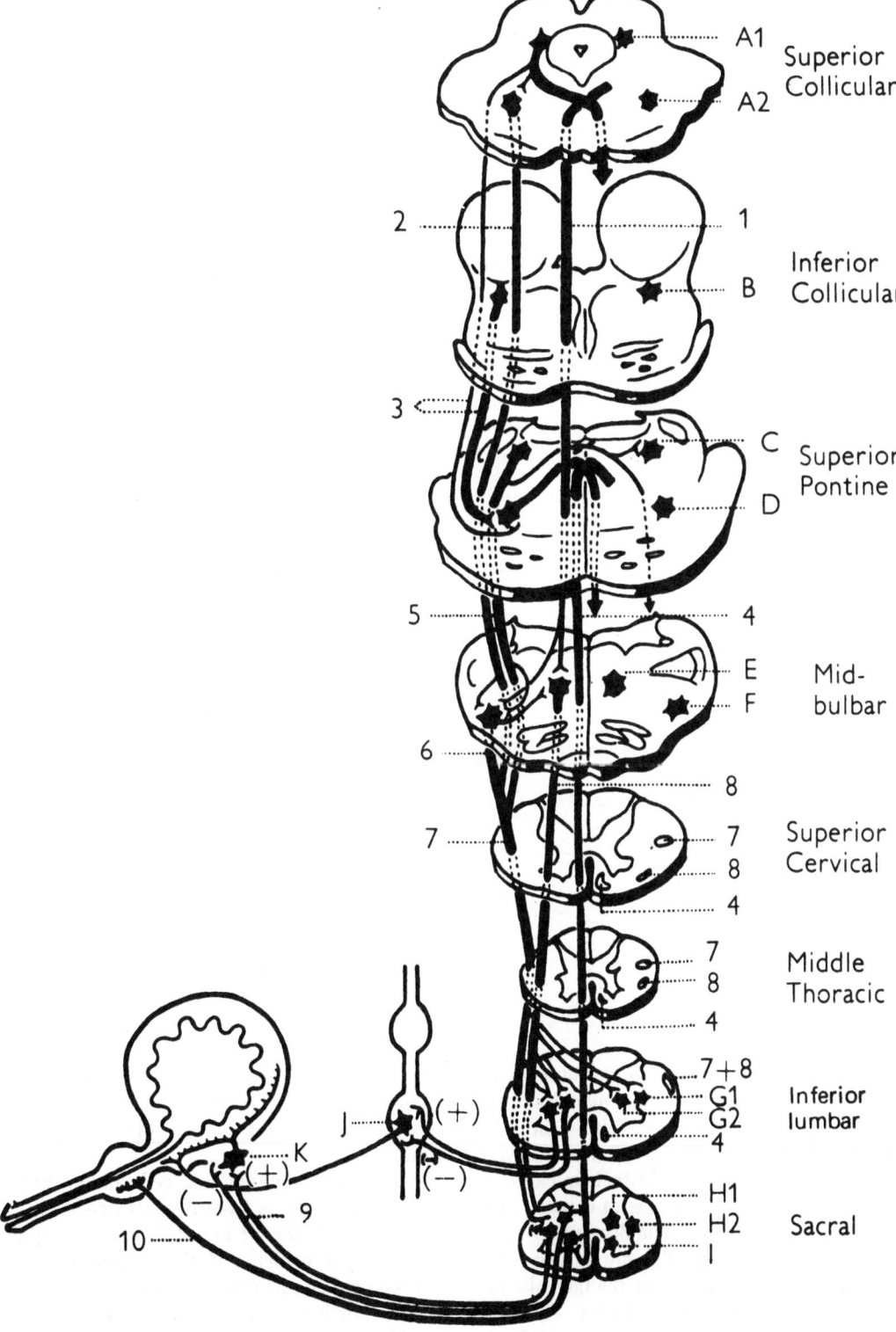

Fig. 7.8. Descending pathways concerned with micturition in the cat. *A1, A2,* mesencephalic micturition facilitatory area; *B,* mesencephalic micturition inhibitory area; *C,* pontine detrusor nucleus (Barrington); *D,* pontine nucleus for external urethral sphincter; *E,* bulbar vesicorelaxer centre; *F,* bulbar vesicoconstrictor centre; *G1,* lumbar vesicomotor centre (excitor); *G2,* lumbar vesicomotor centre (inhibitor); *H1,* sacral vesicomotor centre (inhibitor); *H2,* sacral vesicomotor centre (excitor); *I,* sacral nucleus for external urethral sphincter; *J,* lumbar sympathetic ganglion; *K,* vesical plexus; *1,* tectobulbar tract; *2,* mesencephalobulbosacral tract; *3,* tectopontine tract; *4,* medial reticulospinal tract; *5,* pontobulbosacral tract; *6,* bulbosacral tract; *7,* lateral reticulospinal tract; *8,* ventral reticulospinal tract; *9,* pelvic nerves; *10,* pudendal nerves. (Kuru 1965, quoted in Appenzeller 1982)

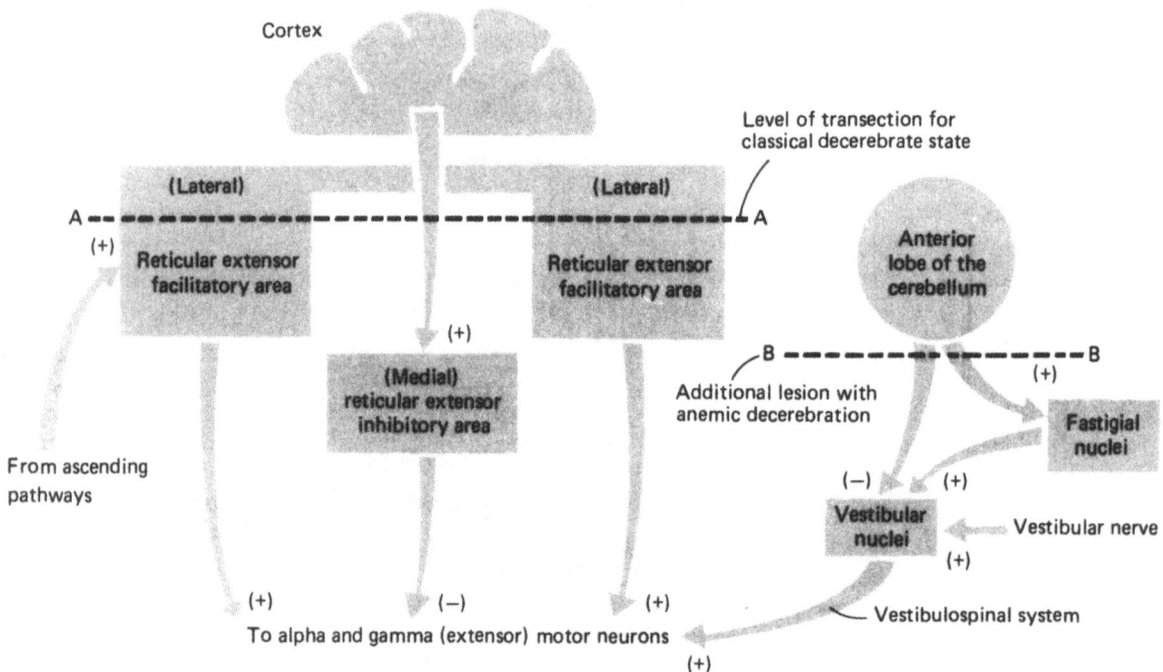

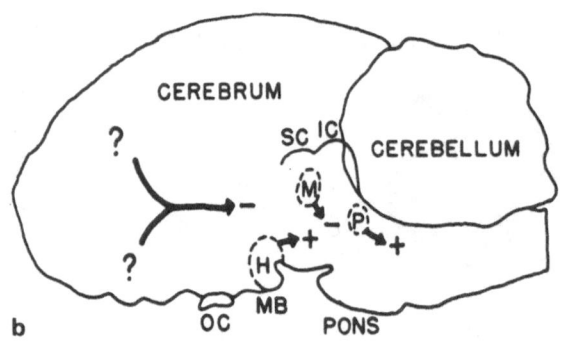

Fig. 7.9a,b. Balance of descending influences on **a** somatic muscle and **b** on the bladder. **a** Major pathways involved in decerebrate and decerebellate rigidity. Mesencephalic transection (*A*) produces decerebrate rigidity; additional disruption of inhibition from the cerebellum onto Deiter's and fastigial nuclei (*B*) produces decerebellate rigidity. **b** Areas concerned with micturition in the cat brain. *H*, posterior hypothalamic facilitatory area; *M*, mesencephalic inhibitory area; *P*, anterior pontine facilitatory area; *OC*, optic chiasma; *MB*, mammillary body; *SC*, superior colliculus; *IC*, inferior colliculus; +, facilitatory area; −, inhibitory area. (**a** from Kandel and Schwartz 1981; **b** from Tang and Ruch 1956)

Summary

Descending pathways from the brain to the autonomic nuclei originate principally from the medulla, pons, hypothalamus and cerebral cortex. Within the reticular formation there are a series of discrete nuclei that have been identified using histochemical methods, and the role of some of these with respect to micturition has been established. The descending pathways from the region of the pontine micturition centre originate from the locus coeruleus, the lateral tegmental nucleus and the parabrachial nuclei. The first two contain noradrenergic neurons; noradrenaline depresses the activity of preganglionic parasympathetic neurons, and it may be that these pathways end on interneurons. Descending pathways that inhibit bladder and sphincteric activity arise from the region of the nucleus gigantocellularis reticularis and the midline raphe nuclei in the medulla; some neurons in the raphe-spinal pathway contain serotonin and substance P, and the inhibitory effects of stimulation of this pathway can be partially blocked by drugs that block the action of 5-hydroxytryptamine.

The sympathetic preganglionic cell column also receives descending inputs from the locus coeruleus, from the A1 and A5, the B1 and B3 monoaminergic neurons, the raphe nuclei, and the paraventricular nucleus. Oxytocin can be demonstrated in the latter pathway. The corticospinal tract and the brain stem also send fibres to the motoneurons controlling the sphincters. Onuf's nucleus contains the terminals of descending noradrenergic and serotonergic axons.

All of these pathways are, in themselves, direct pathways, but all have axon collaterals that pass to other nuclei, many of which also have direct or

indirect connections with the autonomic outflow. The effects of these pathways combine to produce a balance of activities that regulates the activity of the different efferent pathways.

Reflex Integration of Bladder and Sphincteric Function

Normal bladder function can be divided into two phases: (a) concerned with filling and storage of urine; and (b) concerned with voiding. Filling is normally a slow function and few of the investigations on this aspect of bladder function take place over the natural time course of urine production (Klevmark 1977). There is some controversy as to the roles of mechanical accommodation of the tissues, and of neural relaxatory influences mediated by the sympathetic nervous system during this phase. However, the faster voiding process is more fully understood and depends on the parasympathetic and somatic innervation of the lower urinary tract. Micturition is a complex integrated series of acts involving the autonomic innervation of the bladder, the somatic innervation of skeletal muscle, and afferent nerves to provide sensation and reflex coordination. These reflexes are themselves under the control of higher structures in the central nervous system, and can also be influenced by local and systemic processes which may facilitate or depress them. For instance, sexual activity and large bowel function alter the activity of the lower urinary tract, as can changes in motor control such as those which occur in Parkinson's disease.

Micturition is normally a voluntary response initiated by a sensation of bladder fullness, which results from excitation of afferent nerve endings in the bladder wall, the immediately adjacent structures, and possibly from the proximal urethra. The sensation and the response can normally be suppressed for a time, after which the urge to micturate becomes more intense. This ability to suppress lower urinary tract sensation, and to control the reflex voiding response develops during infancy and may be related to the development of neuronal connections between the brain and the spinal cord, or within the brain itself. The reflex which is controlled in this manner, i.e. the micturition reflex, is elicited by bladder distension, and results in bladder contraction.

Much of the experimental work on the neural control of the bladder has been done in cats. While there are undoubtedly species differences in the sexual and voiding behaviour, and in anatomical or physiological details, some of which are relevant to diagnosis and treatment, it seems unlikely that the overall design of the control systems for this viscus are greatly different in the different species. One of the consequences of studying cats, however, is that the investigator is immediately confronted by what appears to be an abnormal situation: the anaesthetized cat appears to have what most urologists would regard as an unstable bladder, i.e. a bladder that gives large spontaneous contractions at moderate volumes. This features is not true of the conscious cat or dog (Klevmark 1980; McMahon 1986), and is best attributed to anaesthesia; anaesthetics are presumed to affect some critical inhibitory synapses in the neural pathways regulating bladder motility so as to make it hyperactive. Should one therefore regard the anaesthetized cat bladder as being so different from the human as to make the data irrelevant? or should one regard it as one model of the "unstable bladder", and therefore worthy of study because it might be clinically relevant? Whether the reader decides it is relevant or irrelevant, there are interesting observations, hypotheses and conclusions to be drawn; nevertheless, throughout this chapter attempts have been made to draw a distinct dividing line between human and animal studies. Barrington (1915, 1921, 1925, 1928, 1931, 1933, 1941) provided a basic scheme of reflexes that control bladder function as a result of investigations on cats between 1915 and 1941. In all, he described seven reflexes, and subsequent work has challenged some and strengthened the position of others. Table 7.1 outlines the main reflexes and their current state.

Barrington's First Reflex: Pathways for Normal Reflex Micturition

Barrington (1915, 1921, 1925) described the reflex contraction elicited by distension of the cat bladder as his first reflex, and the afferent as well as efferent pathways ran in the pelvic nerves; the central pathways which were necessary for this response lay in the spinal cord and brain stem, as high as the pons. He described a pontine "micturition centre", destruction of which resulted in overdistension and absence of bladder contractions. Subsequent workers have confirmed that the rostral pons contains an important area without which bladder emptying is deficient. de Groat (1975) located it near the locus coeruleus in the cat (Fig. 7.10), and Satoh et al. (1978a,b) to the dorsolateral pontine tegmentum in the rat; Westlund and Coulter (1980)

Table 7.1. Barrington's reflexes

	Stimulus	Response	Afferent	Central pathway	Efferent	Comments
I	Bladder distension	Bladder contraction	PN	Pontine	PN	
II	Flow through urethra	Bladder contraction	PUDN	Pontine	PN	Variable in cats; variable or absent in humans
III	Urethral distension	Small bladder contraction	HGN	Spinal	HGN	
IV	Flow through urethra	Urethral relaxation	PUDN	Spinal	PUDN	Absent in cats and humans
V	Bladder distension	Rhabdosphincter relaxation	PN	Spinal	PUDN	
VI	Bladder distension	Relaxation of urethral smooth muscle	PN	Spinal	PN	
VII	Flow through urethra	Bladder contraction	PN	Spinal	PN	Absent in humans

HGN, hypogastric nerve; PN, pelvic nerve; PUDN, pudendal nerve.

doubted that the locus coeruleus gave rise to descending axons which excite parasympathetic preganglionic neurons, on pharmacological grounds. Nevertheless, Elam et al. (1986) have shown that spike activity in the locus coeruleus increases when the rat bladder is distended with small volumes of urine, and when the colon is distended to a variable extent. Whatever the precise location of pontine neurons that elicit bladder contractions, the importance of this part of the brain stem for normal voiding is generally still upheld. Wang and Ranson (1939a,b) showed that electrical stimulation of the rostral pons elicited bladder contractions, and Tang (1955) found that the rostral pons and posterior hypothalamus were excitatory to the bladder.

Barrington's Second and Seventh Reflexes: the Influence of Urine Flow on Micturition

Barrington (1931, 1941) believed that urine flow or mechanical stimulation of the urethra with a catheter could excite afferent endings in the urethra that initiated bladder contractions and helped to maintain the bladder pressure at a high level. These reflexes were called Barrington's second and seventh reflexes and had (in cats) an afferent limb mediated by pudendal or pelvic nerve afferents, and an efferent limb in the pelvic nerves. He believed

that this reflex kept the bladder contracting as long as urine flow through the urethra continued; as such it could be an important pathway for minimizing the residual volume in the bladder. The existence of these reflexes in cats has been confirmed by Langworthy et al. (1940). Todd (1964) provided experimental evidence for the existence of urethral "flow" receptors in cats. In humans, Nathan (1952) examined the effects of catheterization or urinary flow on detrusor contractions and concluded that they had found no evidence in favour of these reflexes. More recent investigations by Brindley et al. (1974) examined the effects of pudendal nerve anaesthesia on the function of the lower urinary tract; they observed that the rate of urine flow during voluntary micturition was reduced to about half if pudendal anaesthesia was complete, and tentatively concluded that the low flow rate was due to the loss of Barrington's urethrovesical reflexes.

Barrington's Third Reflex: Bladder Relaxation During the Filling Phase?

Barrington's third reflex was a spinal reflex which was mediated by hypogastric afferents and efferents; distension of the bladder resulted in a slight contraction of the viscus mediated by the hypogastric efferents (Barrington 1931, 1941). We now know that a transient contraction is the normal initial result of stimulating the noradrenergic nerves

MICTURITION PATHWAY

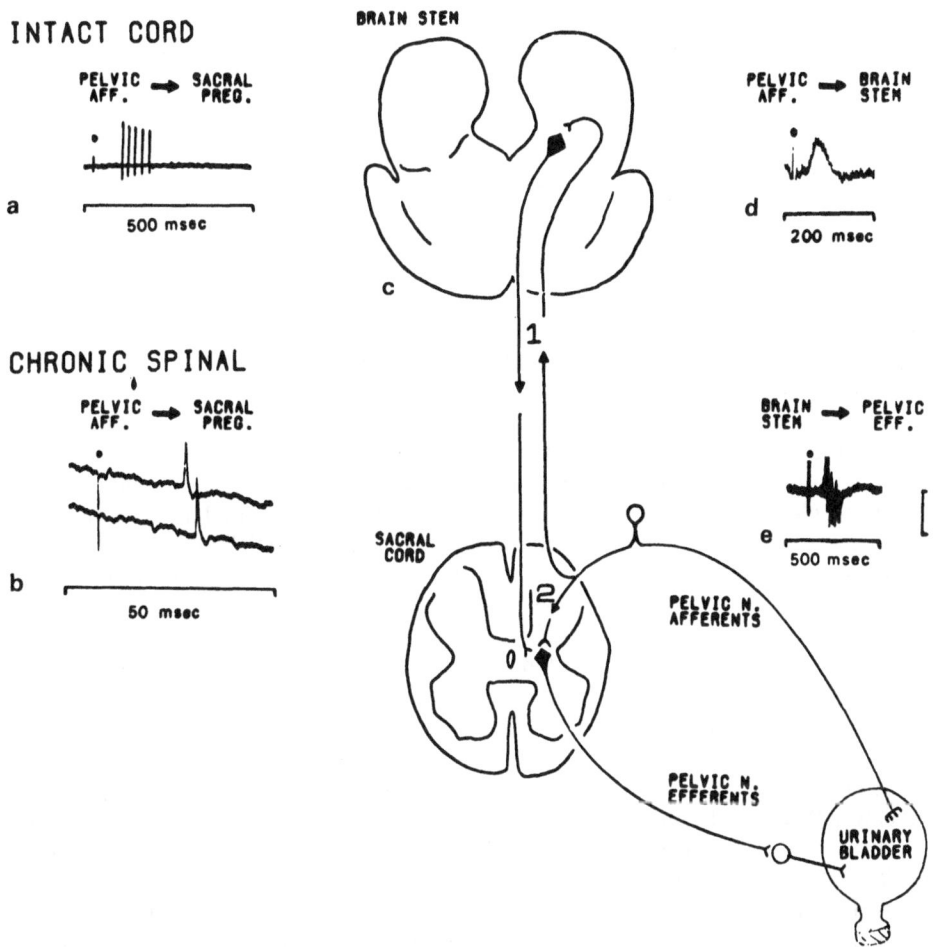

INTACT CORD

PELVIC AFF. → SACRAL PREG.

a

500 msec

CHRONIC SPINAL

PELVIC AFF. → SACRAL PREG.

b

50 msec

BRAIN STEM

c

1

SACRAL CORD

2

0

PELVIC N. AFFERENTS

PELVIC N. EFFERENTS

URINARY BLADDER

PELVIC AFF. → BRAIN STEM

d

200 msec

BRAIN STEM → PELVIC EFF.

e

500 msec

Fig. 7.10a–e. Sacral parasympathetic reflex pathways involved in micturition in normal and chronic spinal cats. **a,b** Responses of sacral preganglionic neurons to stimulation of pelvic nerve afferent fibres in a cat with an intact spinal cord (**a**) and in a chronic spinal cat (**b**). (**c**) Diagram of supraspinal (*1*) and spinal (*2*) reflex pathways. **d** Field potential recorded monophasically in rostral pons (P-2,L-2,H-4) elicited by a single shock to the pelvic nerve. **e** Discharge of a postganglionic parasympathetic fibre on the surface of the bladder elicited by a short train of stimuli (7 shocks, 150 Hz) applied to the rostral pons (P-3,L-2,H-4) via a metal neurological electrode (0.25 mm, outer diameter). Vertical calibration represents 5 mV in **a**, 2 mV in **b**, 50 μV in **d** and 100 μV in **e**, negativity upward in this and all subsequent figures. *Dots* above each record indicate stimulus artefacts. (de Groat 1975)

to the bladder; this response is followed by a longer lasting relaxation. The reflex regulation of the bladder is much more complicated than would seem at first sight, and the third reflex can operate at low bladder pressure, possibly playing a part in the receptive relaxation of the viscus during filling (Gjone 1966; Edvardsen 1967). de Groat and Lalley (1972) showed that the pelvic nerves can influence sympathetic tone to the bladder, and Floyd et al. (1982a) showed that the hypogastric afferents also participate in the reflex; there are no systemic differences between the functional properties of the

pelvic and hypogastric afferents. This reflex is considered in more detail in the section on sympathetic reflexes (see p. 220).

Barrington's Fourth Reflex: the (Doubtful) Effect of Urine Flow on Sphincteric Tone

The fourth reflex was thought to be elicited by running water through the urethra and had an afferent limb in the pudendal nerves (Barrington 1931, 1941). This was a spinal reflex which caused relaxation of the urethral skeletal muscle by a pudendal pathway. Denny-Brown and Robertson (1933) denied that the presence of fluid in the urethra of spinal humans can elicit relaxation of the external urethral sphincter. More recent studies on the cat by Garry et al. (1959) have also failed to confirm the existence of the fourth reflex; instead, they showed that forcing fluid along the urethra always augmented the activity of the external urethral sphincter, except when the bladder was stretched passively or was actively contracting (these have powerful inhibitory influences on the external sphincter—Barrington's fifth reflex). Indeed Garry et al. (1959) explained the fourth reflex by a combination of the second and the fifth; urethral flow induced bladder contraction by the second reflex, and the raised bladder pressure gave rise to urethral relaxation by the second. This would also explain Barrington's observation that the fourth reflex was difficult to elicit if the bladder could no longer contract. Garry et al. (1959) therefore proposed the existence of a guarding reflex which helped to maintain continence.

Barrington's Fifth and Sixth Reflexes: Sphincteric Relaxation During Micturition

Barrington's fifth and sixth reflexes were concerned with relaxation of the external urethral sphincter (the fifth reflex) or of the smooth muscle of the proximal urethra (the sixth reflex) during bladder distension or contraction (Barrington 1931, 1941). The pelvic nerve afferent pathway mediated these spinal reflexes which had efferents in the pudendal and pelvic nerves. The existence of the fifth reflex has been confirmed by Garry et al. (1959) and McMahon et al. (1982). However, during normal voluntary micturition in humans, the relaxation of the external anal sphincter occurs at the start of the rise in bladder pressure, and precedes the start of urine flow (Scott et al. 1964); urethral relaxation is thought to occur with a similar time course (Tanagho 1978), and descent of the bladder, i.e. relaxation of the pelvic floor, occurs at the initiation of micturition. It seems likely that descending spinal pathways concerned with the initiation of voluntary micturition as well as reflex pathways from the bladder are involved in the relaxation of periurethral and pelvic muscles during normal voiding. The urethral relaxation by parasympathetic nerves in the cat described by Barrington in his sixth reflex is known to occur in humans; Torrens (1978) showed that stimulation of the distal cut end of human sacral nerves gives rise to urethral relaxation (Fig. 7.11).

Barrington's Seventh Reflex

Barrington's seventh reflex was discussed along with his second because both have similar effects; the pathways described by Barrington, however, were quite different (Barrington 1931, 1941). In the seventh reflex, the afferents which sensed the flow of water through the urethra travelled in the pelvic nerve, as did the efferent pathway to the bladder; the reflex persisted after transection of the spinal cord.

Summary

Barrington was concerned with describing the mechanisms by which bladder contractions and urethral relaxation were coordinated during voiding. He saw the process as being initiated by bladder distension; reflex bladder contractions elicited by his first reflex were maintained and reinforced by urine flow through the urethra, which elicited his second and seventh reflexes; the raised intravesical pressure during bladder distension or contraction also elicits urethral relaxation by his fifth and sixth reflexes. These reflexes still provide the basis of our understanding of the voiding process.

The second and seventh reflexes are difficult to demonstrate in cats, and probably cannot initiate reflex bladder contractions, but may facilitate or reinforce those already present. In humans there is some controversy as to the existence of the second and seventh reflexes. Nathan (1956) did not observe bladder contractions on flowing water through the human urethra, but Garry et al. (1959) did not find that in cats either. There is no evidence for the fourth reflex in cats or in humans; the pathway of the third reflex is in the sympathetic nerves and the transient contraction described by Barrington in acute distensions may be followed by relaxation, so that this reflex may play a role in receptive relaxation. The sympathetic reflexes are known to be much more complicated than described by Barrington (see p. 206). The sphincteric reflexes are basically spinal reflexes, and sphincteric activity can be facilitated by guarding reflexes, or inhibited at high pressures in the bladder or colon. There is evidence that the sphincteric activity can be inhibited by raphe–spinal and reticulospinal pathways, and these may contribute to the inhibition of the striated muscle sphincters at high visceral pressures (see p. 211). Finally, there are a number of other reflexes, either initiated from the bladder or affecting the viscus, that may be of more peripheral interest, but nonetheless explain a number of phenomena.

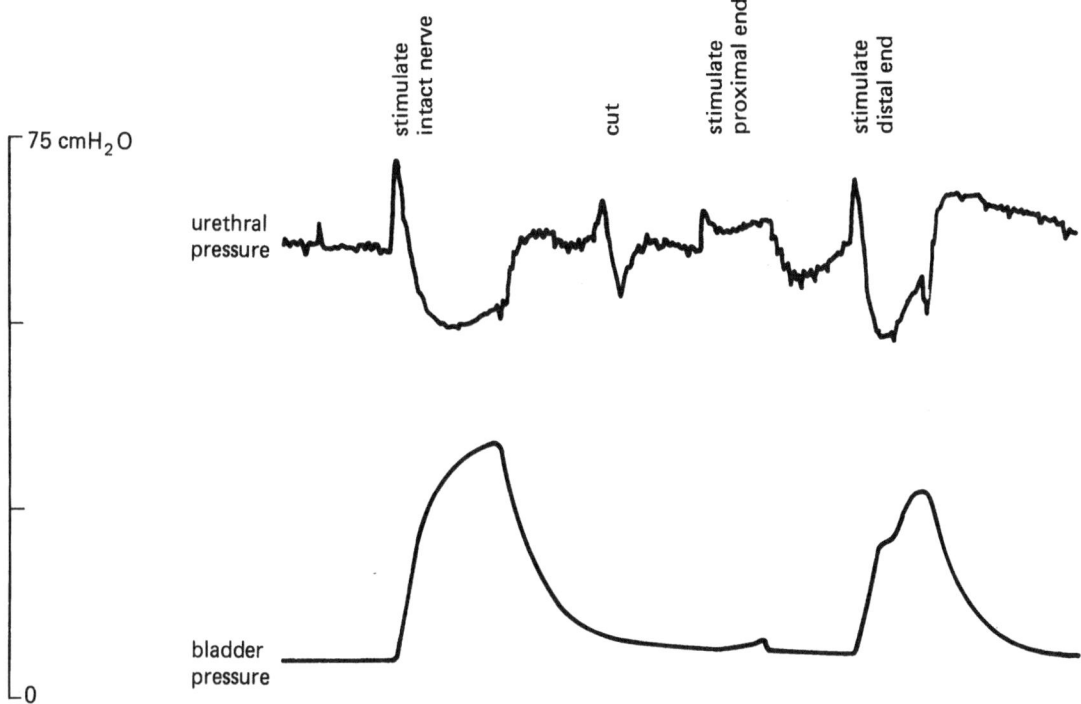

Fig. 7.11. The effects of stimulation of the third sacral nerve on the bladder and urethra in the human female (Torrens 1978)

Contemporary electrophysiological studies of these reflexes have tended to group together: (a) reflexes concerned with regulating the force of bladder contraction; (b) reflexes concerned with the control of sphincters; and (c) sympathetic reflexes concerned with the filling phase. The spirit of Barrington can still be seen in these studies, but the recent emphasis has been on how they function.

Reflexes Regulating the Force of Bladder contractions

The size of bladder contractions, as measured by the peak pressure, increases as the resting pressure is increased; the frequency of spontaneous contractions is related to the peak of bladder pressure in an inverse manner over the range 40–80 cmH$_2$O, which suggests that some inhibitory process affecting cycle length may be initiated at high pressures (Fig. 7.12; McPherson 1966a). A number of other inhibitory events accompany the high pressures that occur during bladder contractions: inhibition of sympathetic reflexes, inhibition of flexor reflexes, and inhibition of nociceptive transmission have been described (Evans and McPherson 1959; de

Groat 1975; Cadden and Morrison 1984), and all have been attributed to descending pathways of supraspinal origin. Most studies during the last two decades, however, have concentrated on the reflex changes that elicit the contractions, rather than on any inhibitory processes that may accompany them.

Electrophysiological experiments reinforce the belief that one of the major pathways which brings about reflex bladder contractions runs between the brain stem and the sacral cord. Bradley and Teague (1968) and de Groat and Ryall (1969) showed that efferent firing in parasympathetic nerves to the bladder occurred at long (80–120 ms) latency following electrical stimulation of fine myelinated fibres in other pelvic nerve branches from the bladder, provided the bladder was distended, and was abolished by acute spinalization. Their conclusions, that the micturition reflex had a spino-bulbospinal pathway was in keeping with the earlier conclusions of Barrington. In chronic spinal animals however, the reflex response following stimulation of the A-delta fibres was not only absent, it was replaced by a response that only occurred when ipsilateral C-fibre afferents were stimulated in the presence of bladder distension; the latter response was relatively weak or absent in intact animals, and was not seen in all the efferent neurons. There is no doubt, however, that reflex bladder contractions can occur in spinal humans

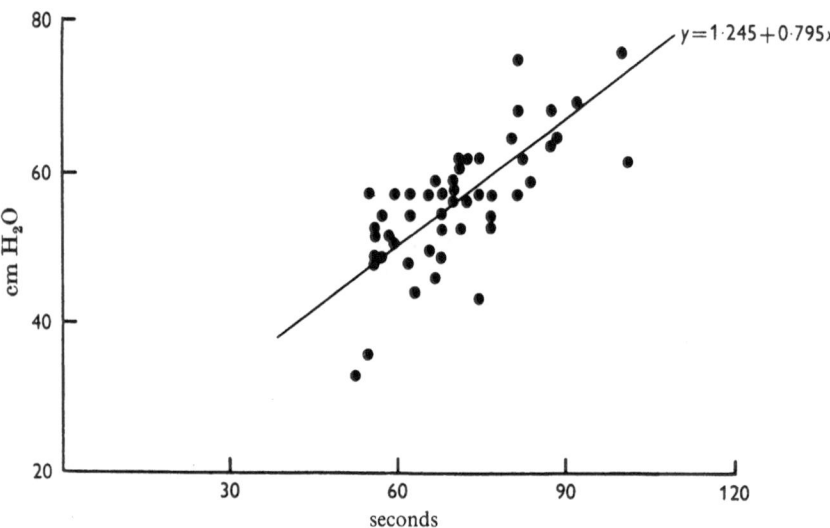

Fig. 7.12. The effect of distension on the size of bladder contractions, and the relationship between amplitude and frequency. (McPherson 1966)

and animals after spinal cord shock is over, and these may be due to the C-fibre afferent pathway as suggested by de Groat et al. (1981, 1982); however, the C-fibre afferents in the pelvic nerve do not respond to pressures of less than 70 mmHg (see Chap. 4, p. 103). This reflex is discussed in the section on the effects of spinal transection (see p. 224).

However, the spinobulbospinal reflex response to electrical stimulation of the pelvic and other nerves in the intact animal is difficult to evaluate; it is not elicited by natural stimulation, and the electrical stimuli employed may excite nerve fibres innervating other viscera or structures in the pelvis and perineum in the pelvic and perineum in addition to the bladder. While the response to an electric shock is undoubtedly mediated by this route, the asynchronous and more discrete afferent activity elicited by distension may or may not. The experiments using electrical stimuli suggest that the reflex pathway for micturition loops through the brain stem, and it is conventional to interpret the data to suggest that every synapse in this path is excitatory.

One piece of evidence that is at variance with the model presented by de Groat (1975) is the fact that neurons which carry bladder information up to the brain show viscerosomatic convergence and viscerovisceral convergence, and usually receive information from the pelvic, pudendal and hypogastric nerves (McMahon and Morrison 1982a). This suggests that the neurons that form the ascending limb of the spinobulbospinal loop do not carry specific information about the activities of the bladder, but mix signals from this viscus with discharges from other sources that are normally incapable of inducing a micturition reflex. McMahon and Morrison (1982b,c) and McMahon (1986) argue that the specific nature of the micturition reflex (i.e. that it can be elicited only by bladder distension and not by other afferent pathways) can be attributed to a spinal circuit which controls the excitability of the parasympathetic efferent neurons in such a way that they are excited only when the bladder pressure is raised (Fig. 7.13; see also Fig. 8.18, p. 263, Fig. 8.21, p. 267). In these conditions many different sources may be allowed to excite the parasympathetic efferents, including the "nonspecific" spinobulbospinal loop (see Chap. 8, p. 265). In McMahon and Morrison's model therefore the descending pathway from the pons could be regarded simply as a source of facilitation that can be excited by a number of afferent sources. The effectiveness of this pathway in eliciting efferent discharges depends upon the state of the spinal loop. The results of this study suggest that the parasympathetic efferent activity that regulates the force of bladder contractions is dependent on a feedback of afferent impulses elicited by bladder pressures in the range 5–30 mmHg. This conclusion reinforces the results of earlier work which shows that the size of bladder contractions can be increased by raising resting pressure within this same range (McPherson 1966a).

The absence of a spinal reflex in normal animals may be due to inhibition, because de Groat and Ryall (1969) found evidence of early inhibitory postsynaptic potentials in parasympathetic efferents following stimulation of the pelvic nerve. Tonic descending inhibitory pathways may inhibit the spinal reflex, and at the same time facilitate "in parallel" inhibitory systems. The short latency of the IPSPs in de Groat and Ryall's (1969) study could be attributed to a pelvic nerve input to a spinal inhibitory pathway that was switched on by tonic descending activity. Unfortunately, we are unable to provide precise answers to these problems at the moment.

A micturition threshold is present in spinal and intact animals and can be modulated by factors such as raised colonic pressure and vaginal stimulation. McMahon and Morrison (1982b,c) attribute the control over micturition threshold to spinal interneurons which show antagonistic influences from bladder and bowel distension. These neurons are regarded as performing a switching or gating function (see chap. 8, p. 265), which allows activity from other pathways to be passed to the efferent system in a controlled manner, the most obvious

feature of which is the presence of a threshold. They may not actually perform a reflex function, and in the railway analogy described in Chapter 8 they could function as signals rather than as trains.

Inhibitory Reflexes Affecting the Force of Bladder Contraction

Since Barrington's time, a number of workers have found that other local influences may inhibit detrusor function. Touching the perianal skin in humans and animals reduces or stops detrusor activity (Kock and Pompeius 1963); distension of the anal canal or rectum or vaginal stimulation also inhibit the micturition reflex (Fig. 7.14; de Groat 1971; de Groat et al. 1981; Floyd et al. 1982b;

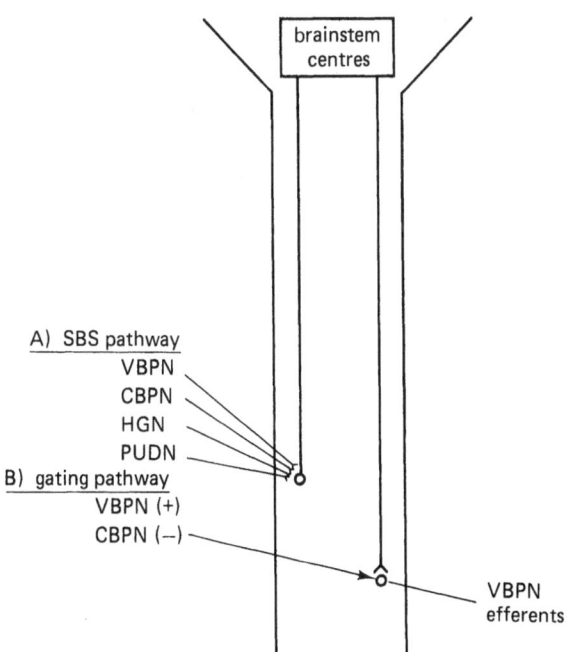

Fig. 7.13. Summary of the neuronal pathways controlling firing in pelvic efferents to the bladder (*VBPN efferents*). A long spinobulbospinal reflex actually activates the efferents. This pathway shows extensive convergence from afferents in a number of nerves (*VBPN*, vesical branches of the pelvic nerve; *CBPN*, colonic branches of the pelvic nerve; *HGN*, hypogastric nerve; *PUDN*, pudendal nerve). However, the ability of discharges in the descending limb of this path to activate the efferents is regulated or gated by a second, sacral pathway. This cannot itself normally drive the efferents. It is switched on by activity in pelvic afferents from the bladder, and off by activity in pelvic afferents from the colon. It does not show the convergence seen in the spinobulbospinal pathway, and it is this property which confers specificity on micturition. (McMahon 1986)

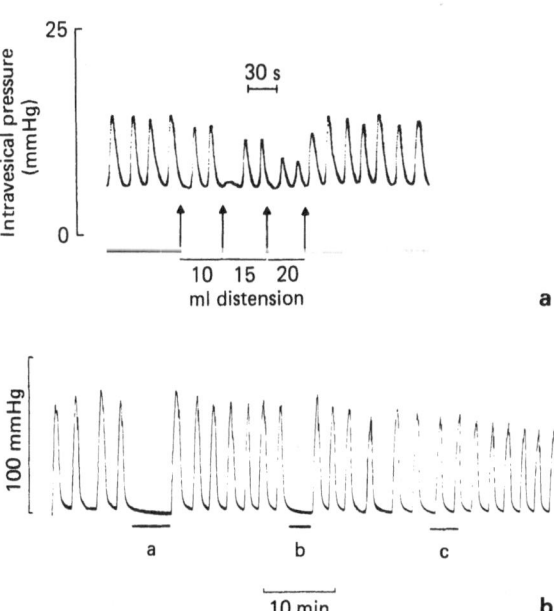

Fig. 7.14a,b. The effects of **a** colonic distension and **b** vaginal stimulation on bladder motility. **a** A recording of intravesical pressure changes during spontaneously occurring bladder contractions and the effects of distension of the descending colon to volumes of 10, 15 and 20 ml followed by release of this distension. The sympathetic and pudendal nerves were sectioned in this experiment. **b** Recording of intravesical pressure showing the rhythmical contractions of the bladder in an anaesthetized cat. At the signals (*a–c*) a vaginal probe was rotated by means of a small electric motor. At the start of the trace the afferent innervation of the genital tract was intact; the pudendal nerve afferents from the perineum were sectioned bilaterally between *a* and *b*; the branches of the pelvic nerves innervating the internal genitalia were sectioned between *b* and *c*. The recording shows that vaginal stimulation inhibits bladder motility and that the afferent pathway for the effect runs in the pelvic nerves (**a** from Floyd et al. 1982b; **b** from Morrison 1982)

Morrison 1982). These effects are mediated by inhibition of parasympathetic reflexes and probably also by peripheral inhibition of the bladder through sympathetic pathways, either at their ganglia or smooth muscle (Bahr et al. 1986a–c; Bartel et al. 1986). From a scientific point of view, these inhibitory interactions are useful in differentiating neurons that are involved in mediating these functions from others that respond in the same manner to these stimuli. In addition, a number of somatovesical reflexes have been studied, and these suggest that stimulation of muscle nerves is generally inhibitory to the bladder (McPherson 1966a). Cutaneous stimuli may excite bladder contractions if the bladder pressure is low (McPherson 1966a; Sato et al. 1977), but if the pressure is above the micturition threshold, noxious mechanical or thermal stimuli to the skin are inhibitory (Sato et al. 1977, 1980).

McMahon and Spillane (1982) and Spillane (1982) found that stimulation of the raphe–spinal and reticulospinal tracts inhibited bladder contractions in cats. In rats, these descending pathways can be brought into activity by noxious thermal and mechanical stimuli to the skin and by high bladder pressures (Lumb and Morrison 1984; Lumb 1986). It may be that these looped pathways, more commonly associated with antinociception, also have some inhibitory effect on bladder contractions (see Chap. 8, p. 248); multiple functions are possible given the extensive ramifications of raphe–spinal axon terminals in the lumbosacral cord (Light 1985). One phenomenon that remains unexplained is the slowing of the rate of spontaneous contractions when the peak pressure in spontaneous bladder contractions is increased; whether this is related to descending inhibition via raphe–spinal or reticulospinal pathways has still to be resolved.

Bladder Reflexes During Inflammation

There has been some work using irritant substances in experimental animals to cause bladder inflammation; little is known about sensation in these circumstances, although irritation can be presumed, and the reflex changes observed may give some indication as to the type of effects on sensory endings. Capsaicin, a substance that can be extracted from red peppers, is known to be acutely irritant to cutaneous and other sensory endings when administered acutely, largely because it releases substance P from nerve endings that contain this peptide. Capsaicin also causes a failure of development of small afferent nerve fibres when given neonatally to rats. Sharkey et al. (1983) showed that neonatal administration of capsaicin causes the development of enlarged bladders, an observation that was confirmed by Santicioli et al. (1985); the latter group also showed that there was an increase in micturition threshold, consistent with a loss of sensory fibres. In addition, the bladders exhibited a hypertonic response to filling with saline, the flat phase of the cystometrogram being absent. In adults, pretreatment of the animals with capsaicin gave rise to an increase in micturition threshold for about 1 month; acute administration of the drug into the bladder gave rise to an acute excitatory effect, and a transient increase in sensitivity of the bladder to distension. In these experiments, the acute excitatory effect of intravesical administration may have been due to the release of substance P from nerve endings, and the decreased responsiveness in the pretreated animals may have been due to depletion of substance P. Maggi et al. (1986) applied capsaicin to the serosal surface of the rat bladder, and reported the development of tonic and rhythmic bladder contractions that were dependent on the extrinsic nerve supply to the viscus. These contractions were also abolished by spinal cord transection, indicating a dependence on supraspinal pathways. The authors suggest that information carried by capsaicin-sensitive afferent fibres is relevant for initiation of a supraspinally mediated micturition reflex. Recently, McMahon and Abel (1987) have shown that sterile inflammation of the rat bladder, induced by instillation of turpentine, mustard oil or croton oil, gives rise to a lowering of reflex threshold in the viscus. This procedure, which is performed on chronic decerebrate rats, is being used as a model for the study of pain arising from the bladder; in addition to having hyperexcitable reflexes, these animals are hypersensitive to noxious somatic stimuli given to the tail, perineum and lower abdomen, and should prove to be useful in the study of central pain pathways from the bladder.

Other Bladder Reflexes: Vesicorenal, Vesicosomatic and Somatovesical Reflexes

The reflex influence of bladder distension on the functioning of other viscera and the somatic nervous system has also been studied by a number of authors; a vesicorenal reflex has been described (Lytton et al. 1967; Langley and Kimura 1959); rapid filling of the bladder is thought to diminish urine output. Changes in the tone of the abdominal and limb muscles have been reported by Evans and McPherson (1959) during bladder distension. In the latter study bladder distension inhibited flexor and extensor reflexes, and a major part of this (non-reciprocal) inhibition was mediated by supraspinal looped pathways; the afferent limbs were in the pelvic and hypogastric nerves (Koley et al. 1983). The inhibitory effect was partly mediated through the γ efferent system (Evans 1963; McPherson 1966b). In humans, bladder distension can influence the excitability of the H-reflex (McPherson and Skorpil 1966; Porter and Krell 1976), and Marsden et al. (1969) suggested that micturition had inhibitory effects on tonic vibration reflexes.

McPherson (1966a) found that stimulation of cutaneous nerves could sometimes induce bladder contractions or augment spontaneous contractions; in contrast, stimulation of muscle nerves produced inhibition. In the chronic spinal animal, cutaneous stimuli produced reflex contractions, but muscle nerve stimulation had no effect. Sato et al. (1980) found that electrical stimulation of finely myelinated and unmyelinated afferents in cutaneous and muscle nerves induced contractions if the bladder was nearly empty and quiescent, but the same stimuli inhibited the bladder if the volume was high and spontaneous contractions were present. One interpretation of this is that the micturition threshold not only allows coordinated bladder reflexes to occur—it can completely modify the reflex responses to other stimuli.

Summary

There have been a series of investigations during the last 20 years which have attempted to elucidate how Barrington's first reflex is mediated. In anaesthetized or decerebrate cats, the rate of bladder contractions increases above the micturition threshold, but the frequency is slowed progressively when the peak pressure becomes high (40–80 cmH$_2$O). Most subsequent work has concentrated on the micturition reflex pathway. The work of de Groat (summarized by de Groat et al. 1981) suggests the existence of a spinobulbospinal reflex that is activated by bladder distension and is excitatory at all synapses; however, McMahon (1986) suggests that this loop can be activated by a number of stimuli that do not elicit reflex contraction of the bladder, and that the excitability of the reflex is regulated by a spinal loop. This spinal circuit appears to act as a switch or gate, concerned with regulation rather than reflex traffic; it probably sets the micturition threshold, above which many types of stimuli can influence parasympathetic efferent activity to the bladder in normal and spinal animals. This gating mechanism utilizes afferent information from the bladder to regulate the force of contractions generated by the parasympathetic efferent supply.

Reflex Control of Sphincters: Connections Within Onuf's Nucleus

Some of the dendrites of pudendal motoneurons extend dorsally towards the intermediolateral column and the dorsal grey commissure, near the central canal, and here some contacts with primary afferent fibres may occur. Mackel (1979) found that only 29% of urethral motoneurons had primary afferent inputs, and the areas of cord which contain the dendrites of these cells are known to be the site of terminations of primary afferent fibres, from the pudendal (Roppolo et al. 1985) and the pelvic nerves (Morgan et al. 1981). Fibre terminals in Onuf's nucleus and the intermediolateral grey column and dorsal grey commissure are rich in somatostatin (Schrøder 1984), Met-enkephalin and avian pancreatic polypeptide (Hunt et al. 1981), Leu-enkephalin (Glazer and Basbaum 1980), vasoactive intestinal polypeptide (VIP) (Honda et al. 1983; Gibson et al. 1984), catecholamine, probably noradrenaline and a material, probably containing zinc, that stains with the Timm's reaction (Schrøder 1985), and these histochemical features distinguish them from other parts of the cord. The distribution of these chemicals does not change following transection of the cord, which suggests that these endings arise from spinal interneurons or from primary afferents. In the cat, 90% of the VIP is present in primary afferents, and the noradrenaline arises from nerve cells originating in the brain stem (Gibson et al. 1984; Hancock and Fougerousse 1976).

Vesicourethral Reflexes

Vesicourethral reflexes have been described already under the heading of Barrington's fifth and sixth reflexes. In addition a guarding reflex is thought to be present normally, with an increase in rhabdosphincteric contraction induced by bladder distension (Garry et al. 1959). A similar reflex is present in the terminal gastrointestinal tract. More information is available about the anal than the urethral sphincter, in terms of reflex control and the mechanisms involved; furthermore, more information is available concerning the skeletal muscle components than the smooth muscle at either site. The parasympathetic efferents can inhibit the urethral smooth muscle (Torrens 1978) and the internal anal sphincter (Garrett et al. 1974), and the reflex relaxation of the urethral smooth muscle is described as Barrington's sixth reflex. Few modern workers have studied this reflex, and the control of smooth muscle sphincters is discussed later in this chapter under the heading of sympathetic reflexes (see p. 215).

Motoneurons in the pudendal nerves exhibit a resting discharge, and the resting activity that can be recorded from the skeletal muscle of the sphincters is a consequence of this. This activity survives spinal transection provided a cauda equina lesion is not present. The reflexology of pudendal motoneurons differs from that of limb nerves in that Renshaw cell inhibition and crossed disynaptic inhibition are absent (Jankowska et al. 1978; Mackel 1979). In addition, there is evidence of crossed primary afferent, and uncrossed corticospinal tracts in the sacral cord (Nadelhaft et al. 1983; Matsushita and Tanami 1983; Sherrington 1892). The easiest way to excite pudendal motoneurons is to touch the skin in the perianal and perineal regions, which will evoke an anal reflex. Electrical stimulation of the contralateral pudendal nerve and the posterior cutaneous nerve of thigh both elicit reflex discharges in pudendal efferents after a short latency. These responses are inhibited by bladder contractions or by raising bladder or colonic pressure, and by stimulation of the raphe nucleus or the nucleus gigantocellularis reticularis of the medulla (Mackel 1979; McMahon et al. 1982). Bladder pressures of 5–30 mmHg were necessary to produce a range of inhibitions in the sphincteric reflexes. It was suggested that some interneurons with the properties described by McMahon and Morrison (1982b), McMahon et al. (1982) and de Groat et al. (1981) may have been responsible for these effects. In this system, as well as in the micturition reflex, there appears to be some sort of gating mechanism which, once switched on, allows or disallows stimuli from a number of sources to excite the motoneurons. While sphincter inhibition during bladder contractions or distensions are mediated by a spinal reflex (Barrington's fifth and sixth reflexes), possibly involving sphincteric gating interneurons (Fig. 7.15; McMahon et al. 1982), we cannot exclude a looped pathway through the medulla as playing some additional role, particularly if the stimuli are noxious. Indeed, the long time course of some of the inhibitory effects reported by McMahon et al, (1982) could be attributable to supraspinal mechanisms.

Attempts to demonstrate changes in anal reflex latency in paraplegic men when bladder or rectal

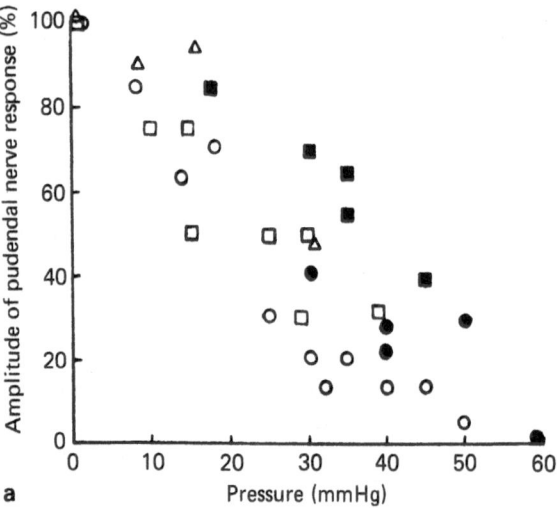

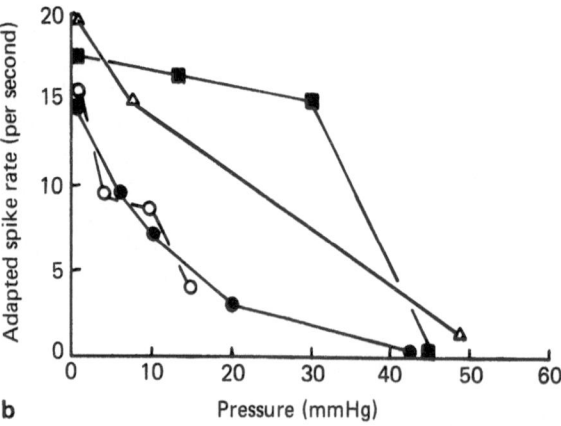

Fig. 7.15. a The effects of intravesical (*open symbols*) and intracolonic pressure (*solid symbols*) on the amplitude of the mass response evoked in the anal branches of the pudendal nerve by electrical stimulation of the afferents in the contralateral pudendal nerve (*squares*; 4 mA, 0.2 ms), the posterior cutaneous nerve of thigh (*circles*; 0·5 V, 0.2 ms) and the colonic branches of the pelvic nerve (*triangles*; 10 V, 1 ms). **b** The effects of intravesical (*open symbols*) and intracolonic pressure (*solid symbols*) on the rate of discharge in three sacral sphincteric interneurons with receptive fields in the bladder and colon. (McMahon et al. 1982)

pressure is raised have proved successful on occasions, but the necessary protocol is difficult to complete as the single-fibre electromyographic (EMG) recording is not sufficiently stable (A. L. Wright and J. F. B. Morrison 1984, unpublished work). The resting activity of the external anal sphincter is commonly inhibited during spontaneous bladder contractions in paraplegic subjects, but may increase at high intravesical pressure in some individuals; if a similar effect occurs in the urethral sphincter, this might explain part of the phenomenon of vesicosphincteric dyssynergia (see later).

Electromyographic studies on the anal reflex in humans have led to suggestions that the anal reflex latency is lengthened in patients with fecal incontinence. Parks et al. (1977) recorded EMG activity from the external anal sphincter and studied the latency of the first evoked discharge following an electrical stimulus to the perianal skin. This evoked response was thought to be the electrical correlate of the anal reflex, induced by touching or pricking perianal skin. The validity of their interpretation has recently been challenged, partly on the grounds that the EMG–evoked response that was recorded is not a reflex response, and that there is a rather larger range of latencies in the normal continent population, than was evident in the Henry and Swash paper (Wright et al. 1985). The early evoked response is believed to be due to activation by the electrical stimuli of motoneuron axon collaterals, and persists during spinal anaesthesia. The response is also present in paraplegic patients without cauda equina lesions (A. L. Wright and J. F. B. Morrison 1984, unpublished work).

Summary

The interneurons that modulate the activity of pudendal motoneurons in Onuf's nucleus differ from interneurons in other somatic motor nuclei in a number of respects. First, there appear to be no Renshaw cells associated with these nuclei. Indirect evidence has come from studies of acetylcholinesterase (AChE) which can be demonstrated in electron micrographs of terminals of neurons believed to be Renshaw cells; the coarse AChE-containing particles observed in these studies of somatic motoneuron pools are largely absent from some of the motoneurons in Onuf's nucleus X (Schrøder 1980). More direct evidence has arisen from electrophysiological studies of pudendal motoneurons by Mackel (1979), who demonstrated an absence of recurrent inhibition in the pudendal motoneurons of the cat. The function of Renshaw cells is bound up with inhibition and the asynchronous activation of motoneurons within a motoneuron pool, and these findings may also be one reason why pudendal motoneurons are active at rest. Little is known of the origin of the spontaneous electrical activity in pudendal motor units, but it is known that these neurons are only medium sized (Roppolo et al. 1985), which increases the probability that they will be spontaneously active, if the size principle were to apply in this situation (Henneman 1981).

The function of sphincteric motoneurons is organized quite differently from other groups of α-motoneurons: the neurons innervating each side of the sphincteric apparatus have to work in harmony and synchronously, as distinct from the reciprocal innervation that is common in the limb muscles. Jankowska et al. (1978) and Mackel (1979) found that these neurons do not show crossed disynaptic inhibition as occurs in the limb muscles, and Renshaw cell inhibition was also absent. Reflex activity in the external anal sphincter may be generated by light cutaneous stimuli, by rectal distension and by increasing intra-abdominal pressure; inhibition occurs during micturition and defecation. These motoneurons are subject to inhibitory descending controls from the raphe–spinal and reticulo–spinal pathways, and from the corticospinal tract.

Sympathetic Reflexes

Areas of Influence relevant to the Function of the Lower Urinary Tract

Sympathetic reflexes can be initiated by contraction or distension of the bladder, and can influence the motility of this viscus (Edvardsen 1967; de Groat and Lalley 1972; de Groat and Theobald 1976; Floyd et al. 1982a; Maggi et al. 1985; Jänig 1986; Jänig and McLachlan 1987) as well as the cardiovascular system. Cardiovascular effects include hypertension and vasoconstriction, which are present in normal animals (Mukerjee 1957; Taylor 1965), but can dominate after interruption of the baroreceptor reflex pathway by high spinal transection. The influence of sympathetic reflexes on the bladder may play a part in the filling phase in some species, and the inhibitory effects on the bladder that occur during pain, stress and some other circumstances. The role of the sympathetic in the regulation of bladder motility in normal humans is controversial, but almost certainly plays a part in certain facets of disease states, or as a side effect of treatment.

Cardiovascular Effects of Bladder Distension

The cardiovascular effects of distension of the bladder can be seen most clearly in patients with spinal injuries, where they are not subject to the control of the brain stem, and in conditions where cardiovascular decompensation is present, e.g. the effects of urinary retention on patients in incipient heart failure. Guttmann and his colleagues (see Guttmann 1976) provided a considerable amount of information concerning the reflex cardiovascular changes during distension of the bladder in human paraplegics and quadriplegics. The effects included arterial hypertension, bradycardia, a reduction in peripheral blood flow, fall in skin temperature and pallor below the level of the spinal transection, redness and sweating above that level, and headache associated with the hypertension (Fig. 7.16; see also Fig. 7.25).

These changes are explained by activation of afferents from the bladder during distension or contraction giving rise to reflex sympathetic discharge to skin, muscle and visceral vascular beds; the splanchnic efferent pathway makes a major contribution to these changes (Mukerjee 1957). Figure 7.17 shows the changes in activity in sympathetic efferents in skin and muscle nerves in normal and spinal cats.

Autonomic hyperreflexia occurs to the greatest degree in quadriplegic patients, and can be sensed by them by the symptoms of headache and sweating, as well as other somatic reflex manifestations (Pedersen 1983). The symptoms are troublesome in patients with a complete cord transection above the midthoracic level, an account of removal of the descending controls over the sympathetic reflex pathways. These changes, which are characteristic of autonomic hyperreflexia, are discussed in more detail in the section on spinal injury.

Involvement of the Sympathetic in Visceral Reflexes

It has been known from the time of Langley (1894), that the organization of prevertebral ganglia, such as the inferior mesenteric and coeliac ganglia, differs from the paravertebral ganglia in the sympathetic chain; visceral reflexes can be mediated by synaptic

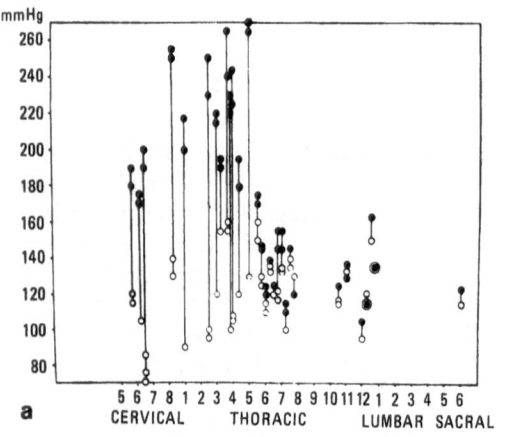

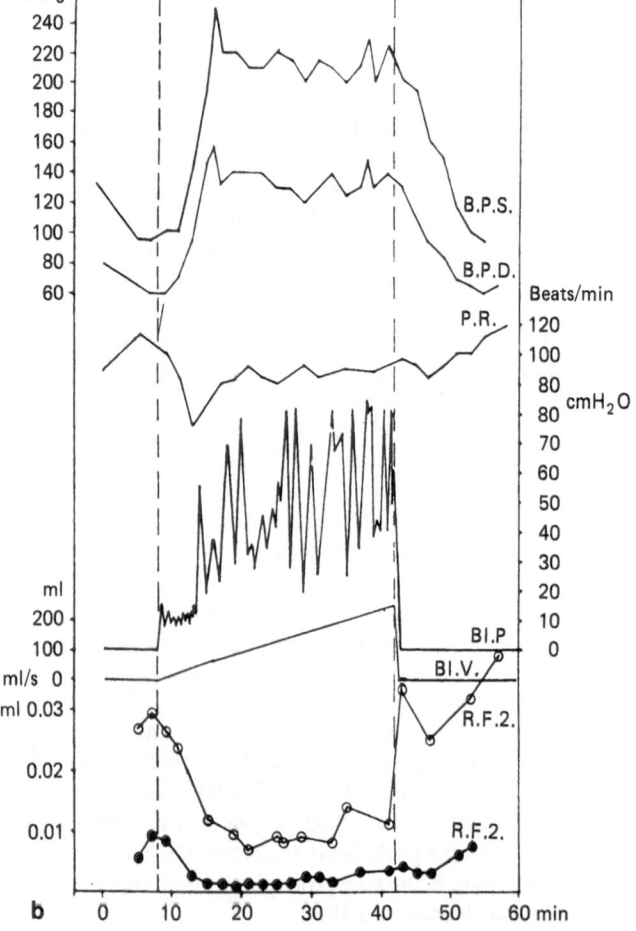

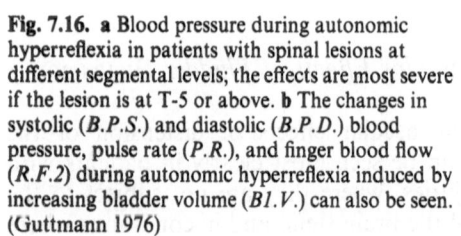

Fig. 7.16. a Blood pressure during autonomic hyperreflexia in patients with spinal lesions at different segmental levels; the effects are most severe if the lesion is at T-5 or above. **b** The changes in systolic (*B.P.S.*) and diastolic (*B.P.D.*) blood pressure, pulse rate (*P.R.*), and finger blood flow (*R.F.2*) during autonomic hyperreflexia induced by increasing bladder volume (*Bl.V.*) can also be seen. (Guttmann 1976)

connections between afferents and efferents in pre-vertebral ganglia, without the intervention of the central nervous system, as well as through central nervous pathways. For many years it was thought that these peripheral pathways only operated in unphysiological circumstances (Kuntz 1946), but more recent studies using electrophysiological techniques, beginning with the work of Crowcroft et al. (1971), has shown that some reflexes mediated by the inferior mesenteric and coeliac ganglia can result from a physiological stimulus. King and Szurszewski (1984) have gone so far as to suggest that the reflexes through prevertebral ganglia are more significant than those through the central nervous system. This view seems to be based entirely on negative evidence from their own experiments, and other workers have shown that central reflexes operate in normal conditions (see Jänig 1986; Jänig and McLachlan 1987). The following two sections therefore deal with reflexes through prevertebral ganglia, and through the central nervous system.

Sympathetic Reflexes Mediated by Prevertebral Ganglia

Types of Connections in Prevertebral Ganglia

In most of the work which shows functional reflexes mediated by prevertebral ganglia, it is not absolutely clear as to the type and origin of the afferent fibres involved. The synapses could be due to contacts between the terminations of primary afferent collaterals and the postganglionic neurons; in this case the postulated site of the afferent cell body is in the dorsal root ganglia. Evidence for this pathway has recently been provided by Aldskogius et al. (1986); injection of HRP conjugated to wheat germ agglutinin (WGA-HRP) into the second and third lumbar dorsal root ganglia was followed by labelled axons being stained in the bladder and colon, and some of these had collaterals with varicosities in close relation to inferior mesenteric ganglion neurons. However, in addition, there is good

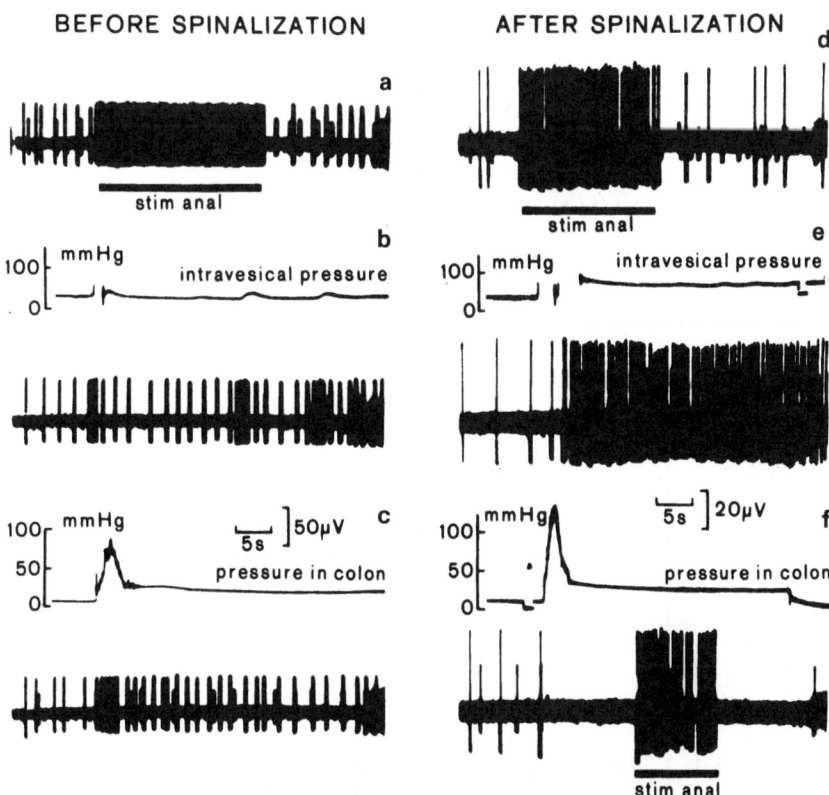

Fig. 7.17a–f. Reaction of two preganglionic motility-regulating neurons to mechanical stimulation of the mucosal skin of the anus, to distension and contraction of the urinary bladder and to colon distension before and after spinalization. One neuron (small signal) reacted neither to stimulation of the afferents from the urinary bladder before spinalization nor to anal stimulation. The bladder was distended by injection of 5 ml fluid through the urethral catheter, the colon by injection of 30 ml. Note change of the gain in fibre records after spinalization. (Bartel et al. 1986)

evidence in the gastrointestinal tract that some intramural neurons have centripetal axons that could reach the prevertebral ganglia, but not a lot is known about the properties of these neurons, which are believed to release bombesin. In support of this pathway, Job and Lundberg (1953) did find that the peripheral reflex still operated after decentralization of the prevertebral ganglia when time was allowed for the connections with the central nervous system to degenerate. However, some of the afferent synaptic effects on prevertebral ganglia appear to operate by means of substance P, which might suggest the involvement of primary afferent collaterals (Otsuka and Konishi 1983).

Sokownin (1874) showed that after central denervation of the inferior mesenteric ganglion, stimu-

lation of the central end of one hypogastric nerve caused contraction of the bladder on the opposite side; these observations have sometimes been called the Sokownin reflex and were extended by Langley and Anderson (1894). They showed that a number of other changes occurred, including contraction of the internal anal sphincter and pallor of the mucous membrane of the rectum, all of which were abolished by nicotine, and concluded that there must be a synapse in the inferior mesenteric ganglion. Later experiments suggested that the reflex was mediated by C-fibres (Job and Lundberg 1953). All of these effects were the result of electrical stimulation of the central end of one hypogastric nerve, and may have been mediated by a variety of different pathways; these include excitation of sympathetic pre-

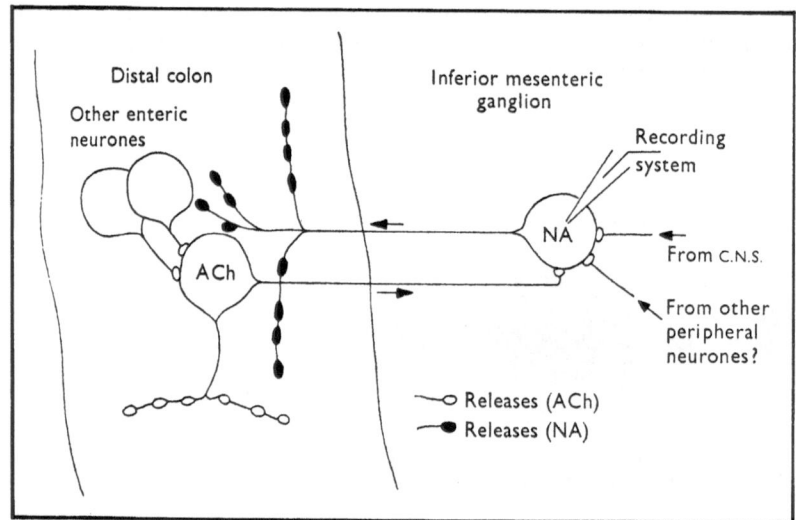

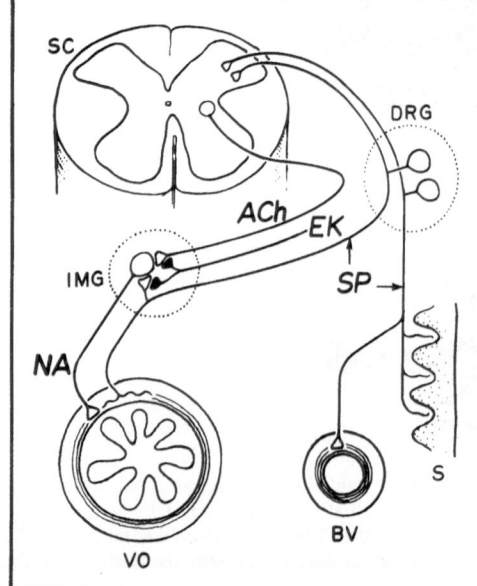

Fig. 7.18a,b. Reflex pathways through prevertebral ganglia. **a** Possible origins of afferent synaptic input to the inferior mesenteric ganglion. **b** Schematic representation of somatic and visceral primary afferent neurons and the sympathetic nervous system. *ACh*, acetylcholine; *NA*, noradrenaline; *EK*, enkephalin; *SP*, substance P; *BV*, blood vessel; *DRG*, dorsal root ganglion; *IMG*, inferior mesenteric ganglion; *S*, skin; *SC*, spinal cord; *VO*, visceral organ. (**a** from Crowcroft et al. 1971; **b** from Otsuka and Konishi 1983)

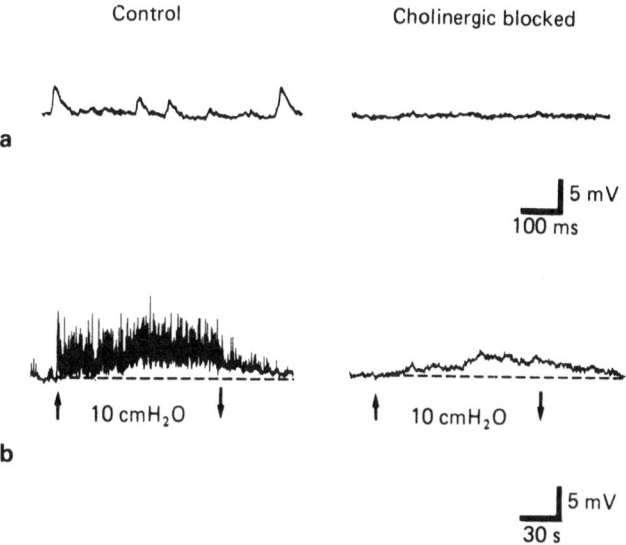

Fig. 7.19a,b. Mechanosensory input to the inferior mesenteric ganglion (IMG) from the distal colon and its cholinergic and noncholinergic components. In **a**, asynchronous synaptic activity recorded from an IMG neuron with the colon in the nondistended state (0 cmH₂O) was abolished after superfusion of the ganglion with hexamethonium ($1 \times 10^{-6}M$) and atropine ($2 \times 10^{-6}M$) for 3 min. In **b** (slower sweep speed), the colon was distended to an intraluminal pressure of 10 cmH₂O (*between arrows*). Distension produced an increase in asynchronous synaptic activity superimposed over a slow depolarization. Cholinergic antagonists abolished the asynchronous fast EPSPs but not the slow depolarization. *Dashed line* represents the resting membrane potential before distension. (Kreulen and Peters 1986)

ganglionic efferents that had collaterals synapsing in the inferior mesenteric ganglion and in the bladder, intramural neurons that have centripetal projections, other neurons running in a rostral direction via the pelvic plexus (such as parasympathetic efferents in the hypogastric nerve described by McLachlan 1985; see Chap 3, p. 82), or hypogastric afferents with collaterals that synapse in the inferior mesenteric ganglion (Fig. 7.18). In addition to the classical cholinergic transmission, blocked by nicotine or hexamethonium, there is evidence from microelectrode penetrations of postganglionic cells in the inferior mesenteric ganglion for a slow excitatory postsynaptic potential (slow EPSP) that is not affected by cholinergic ganglion blockers, and which could modulate the excitability of synaptic transmission at this site (Fig. 7.19; Kreulen and Peters 1986). Otsuka and Konishi (1983) and Kreulen and Peters (1986) suggest that substance P may be involved in producing this EPSP, and that this peptide may originate from primary afferent terminals; in addition they suggest that enkephalin may be present in axon terminals in the inferior mesenteric ganglia, and have an inhibitory action, possibly on other axonal terminals.

There is good evidence that prevertebral ganglia contain different types of neurons; one basis for the classification is the discharge patterns and the axonal projections of the neurons. The discharge patterns of the neurons recorded with microelectrodes in vitro and described below can also be correlated with those recorded by nerve dissection in the intact preparation (see the next section). In an in vitro preparation of the inferior mesenteric ganglion of the cat, Julé et al. (1983) found sympathetic postganglionic neurons with different types of discharge patterns. Many of the irregularly discharging and nonspontaneous neurons in the rostal and caudal poles of the ganglion had synaptic inputs from the hypogastric nerve, and the nonspontaneous neurons in the caudal lobes, in contrast to those in the rostral lobes, received peripheral synaptic input primarily from the hypogastric nerve; most of these neurons had axons that projected to the viscera also through the hypogastric nerves. They conclude that one of the main physiological functions of the caudal lobe of the inferior mesenteric ganglion is to transact and integrate neural signals derived from or destined for the pelvic abdominal viscera innervated by the hypogastric nerves, including the bladder, the genitalia and the internal anal sphincter. This function is also suggested by Bouvier and Grimaud (1984), who found that vesical distension causes contraction of the internal anal sphincter by a hypogastric pathway, mediated mainly in the periphery by connections in the inferior mesenteric ganglion. This response was blocked by phentolamine. They also found that distension of the bladder inhibited

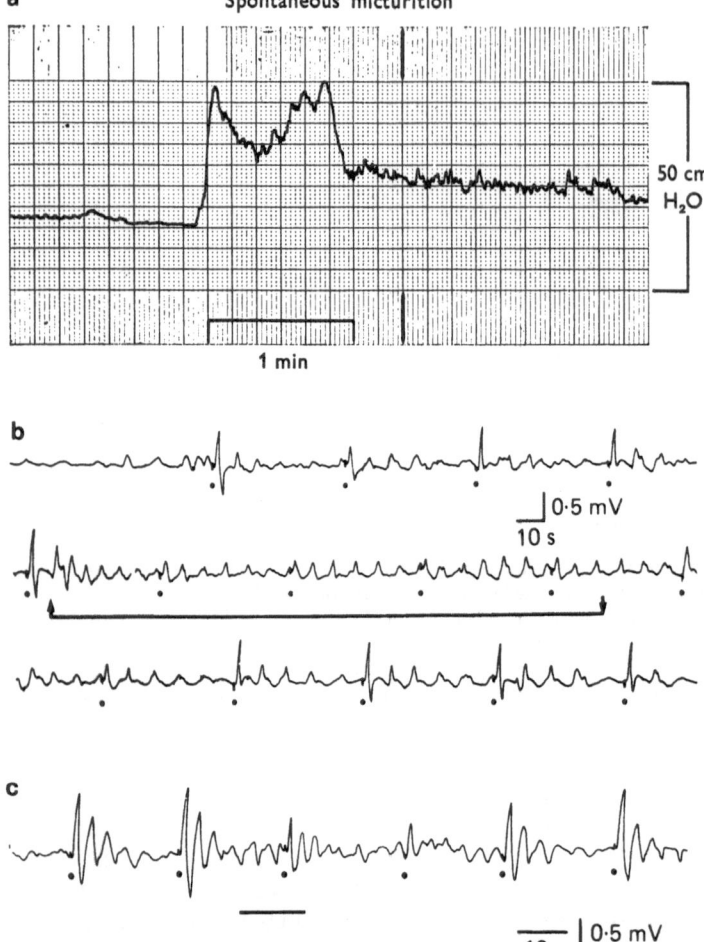

Fig. 7.20a–c. Reflex effects of bladder contractions and distension on the smooth muscle of the internal anal sphincter. **a** The effects on the pressure exerted by the internal anal sphincter during a spontaneous bladder contraction. **b,c** Inhibitory effects of vesical afferents mediated via prevertebral ganglia. *Dots* under each tracing indicate hypogastric nerve stimulation (2 shocks, 20 Hz, 2 ms, 10 V). **b** Inhibition of the response of the internal anal sphincter, evoked by hypogastric nerve stimulation, during bladder distension. *Black bar* under middle tracing indicates the duration of distension (volume: 10 ml). Continuous recording. **c** Inhibitory effect obtained by electrical stimulation of vesical afferents. *Black bar* indicates the duration of stimulation (20 Hz, 2 ms, 10 V). (**a** from Garrett et al. 1974; **b,c** from Bouvier and Grimaud 1984)

EMG activity in the internal anal sphincter evoked by hypogastric nerve stimulation (Fig. 7.20); this interaction was blocked by naloxone, whose effect may have been due to blockade of enkephalins released from visceral afferent fibres. Some visceral afferents contain enkephalins (see Chap. 3, p. 65), and the release of these peptides at the sites suggested in Fig. 7.18b might explain these observations. Both responses were unaffected by transection of the sacral dorsal roots and only slightly affected by transection of the lumbar dorsal roots. The reflexes were strongly affected by transection of the lumbar ventral roots, which suggested that they were facilitated by the efferent sympathetic tone. The responses were entirely dependent on the integrity of the remaining hypogastric nerve, which constituted the efferent limb of the reflex. The effects of enkephalins on ganglionic transmission were considered in Chapter 5 (see p. 137).

Sympathetic Reflexes Mediated by the Central Nervous System

Effects of Bladder Distension and Contraction on Cardiovascular Sympathetic Efferents

Jänig (1986) has recently reviewed the properties of sympathetic preganglionic neurons concerned with

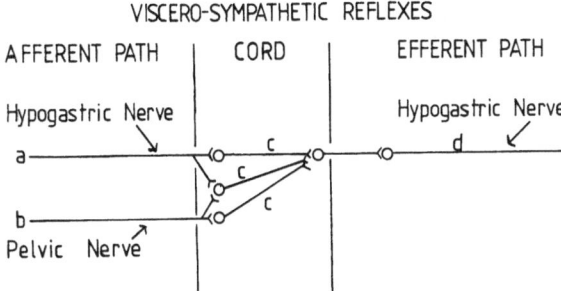

VISCERO-SYMPATHETIC REFLEXES

AFFERENT PATH | CORD | EFFERENT PATH

Hypogastric Nerve | | Hypogastric Nerve

Fig. 7.21. Afferent and central pathways that are operative in viscero-sympathetic reflexes. Identified neurons have been described in the following papers: *a* Floyd et al. (1976); *b* Iggo (1955), Floyd and Lawrenson (1979); *c* McMahon and Morrison (1982b); *d* de Groat and Lalley (1972), Floyd et al. (1982b). (Morrison 1982)

cardiovascular function, and the effects that visceral stimuli may have on them. In respect of the sympathetic efferents that project in the lumbar splanchnic nerves, the vasoconstrictor neurons show ongoing activity that had marked cardiovascular rhythmicity and modulation during the respiratory cycle, inhibition when carotid sinus pressure is raised, weak excitation during stimulation of the arterial chemoreceptors, and (usually) no influence of distensions and contractions of the bladder or colon (Bahr et al. 1986b).

Effects of Bladder Distension and Contraction on Sympathetic Efferents to the Viscera

Figure 7.21 shows the main pathways responsible for reflex effects from the bladder and colon on the discharge of sympathetic efferents (Morrison 1982). Garrett et al. (1974) also found that rectal or

bladder distension gave rise to reflex contractions of the internal anal sphincter, but did not investigate whether the site of the synapses were central or peripheral. They concluded that excitatory motor pathways involved α-adrenergic and cholinergic mechanisms, and that inhibitory pathways were also present which depended on β-adrenergic and nonadrenergic, noncholinergic mechanisms. The cholinergic contractions were dependent on connections that were blocked by α-adrenoceptor antagonists.

However, not all reflex activities are mediated by peripheral reflex loops: there is good evidence that preganglionic neurons that project to the inferior mesenteric ganglion alter their activity during visceral reflexes (Jänig 1986). Bahr et al. (1986a–c) and Bartel et al. (1986) have recently described sympathetic preganglionic neurons in the cat which they call motility-regulating (MR) neurons (Table 7.2). These neurons can be divided into two types. Both types have reciprocal inhibitory inputs from bladder and colonic afferents and usually had inputs also from the perineum and perianal skin; in MRI neurons the influence of bladder distension and contractions are excitatory. They could be differentiated from vasoconstrictor neurons by the lack of baroreceptor and chemoreceptor inputs, and had different axonal conduction velocities, different patterns of discharge, different reflex latencies and different segmental origins within the spinal cord. The reflexes elicited in these preganglionic neurons by natural stimuli, such as bladder distension, are probably mediated by sacral afferents and by spinal pathways, which are under the influence of descending inhibitory and excitatory pathways from the brain stem.

de Groat and Lalley (1972) also investigated the reflex influence of pelvic nerve afferents on sympathetic efferents in the lumbar splanchnic and

Table 7.2. Properties of sympathetic vasoconstrictor and motility-regulating neurons in cats (Bahr et al. 1986c)

Parameters	Vasoconstrictor	Motility-regulating (MR)	
		MR1	MR2
Baroreceptors	Inhibit cardiac rhythm	0	0
Chemoreceptors	Excite	0	0
Bladder distension	0 (few +)	+	−
Colonic distension	0 (few +)	0	+ or 0
Anal stimulation	0 (few +)	+ (few −)	+ (few −)
Mean conduction velocity (m s⁻¹)	2.8	8.5	6.9
Mean resting discharge (s⁻¹)	1.6	0.7	0.9
Location of white rami	L-1 to L-4	L-3 to L-4	L-3 to L-4

hypogastric nerve. Electrical stimulation of the pelvic nerve gives rise to a burst of activity in sympathetic efferents, at latencies that were consistent with a spinal reflex. The amplitude of this evoked response was reduced at high bladder pressure because of supraspinal inhibition. Reflex activity could also be elicited by bladder contraction, and the majority of sympathetic efferents were excited at low bladder pressure, whereas a smaller proportion were inhibited at the same pressures. These authors hypothesize that the sympathetic efferents should be activated during bladder distension (i.e. when accommodation and urethral sphincter contraction are required during the filling phase) and should be inhibited during contraction. There is no evidence, however, that distension and contraction differ in their effects on the discharge of single sympathetic pre- or postganglionic fibres (Floyd et al. 1982a; Bahr et al 1986a; Jänig and McLachlan 1987).

de Groat (1975) provided some evidence that the sympathetic reflexes are more excitable at low intravesical pressure during the filling phase; electrically evoked sympathetic reflexes were inhibited at high bladder pressures, and were most active when the bladder pressure was low. The difference in the excitability of these reflexes was dependent on supraspinal descending inhibitory pathways that become active when bladder pressure is high. The interpretation of this data depends on how closely one believes that electrical stimuli simulate natural activation of the afferent pathway. Maggi et al. (1985) have investigated the effects of the sympathetic on bladder tone during filling at physiological rates in rats, and have concluded that sympathetic reflexes are normally active during the filling phase. Propranolol or hypogastric nerve section both increased the intravesical pressure response to saline filling in spinal rats. Propranolol also increased the pressure at which rhythmical contractions of the bladder were elicited.

Floyd et al. (1982a) studied the influence of lumbar visceral afferents on the discharge of sympathetic efferents, during distension of the bladder or colon (Fig. 7.22). The threshold intravesical pressures that elicited these responses were similar to those of the de Groat and Lalley study. The effects of bladder and colonic stimuli were often seen in the same sympathetic neuron and were always of the same sign, i.e. both were excitatory or both were inhibitory; furthermore, distension and contraction of a viscus produced the same effect. These reflexes are therefore not specifically vesicosympathetic, but viscerosympathetic, elicited by distension of more than one viscus. This result contrasts with that of Bahr et al. (1986a,b) and Bartel et al. (1986) in that reciprocal interactions between the viscera are not

seen when the influence of the pelvic nerves is removed. This is consistent with the results of McMahon and Morrison (1982b), who found that central neurons that showed reciprocal properties always have pelvic nerve afferent inputs, and do not project above the thoracolumbar junction, i.e. not above the level of the lumbar splanchnic outflow.

Floyd et al. (1982a) studied the properties of sympathetic efferents in the hypogastric nerve, and looked at their reflex activities in an anaesthetized cat preparation in which the only afferent input from the bladder was from the lumbar pathway. They found that some neurons did not respond to any afferent stimulus (visceral or somatic) they tried, and that many of these neurons showed a regular spontaneous discharge. Indeed, there was a highly significant association between the occurrence of a regular spontaneous discharge and the absence of reflex activity. Julé and Szurszewski (1983) studied the cat inferior mesenteric ganglion in vitro and found that some of the neurons they studied with intracellular electrodes showed a regular spontaneous discharge. They could find no evidence either for synaptic input from the hypogastric nerves, or for hypogastric axonal projections, in their population of regularly discharging neurons (Julé et al. 1983). They thought that the regularly discharging neurons were intraganglionic neurons that were endogenously active, and provided synaptic input to the irregularly discharging neurons. Floyd et al. (1982a), however, did have evidence of a hypogastric projection of their neurons (whose properties otherwise fitted the description given by Julé et al. 1983), and this might suggest that these cells provide an endogenous resting discharge to other neurons in the vesical plexus.

Summary

Reflexes through prevertebral ganglia are known to exist, but there is little evidence as to their function in humans. The evidence of the last 15 or so years has suggested that these reflexes operate in physiological conditions. The reflex contraction of the internal anal sphincter during bladder distension appears to be largely due to activity in this prevertebral pathways. The afferent limb of these reflexes may be mediated by axon collaterals of primary afferents or by the centripetal processes of enteric neurons. The efferent limb of the reflex involves postganglionic neurons with cell bodies in the prevertebral ganglia; these ganglia may also contain other neurons, some of which are thought to discharge spontaneously and

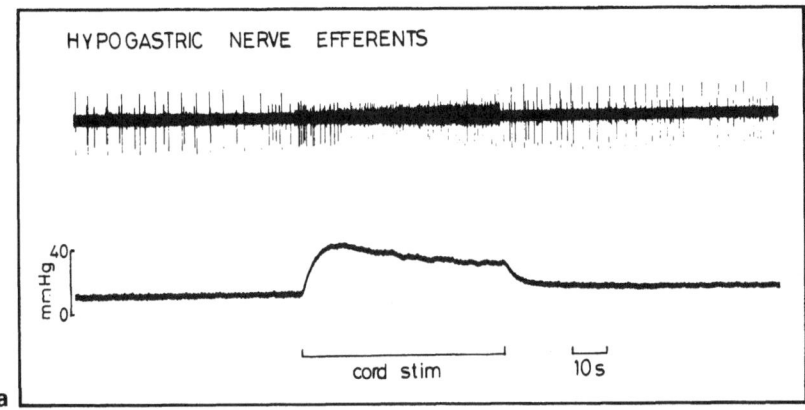

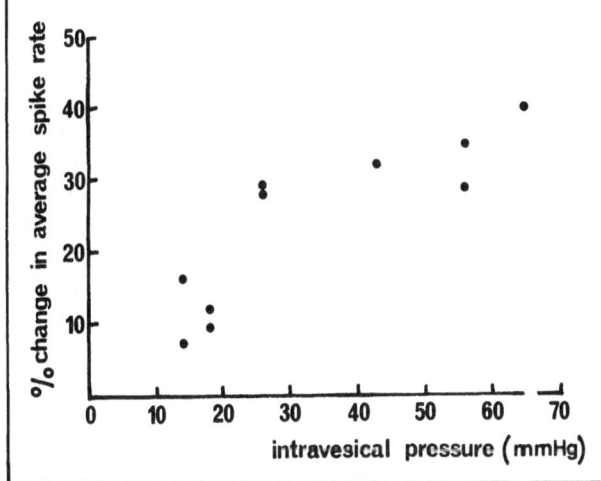

Fig. 7.22. a The discharges recorded from two hypogastric nerve efferent units (*top*) during a bladder contraction. Bladder pressure (*bottom*) is increased by electrical stimulation of the parasympathetic efferents in the sacral cord below a complete transection at the sixth lumbar segment. The bladder contraction induces a decrease in the firing of the large unit and an increase in that of the smaller unit. (Recording by courtesy of Dr. Verity E. Hick) **b** The relationship between adapted discharge rate in a sympathetic efferent unit and intravesical pressure during the plateau phase of a series of bladder contractions. (**a** from Morrison 1982; **b** from Floyd et al. 1982a)

regularly, to provide an endogenous input to other postganglionic neurons that receive synaptic inputs from the viscera and from the central nervous system.

Reflexes through the central nervous system appear to mediate many of the sympathetic phenomena responsible for cardiovascular adjustments during micturition or bladder distension, particularly following spinal transection. In humans, the effects of spinal transactions rostral to the midthoracic segments result in greater autonomy in the sympathetic nervous system than when lower lesions are present, as a result of the loss of the baroreceptor reflex input to the autonomic preganglionic neurons. The reflexes mediated by the separated spinal segments are then no longer subject to inhibition from the brain stem, and autonomic hyperreflexia is more likely to occur in quadriplegics or paraplegics with spinal lesions above the fifth thoracic segment. The afferent limb of these spinal sympathetic reflexes is in the pelvic and hypogastric nerves.

Other sympathetic reflexes may act on the lower urinary tract, e.g. to modulate the activity of urethral smooth muscle, or to produce bladder relaxation.

Whether or not they contribute to bladder relaxation during normal filling in humans has been questioned; the effect of this reflex in unphysiological conditions would nevertheless be to relax the bladder. It has been suggested that sympathetic reflexes are activated by distension so as to aid accommodation, but are inhibited during bladder contractions, so as to facilitate voiding. There is evidence that sympathetic reflexes may operate to facilitate bladder relaxation during filling in certain species; there is no evidence however for sympathetic efferent (or afferent) single units that respond to bladder distensions and contractions in opposite ways, as would be necessary if the hypothesis were correct. Some efferent sympathetic units are inhibited by bladder distensions and by bladder contractions; others (the majority) are excited by these stimuli. The functional characteristics of these two types of MR neurons have now been defined, and differ in many respects from those of the vasoconstrictor neurons in the hypogastric nerve (Table 7.2); in particular they respond to bladder distensions and contractions in the same way, and show reciprocal changes during colonic activity.

Effects of Spinal Transection on the Function of the Lower Urinary Tract

Neurological Changes in Human Paraplegics and Quadriplegics

Lesions at different levels can affect the physiology of the individual in different ways. Thus, cervical lesions have immediate effects on blood pressure, and ventilation can be compromised; spinal transection above the fourth cervical segment disconnects the phrenic nerve as well as the intercostal muscles from the medullary respiratory centres, and is not compatible with life.

The immediate neurological effect of experimental spinal transection is a period of "spinal shock" characterized by flaccid paralysis and sensory loss below the level of the lesion; its duration is a few days or less in laboratory animals and the analagous period following traumatic spinal transection in humans can last for only a few hours, but usually persists for several weeks. During this phase, there is depression of all reflex activity below the lesion, and the bladder is flaccid. After recovery from spinal shock, laboratory animals regain somatic reflexes and usually show spontaneous reflex bladder activity, but emptying is incomplete; reflex bladder contractions can often be initiated by cutaneous or other stimulation below the level of the lesion (de Groat and Ryall 1969; Jänig 1986). It is not clear why some recovery of reflexes should occur, often months after the spinal lesion. In the shorter term, it would seem likely that the loss of descending pathways will unmask reflexes that are normally suppressed, and alter the excitability of spinal reflexes that are present in the normal individual. It is much more difficult to explain the development of reflex functions that appear over the course of months, and one interesting explanation might be that new synaptic connections are formed on neurons that have suffered a loss of synaptic inputs by sprouts from neurons that are active. The principle involved is supported by some evidence, particularly within the brain; however, more work needs to be done before we can conclude that mechanisms involving plasticity of neuronal connections are operating in the regulation of the lower urinary tract in spinal humans.

Cauda equina lesions, which interrupt the afferents and efferents running from the sacral cord to the bladder and urethra, result in urinary retention, and overflow dribbling. In this condition the bladder is disconnected from the sacral cord, and, in humans, the bladder can only rarely produce effective detrusor contractions by means of its own intramural innervation (Guttmann 1976).

The effects of spinal transection are generally referred to in relation to the level of the lesion; the vertebral level of a fracture is different from the spinal segment lesioned by that trauma if transections occur in the lower thoracic and upper lumbar cord. Thus, the twelfth thoracic vertebra in humans overlies the fifth lumbar segment of the spinal cord. In humans, the return of some function to the lower urinary tract can take months and there is great variation in the degree of functional recovery. This appears to be due partly to the fact that acute trauma to the spinal cord can give rise to a variety of pathology below the level of the lesion. Haemorrhagic necrosis and oedema affect the cord segments adjacent to the transection, and later changes include gliosis along the degenerating tracts which normally connect the sacral cord and brain. Some patients with complete lesions well above the level of the sacral cord also have evidence of trauma or infarction of the lumbosacral cord. Such patients have a flaccid bladder, and little sphincteric tone. These pathological changes explain the variation that can be seen in the urodynamic findings of groups of paraplegic patients in some studies (Fig. 7.23).

Somatovesical Interactions in Spinal Humans and Animals

In humans, the anal and bulbocavernosus reflexes are the first reflexes to reappear following spinal shock (if in fact they disappeared), followed by the return of more proximal somatic reflexes; bladder reflexes usually return to some extent, possibly months later, if at all. When they do return, bladder contractions may be elicited in certain individuals by one or more stimuli that are not normally associated with micturition; these include touching or pulling hairs on the skin of the thigh, anal stimulation, tapping on the lower abdominal wall, and an association with spontaneous events such as flexor spasms. Figure 7.24 shows the effects of electrical stimulation of the common peroneal nerve 11 days after a complete cord lesion at the seventh thoracic segment. The bladder contracts and the urethra reflexes reflexly.

Distension of the bladder is another stimulus that may also give rise to bladder contractions, mediated by reflex pathways which are released from descending inhibitory influences. Figure 7.25 shows such spontaneous contractions in a male paraplegic with a complete lesion at T-7 to T-8, recorded 6 months after the injury. These bladder reflexes may

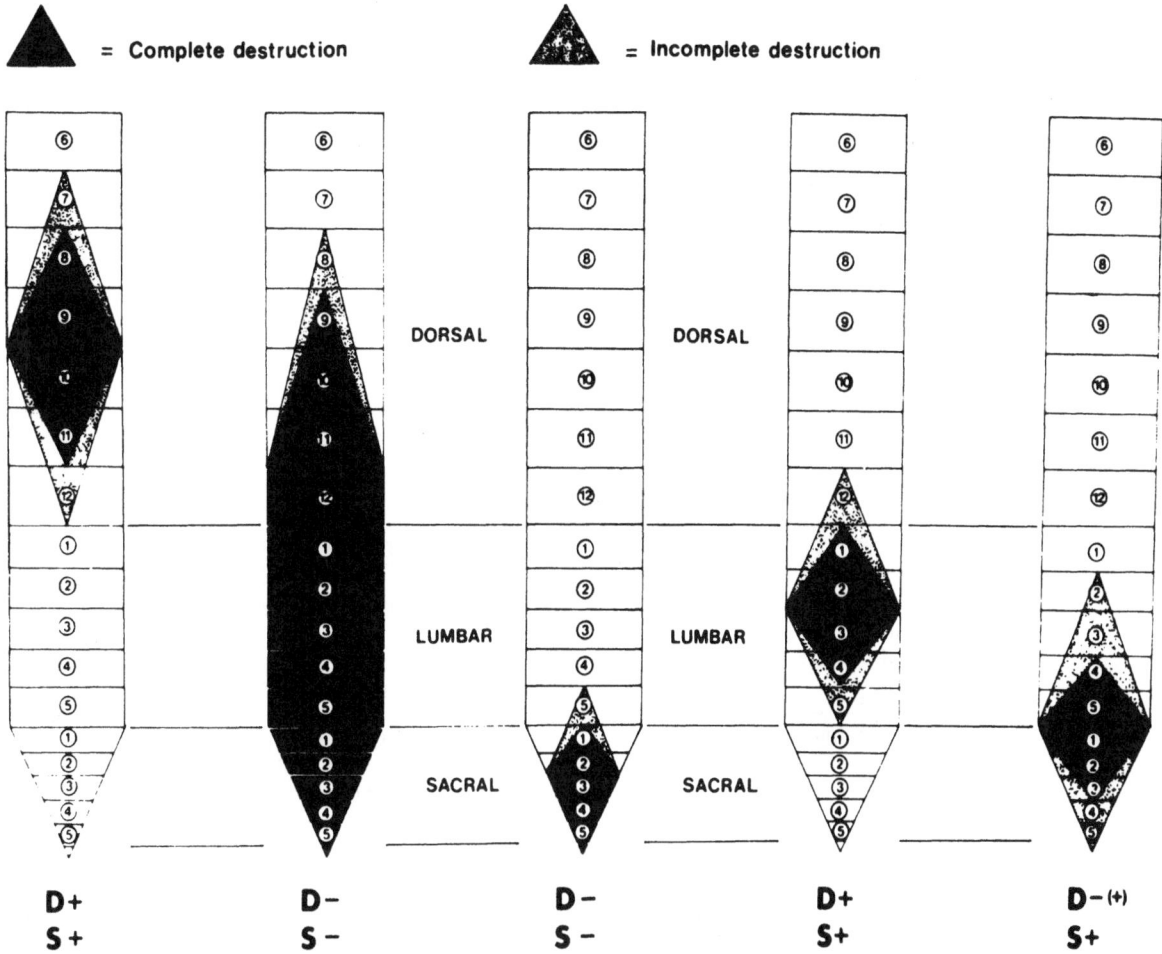

Fig. 7.23. The effects of different levels and degrees of cord damage on bladder and sphincteric function. $D+$, reflex detrusor activity; $D-$, absent reflex detrusor activity; $S+$, reflex activity of external urethral sphincter; $S-$, absent external urethral sphincter activity. (Thomas 1982)

be weak or powerful, and may not be maintained throughout voiding, so that emptying is incomplete; Barrington might have attributed this partly to the loss of his second reflex. This limited detrusor activity may be opposed by sphincteric contractions because of a failure of the normal reflex mechanisms that lead to urethral relaxation (i.e. Barrington's fifth and sixth reflexes). In addition, an increase in the tone of the urethral sphincter caused by spasticity may oppose emptying, and in some patients there is a reflex increase in sphincteric tone during attempts to void, or during contractions of the bladder.

In Fig. 7.25, the anal sphincter EMG is shown, and the units become active at high intravesical pressures, often near the peak of the contractions. If similar events occurred in the urethral sphincter, then emptying would have been actively opposed by such reflex activity. In some patients, this increase in

urethral tone at rest (Fig. 7.26) as well as during bladder contractions may be one component of a general increase in somatic muscle tone which may be associated with flexor spasms. The increase in sphincteric tone during straining is a normal response, and can be seen in paraplegic patients; it is assumed to be a somatic reflex initiated by receptors in abdominal or pelvic floor muscles. The difficulty in voiding during detrusor activity that is opposed by contractions of the urethral sphincter is commonly called detrusor–sphincteric dyssynergia. The pattern of bladder reflexology that exists in an individual patient with spinal injuries can change from time to time, and is influenced by, amongst other factors, the presence of inflammation or infection, and the occurrence of episodes of overdistension.

McPherson (1966a,b) studied the changes in bladder activity that could occur in chronic spinal

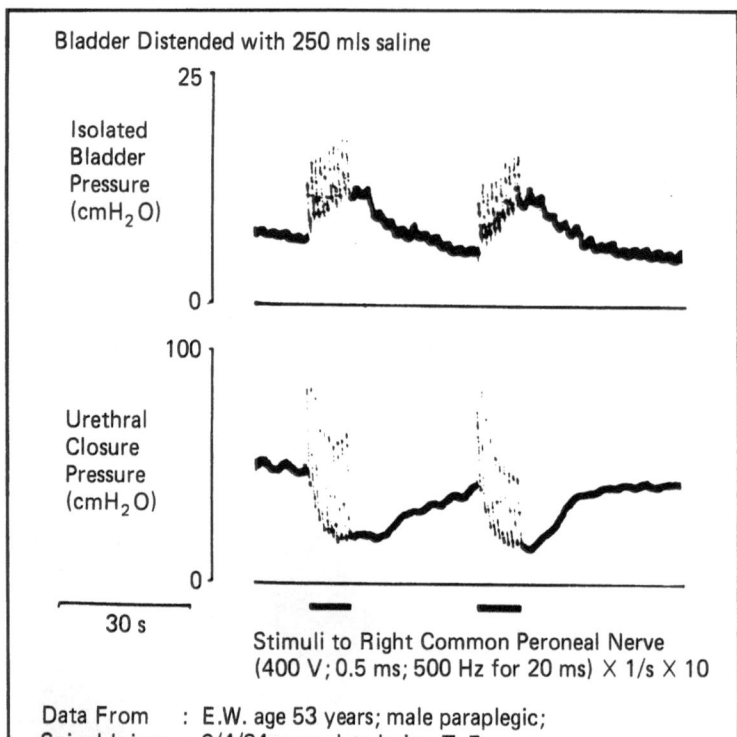

Bladder Distended with 250 mls saline

Isolated Bladder Pressure (cmH₂O)

Urethral Closure Pressure (cmH₂O)

30 s

Stimuli to Right Common Peroneal Nerve
(400 V; 0.5 ms; 500 Hz for 20 ms) × 1/s × 10

Data From : E.W. age 53 years; male paraplegic;
Spinal Injury : 2/4/84; complete lesion T-7.
Recorded : 13/4/84

Fig. 7.24. The effects of electrical stimulation of the common peroneal nerve on bladder and urethral pressures following spinal transection. (A.L. Wright and J.F.B Morrison 1984, unpublished work)

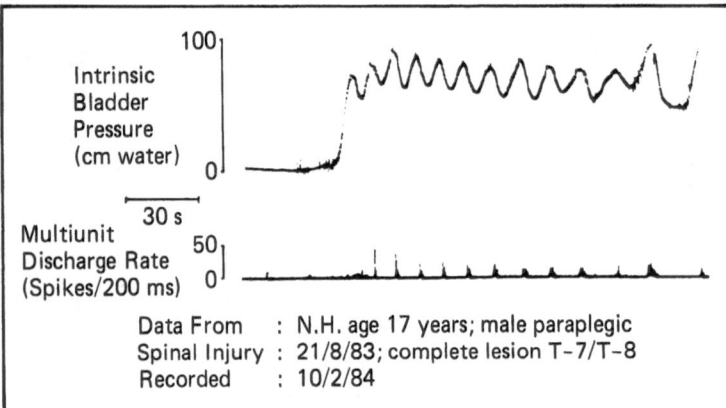

Intrinsic Bladder Pressure (cm water)

30 s

Multiunit Discharge Rate (Spikes/200 ms)

Data From : N.H. age 17 years; male paraplegic
Spinal Injury : 21/8/83; complete lesion T-7/T-8
Recorded : 10/2/84

Fig. 7.25. Spontaneous bladder activity 6 months after a spinal cord transection. The raised bladder pressure is associated with increasing EMG activity in the external anal sphincter. (A.L. Wright and J.F.B. Morrison 1984, unpublished work)

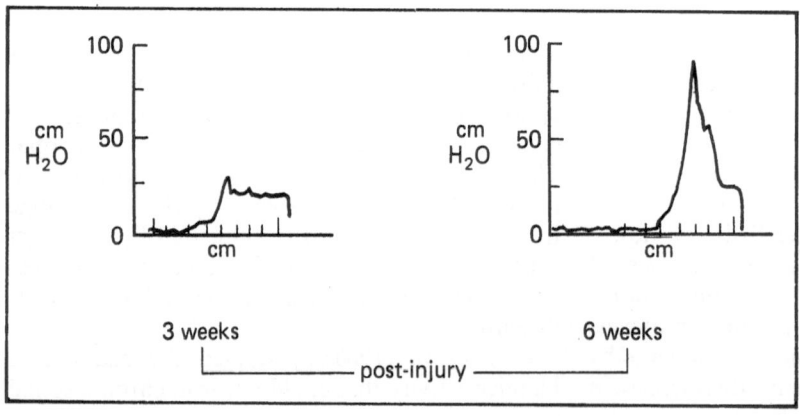

cm H₂O

cm H₂O

cm   cm

3 weeks   6 weeks

——————— post-injury ———————

Fig. 7.26. Recovery of the urethral pressure profile following spinal shock in a patient with a complete lesion at the fourth thoracic level. (Thomas et al. 1975)

cats during somatic stimulation. Cutaneous stimuli produced reflex bladder contractions or an augmentation of existing contractions; however, stimulation of muscle nerves was without effect. Spontaneous bladder contractions occur in spinal animals (de Groat and Ryall 1969), and the frequency and amplitude of the contractions increases when the bladder is filled. At baseline pressures of around 10 cmH$_2$O, large bladder contractions could be induced; injections of small volumes of fluid into the bladder produced reflex contractions. Tactile stimulation of the perineum was also a potent stimulus that induced firing of these neurons and reflex bladder contractions. de Groat et al. (1975) compared the somatovesical reflexes of chronic spinal cats with those present during postnatal development in kittens. An excitatory reflex can be elicited from the perineal skin of young cats and other nonprimate species which results in contraction of the bladder and voiding; the reflex is organized in the lumbosacral cord and utilizes a pelvic nerve efferent pathway. This reflex is essential for existence in these neonates: the pups micturate and defecate when the perineal region is licked by the mother. The reflex becomes nonfunctional at the age of 5–7 weeks in cats, and the latency of the response more than doubles by the time the response disappears, probably because of the development of myelinated bulbospinal pathways. This reflex can only be elicited in adult cats following spinal transection, and its absence in normal adults is presumed to be because of inhibitory descending pathways that suppress it. Reflex activity in pelvic nerve efferents can be elicited at short latency (3–5 ms) following electrical stimulation of somatic afferents in spinal afferents (de Groat and Ryall 1969; de Groat 1975).

Bladder Reflexes Following Spinal Transection

Following spinal shock in animals in which cord transection has been performed by a cut which seeks to minimize injury to the adjacent tissues, bladder reflexes recover and exhibit a micturition threshold, reciprocal activity between bladder and colon, and somatovisceral interactions (de Groat et al. 1981, 1982; Kümmel 1983; Bartel et al. 1986).

de Groat et al. (1981) found that normal animals sometimes show a long latency reflex that can be recorded from fine nerve bundles on the bladder surface following stimulation of C-fibre afferents in the pelvic nerve. This reflex disappears during spinal shock, but becomes progressively larger during the development of automatic micturition. Further-

more, the reflex only occurs when the bladder is distended; it is unlikely that the effects of distension are mediated by C-fibre afferents in the pelvic nerve, as none of these appear to respond to pressures less than 70 mmHg (Floyd and Lawrenson 1979; Bahns et al. 1986). It is probable that the effects of raising bladder pressure are mediated by A-delta fibres acting on sacral local interneurons that are believed to be responsible for setting the micturition threshold (McMahon and Morrison 1982b,c; see Chap. 8, p. 265). These properties of C-fibre afferents in the pelvic nerve contrast with those of the hypogastric nerve, which are mechanosensitive (Floyd et al. 1976; Bahns et al. 1986; Jänig and Morrison 1986), but not involved in the regulation of the micturition threshold (McMahon and Morrison 1982c; McMahon 1986). The effect of bladder distension in spinal animals was to gate the C-fibre reflex in much the same way as the spinobulbospinal reflex was gated in intact animals (McMahon and Morrison 1982c; see Chap. 8, p. 262). It is not entirely clear whether the responses of postganglionic efferents recorded from the bladder base when C-fibre afferents are stimulated in these experiments are also seen in parasympathetic preganglionic neurons in the cord. If the evidence for the C-fibre reflex depends entirely on recordings from the pelvic nerves at the bladder base, then the question has to be raised whether the recordings were from parasympathetic or sympathetic efferents (Kuo et al. 1984). The interpretation that de Groat et al. (1981) place on these experiments is that the efferents are parasympathetic and that the C-fibre reflexes only become dominant following spinal transection. A possible alternative interpretation of the data could be that the efferent activity is in sympathetic efferents (Kuo et al. 1984) and appears at high bladder pressure in spinal animals because it is normally suppressed by descending inhibitory pathways from the brain stem. There are a number of such pathways (see the start of this chapter, and Chap. 8, p. 248 and two possibilities have been described by Dembowsky et al. (1981) and Gilbey et al. (1981). Clearly, the situation is a complex one, and most of the discussion has focused on the changes in the spinal cord; whether or not there are associated changes in peripheral ganglia remain to be seen, as sympathetic pathways appear to be able to excite pelvic postganglionic neurons in certain circumstances (Sundin 1972; de Sy et al. 1974; see earlier discussion, p.135).

Micturition can be induced in paraplegic cats by administration of naloxone. The time course of this response lasted about 1 h, and was accompanied by hind limb movements, which were attributed to the reflex activity of bladder afferents as they did not occur when the bladder was empty. This suggests that endogenous opiates may have a tonic inhibitory role to play in the regulation of micturition (Thor et al. 1983). Hisamitsu and de Groat (1984) later found that enkephalins appear to work in at least two sites in the micturition reflex: in the brain the effects are mediated by δ- and μ-receptors and control the frequency of bladder contractions and the micturition threshold, and in the spinal cord, where δ-receptors seem to control the amplitude of bladder contractions. This is in addition to an inhibitory effect of opioid peptides in bladder ganglia (de Groat et al. 1983). In humans, the effects of opiate receptor blockade on detrusor and urethral function has been studied by Murray and Feneley (1982), who found that naloxone increased the detrusor pressure and reduced the bladder capacity at the first sensation and the sensation of urge to micturate; in addition urethral closure pressure was reduced.

The bladder may be made to contract by electrical stimulation of perineal skin or of the lateral popliteal nerve in some paraplegic men. The effect occurs at resting pressures just below the micturition threshold, and usually occurs in patients who can generate spontaneous bladder contractions. The size of the bladder contractions elicited by nerve stimulation increases as the resting pressure is increased. Raising rectal pressure can abolish or reduce the size of spontaneous contractions in these patients (A. L. Wright and J. F. B. Morrison 1984, unpublished work). Figure 7.24 shows one example from a patient 11 days after the initial injury, and the reflex bladder contraction can be seen in some paraplegic patients 6–12 months after the lesion.

Sphincteric Reflexes Following Spinal Transection

Barrington found that the pathways that elicited sphincteric relaxation during micturition were at least partly within the spinal cord. Melzak and Porter (1964), who studied spinal men, found that the external anal sphincter showed continuous tonic activity, which could be increased by skin stimulation and raised intra-abdominal pressure. Distension of the anal canal inhibited contractions,

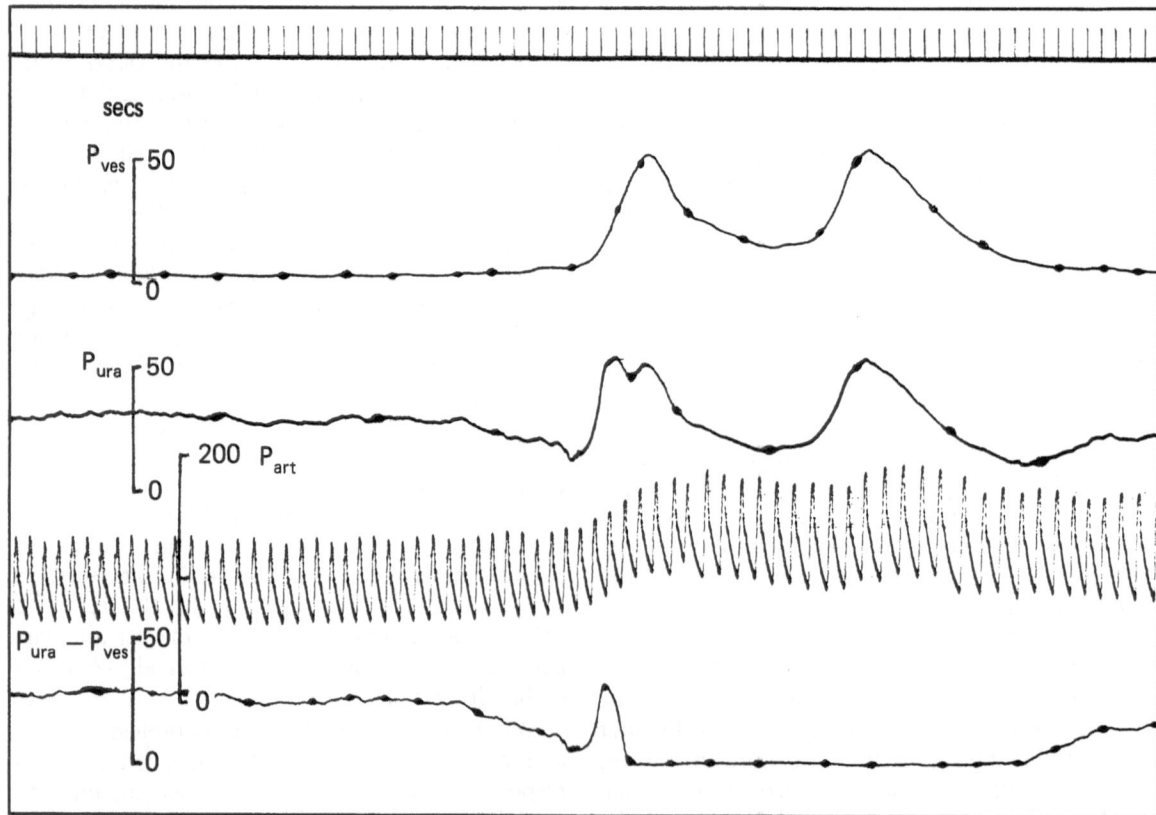

Fig. 7.27. Changes in bladder and urethral pressures during autonomic hyperreflexia. (Courtesy of M.J. Torrens)

whereas rectal distension could often excite EMG activity. A. L. Wright and J. F. B. Morrison (1984, unpublished work) found that the anal reflex latencies did not differ significantly from normal in paraplegics with complete transverse lesions of the cord at low and midthoracic levels, unless there was evidence of cord damage below the level of the transection; in the latter situation, the response was usually absent.

Coughing can evoke electrical activity in the external anal sphincter of paraplegic men, as can raising intra-abdominal pressure by a Valsalva manoeuvre or by abdominal compression (N. Womack and J. F. B. Morrison 1986, unpublished work). Figure 7.27 shows the effects of autonomic hyperreflexia on the urethral sphincter in humans following a spinal cord lesion.

Sympathetic Reflexes Following Spinal Transection

Patients who survive spinal transections above the fifth thoracic segment may be subject to abnormal sympathetic reflexes mediated by the isolated spinal segments and initiated by afferent inputs from the viscera, and particularly the bladder; the symptoms and signs elicited in this state are usually referred to as autonomic dysreflexia or hyperreflexia. Patients who survive a low cervical transection may be severely affected by autonomic dysreflexia, but may also suffer from pooling of blood in the lower limbs and a redistribution of fluid which leads to retention when they are in the erect or semierect posture. This fluid is excreted as urine when they assume the recumbent position, and the diuresis may precipitate some of the abnormal reflexes.

In addition to orthostatic hypotension, another effect of separation of the sympathetic outflow from the brain is a defect in thermoregulation caused by the lack of control over cutaneous vasoconstrictor nerves, and the paralysis of somatic muscle that normally takes part in shivering thermogenesis. The vasodilatation may also give rise to passive engorgement of the penis, that differs in its mechanism from true erection.

In patients with lesions above the fifth dorsal segment, a major proportion of the sympathetic outflow is disconnected from higher levels of the central nervous system, and reacts to afferent inputs below the lesion. One of the most potent of these afferent inputs is from the bladder, and overdistension results in hypertension, because of reflex excitation of noradrenergic sympathetic efferents to blood vessels. Diastolic pressure can rise by more than 50 mmHg, and the systolic pressure may reach more than 250 mmHg. Figure 7.11 shows the changes in arterial, intravesical and intraurethral pressures during an episode of autonomic hyperreflexia. The vasoconstriction is confined to skin, muscle and viscera innervated by the isolated segments of spinal cord, and the cutaneous vascular changes give rise to pallor, and a reduction in skin temperature. The arterial pressure is raised and gives rise to vasodilatation and sweating above the lesion, and headache, caused by excitation of sensory endings in the meningeal vessels. Changes in heart rate occur, however, because the vagal supply to the sinoatrial node is intact; thus, bradycardia is commonly seen during hypertensive episodes (see Fig. 7.16). The reflex basis of autonomic hyperreflexia is present in normal individuals, and these pathways are described more fully in the section on sympathetic reflexes (see p. 215); the special features in the paraplegic or quadriplegic patient are due to removal of descending controls by the cord lesion. In patients with cord transections rostral to the fifth thoracic segment, the loss of descending control becomes critical, and the degree of hypertension becomes more severe (see Fig. 7.16a). The reader is referred to reviews by Guttmann (1976) and Mathias and Frankel (1983).

Recordings from Sympathetic Efferents in Normal and Spinal Humans and Chronic Spinal Animals

Kümmel (1983) studied the changes in activity of postganglionic sympathetic vasoconstrictor neurons in skin and muscle nerves, and the activity of sudomotor neurons before and after spinal transection in cats. In anaesthetized cats with an intact neuraxis, muscle vasoconstrictor and sudomotor neurons are excited by skin and visceral stimulation, whereas cutaneous vasoconstrictors are inhibited. In chronic spinal cats, however, cutaneous vasoconstrictor neurons are excited, and noxious stimulation of skin leads to long-lasting aftereffects in these and the sudomotor neurons.

Wallin and Stjernberg (1984) have recorded from fascicles of skin and muscle nerves containing sympathetic efferents in spinal men. Spontaneous activity was sparse, but pressure over the bladder, electrical stimulation of skin, or pinching the skin below the level of the lesion gave rise to bursts of impulses, which lasted longer than in normal subjects (Figs. 7.28, 7.29). These results correlate well with the single-unit recordings of Kümmel (1983) in cats.

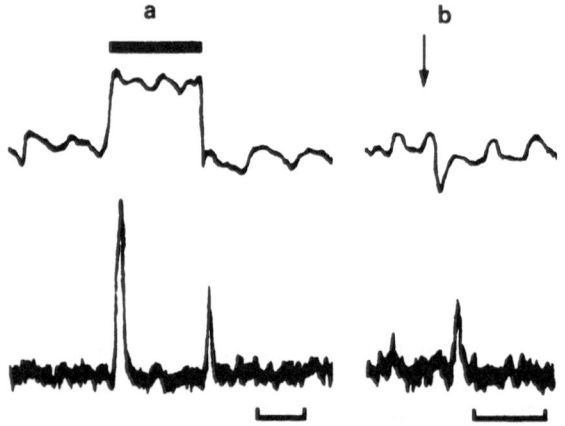

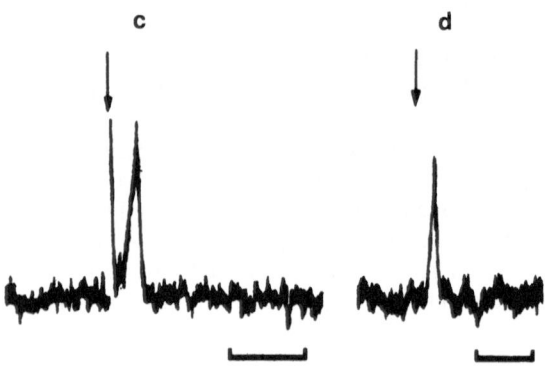

Fig. 7.28a–d. Peroneal mean voltage neural bursts evoked by pressure over the bladder (**a**), a deep breath (**b**), a short train of electrical stimuli delivered via intracutaneous needle electrodes to the abdominal skin over the bladder (**c**) and a skin pinch applied over the bladder in patients with cervical spinal cord transection (**d**). *Upper traces* in **a** and **b** show respiratory movements (inspiration upwards. Manoeuvres indicated by *horizontal bar* (**a**) and *arrows* (**b–d**). Time calibration: 5 s. (Wallin and Stjernberg 1984)

that this inhibitory effect is mediated via spinal pathways.

Summary

Immediately following complete spinal transection there is a period of spinal shock that lasts about 3 weeks or longer in humans, and a considerably shorter time in most laboratory animals. This period is characterized by muscular flaccidity and a loss of spinal reflexes. The recovery of reflex activities below the level of the lesion occurs at different times in different individuals and sometimes fails to occur at all, depending on a number of factors, such as the

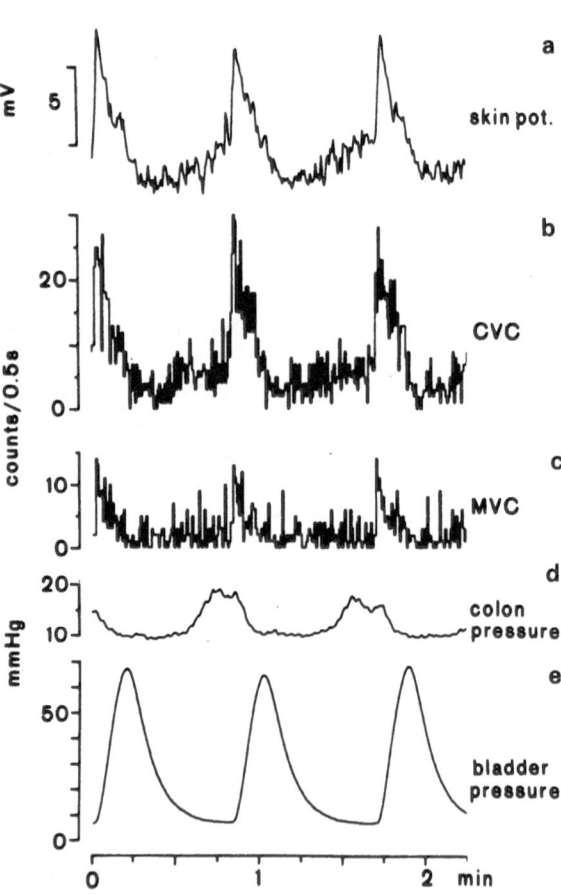

Fig. 7.29a–e. Reactions of sudomotor system (skin potential; *skin pot*) and vasoconstrictor systems supplying hairy skin (*CVC*) and skeletal muscle (*MVC*) during alternating contractions of urinary bladder and colon. The intraluminal pressures in the organs were measured. The postganglionic vasoconstrictor activity was recorded from multifibre bundles isolated from the superficial peroneal (CVC) and from branches of the deep peroneal (MVC) nerve. Chronic spinal cat. (Kümmel 1983)

Bartel et al. (1986) studied bladder motility and reflexes elicited in sympathetic MR neurons in the lumbar splanchnic nerves of spinal cats. Two types of activity were described in different types of neurons. Inhibitory phenomena that are present in the intact animal were uncommon following spinal transection; they conclude that the excitatory effects are mediated via spinal cord mechanisms, whereas the inhibitory effects were mediated via the brain stem and hypothalamus; these conclusions are in keeping with those of de Groat and Lalley (1972), de Groat (1975) and de Groat and Theobald (1976). The effects of colonic stimulation in inhibiting the discharge of certain MR neurons, however, is exaggerated following spinal transection, which suggests

condition of the isolated segments of the spinal cord, and the occurrence of cauda equina lesions. Recovery of reflexes in limb muscles generally occurs before the recovery of bladder reflexes, and when these do reappear they can differ in some important respects from those in the normal individual. The "automatic bladder" is one in which the bladder contracts in response to distension, but the power is rarely that of the normal, such that the residual volume is increased; also reflex activity can be elicited from the skin and during muscle spasms, indicating that the skin and muscle afferents make contact with spinal neuronal circuits that elicit bladder contractions. This is not surprising, given the degree of somatovisceral convergence that is present in these neuronal paths (see Chap. 4, p. 116). It is likely that these somatic influences are normally suppressed by tonically active descending pathways from the brain that appear during the development of bladder control (see Chap. 8, p. 261). Another reason for an increase in the residual volume is the presence of detrusor–sphincteric dyssynergia, which describes the loss of sphincteric relaxation during bladder contractions. The reason for this is little understood, and it is associated with increased resting tone in the urethral sphincter which may be demonstrated by the urethral pressure profile.

Sympathetic reflexes alter their activities following spinal transection, particularly if the lesion is above the midthoracic level. The loss of the baroreceptor reflex contributes to the occurrence of orthostatic hypotension and to the development of hypertensive episodes during autonomic hyperreflexia associated with bladder distension. It has been suggested that some of the main inhibitory influences on the spinal sympathetic reflexes originate from the brain stem and hypothalamus, and that release from these descending inhibitory influences unmasks the predominantly excitatory spinal reflexes. One exception is the inhibitory effect of colonic distension on sympathetic reflexes, which is exaggerated in spinal animals.

Following spinal cord transection it is possible to demonstrate reflex activity elicited by stimulation of C-fibre afferents in fine bundles of nerve fibres at the bladder base; this activity can only be demonstrated when the bladder pressure is high, and increases in amplitude as automatic micturition develops. It is not clear whether or not these recordings are from the parasympathetic or sympathetic pathways; sympathetic reflexes are normally suppressed by descending inhibitory systems in the intact animal, and their presence following spinal transection might be expected. The idea that C-fibre afferents mediate automatic micturition via parasympathetic reflexes in the spinal animal is also in doubt because pelvic nerve unmyelinated afferents appear to be insensitive to bladder pressure.

References

Aldskogius H, Elfin L-G, Andersson-Forsman C (1986) Primary sensory afferents in the inferior mesenteric ganglion and related nerves of the guinea pig. J Auton Nerv Syst 15:179–190

Appenzeller O (1982) The autonomic nervous system: an introduction to basic and clinical concepts, 3rd edn. Elsevier, New York, p 524

Bahns E, Halsband V, Jänig W (1987) Functional characteristics of sacral visceral afferents from the lower urinary tract, colon and anus. Pflügers Arch (in press)

Bahr R, Bartel B, Blumberg H, Jänig W (1986a) Functional characterisation of preganglionic neurons projecting in the lumbar splanchnic nerves: neurons regulating motility. J Auton Nerv Syst 15:109–130

Bahr R, Bartel B, Blumberg H, Jänig W (1986b) Functional characterisation of preganglionic neurons projecting in the lumbar splanchnic nerves: vasoconstrictor neurons. J Auton Nerv Syst 15:131–140

Bahr R, Bartel B, Blumberg H, Jänig W (1986c) Secondary functional properties of lumbar visceral preganglionic neurons. J Auton Nerv Syst 15:141–152

Barrington FJF (1915) The nervous control of micturition. Q J Exp Physiol 8:33–71

Barrington FJF (1921) The relation of the hind brain to micturition. Brain 44:23–53

Barrington FJF (1925) The effect of lesions of the hind and midbrain on micturition in the cat. Q J Exp Physiol 15:181–202

Barrington FJF (1928) The central nervous control of micturition. Brain 51:209–220

Barrington FJF (1931) The component reflexes of micturition in the cat. Parts I and II. Brain 54:177–188

Barrington FJF (1933) The localisation of the paths subserving micturition in the spinal cord of the cat. Brain 56:126–148

Barrington FJF (1941) The component reflexes of micturition in the cat. Part III. Brain 64:239–243

Bartel B, Blumberg H, Jänig W (1986) Discharge patterns of motility-regulating neurons projecting in the lumbar splanchnic nerves to visceral stimuli in spinal cats. J Auton Nerv Syst 15:153–163

Bouvier M, Grimaud JC (1984) Neuronally mediated interactions between the urinary bladder and internal anal sphincter motility in the cat. J Physiol 346:461–470

Bradley WE, Teague CT (1968) Spinal cord organisation of micturition reflex afferents. Exp Neurol 22:504–516

Brindley GS, Rushton DN, and Craggs MD (1974) The pressure exerted by the external sphincter of the urethra when its motor nerve fibres are stimulated electrically. Br J Urol 46:453–462

Cadden SW, Morrison JFB (1984) The effects of visceral distension on the activities of lumbar dorsal horn neurons in the rat. J Physiol 350:71P

Coote JH, Fleetwood-Walker SM, Gilbey MP, Peterson DF (1982) Possible role of oxytocin, vasopression and the paraventriculo-spinal pathway in the regulation of sympathetic preganglionic neurons. J Physiol 324:81P

Crowcroft PJ, Holman ME, Szurszewski JH (1971) Excitatory input from the distal colon to the inferior mesenteric ganglion in the guinea-pig. J Physiol 219:443–461

Cuello AC, Priestley JV, Sofroniew MV (1983) Immunocytochemistry and neurobiology. Q J Exp Physiol 68:545–578

Dahlström A, Fuxe K (1964) Evidence for the existence of monoamine-containing neurons in the central nervous system. I. Demonstration of monoamines in the cell bodies of brainstem neurons. Acta Physiol Scand [Suppl] 232:1–55

de Groat WC (1971) Inhibition and excitation of sacral parasympathetic neurons by visceral and cutaneous stimuli in the cat. Brain Res 33:499–503

de Groat WC (1975) Nervous control of the urinary bladder of the cat. Brain Res 87:201–211

de Groat WC (1976) Mechanisms underlying the recurrent inhibition in the sacral parasympathetic outflow to the urinary bladder. J Physiol 257:503–514

de Groat WC, Lalley PM (1972) Reflex firing in the lumbar sympathetic outflow to activation of vesical afferent fibres. J Physiol 226:289–309

de Groat WC, Ryall RW (1968) Recurrent inhibition in sacral parasympathetic pathways to the bladder. J Physiol 196:579–591

de Groat WC, Ryall RW (1969) Reflexes to the sacral parasympathetic neurons concerned with micturition in the cat. J Physiol 200:87–108

de Groat WC, Theobald RJ (1976) Reflex activation of sympathetic pathways to vesical smooth muscle and parasympathetic ganglia by electrical stimulation of vesical afferents. J Physiol 259:223–238

de Groat WC, Douglas JW, Glass J, Simonds W, Weimer B, Werner P (1975) Changes in somato-vesical reflexes during postnatal development in the kitten. Brain Res 94:150–154

de Groat WC, Nadelhaft I, Milne RJ, Booth AM, Morgan C, Thor K (1981) Organisation of the sacral parasympathetic reflex pathways to the urinary bladder and large intestine. J Auton Nerv Syst 3:135–160

de Groat WC, Booth AM, Milne RJ, Roppolo JR (1982) Parasympathetic preganglionic neurons in the sacral spinal cord. J Auton Nerv Syst 5:23–43

de Groat WC, Kawatani M, Hisamitsu T, Lowe I, Morgan C, Roppolo J, Booth AM, Nadelhaft I, Kuo D, Thor K (1983) The role of neuropeptides in the sacral autonomic reflex pathways of the cat. J Auton Nerv Syst 7:339–350

Dembowsky K, Lackner K, Czachurski J, Seller H (1981) Tonic catecholaminergic inhibition of the spinal somato-sympathetic reflexes originating in the ventrolateral medulla oblongata. J Auton Nerv Syst 3:277–290

Denny-Brown D, Robertson EG (1933) On the physiology of micturition. Brain 56:149–190

de Sy W, Lacroix E, Leusen I (1974) An analysis of the urinary bladder response to hypogastric nerve stimulation in the cat. Invest Urol 11:508–516

Edvardsen P (1967) Nervous control of urinary bladder in cats. Acta Neurol Scand 43:543–563

Elam W, Thorén P, Svensson TH (1986) Locus coeruleus neurones and sympathetic nerves: activation by visceral afferents. Brain Res 375: 117–125

Evans MH (1963) Alterations in activity of gamma efferents during distension of the bladder in the cat. J Physiol 165:358–367

Evans MH, McPherson A (1959) The effects of distension of the bladder on somatic reflexes in the cat. J Physiol 146:438–458

Evans MH, McPherson A (1960) The effects of electrical stimulation of visceral afferent nerve fibres on monosynaptic and polysynaptic reflex responses. J Physiol 150:105–113

Floyd K, Lawrence G (1979) Mechanosensitive afferent units in the cat pelvic nerve. J Physiol 290:51–52P

Floyd K, Hick VE, Morrison JFB (1976) Mechanosensitive afferent units in the hypogastric nerve of the cat. J Physiol 259:457–471

Floyd K, Hick VE, Morrison JFB (1982a) The influence of visceral mechanoreceptors on sympathetic efferent discharge in the cat. J Physiol 323:65–75

Floyd K, McMahon SB, Morrison JFB (1982b) Inhibitory interactions between colonic and vesical afferents in the micturition reflex of the cat. J Physiol 322:45–52

Garrett JR, Howard ER, Jones W (1974) The internal anal sphincter of the cat: a study of nervous mechanisms affecting tone and reflex activity. J Physiol 243:153–166

Garry RC, Roberts TDM, Todd JK (1959) Reflexes involving the external urethral sphincter in the cat. J Physiol 149:653–665

Gibson SJ, Polak JM, Anand P, Blank MA, Morrison JFB, Kelly JS, Bloom SR (1984) The distribution and origin of VIP in the spinal cord of six mammalian species. Peptides 5:201–207

Gilbey MP, Coote JH, Macleod VH, Peterson DF (1981) Inhibition of sympathetic activity by stimulating in the raphe nuclei and the role of 5-hydroxytryptamine in this effect. Brain Res 226:131–142

Gjone R (1966) Peripheral autonomic influence on the motility of the urinary bladder in the cat. Acta Physiol Scand 66:72–80

Gjone R, Setekleiv J (1963) Excitatory and inhibitory responses to stimulation of the cerebral cortex in the cat. Acta Physiol Scand 59:337–349

Glazer EJ, Basbaum AL (1980) Leucine enkephalin: localisation in and axonal transport by sacral parasympathetic preganglionic neurons. Science 208:1479–1480

Guttmann L (1976) Spinal cord injuries. Blackwell, Oxford, p 731

Hancock MB, Fougerousse CL (1976) Spinal projections of the nucleus locus coeruleus and nucleus subcoeruleus in the cat and monkey as demonstrated by the retrograde transport of horseradish peroxidase. Brain Res Bull 1:229–234

Henneman E (1981) Recruitment of motoneurones: the size principle. In: Desmedt J (ed) Motor unit types, recruitment and plasticity in health and disease. Karger, Basel, pp 26–60 (Progress in clinical neurophysiology, vol 9)

Henry MM, Swash M (1978) Assessment of pelvic floor disorders and incontinence by electrophysiological recording of the anal reflex. Lancet 1:1290–1291

Hisamitsu T, de Groat WC (1984) The inhibitory effect of opioid peptides and morphine applied intrathecally and intra-cerebroventricularly on the micturition reflex in the cat. Brain Res 298:51–65

Hökfelt T, Terenius L, Kuypers HGJM, Dann O (1979) Evidence for enkephalin immunoreactive neurons in the medulla oblongata projection to the spinal cord. Neurosci Lett 14:55–60

Hökfelt T, Everett B, Meister B, Melander T, Schalling M, Johansson O, Lundberg JM, Hulting A-L, Werner S, Cuello C, Hemmings H, Ouimei C, Walaas I, Greengard P, Goldstein M (1986) Neurons with multiple messengers, with special reference to neuroendocrine systems. Recent Prog Horm Res 42:1–70

Holstege G, Kuypers HGJM (1982) The anatomy of brain stem pathways to the spinal cord in cat. A labeled amino acid tracing study. In: Kuypers HGJM, Matin GF (eds) Descending pathways to the spinal cord. Elsevier, Amsterdam, pp 145–175 (Progress in brain research, vol 57)

Honda CN, Rethelyi M, Petrusz P (1983) Preferential histochemical location of vasoactive intestinal polypeptide (VIP) in the sacral spinal cord of the cat: light and electron microscopic observations. J Neurosci 3:2183–2196

Hunt SP, Emson PC, Gilbert R, Goldstin M, Kimmel JR (1981) Presence of avian pancreatic polypeptide-like immunoreactivity in catecholamine and methionine-enkephalin-containing neurones within the central nervous system. Neurosci Lett 21:125–130

Iggo A (1955) Tension receptors in the stomach and the urinary bladder. J Physiol 128:593–607

Jänig W (1985) Systemic and specific autonomic reactions in pain: efferent, afferent and endocrine components. Eur J Anaesthesiol 2:319–346

Jänig W (1986) Spinal cord integration of visceral sensory systems and sympathetic nervous system reflexes. In: Cervero F, Morrison JFB (eds) Visceral sensation. Elsevier, Amsterdam, pp 255–277 (Progress in brain research, vol 67)

Jänig W, McLachlan E (1987) Organisation of the lumbar sympathetic and afferent supply of colon and pelvic organs. Physiol Rev (in press)

Jänig W, Morrison JFB (1986) Functional properties of spinal visceral afferents supplying abdominal and pelvic organs, with special emphasis on visceral nociception. In: Cervero F, Morrison JFB (eds) Visceral sensation. Elsevier, Amsterdam, pp 87–114 (Progress in brain research, vol 67)

Jankowska E, Padel Y, Zarzecki P (1978) Crossed disynaptic inhibition of sacral motoneurones. J Physiol 285:425–444

Job C, Lundberg A (1953) Facilitation and inhibition in the sympathetic ganglion of the cat. Acta Physiol Scand 28:14–28

Julé Y, Szurszewski JH (1983) Electrophysiology of neurones of the inferior mesenteric ganglion of the cat. J Physiol 344:277–292

Julé Y, Krier J, Szurszewski JH (1983) Patterns of innervation of neurones in the inferior mesenteric ganglion of the cat. J Physiol 344:293–304

Kandel ER, Schwartz JH (1981) Principles of neural science. Arnold, London

Katayama Y, deWitt DS, Becker DP, Hayes RL (1984) Behavioural evidence for a cholinoceptive pontine inhibitory area: descending control of spinal motor output and sensory input. Brain Res 296:241–262

Kerr FWL, Alexander S (1964) Descending autonomic pathways in the spinal cord. Arch Neurol 10:249–261

King B, Szurszewski JH (1984) Mechanoreceptor pathways from the distal colon to the autonomic nervous system in the guinea-pig. J Physiol 350:93–107

Klevmark B (1977) Motility of the urinary bladder in cats during filling at physiological rates. II. Effects of extrinsic bladder denervation on intraluminal tension and intravesical pressure patterns. Acta Physiol Scand 101:176–184

Klevmark B (1980) Motility of the urinary bladder in cats during filling at physiological rates. III. Spontaneous rhythmic bladder contractions in the conscious and anaesthetised animal. Influence of distension and innervation. Scand J Urol Nephrol 14:219–224

Kock NG, Pompeius R (1963) Inhibition of vesical motor activity induced by anal stimulation. Acta Chir Scand 126:244–250

Koley BN, Das AK, Koley J (1983) Viscerosomatic reflexes following distension of the urinary bladder in the cat. Indian J Physiol Allied Sci 37:84

Kreulen DL, Peters S (1986) Non-cholinergic transmission in a sympathetic ganglion of the guinea-pig elicited by colon distension. J Physiol 374:315–334

Kümmel H (1983) Activity in sympathetic neurons supplying skin and skeletal muscle in spinal cats. J Auton Nerv Syst 7:319–327

Kuntz A (1946) The autonomic nervous systems, 2nd edn. Ballière, Cox and Tindall, London

Kuo DC, Hisamitsu T, de Groat WC (1984) A sympathetic projection from sacral paravertebral ganglia to the pelvic nerve and to postganglionic nerves on the surface of the urinary bladder and large intestine of the cat. J Comp Neurol 226:76–86

Kuru M (1965) Nervous control of micturition. Physiol Rev 45:425–494

Langley JN, Anderson HK (1894) On reflex action from sympathetic ganglia. J Physiol 16:410–440

Langley LL, Kimura K (1959) The afferent pathway of the vesico-renal reflex. J Urol 82:476–480

Langworthy OR, Kolb LC, Lewis LG (1940) Physiology of micturition. Williams and Wilkins, Baltimore

Light AR (1985) The spinal terminations of single, physiologically characterised axons originating in the ponto-medullary raphe of the cat. J Comp Neurol 234:536–548

Loewy AD, Neil JJ (1981) The role of descending monoaminergic systems in central control of blood pressure. Fed Proc 40:2778–2785

Loewy AD, Wallach JH, McKellar S (1981) Efferent connections of the ventral medulla oblongata in the rat. Brain Res Rev 3:63–80

Lumb BM (1984) Neural basis of stimulation-produced analgesia. In: Holder AV, Winlow W (eds) Neurobiology of pain. Manchester University Press, Manchester pp 188–196

Lumb BM (1986) Brain stem control of visceral afferent pathways in the spinal cord. In: Cervero F, Morrison JFB (eds) Visceral sensation. Elsevier, Amsterdam, pp 279–294

Lumb BM, Morrison JFB (1984) Convergence of visceral and somatic information on to identified reticulo- and raphe-spinal neurons in the rat. J Physiol 357:33P

Lumb BM, Morrison JFB (1987) Functions of the pathway from the parabrachial region to nucleus raphe magnus in the rat. J Physiol 386:30P

Lytton B, Kupfer DJ, Traurig AR (1967) The vesico-renal reflex. Invest Urol 4:521–530

Mackel R (1979) Segmental descending control of the external urethral and anal sphincters in the cat. J Physiol 294:105–122

Maggi CA, Santicioli P, Meli A (1985) Sympathetic inhibition of reflex activation of bladder motility during filling at a physiological-like rate in urethane-anaesthetised rats. Neurourol Urodyn 4:37–46

Maggi CA, Santicioli P, Borsini F, Giuliani S, Meli A (1986) The role of capsaicin-sensation innervation of the rat urinary bladder in the activation of micturition reflex. Naunyn-Schmiedebergs Arch Pharmacol 332:276–283

Marsden CD, Meadows JC, Hodgson HJF (1969) Observations on the reflex response to muscle vibration in man and its voluntary control. Brain 92:829–846

Mathias CJ, Frankel HL (1983) Clinical manifestations of malfunctioning sympathetic mechanisms in tetraplegia. J Auton Nerv Syst 7:303–312

Matsushita M, Tanami T (1983) Contralateral termination of primary afferent axons in the sacral and caudal segments of cat, as studied by anterograde transport of horseradish peroxidase. J Comp Neurol 220:206–218

McLachlan Elspeth M (1985) The components of the hypogastric nerve in male and female guinea-pigs. J Auton Nerv Syst 13:327–342

McMahon SB (1986) Sensory-motor integration in urinary bladder function. In: Cervero F, Morrison JFB (eds) Visceral sensation. Elsevier, Amsterdam; pp 245–253 (Progress in brain research, vol 67)

McMahon SB, Abel C (1987) A model for visceral pain states: chronic inflammation of rat urinary bladder by irritant chemicals. Pain 28:109–127

McMahon SB, Morrison JFB (1982a) Spinal neurones with long projections activated from the abdominal viscera of the cat. J Physiol 332:1–20

McMahon SB, Morrison JFB (1982b) Two groups of spinal

interneurones that respond to stimulation of the abdominal viscera of the cat. J Physiol 332:21–34

McMahon SB, Morrison JFB (1982c) Factors that determine the excitability of parasympathetic reflexes to the bladder. J Physiol 332:35–43

McMahon SB, Spillane Kathy (1982) Brain stem influences on the parasympathetic supply to the urinary bladder of the cat. Brain Res 234:237–249

McMahon SB, Morrison JFB, Spillane Kathy (1982) An electrophysiological study of somatic and visceral convergence in the reflex control of the external sphincters. J Physiol 328:379–387

McPherson A (1966a) The effects of somatic stimuli on the bladder of the cat. J Physiol 185:185–196

McPherson A (1966b) Vesico-somatic reflexes in the chronic spinal cat. J Physiol 185:197–204

McPherson A, Skorpil V (1966) Effects of micturition on leg reflexes. Lancet II:309–312

Melzak J, Porter NH (1964) Studies of the reflex activity of the external sphincter ani in spinal man. Paraplegia 1:277–296

Mokha SS, McMillan JA, Iggo A (1985) Descending control of spinal nociceptive transmission. Actions produced on spinal multireceptive neurons from the nuclei locus coeruleus and raphe magnus. Exp Brain Res 58:213–226

Morgan C, Nadelhaft I, de Groat WC (1981) The distribution of visceral primary afferents from the pelvic nerve within Lissauer's tract and the spinal gray matter and its relationship to the sacral parasympathetic nucleus. J Comp Neurol 201:415–440

Morrison JFB (1982) The neural control of the bladder. In: Bloom SR, Polak JM, Lindenlaub E (eds) Systemic role of regulatory peptides. Schattauer, Stuttgart, pp 381–396

Morrison JFB, Spillane Kathy (1986) Neuropharmacological studies on descending inhibitory controls over the micturition reflex. J Auton Nerv Syst (suppl) pp 393–397

Mukerjee SR (1957) Effect of bladder distension on arterial blood pressure and renal circulation: role of splanchnic and buffer nerves. J Physiol 138:307–325

Murray KHA, Feneley RCL (1982) Endorphins—a role in lower urinary tract function? The effect of opioid blockade on the detrusor and urethral sphincter mechanisms. Br J Urol 54:638–640

Nadelhaft I, Roppolo J, Morgan C, de Groat WC (1983) Parasympathetic preganglionic neurons and visceral primary afferents in the monkey sacral spinal cord revealed following the application of horseradish peroxidase to pelvic nerve. J Comp Neurol 216:36–52

Nathan PW (1952) Thermal sensation in the bladder. J Neurol Neurosurg Psychiatry 15:148–149

Nathan PW (1976) The central nervous connections of the bladder. In: Williams DI, Chisholm GD (eds) Scientific foundations of urology, vol II. Heinemann, London, pp 51–58

Nathan PW, Smith MC (1958) Centrifugal pathway for micturition within the spinal cord. J Neurol Neurosurg Psychiatry 21:177–189

Otsuka M, Konishi S (1983) Substance P—the first peptide neurotransmitter? TINS 6:317–320

Parks AG, Swash M, Urich H (1977) Sphincter denervation in anorectal incontinence and rectal prolapse. Gut 18:656–665

Pedersen E (1983) Regulation of bladder and colon-rectum in patients with spinal lesions. J Auton Nerv Syst 7:329–338

Porter RW (1967) A pallidal response to detrusor contraction. Brain Res 4:381–383

Porter RW, Krell M (1976) Alterations in the H-reflex in the paraplegic induced by bladder distension. Paraplegia 14:105–114

Porter RW, Pazo JH, Dillard GV (1971) Triphasic brain stem response to detrusor contraction. Brain Res 35:119–126

Roppolo JR, Nadelhaft I, de Groat WC (1985) The organisation of pudendal motoneurones and primary afferent projections in the spinal cord of the Rhesus monkey revealed by horseradish peroxidase. J Comp Neurol 234:475–488

Ryall RW, de Groat WC (1972) The microontophoretic administration of noradrenaline, 5-hydroxytryptamine, acetylcholine and glycine to sacral parasympathetic preganglionic neurones. Brain Res 345–348

Santicioli P, Maggi CA, Meli A (1985) The effect of capsaicin pretreatment on the cystometrograms of urethane anaesthetised rats. J Urol 133: 700–703

Sato A, Sato Y, Schmidt RF (1980) Reflex bladder activity induced by electrical stimulation of hind limb somatic afferents in the cat. J Auton Nerv Syst 1:229–241

Sato A, Sato Y, Sugimoto H, Terui N (1977) Reflex changes in the urinary bladder after mechanical and thermal stimulation of the skin at various segmental levels in cats. Neuroscience 2:111–117

Satoh K, Tohyama M, Tetsuro S, Yamamoto K, Shimizu N (1978a) Descending projection of the nucleus tegmentalis laterodorsalis to the spinal cord; studied by the horseradish peroxidase method following 6-hydroxy-dopa administration. Neurosci Lett 8:9–15

Satoh K, Shimizu N, Tohyama M, Maeda T (1978b) Localisation of the micturition reflex at dorsolateral pontine tegmentum of the rat. Neurosci Lett 8:27–33

Schrøder HD (1980) Organisation of the motoneurons innervating the pelvic muscles of the male rat. J Comp Neurol 192:567–587

Schrøder HD (1984) Somatostatin in the caudal spinal cord: an immunohistochemical study of the spinal centres involved in the innervation of pelvic organs. J Comp Neurol 223:400–411

Schrøder HD (1985) Anatomical and pathoanatomical studies on the spinal efferent systems innervating pelvic structures. J Auton Nerv Syst 14:23–48

Scott FB, Quesada EM, Cardus D (1964) Studies on the dynamics of micturition: observations on healthy men. J Urol 92:455–463

Sharkey KA, Williams RG, Schultzberg M, Dockray GJ (1983) Sensory Substance P-innervation of the urinary bladder: possible site of action of capsaicin in causing urine retention in rats. Neuroscience 10:861–868

Sherrington CS (1892) Notes on the arrangement of some motor fibres in the lumbo-sacral plexus. J Physiol 13:621–772

Sokownin (1874) Arch Physiol (Bonn) 8:S600. Quoted by Langley JN (1900) The sympathetic and other related systems of nerves. In: Schäfer EA (ed) Textbook of physiology, vol II. Pentland, Edinburgh, pp 616–696

Spillane Kathy (1982) Brain stem influence on pelvic parasympathetic reflexes. PhD Thesis, University of Leeds

Sundin T (1972) Reinnervation of the urinary bladder. Scand J Urol Nephrol [Suppl] 17:1–25

Swanson LW, Sawchenko PE (1983) Hypothalamic integration: organisation of the paraventricular and supraoptic nuclei. Annu Rev Neurosci 6:269–324

Tanagho EA (1978) The anatomy and physiology of micturition. Clin Obstet Gynaecol 5:3–26

Tang PC (1955) Levels of brainstem and diencephalon controlling micturition reflex. J Neurophysiol 18:583–595

Tang PC, Ruch TC (1956) Localisation of brainstem and diencephalic areas controlling the micturition reflex. J Comp Neurol 106:213–245

Taylor DEM (1965) Reflex effects of slow bladder filling on the blood pressure of cats. Q J Exp Physiol 50:263–270

Thomas DG (1982) The urinary tract following spinal cord injury. In: Chisholm GD, Williams DI (eds) Scientific foundations of urology, 2nd edn. Heinemann, London, pp 421–433

Thomas DG, Smallwood R, Graham D (1975) Urodynamic observations following spinal trauma. Br J Urol 47:161–175

Thor KB, Roppolo JR, de Groat WC (1983) Naloxone induced micturition in unanaesthetised paraplegic cats. J Urol 129:202–205

Todd JK (1964) Afferent impulses in the pudendal nerves of the cat. Q J Exp Physiol 49:258–267

Torrens MJ (1978) Urethral sphincteric responses to stimulation of the sacral nerves in the human female. Urol Int 33:22–26

Wallin G, Stjernberg L (1984) Sympathetic activity in man after spinal cord injury. Brain 107:183–198

Wang SC, Ranson SW (1939a) Autonomic responses to stimulation of the lower brainstem. J Comp Neurol 71:437–455

Wang SC, Ranson SW (1939b) Descending pathways from the hypothalamus to the medulla and spinal cord. Observations on blood pressure and bladder responses. J Comp Neurol 71:457–472

Westlund KN, Coulter JD (1980) Descending projections of the locus coeruleus and subcoeruleus/medial parabrachial nuclei in the monkey: axonal transport studies and dopamine-*beta*-hydroxylase immunocytochemistry. Brain Res Rev 2:235–264

Wright AL, Williams NS, Gibson JS, Neal DE, Morrison JFB (1985) Electrically evoked activity in the human external anal sphincter. Br J Surg 72:38–41

Yamamoto S, Araki K (1962) Intra-abdominal pressure response to medullary stimulation in cats. Exp Neurol 5:110–119

Yamamoto S, Araki K, Kikuchi M (1962) Electromyographic study of reflex activity in abdominal wall muscles and diaphragm following pelvic afferent excitation in cats. Tohoku J Exp Med 75:375–383

8 · Bladder Control: Role of Higher Levels of the Central Nervous System

John F. B. Morrison

Methods, Conclusions and Concepts
Bladder Control by the Central Nervous System
Functional Changes Mediated by Different Anatomical Sites in the Central Nervous System
Micturition Threshold and Factors that Influence It
Role of Feedback in Voluntary Motor Control of the Lower Urinary Tract

Methods, Conclusions and Concepts

During the last 150 years many investigators have attempted to study the central pathways which control micturition in humans and in animals; these results suggest that control over the bladder and sphincters is organized in different areas of the cortex and at various levels of the nervous system, the lowest of which is the spinal cord. The precise roles of many supraspinal sites in the regulation of the bladder are uncertain, and different workers interpret their results in different ways.

Classical Neuroanatomical Studies

Investigations on the anatomical connections between different structures in the central nervous system that are involved in the control of the lower urinary tract were summarized by Kuru (1965), Bors and Comarr (1971) and by Appenzeller (1982). The neuroanatomical work of Kuru and his col-

leagues was mainly based on the method of creating electrolytic lesions at sites, which, when stimulated electrically, gave rise to contraction or relaxation of either the bladder or the urethral sphincter; after time had been allowed for the degeneration of myelin, the Marchi method was used to trace the lesioned pathways. The method is likely to provide a large number of false positives, because fibres of passage that have nothing to do with the bladder are also lesioned and stained by these methods.

Kuru (1965) concluded that there are three descending pathways subserving reflex micturition: the lateral, ventral and medial reticulospinal tracts; these were concerned respectively with bladder contraction, bladder relaxation and the control of the urethral sphincter (see Fig. 8.8). The lateral reticulospinal pathway originates from the midbrain, pons and medulla and subserves bladder contraction. The ventral reticulospinal pathway is believed to subserve bladder relaxation and originates from the dorsomedial reticular formation of the medulla, and may correspond to the descending pathways from the nucleus reticularis gigantocellularis or the raphe nuclei (McMahon and Spillane 1982; Spillane

1982; Morrison and Spillane 1986). The medial reticulospinal tract originates from the rostral pons and is concerned with the control of the external sphincter, according to Kuru (1965). Kuru (1965) also recognized the existence of sites rostral to the brain stem that were concerned with micturition, such as the hypothalamus, the amygdala and the cerebral cortex, but did not describe the precise connections between these structures.

Retrograde Tracing and Autoradiographic Studies

Retrograde tracing was mentioned in Chapter 3 with respect to the situation of the cell bodies of parasympathetic efferents (see p. 69). Retrograde tracers such as HRP and True Blue have also been used to elucidate central anatomical pathways, and they have the advantage that they are a fast and efficient method for determining the sites of origin of axons which pass or terminate in the area injected with the tracer. Thus, injections of HRP into the lumbosacral cord or localized regions of it can be used to elucidate pathways that connect directly with this part of the nervous system. We know that this region receives descending inputs from, for instance, the dorsal column nuclei, the raphe nuclei, the nucleus reticularis gigantocellularis and the vestibular nuclei of the medulla, the locus coeruleus, the dorsolateral tegmental nuclei and the reticular nuclei of the pons, and the cerebral cortex. In addition, the autonomic cell columns at all levels of the spinal cord receive an oxytocin-containing pathway from the anterior hypothalamus. These pathways can also be demonstrated by localized microinjection of tritiated amino acids into the region containing the neuronal cell bodies, following which the radioactive label appears in proteins along the path of the nerve and at its terminals. Figure 8.1 shows the distribution of descending axons from the hypothalamus, and fibres from neurons in this localized region of the brain project to a number of structures which in turn are connected to the autonomic efferent outflows. Serotonin and substance P are present in the raphe–spinal pathway and noradrenaline can be demonstrated in the fibres originating from the locus coeruleus (Jänig 1983). Not all of these are involved in the control of micturition, and in particular, descending pathways from the dorsal column nuclei and the vestibular nuclei have not so far been implicated in the control of the lower urinary tract.

It is probable that most of the sites in the central nervous system that influence the activity of the lower urinary tract have direct as well as indirect connections with the efferents in the lumbosacral cord. The most familiar example will be the connections between the cortex and the lumbosacral cord; the pyramidal (corticospinal) tract provides a direct connection, but other descending pathways which synapse in the brain stem also exert important influences. Indeed, in the case of the connections between the cortex and the spinal cord, the direct (corticospinal) pathway is quite slow, and the polysynaptic pathways play an important part. The clinical concept of "the upper motoneuron" therefore glosses over many of the different functional constituents that give rise to the signs of spasticity; this is quite apart from the fact that no neurons in this pathway are actually motoneurons, and that some are not simply concerned with motor functions, viz. the pyramidal tract can also modulate the transmission of sensory information in the dorsal horn. Another example can be seen if we consider the relationship between the locus coeruleus/medial parabrachial region of the rostral pons—the region of Barrington's "micturition centre" and the sacral cord. In addition to direct spinal projections from the locus coeruleus and the parabrachial nuclei to the sacral cord, there are reciprocal connections between the raphe nucleus and the reticular nuclei, the locus coeruleus, the parabrachial nuclei, and the periventricular and periaqueductal grey matter (to mention but a few). Thus the locus coeruleus/medial parabrachial region of the pons may exert its action on the sacral cord by direct connections as well as through the raphe–spinal, and possibly the reticulospinal pathways. Similarly, the anterior hypothalamus has direct connections with the spinal cord, and indirect ones via the raphe–spinal pathway and possibly via the locus coeruleus–medial parabrachial complex.

Clearly, the anatomy is complex, and it is probably not possible to ascribe the effects of electrical stimulation at one site in the nervous system to a direct pathway connecting that site with an effector system; instead, the effects are likely to be due to the combination of direct and indirect pathways. In addition, these pathways may have axon collaterals that terminate in different regions of the cord; there is a possibility that some axons may subserve more than one function, and, for instance, the degree of separation of neural controls over noxious sensory transmission and over the autonomic phenomena associated with pain has not yet been worked out. In a later section, a distinction will be drawn between neural pathways concerned with: (a) signal transmission, for instance in sensory or reflex pathways; and (b) with control over information transfer at synapses in those pathways; an analogy will be made to the operation of a rail network, in which

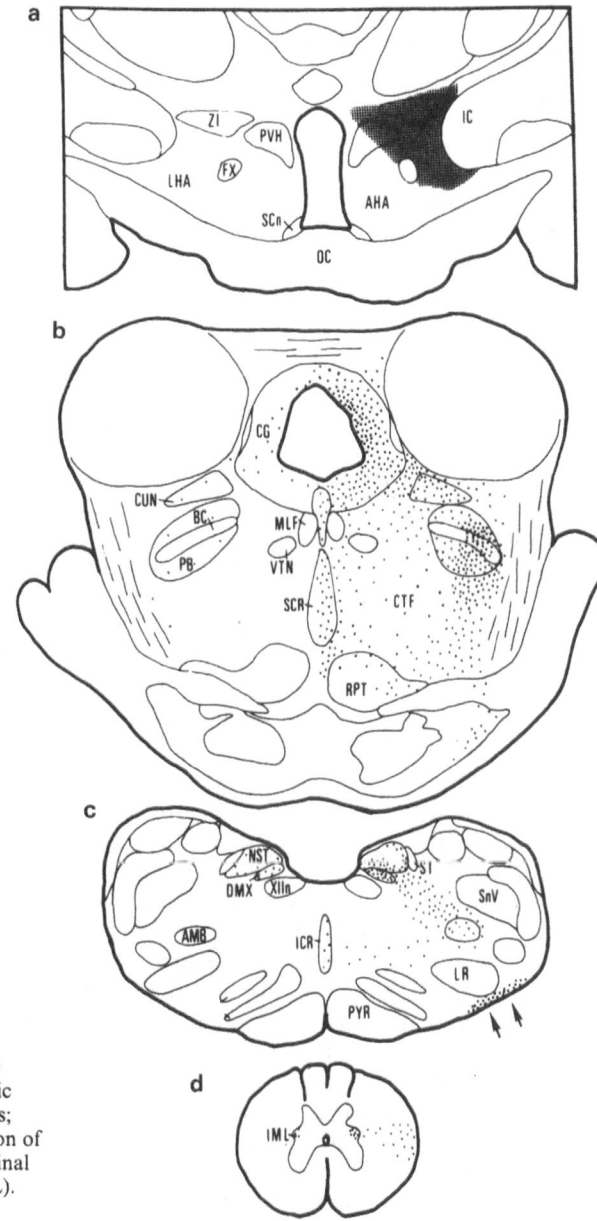

Fig. 8.1a–d. Following injections of [³H]amino acids into the hypothalamus, descending projections to the parasympathetic preganglionic nuclei of the medulla (*AMB*, nucleus ambiguus; *DMX*, dorsal motor nucleus). The *arrows* indicate the position of the descending autonomic pathway to the spinal cord. Terminal labelling is present in the intermediolateral cell column (*IML*). (Saper et al. 1976)

the track carries trains, the onward movement of which is regulated by a network of controls over signals and points. In the central nervous system it is not surprising to find that the passage of information along a particular pathway can be controlled, and that some of the information about information transfer, is diverted to sites which regulate signal transmission. One instance is to be seen in a study by Dembowsky et al. (1981); reversible cold block of cell groups in the ventral brain stem increases the size of spinal sympathetic reflexes and

abolishes supraspinal reflexes. The inference is that afferent information is normally diverted from spinal to supraspinal pathways, and this may be under dynamic control by the brain. Perhaps it would be a mistake to view the complex networks of neurons which control the lower urinary tract as simple motor pathways. I hope the railway analogy will help to emphasize the different roles that might be played by different parallel pathways connecting one site in the central nervous system with the effector organs. It may be of some use in describing how a control system can operate on the basis of switching between stop and start modes, but does not readily explain how oscillatory activity or feedback occurs: these will be dealt with later in this chapter.

Railway Analogy

One of the significant conclusions that one can make from a general survey of modern neurophysiology is that certain anatomical pathways are concerned with the control of information transfer. Thus, an attempt to separate a particular reflex pathway from other pathways that modulate it could be useful. The process is akin to separating the lines from the signals in a railway network. The passage of locomotives along the track is essential for transfer of goods; however the switching of signals and of points is a major factor in determining the speed at which the goods travel, and their destination. Furthermore, the switching of signals and of points occurs in a sequence determined by the passage of the locomotive. In a polysynaptic reflex pathway such as the micturition reflex, the passage of nerve impulses round the basic nerve cell network is analogous to the locomotive; however, the controlling pathways, i.e. the signals and points of the rail network, also function by means of nerve action potentials. The recording microelectrode cannot distinguish directly between action potentials that mediate the transfer of information from impulses that perform some control function; similarly, electrical stimuli do not distinguish between neurons of different functions. Just as the sequence of controls over railway signals and points depends on the progress of a train along the track, there is likely to be some form of feedback of information from the reflex pathway to pathways which are responsible for control in certain circumstances. Electrical stimuli at different sites can produce excitation or inhibition of bladder motility (usually concomitantly with some other physiological changes), but we are still a long way from knowing the exact functional role of such pathways, and the circumstances in which they have a special role to play.

Interpretation of Lesions, and Their Effects

Studies of the effects of lesions have provided a considerable body of evidence concerning the roles of different structures within the CNS. Sometimes the effects of lesions are very difficult to interpret because of a number of factors: the nature of the pathological process (which may not be as localized as one might hope) or the lack of adequate neuropathological checks; the changes in the physical or mental state of the patient (frontal lobe lesions have many more serious effects than a reduced awareness of the bladder and its function, such as generalized hypoactivity and sensory inattention); the extent of the lesions in different subjects in a series, which are rarely in exactly the same place, or all neatly circumscribed. The interpretation of circumscribed lesions, the result of discrete accidental trauma or of neurosurgical intervention are often easiest to interpret, but even then one cannot be sure how much of the effect is due to removal of cells at the site of the lesion, or to interruptions of axons of passage. Thus, some studies of the effects of lesions are of little value, but others, performed by critical, careful investigators, with adequate neuropathological control, can be invaluable.

Studies Using Electrical Stimulation of the Brain

Studies of the effects of electrical stimulation of the central nervous system have most commonly been performed on animals, and the interpretation of this type of result depends on the stimulus parameters used, the type of electrode used, and checks on the position of the electrode tip. The problem with stimulus parameters is that the stimulus can spread to affect neurons distant from the electrode tip, particularly if current is passed between an electrode and some remote earth. Figure 8.2 shows the spread of current around the tip of an electrode; currents of around $100\,\mu A$ spread around the tip of a bipolar electrode for about 1 mm (Bagshaw and Evans 1976). It is often possible to excite pathways at the electrode tip with currents of $10-20\,\mu A$ or less.

If voltages are known, rather than currents, they may be related to the current passing if the resistance of the electrode is known, by using Ohm's law. Constant voltage stimulation has the disadvantage that the current delivered to the tissue can vary if the resistance varies, and this may happen if the electrode becomes polarized, with the formation of a hydrogen bubble on its end; in this circumstance most of the voltage drop is across the electrode tip, rather than the tissue. Histological checks of the electrode position are essential for the proper evaluation of the site of the stimulus, and in animal studies the location of the electrode tip is often marked by microinjection of a dye such as pontamine sky blue or fast green; alternatively large

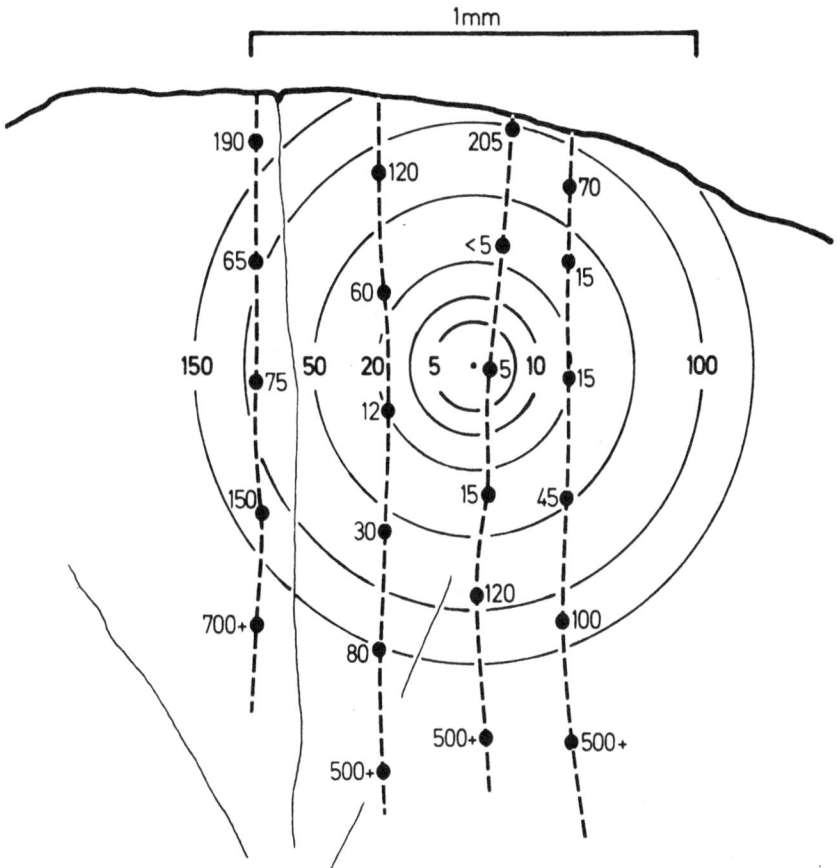

Fig. 8.2. Diagram of 4 microelectrode tracks penetrating the dorsal columns of a cat spinal cord. The surface of the cord is indicated by the *heavy line*, and the *fine lines* show the position of the midline and part of the boundaries between the gracile and cuneate fasciculi. The *dashed lines* represent microelectrode tracks seen in histological sections, with the sites of stimulation marked by *solid circles*. Next to these sites are marked the threshold currents for activation of one axon with a peripheral process in the sural nerve. The circles are centred on the probable location of this axon in the cord. The dimensions of these circles, which have been determined theoretically, give an indication of the probable mean spread of excitation from a monopolar microelectrode tip placed at the centre, when passing currents of 5–150 μA as marked on the circles. (Bagshaw and Evans 1976)

currents may be passed through the electrode at the end of an experiment to create a lesion that can later be identified histologically. Clearly, the histological checks cannot be done on patients except in unforeseen circumstances, but it is worthwhile looking at the stimulus parameters to try to evaluate the amount of spread of current from the electrode tip.

Electrical stimulation within the nervous system rarely gives rise to bladder contractions alone; usually these occur concomitantly with other events, such as changes in blood pressure, heart rate, intestinal motility, and the secretion of hormones. Furthermore, electrical stimuli do not distinguish between cells and axons of passage, exciting both indiscriminately. One might ask whether these combinations of effects are really attributable to integration within nuclei at or near the electrode tip, or due to excitation of fibres of different origins that have in common the chance property that they pass close to the tip of a stimulating electrode strategically placed by some neuroscientist? Is it justifiable to describe a "centre" which supposedly integrates these activities whenever the combination of effects desired by the observer is recorded? Thus, many scientists question the interpretation of experiments using electrical stimuli, the spread of which from the electrode tip is uncertain, to label anatomical sites which are purported to be "integrating centres". One has to be prepared to doubt some of these interpretations; structure can be related to function, but too great a dependence on the localization of function can make it difficult to interpret the existence of anatomically diffuse, polysynaptic, parallel pathways such as those responsible for the normal functioning of the autonomic nervous system. Concepts such as "the upper motoneuron" in the control of the bladder may in reality mean a series of descending pathways, some excitatory, some inhibitory, originating from a variety of interconnected sites in the brain, and ending on different components of a simpler reflex pathway.

Studies of the Activity of Neurons

Similarly, when neurophysiological recordings have shown the existence of neurons that respond to bladder distension or contraction, can one infer that these neurons are concerned specifically with the control of the bladder? When physiologists have set out to study systematically the properties of these neurons, it has been found that they respond not only to excitation of sensory inputs from the bladder, but that they can respond to stimulation of a number of afferent systems, from skin, muscle or other viscera. Somatovisceral convergence is so common that one has to recognize that specific pathways concerned only with bladder information may not even exist. One might ask whether these neurons are concerned with the effects of bladder distension on the bladder itself, or on some other function, say sensation, cardiovascular or respiratory control, or postural adjustment, or possibly more than one of these. The railway analogy may also be useful in considering the function of neurons which handle information from several sources: trains from different places can travel along the same stretch of line and be routed to different destinations, by means of points, etc. In the biological situation, information carried in the ascending pathways in the spinal cord may have originated in the bladder, the skin, or in muscles or joints (see Chap. 4, p. 116). If some physiological mechanism were to require purely visceral information, that might be achieved by comparing the signals in somatic neurons with those in viscerosomatic neurons; activity in the latter, but not in the former could be used as an indicator of the visceral origin of the message. Nothing is known as to whether somatic neurons can divert the information in viscerosomatic neurons through one collateral pathway rather than another. Nevertheless, control theory can cope reasonably successfully with the movement of trains of different types from different origins to predetermined destinations down the same stretch of track. It is not impossible that the central nervous system can use the same neurons to carry messages concerning different functions from somatic or visceral structures to sites where they may be diverted into appropriate functional pathways. These ideas are noted in order to emphasize

that the anatomical view of complex connections between the bladder and different sites in the nervous system may not be the only or even the best way to view the control of the lower urinary tract. The misunderstanding that can be introduced by trying to think too simply, in terms of "the anatomical pathway" of the reflex, is akin to the chaos that would result if a railway did not have signals or points. We will consider the ways in which control can be exerted over reflex pathways later in this chapter, and some of the situations in which reflex pathways can be switched on or off, and their threshold and gain regulated.

Summary

An appreciation of the role of the nervous system in the control of the lower urinary tract is difficult to achieve without an understanding of the limitations as well as the strengths of modern electrophysiological techniques, and the differences between the state of the nervous system in an anaesthetized animal, a normal animal, and an animal with neurological lesions. Simplistic ideas about pathways have to be treated with caution, as no part of the CNS exists in isolation. It is possible to consider the functional control of the bladder without recourse to the details of neuroanatomy, and the accounts of the control of the lower urinary tract which follow indicate the variety of interpretations and emphasis that can be brought to bear on this subject. Most of the recent knowledge concerns the peripheral and spinal cord controls over the lower urinary tract, and the effects of higher levels of the central nervous system may be best thought of in terms of a series of descending pathways which facilitate or inhibit a basic reflex pathway. The functional balance between these controlling pathways appears to be responsible for the state of the reflexes that control the lower urinary tract; in addition to the descending pathways, however, there are local factors that also contribute to reflex excitability (as outlined in Chap. 7, see p. 209).

Bladder Control by the Central Nervous System

Four Viewpoints

I have chosen four sets of ideas that are in the recent literature to provide an impression of the variety of ways in which different people think about this subject. Some are based almost entirely on anatomical pathways, while functional considerations assume greater importance in the reviews by de Groat (1975), Nathan (1976) and McMahon (1986).

Bors and Comarr (1971) described a series of descending pathways that converged on the pelvic and pudendal efferents to the bladder and sphincters, respectively. Efferent impulses from various loci of the brain (including the premotor cortex, basal ganglia, preoptic, supraoptic, and septal hypothalamic areas, the limbic system, the reticular formation of the pons and medulla and the cerebellum) are known to have some influence on the sacral efferent pathways by mainly extrapyramidal pathways. The principles of reciprocal innervation, facilitation, feedback, and inhibition were all thought to play their parts in the storage, initiation, continuation and interruption of micturition. Kuru (1965) also gives an anatomically based account, ascribing different functions to different fibre tracts

and brain stem nuclei. Some of these tracts are given rather quaint names, such as the portion of the dorsal columns called the pelvic sensory vagus; the existence of such a pathway has nevertheless been substantiated by modern neuroanatomical techniques. There is some disagreement between Kuru (1965) and Nathan (1976) on some of the pathways.

Ruch (1965) and Nathan (1976) concluded that there are a number of pathways involved in the coordinated process of micturition, that these are polysynaptic in nature, and the basic reflex phenomena involve pathways in the spinal cord and brain stem (Fig. 8.3). More complicated patterns of behaviour are grafted on to the reflex control of the bladder by regions of the anterior hypothalamus, but the highest levels of control, i.e. voluntary, socially acceptable behaviour, are imposed by the medial frontal lobe. At each anatomical level within the brain, evidence of facilitation and inhibition can be seen, and tonic activities in these different pathways result in the overall activity of the cortex

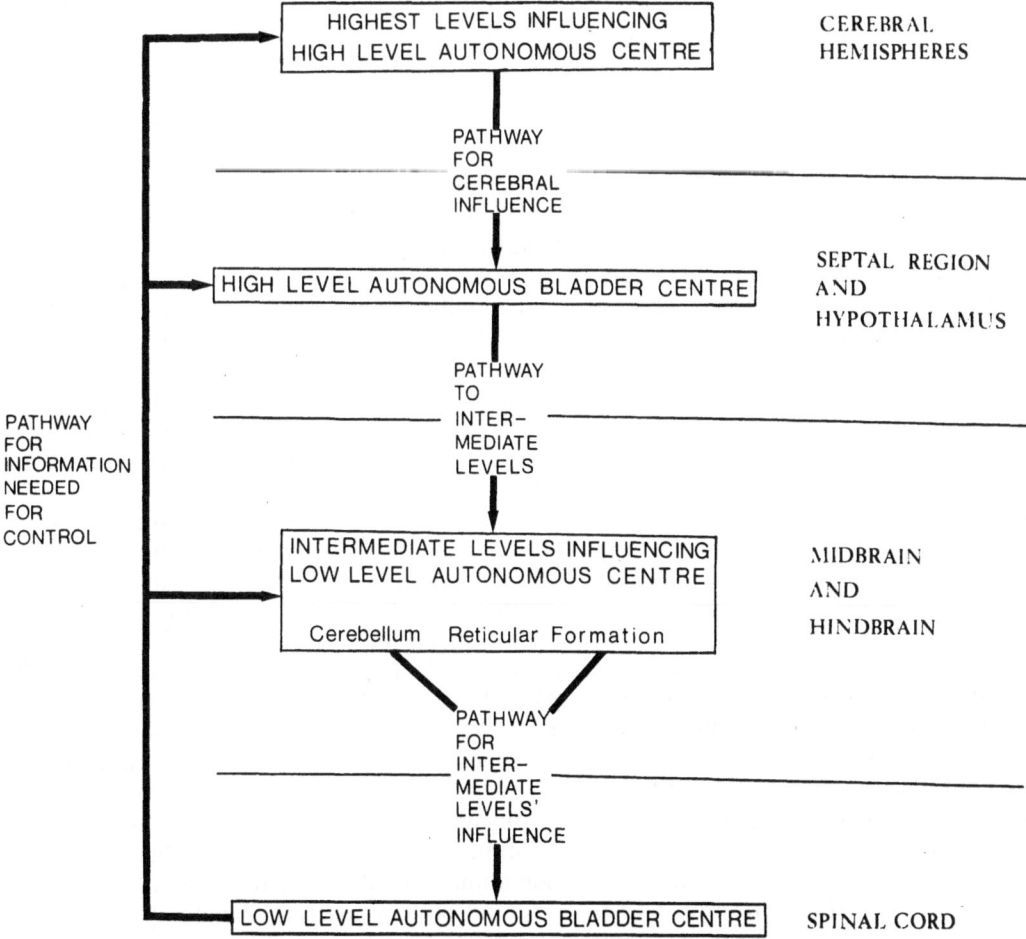

Fig. 8.3. The relationship between structures concerned with the regulation of micturition. (Nathan 1976)

and midbrain being inhibitory, and the hypothalamus and anterior pons being excitatory. The external sphincters and pelvic floor muscles can also be under voluntary control from the paracentral lobule of the cerebral cortex. These reviews, by two of the major contributors to our knowledge of the neural control of the human bladder, are clear, interesting statements of what is known about the functional roles of different structures in the human brain in the integration of micturition behaviour; the concepts they utilize involve the hierarchical view of levels of control in the central nervous system (see below).

In contrast, Bradley and Scott (1978) and Bradley (1984) emphasize a "loop concept", with four different loops. Loop I consists of reciprocal connections between the dorsolateral tegmentum of the brain stem and the frontal lobe; it is hypothesized that this loop is the neural substrate for voluntary control over the bladder. The assertion that it can be tested by making a cystometrogram (Bradley and Scott 1978) has to be viewed with the eye of faith, and changes in the cystometrogram cannot be entirely ascribed to events in pathways involving the frontal and cingulate cortices, the hypothalamus and limbic system, and the midbrain and upper pons. Nevertheless, there is no doubt that lesions in the frontal lobes can influence the cystometrogram (see Fig. 8.20). Loop II is regarded as a pathway carrying information concerning the activity of the detrusor from the sacral cord to the brain stem and back down to the sacral grey matter, and has no peripheral sensory or motor activities, but it is hypothesized that it is concerned with the "gain or amplification of the time course of the detrusor reflex" (Bradley 1984), a rather curious concept that is probably indefinable in terms of function, but presumably includes both the power and the duration of bladder contractions. Loop III includes the peripheral afferents and the intraspinal portion of their connections related to the control of the pudendal motoneurons, and hence the skeletal muscle sphincters; presumably the afferent part of this loop is the same as the sensory neurons that transmit information in a rostral direction, and are also involved in sensation, the micturition reflex and sympathetic reflexes. Loop IV is regarded as having two components: (A) the supraspinal innervation of pudendal motoneurons, and (B) the segmental innervation of periurethral muscles. Bradley appears to believe that detrusor–sphincteric dyssynergia can be explained by abnormalities of loop IV (Bradley 1984), but does not offer any more positive approach to this problem. Conceptualization of this sort may help some people and be viewed as rather sterile by others; the con-

cepts are based on selected ideas from Bradley's and other people's work, but the selection of ideas and the way in which the hypothesis is put together is somewhat idiosyncratic and open to certain criticisms. The ideas are generally vague and superficial, leaving out important anatomical and functional details, the anatomical diagrams that accompany the descriptions of the loops are sometimes imprecise and one might question whether these concepts aid the active researcher or the enquiring clinician to elucidate or understand the causes of changes in the pattern of behaviour of the lower urinary tract. Loops will also be mentioned later in this chapter; they involve descending pathways, such as those in Fig. 7.1 (p. 194), and differ from the Bradley model in that they include some recently defined pathways (such as the hypothalamicospinal pathway and the aminergic descending spinal pathways) that act *in parallel* on efferent systems. Furthermore, emphasis will be placed on function, rather than the effect of lesions.

It was a common approach in the first half of this century to think of the central nervous system in terms of a hierarchy of levels of functional organization, with increasing complexity being added, the more rostral the structure under consideration. Thus, in the organization of the control of ventilation, or of postural reflexes, more complex facets of behaviour appear to be added at different sites between the medulla and the cortex. Ruch's (1965), and Nathan's (1976) views of the control of the bladder are organized along these lines, with different levels of autonomous integrative centres in the pons, midbrain, hypothalamus and cortex. Bradley's (1984) loop concept also involves a hierarchy of structures, and one that might soon be replaced by more definite descriptions of direct pathways between "higher centres" and the efferents themselves, as well as polysynaptic pathways involving different "lower centres".

Today, the concept of this hierarchy is less secure, largely as a result of new knowledge about the complexity of function subserved by the spinal cord, which can no longer be regarded as just a transmission cable that connects the brain with the periphery. A couple of examples should make the point. It has been known for many years that spinal animals can be made to produce coordinated, active walking movements if stimulated appropriately, but it is only relatively recently that such integration of function at a spinal level has been commonly incorporated in the main line of theoretical consideration. The idea that there exists in the cord a central pattern generator (anatomy unspecified) whose function is to coordinate the sequence of basic movements used in walking is widely supported (Kandel and Schwartz 1981), and the idea in fact dates from early in this century. The existence of powerful controls over sensory input, particularly nociceptive input, in the spinal dorsal horn are widely appreciated, and put to practical use. This important function is situated low down in the anatomical hierarchy, yet it is believed to be a major site for control over sensory pathways. The common idea that the more rostral centres are "integrative" is itself open to much questioning, as it tends to fudge issues rather than clarify them. If these concepts are at least partly inadequate, what alternatives are there to explain in functional, rather than in anatomical terms, the behaviour of the nervous system in the control of the lower urinary tract? Control theory analysis of the relationship between output and input in a complex system, the details of which are not fully known, is one other

approach that has been tried successfully in a number of physiological systems (see Milhorn 1966); functional concepts such as the gain of the reflex, the threshold, and the filtering capacities of different components of the reflex arc, will be discussed later in the context of the micturition reflex.

The introduction of more functionally based ideas has occurred mainly in the last 20 years. Reviews by de Groat (1975) and de Groat et al. (1981, 1982) provide a more solid basis for understanding how the filling and voiding functions of the bladder work; there is less concern for anatomical pathways, and greater consideration for the functions of the peripheral nerves, spinal cord and brain stem during different phases of bladder activity (Fig. 8.4). de Groat (1975) states that the primary stimulus for micturition is bladder distension, which

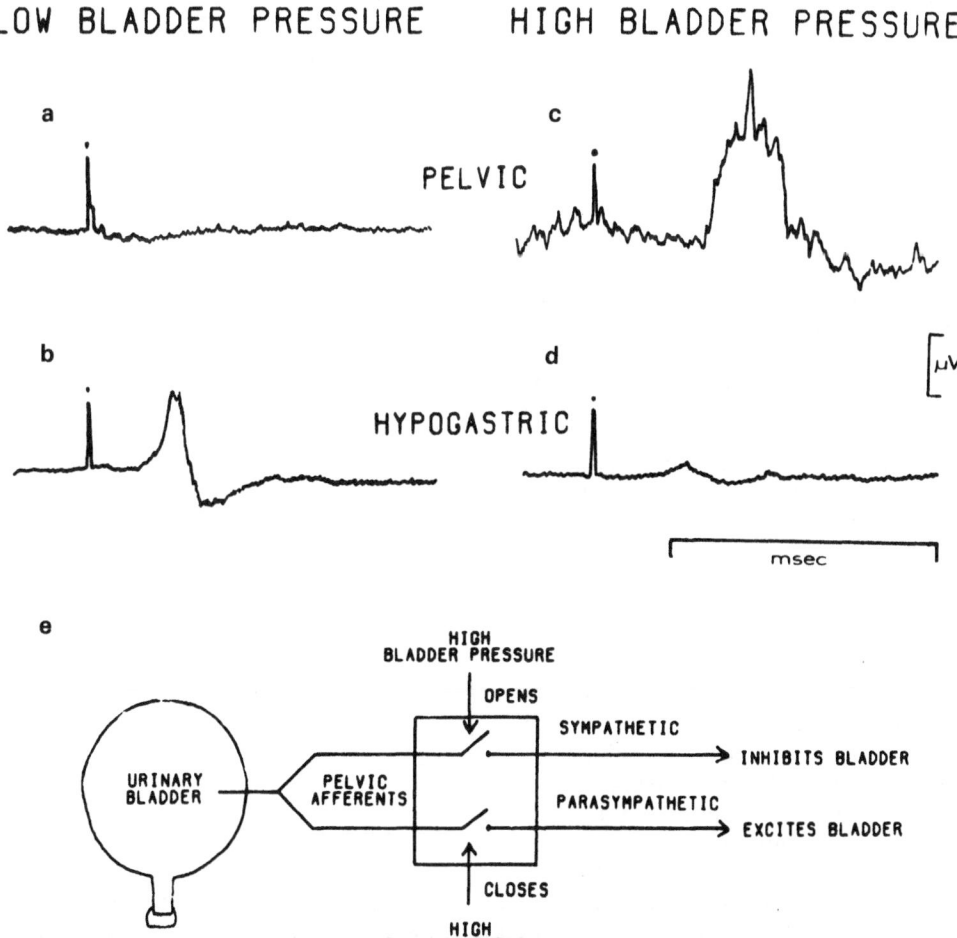

Fig. 8.4a–e. Relationships between bladder activity and reflexes evoked in pelvic (**a, c**) and hypogastric (**b, d**) nerves by electrical stimulation of afferents in the pelvic nerve. Records **a–d** are computer-averaged responses (10 responses) recorded between (**a** and **b**) and during (**c** and **d**) isovolumetric bladder contractions. Vertical calibration represents 30 μV in **a** and **b** and 60 μV in **c** and **d**. **e** Diagram depicting the effects of bladder pressure on the autonomic outflow to the bladder. (de Groat 1975)

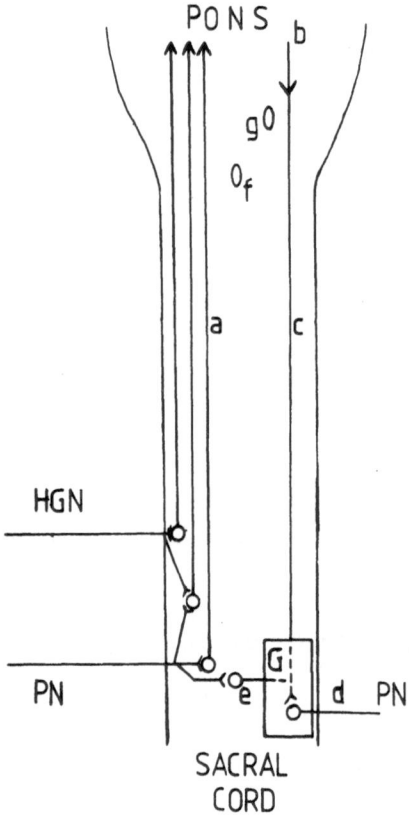

Fig. 8.5. Pathways active during the micturition reflex. The neurons labelled *a* form the ascending limb of the reflex from the lumbosacral cord to the pons (McMahon and Morrison 1982a). The activities of cells in the pontine micturition centre (*b*) and the bulbospinal limb of the reflex (*c*) were described by McMahon and Morrison (1982a). The activities of the pelvic nerve efferents (*d*) and sacral interneurons (*e*) were described by Bradley and Teague (1968), de Groat and Ryall (1969) and McMahon and Morrison (1982c); inhibitory descending pathways which influence the micturition reflex pass through the raphe nucleus (*f*) and lateral reticular formation (*g*) (McMahon and Spillane 1982). The box labelled *G* indicates the site of gating mechanism that regulates the excitability of the reflex (McMahon and Morrison 1982c). (Morrison 1982)

induces reflex activation of the parasympathetic excitatory outflow to the bladder, depression of the sympathetic inhibitory outflow, and depression of the somatic efferent output to the external sphincter; secondary reflexes elicited by the passage of urine through the urethra may reinforce these primary reflexes and facilitate the complete emptying of the bladder. Different patterns of reflex activity occur above and below the micturition threshold, and the parasympathetic reflex pathway through the brain stem is active, while the sympathetic spinal reflex is suppressed when the micturition reflex is exceeded. The role of the sympathetic nervous system in the control of the

bladder in humans is far from clear, and there are some who would argue that, even in cats, it has only a small part to play. A number of other authors agree in essence with the major parts of his story, although inevitably there are differences in the detailed interpretation of the data. McMahon and Morrison (1982a–c), Morrison (1982) and McMahon (1986) used a similar approach, but attempted to separate functional pathways concerned with the reflex pathway and with the control of micturition threshold (Fig. 8.5). The studies by de Groat's and Morrison's groups are much more concerned with function than the others, which are based more on functional anatomy. Mundy (1984) follows the ideas expressed by Nathan (1976) and by de Groat (1975).

The ability to suppress detrusor activity is an important physiological function and an essential part of social existence. Spillane (1982) and McMahon and Spillane (1982) studied the location of descending pathways that could suppress spontaneous bladder contractions and inhibit reflex activity in parasympathetic efferents. Two descending inhibitory pathways were found, one of which

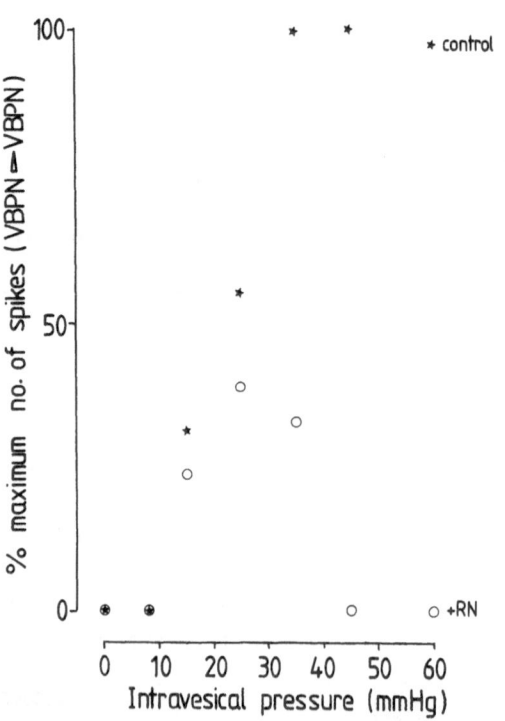

Fig. 8.6. Plot of the relationship between the magnitude of a micturition reflex unitary response (spikes/burst) and intravesical pressure. *Solid symbols* indicate the test response alone. *Open symbols* indicate the test response conditioned by stimulation (introduced 100 ms prior to the test stimulus) of the raphe nucleus (*RN*). (Courtesy of K. Spillane)

originated from the raphe nucleus in the caudal medulla, and the other descending from the region of the nucleus reticularis gigantocellularis. The former appears to utilize serotonin as a neuro-transmitter, whereas the reticulospinal pathway does not (Morrison and Spillane 1986). The inhibitory effects can be seen in neurons which carry bladder information up the cord as well as in local interneurons and in the efferent pathway (Spillane 1982). The effects of these pathways interact with the effects of bladder pressure (Fig. 8.6) on the gain and threshold of the micturition reflex, and can be regarded as switching circuits that modulate the influence of the pelvic nerve input from the bladder, and the efferent pathway in the parasympathetic nerves. The efferent output initiated by stimulation of the pelvic nerve is greatly reduced or abolished by stimulation of the raphe nucleus, showing the reduction in gain that can be achieved by activation of this pathway.

Application of Control Systems Ideas to the Function of the Lower Urinary Tract

The hypotheses developed by the authors referred to in the previous section have incorporated anatomical ideas of a hierarchy of function of increasing complexity, the further rostral one goes in the central nervous system. Functional ideas of switching or gating have been incorporated by de Groat (1975), de Groat et al. (1981, 1982) and McMahon and Morrison (1982c), Morrison (1982) and McMahon (1986). The latter regard the changes in micturition reflex excitability between intravesical pressures of 5 and 30 mmHg as being due to changes in the gain of the reflex. None of them approach the problem of why spontaneous activity should occur in the bladder of the anaesthetized cat or rat, or the correlate in the human, viz. the unstable bladder. Control systems analysis has not been seriously applied to this system, but offers some interesting prospects; the occurrence of spontaneous activity could be regarded as an oscillation generated by feedback, i.e. akin to "hunting" in an engine. Similarly, the gain of the reflex, i.e. the ratio of output could be a valuable measure of the overall activity of the reflex control in different conditions. One of the best treatises on the application of control systems to physiological systems was described in Milhorn's (1966) book on the subject; readers are referred to this book, which is mathematical, but not to such an extent that the non-mathematician cannot obtain the gist of the argument. One or two of the concepts will be mentioned at this point, as they will be referred to later.

Negative and Positive Feedback

The concept of negative feedback is well known to physiologists: afferent signals concerned with a particular parameter, such as body temperature or arterial blood pressure are used to regulate heat loss and heat gain, or signals from the heart and peripheral vessels are used to maintain an approximately constant temperature or pressure. Temperature and pressure are sensed and controlled, and this is achieved by measuring deviations from a set point (an "error signal") that are used to induce proportional or roughly proportional changes in the systems that affect the controlled parameter. Some sort of comparison is required to extract the error signal, and one means of doing this is by subtraction of the afferent input from the set point; biologically, this might be achieved by the algebraic sum of inhibitory and excitatory synaptic inputs. If the effectiveness of the feedback is inadequate, the parameter over which control is desired in fact varies widely. If the gain of the feedback loop is to high, "hunting" may occur, and the output of the system may oscillate. A similar oscillation may occur if some lag time or dead time is introduced into the system.

Positive feedback may also occur in biological systems: the feedback signal is used to reinforce mechanisms that increase the feedback signal. An increment in the afferent signal generates efferent activity that, in turn, increases the afferent signal further. The end result depends on the gain of the feedback loop, i.e. the ratio of output to input, and a new steady state may be achieved if the gain is less than unity, and the extreme limits of the system may be reached if the gain is greater than unity (Milhorn 1966). Figure 8.7 shows diagrammatically the effects of negative feedback.

It is interesting to consider the way in which micturition is switched on, with reference to these control systems. Accommodation of the bladder helps to keep it quiescent, and the pressure low; the sympathetic reflexes which appear to be involved in this process tend to minimize the changes in intravesical pressure, and this may involve some negative feedback. In this case, a rise in bladder pressure initiates relaxation of bladder muscle. In contrast, when micturition occurs the pressure appears to tend towards the maximum that is achievable. Theoretically, this could occur without any feedback whatsoever, just as a car engine responds to pressing the throttle. However, there is a possibility that some positive feedback may be involved, since the switching mechanism referred to by de Groat (1975) and the gating mechanisms of McMahon and Morrison (1982c) imply that the

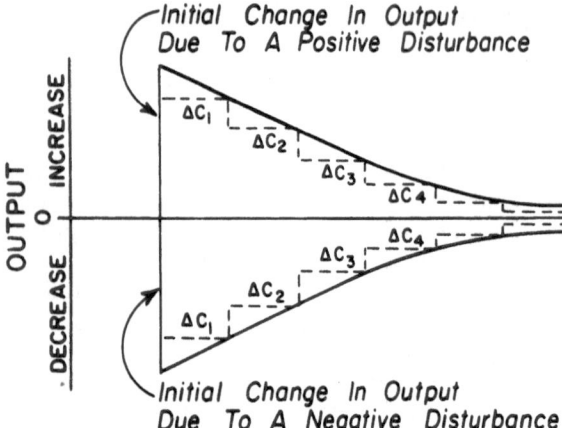

Fig. 8.7. The effects of negative feedback. The figure shows the initial change in output caused by either a negative or a positive disturbance. This disturbance is sensed by an error detector, which brings the output back to within a small deviation from the initial output level. At each point in time following the initial disturbance, the control system exerts a change on the output, and these contributions of the control system are represented by $\varDelta C1$, $\varDelta C2$, etc. (Milhorn 1966)

output of the micturition reflex pathway is reinforced at the high pressures produced during bladder contractions by means of afferent feedback in pelvic nerve afferents. This matter will be considered further near the end of this chapter, after certain functional aspects of the control system, such as the gain of the micturition reflex, have been discussed. For the moment, let it suffice to say that the control systems which govern the activities of the lower urinary tract during the filling and voiding phases present some contrasting properties. Switching between one state and the other is normally accomplished during voluntary micturition, and abnormally, in bladder instability and in anaesthetized animals. In order to understand the factors that might contribute to this switching between states, we have to discuss the anatomical pathways in the central nervous system that influence the function of the lower urinary tract, and the effects these may have on the threshold and gain of the micturition reflex.

Functional Changes Mediated by Different Anatomical Sites in the Central Nervous System

The following paragraphs present a summary of the anatomical information that has been presented by previous reviewers, together with recent additions to the literature. This section is concerned with an anatomical dissection and functional analysis of components of the central nervous system that are influenced by or exert some action on the lower urinary tract. The evidence on which different models and hypotheses concerning how the bladder is controlled is provided in this section, so that our later discussion can concentrate on functional interpretation and a synthesis of how the whole system works. Much of the data was available before 1970, and the reader is referred to the works of Appenzeller (1982), Bors and Comarr (1971) and Kuru (1965) for a fuller bibliography.

Medulla

Kuru (1965) and Kuru et al. (1961) concluded that the ventral reticulospinal tract, originating from the dorsomedial reticular formation of the medulla, is a pathway that inhibits the micturition reflex, and recent work by McMahon and Spillane (1982) confirms this view (Figs. 8.8, 8.9). The latter workers localized the pathway to the region of the nucleus gigantocellularis reticularis. In addition, Spillane (1982) found that the raphe–spinal pathway, originating from the midline of the medulla, was another pathway that had powerful inhibitory effects on the micturition reflex. Kuru (1965) also thought that the lateral reticulospinal tract was responsible for the descending excitation in the micturition reflex and that it originated partly from the medulla, as well as the pons and midbrain. Spillane traced the descending excitatory pathway to the rostral pons, but had no evidence that could be brought to bear on the question of whether the pathway synapsed in the medulla. From a functional point of view, the lesioning experiments of Tang (1955) and Tang and Ruch (1956) suggest that the integrity of the rostral pons is essential for the micturition reflex. Satoh et al. (1978a) have shown that there is a direct connection between the lateral tegmental nucleus of the rostral pons and the lumbar cord in the rat, but polysynaptic connections involving synapses in the medulla could still be involved in this pathway, as suggested by Kuru (1965).

Nathan (1976) takes up Kuru's observation on cats that suggests an involvement for the reticulospinal pathways in the control of the bladder, by saying that no involvement of the bladder is seen if the reticulospinal tracts are lesioned in humans. Mackel (1979) agrees with Kuru et al. (1963), and has localized the source of the inhibitory and excit-

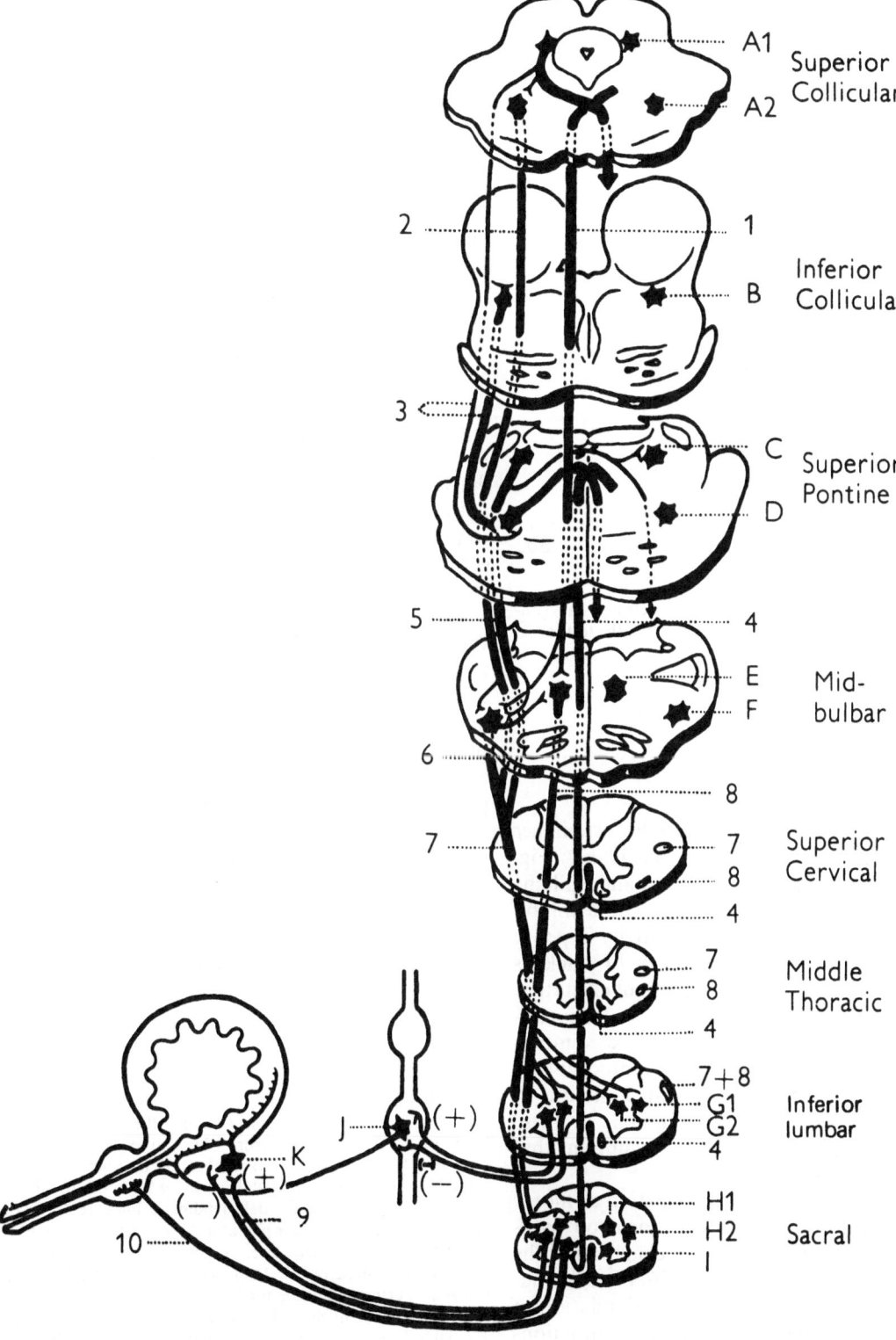

Fig. 8.8. Descending pathways concerned with micturition in the cat. *A1*, *A2*, mesencephalic micturition facilitatory area; *B*, mesencephalic micturition inhibitory area; *C*, pontine detrusor nucleus (Barrington); *D*, pontine nucleus for external urethral sphincter; *E*, bulbar vesicorelaxer centre; *F*, bulbar vesicoconstrictor centre; *G1*, lumbar vesicomotor centre (excitor); *G2*, lumbar vesicomotor centre (inhibitor); *H1*, sacral vesicomotor centre (inhibitor); *H2*, sacral vesicomotor centre (excitor); *I*, sacral nucleus for external urethral sphincter; *J*, lumbar sympathetic ganglion; *K*, vesical plexus; *1*, tectobulbar tract; *2*, mesencephalobulbosacral tract; *3*, tectopontine tract; *4*, medial reticulospinal tract; *5*, pontobulbosacral tract; *6*, bulbosacral tract; *7*, lateral reticulospinal tract; *8*, ventral reticulospinal tract; *9*, pelvic nerves; *10*, pudendal nerves. (Kuru 1965, quoted in Appenzeller 1982)

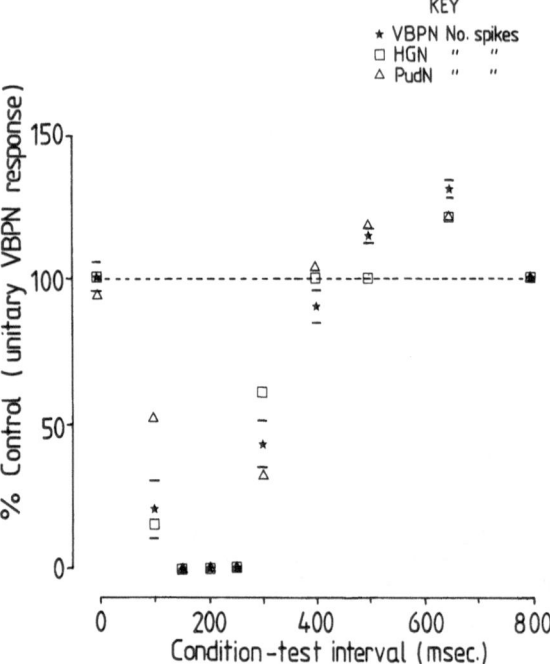

KEY
★ VBPN No. spikes
□ HGN   "   "
△ PudN   "   "

Fig. 8.9. Shows how varying the interval between the conditioning and test stimuli affects the amplitude of the test reflex. (Courtesy of K. Spillane)

Pons

Normal micturition reflex activity depends on links between the sacral cord and the brain stem, where Barrington and others have located two regions in the rostral pons concerned with regulation of the bladder; the first was in the dorsolateral reticular formation, and bilateral lesions of this region result in the animals not being able to empty their bladders. Bladder contractions can be elicited by moderate distensions or by flowing water through the urethra (Barrington's first and second reflexes), and the normal reflex activity that can be seen with electrophysiological techniques depends on these connections being intact (de Groat and Ryall 1969; de Groat 1975). Electrical stimulation within this region elicits large contractions of the bladder, and some workers have called this area the "pontine detrusor nucleus". From this region fibres descend in the lateral reticulospinal tracts, and some appear to make contact with neurons in the medulla concerned with detrusor contraction. The second area was more medial, extending from the bottom of the aqueduct to the lateral part of the floor of the fourth ventricle, and lesions in this site caused the animal to appear to lose the urge to micturate. Whether this loss of sensation is related to damage in the medial lemniscus is not clear. Spillane (1982) plotted the pontine course of pathways that influence bladder motility and electrical activity in the pelvic nerves, and found a lateral excitatory and a more medial inhibitory pathway. de Groat (1975) has also observed efferent activity in the pelvic nerves following stimuli to the pons. In rats, Satoh et al. (1978b) have localized the detrusor centre to the dorsolateral tegmental nucleus, using retrograde tracers and lesioning techniques (Fig. 8.11).

atory controls over the external urethral sphincter to the nucleus gigantocellularis reticularis (Fig. 8.10). Spillane (1982) and McMahon and Spillane (1982) also found that this nucleus could inhibit bladder motility and the electrical activity in pelvic nerve efferents in the micturition reflex. In Mackel's and in Spillane's work, a combination of excitatory and inhibitory effects can be seen in the electrophysiological changes following a brief stimulus to the raphe nucleus or the nucleus gigantocellularis reticularis; this relatively complicated picture of mixed effects may be why gross lesions of these tracts in humans do not give rise to symptoms that point to the lower urinary tract.

One further aspect of the organization of motor control during micturition is dependent on pathways through the lateral reticular formation of the medulla. Yamamoto and Araki (1962) and Yamamoto et al. (1961, 1962) found that stimulation of the pelvic nerve afferents can elicit contractions of abdominal muscles and relaxation of perineal muscles at the same time as the reflexly induced bladder contraction. The perineal relaxation could well be a spinal reflex elicited by the bladder contraction (see Chap. 7, p. 206).

Fig. 8.10a–d. Sections through four levels of the ponto-medullary reticular formation showing the effective stimulation points and the points from which no responses could be evoked. **a** Section through r.p.c. **b** Section through the rostral r.g. **c** Section through the middle of r.g. **d** Section through the caudal r.g. According to Brodal's landmarks lines are drawn on each side to delineate the r.p.c. and the r.g. Abbreviations: *V, VI, VII,* cranial nuclei; *Br.c.,* superior cerebellar peduncle (bracchium conjunctivum); *M.l.f.,* medial longitudinal fasciculus; *N.c.e.,* external cuneate nucleus; *Ol.s.,* superior olive; *V.m.,* medial vestibular nucleus; *V.d.,* descending vestibular nucleus; *V.l.,* lateral vestibular nucleus; *V.s.,* superior vestibular nucleus; *P.h.,* perihypoglossal nucleus; *c.r.g.,* caudal nucleus reticularis gigantocellularis; *r.g.,* nucleus reticularis gigantocellularis; *r.r.g.,* rostral nucleus reticularis gigantocellularis; *r.p.c.,* nucleus reticularis pontis caudalis; Symbols: ●, monosynaptic EPSPs; ○, polysynaptic EPSPs; △, polysynaptic IPSPs; ×, no response. (Mackel 1979)

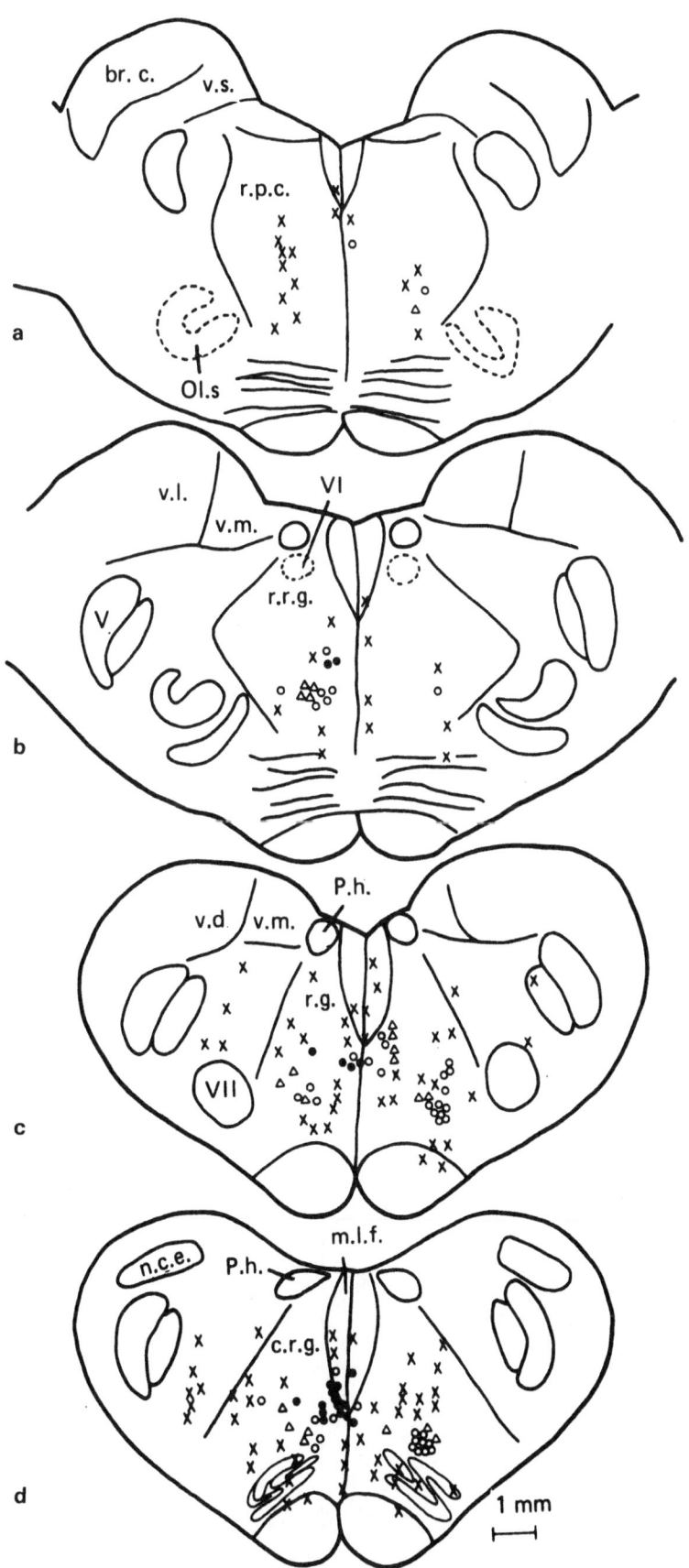

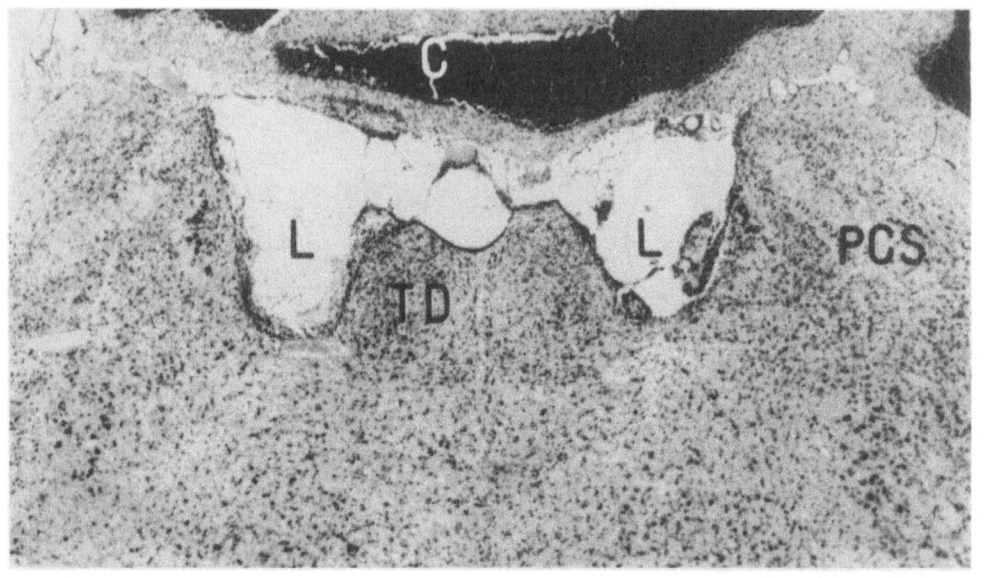

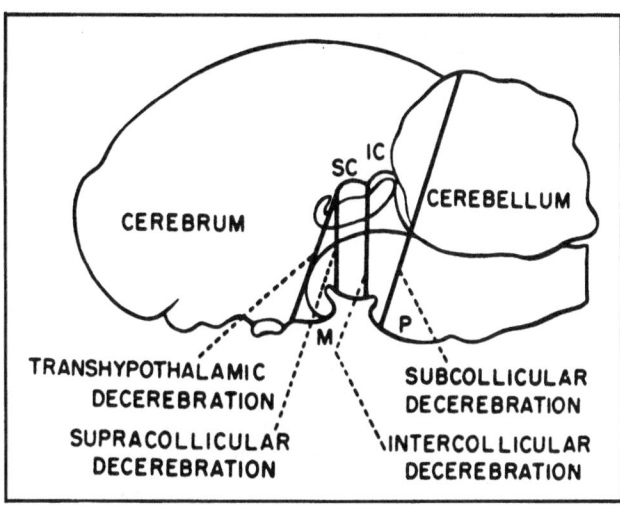

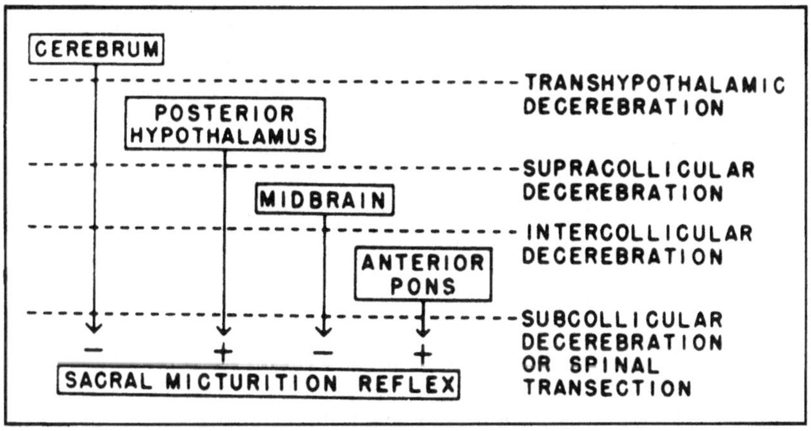

Fig. 8.12. a Schematic sagittal section of cat's brain showing transections employed in studying brain stem control of micturition. *SC, IC*, superior and inferior colliculi; *M*, midbrain, *P*, pons. b Summarizing diagram indicating the net facilitatory or inhibitory action of various levels of the nervous system deduced from surgical procedures shown at the right. For simplicity, the diagram does not take into account the possibility that the descending pathways from the higher structures terminate on lower ones, including the bulbar reticular inhibitory and facilitatory areas of Kuru and co-workers. (See Kuru 1965.) (a from Tang 1955; b from Tang and Ruch 1956)

Midbrain

Transection of the cat neuraxis above the superior colliculi facilitates the micturition reflex, whereas transection below the inferior colliculus inhibits it (Langworthy and Kolb 1933). Skyltety (1959) found that electrical stimulation of the periventricular grey matter resulted in bladder contractions; and Kuru (1965) found that mesencephalic vesicoconstrictor loci near the lateral

margin of the grey matter in the superior and intercollicular area. Gjone (1966) also found that the effects of stimulation of the periaqueductal grey matter and the superior colliculus of the cat gave rise to bladder contractions. In humans, stimulation of the caudal end of the cingulate gyrus or the superior colliculus gave rise to vesical and rectal contractions, and lesions of these areas decreased the micturition and defecation thresholds (Matsumoto 1957; Takebayashi et al. 1957). The superior colliculus is one of a number of areas that receive inputs from the cingulate gyrus (Mukai 1959) and the hypothalamus (Enoch and Kerr 1967a, b); however, Magoun et al. (1938) found that pathways influencing bladder function could be found in many regions of the midbrain. The studies of Tang (1955) and Tang and Ruch (1956) on the effects of lesions and stimulation of mesencephalic and telencephalic structures on micturition threshold and bladder activity showed that the overall effect of the hypothalamus was excitatory and that of the midbrain inhibitory (Fig. 8.12).

Fig. 8.11a,b. Lesions that cause retention of urine in the rat. a A photomicrograph of the focal lesions at the caudal part of nucleus tegmentalis laterodorsalis. *L*, lesion; *C*, cerebellum; *TD*, nucleus tegmentalis dorsalis (Gudden); *PCS*, pedunculus cerebellaris superior. (Gallocyanin staining, × 30.) b Summarizing diagram showing the common lesion of the rats with urinary retention. *Shaded areas* indicate a destroyed area which more than a half of the animals with urinary retention had in common, if destroyed bilaterally. Within this area, caudal part of nucleus tegmentalis laterodorsalis (*TLD*) is situated. *FLM*, fasciculus longitudinalis medialis; *PCS*, pedunculus cerebellaris superior; *TD, TV*, nucleus tegmentalis dorsalis and ventralis (Gudden); *VMT*, nucleus motorius n. trigemini. (Satoh et al. 1978b)

Regions of the anterior hypothalamus and the periventricular and periaqueductal grey matter have been implicated not only in bladder control, but in the regulation of nociceptive input to the spinal cord. Indeed, regions of the brain stem and hypothalamus that are known to be involved in antinociception are also closely associated with bladder control. Regions such as the anterior hypothalamus, periventricular and periaqueductal grey matter, the locus coeruleus, the raphe nuclei and the nucleus gigantocellularis reticularis of the medulla are all locations where cell somata or fibres of passage have a role to play in each of these functions (de Groat 1975; Giesler and Liebeskind 1976; Fields and Basbaum 1978; Hammond and Proudfit 1980; Westlund and Coulter 1980; Carstens 1982; McMahon and Spillane 1982; Spillane 1982; Morrison and Spillane 1981, 1986). The elucidation of these functions, and the extent to which they are separated or linked, is one important task for the future. Autonomic phenomena are elicited by pain, just as antinociceptive mechanisms are; some of the potential interactions between these two systems have been described recently by Lumb (1986).

Recordings have been made from neurons in the substantia nigra, red nucleus, tegmental fields and mesencephalic reticular formation in unanaesthetized paralysed cats by Porter et al. (1971); they showed mixed excitatory and inhibitory responses. Whether or not the changes are due to changes in the bladder, or to some concomitant changes in skeletal muscle control is not clear; the interpretation of neuronal data on cats in this condition is fraught with problems. Elam et al. (1986) have recorded from units in the locus coeruleus in rats, and found that they respond to small distensions of the bladder. These neurons could be involved in the micturition reflex.

Hypothalamus

Contractions of the bladder can be elicited by electrical stimulation of the anterior or lateral hypothalamus of anaesthetized and awake animals, and the effects of stimulation are dependent on pathways through the sacral cord and pelvic nerves (Karplus and Kreidl 1909; Lichternstein 1912; Kabat et al. 1936; Hess and Brügger 1943a,b). In the cat, there are two pathways running through the hypothalamus that initiate bladder contractions; one medial and the other lateral (Enoch and Kerr 1967a,b; Nathan 1976). The medial pathway appears to have its origin in the posterior hypothalamus and to descend in the periventricular

grey and periaqueductal grey matter to a level just below the third nerve nucleus; the pathway divides to end partly in the periaqueductal grey and partly in the lateral half of the superior colliculus, in the intermediate grey stratum. The lateral pathway runs caudally through the medial forebrain bundle in the hypothalamus, to divide into one branch that ends in the periaqueductal grey, and the other branch which terminates in the lateral tegmental nuclei of the midbrain. Magoun et al. (1938) attempted to localize the descending pathways that mediated bladder contractions by observing the effects of lesions in the brain stem on the bladder contractions elicited during hypothalamic stimulation. The descending pathways were widely distributed throughout the tegmentum of the midbrain and pons and were not confined to any one region or tract of fibres. In the spinal cord, there appears to be an uncrossed relationship between the descending pathway and its site of excitation in the hypothalamus: after hemitransection of the spinal cord, bladder contractions can be elicited from the contralateral hypothalamus, but not from the ipsilateral side of the diencephalon (Nishiyama 1959). Wang and Ranson (1939a,b) showed that the spinal pathway was in the ventrolateral columns. The relationship between this descending pathway and the direct hypothalamospinal pathway from the paraventricular nucleus (see Fig. 8.1; see also Chap. 7, p. 201) is unclear; the latter makes contact with the parasympathetic nucleus in the sacral cord, and is uncrossed (Loewy 1982).

Gjone (1966) also recognized two sites in the hypothalamus that influenced bladder function: a dorsolateral vesicoconstrictor pathway and a ventromedial inhibitory pathway were reported. Thus, the lateral pathway of Enoch and Kerr (1967a,b) terminates in the periaqueductal grey and lateral tegmental nuclei, sites which elicit bladder contractions, and stimulation of the descending pathway to these structures in the hypothalamus is also excitatory. The effects of the lateral excitatory pathway from the hypothalamus appear to be the dominant pathway at rest, because in Tang and Ruch's experiments, the overall effect of the hypothalamus was to facilitate the micturition reflex. According to Gjone (1966), all inhibitory effects were mediated by the sympathetic, whereas the excitatory influences involved the parasympathetic efferent pathway.

However, the effects of hypothalamic stimulation are not confined to bladder contractions, but a variety of other cardiovascular and gastrointestinal effects are elicited simultaneously. In addition, somatic movements can be elicited, and the work of Hess and his colleagues has suggested that the hypothalamus actually coordinates the appropriate somatic and autonomic responses characteristic of behaviour during micturition.

Hess (1947) did experiments on conscious cats in which different parts of the hypothalamus and associated structures were stimulated electrically through electrodes which had previously been implanted under anaesthesia. Stimulation within the septal region at the base of the brain elicited a behaviour pattern which included all the normal elements seen when members of that species micturate or defecate; they would dig a hole, assume the posture appropriate for micturition or defecation, voiding would occur, and the normal behaviour patterns that follow these events would then be performed. Micturition of defecation or both could occur, depending on the fullness of these viscera; electrical stimulation of adjacent regions of the septum give rise to erection and ejaculation. In Hess's experiments, stimulation of the neighbouring preoptic areas of the hypothalamus gave rise to changes in bladder motility, without the accompanying postural and behavioural patterns which were characteristic of septal stimulation (Fig. 8.13). In the rabbit, these regions can be subdivided into areas which elicit either voiding or storage activity in the bladder. It should be noted that the paraventricular nucleus in the anterior hypothalamus contains one group of neurons whose axons project to the intermediolateral columns of the cord (see Chap. 7, p. 201).

Bradley (1984) does not include the hypothalamus in his loop concept, despite the interesting results obtained in awake, behaving animals by Hess and his colleagues. Possibly he regards it as a relay linking the frontal cortex and the pontine tegmentum, and this is one of the major differences between Nathan (1976) and Bradley (1984). Nathan regards the anterior hypothalamus as an autonomous group of centres which graft on specific behaviours, in the manner outlined by Hess (1947).

At the current state of knowledge, it is impossible to be certain whether the "integrative" centres in the hypothalamus organize every detail of the complex motor behaviour associated with micturition within the hypothalamus itself; one alternative might be that they alter the processing of information in a subtle or indirect way, possibly by switching or modulating the activities of a series of other pathways, which together are responsible for the coordination of the sequence of motor responses we recognize as a specific behaviour. A small switch can tap great power, but the origin of the power is not within the switch itself. It might be unwise to adopt too fixed an interpretation of the role of the hypothalamus (or other structures) in micturition

behaviour, and, in particular, the term "integrative" can disguise a dearth of knowledge and a lack of understanding, besides perpetuating an idea that has not shown itself to be very discriminating.

Cortical Control of Micturition

The regions of the cerebral cortex concerned with micturition occur in three main areas, the frontal lobe, the paracentral lobule, and the cingulate gyrus, together with the associated regions of the limbic system (the amygdala, hippocampus and hypothalamus). The frontal cortex and cingulate gyrus have connections with the hypothalamus and limbic system (Fig. 8.14), whereas the paracentral lobule has direct connections, via the corticospinal tract, with the pudendal motoneurons controlling the external sphincters.

Voluntary control over micturition in humans appears to depend on the integrity of connections between: (a) the medial surface of the frontal lobe (the superior frontal gyrus and anterior cingulate gyrus) and the septal and preoptic regions of the hypothalamus, or (b) connections between the paracentral lobule and the brain stem and cord (Nathan 1976). These conclusions are based on studies of patients with lesions that produce disturbances of micturition; the lesions included different types of leucotomies, penetrating injuries, thromboses, neoplasms and aneurysms. These lesions appear to remove inhibitory controls over the anterior hypothalamic areas with which they are connected, and which organize voiding behaviour; these lesions can also release bladder reflexes from the inhibitory influence of these cortical areas, thus allowing bladder contractions to occur at low volumes. In addition, these lesions interfere with the processing of sensory information from the bladder, as the affected patients fail to experience a desire to void, but may be aware that micturition is imminent, or of the sensation of urgency. Alternatively, patients may be completely unaware of the state of their bladders, and are surprised by the fact that they have passed urine. Lesions of the same part of the brain may also give rise to loss of bladder sensation, but accompanied by retention or difficulty in passing urine. The difficulty in interpreting many of the results of pathological lesions in this area of the cortex is that those changes in the ability to sense or control events in the lower urinary tract are not isolated changes; they are accompanied by much more widespread effects, such as sensory inattention, and generalized hypoactivity, and the effects on the lower urinary tract may be only one component of these states. For instance, lesions of the

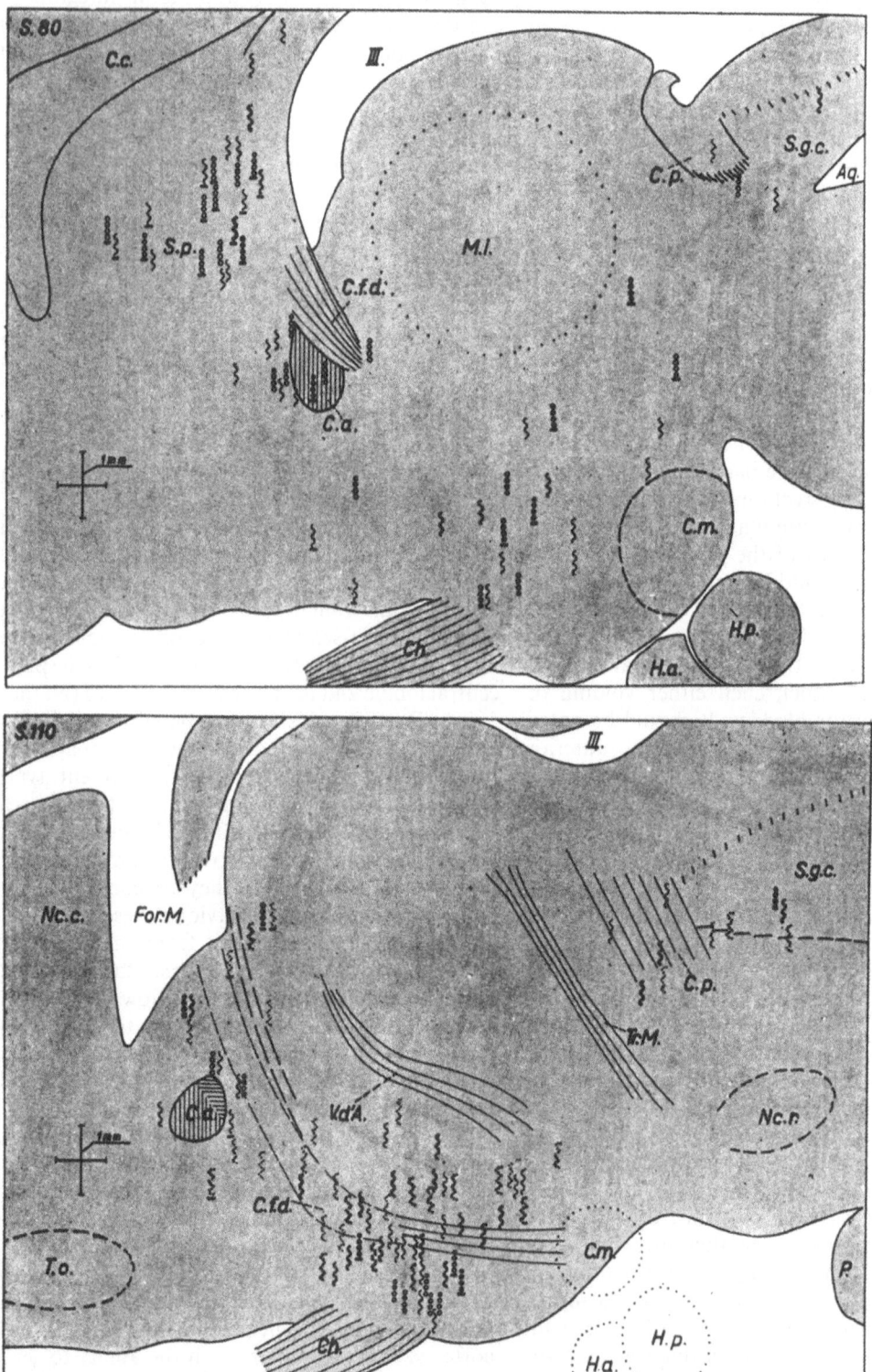

Fig. 8.13a,b. ⚥, Defecation; ⚥, defecation in adequate posture; ⚥, micturition; ⚥, micturition in adequate posture. *Dotted symbols*, points stimulated 1–2 mm more lateralwards than the plane of the section (**b**). *Full line*, first voltage stage. *Light line*, second voltage stage. Note the numerous responses of defecation and micturition in adequate posture on stimulation of the septum pellucidum, the adequate posture not being usually assumed when the hypothalamus was excited.

Defecation and micturition were never elicited by stimulation of the thalamus. (Hess 1947)

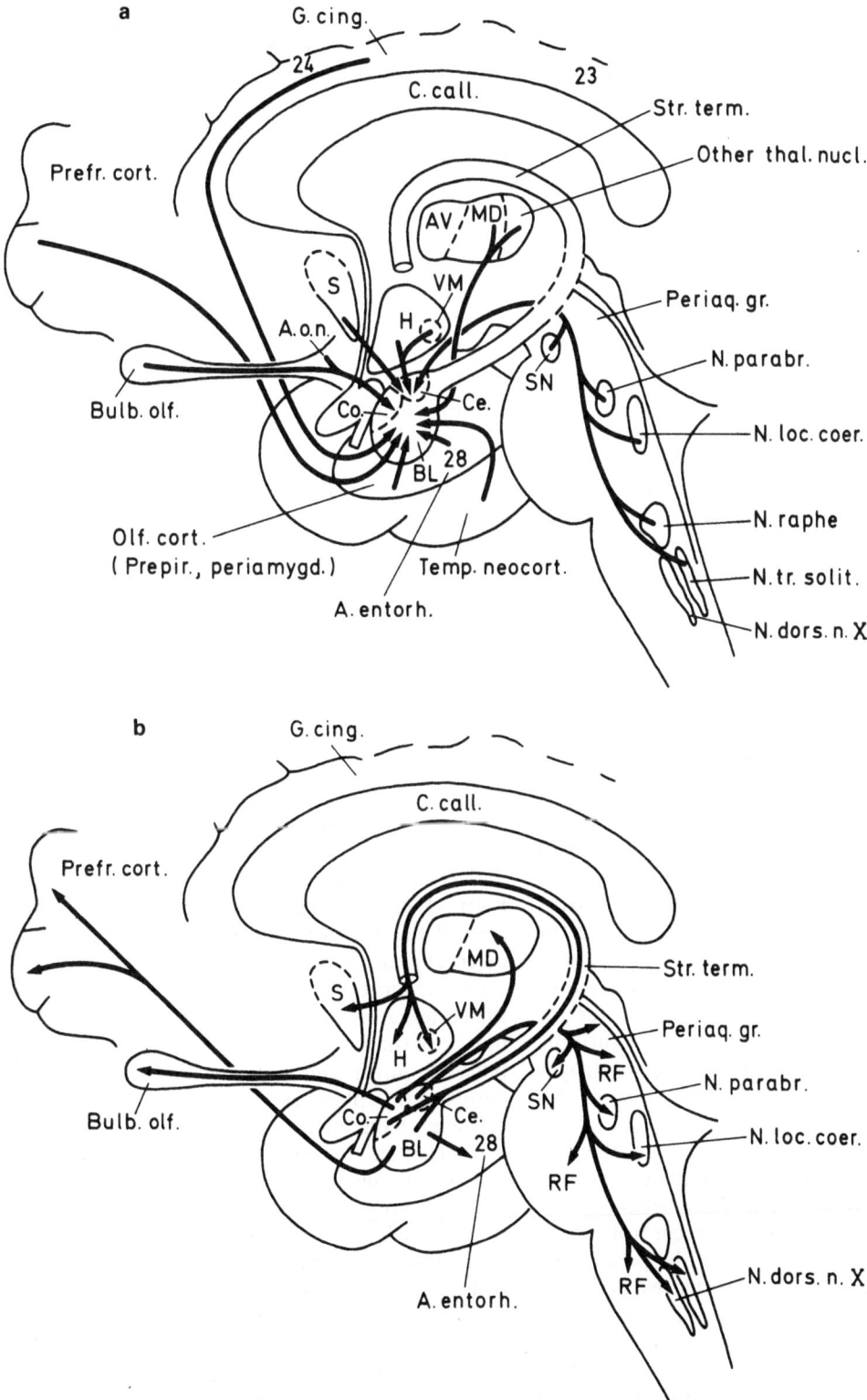

Fig. 8.14a,b. Very simplified diagrams of the main afferent (**a**) and efferent (**b**) connections of the amygdaloid complex. To some extent sites of origin and endings of fibres within particular nuclei of the complex are shown. Note the many reciprocal connections. Not all connections are shown (for example, in **b**, the fibres to the subiculum). *A.o.n.*, anterior olfactory nucleus; *AV*, anteroventral thalamic nucleus; *BL*, basolateral amygdaloid nucleus; *Ce.*, central amygdaloid nucleus; *Co.*, cortical amygdaloid nucleus; *H*, hypothalamus; *MD*, dorsomedial thalamic nucleus; *RF*, reticular formation; *S*, septum; *SN*, substantia nigra; *VM*, ventromedial hypothalamic nucleus. (Brodal 1981)

anterior cingulate gyrus lead to akinetic mutism, indolence of painful stimuli, and indifference to urinary incontinence (Barris and Lehuman 1953). One might wonder to what extent the urinary abnormalities seen in these patients are a reflection of their general state, and to what extent it is really possible to make specific conclusions concerning the role of these areas of the cortex in the voluntary control of micturition. Nevertheless, there is some agreement from animal studies which suggest that the frontal lobe may be important in the regulation of micturition. The older literature suggests that the middle third of the precentral gyrus contains a vesicomotor centre (Pfeifer 1919). In this study of patients with war injuries, no disturbances of micturition were observed in patients with temporal, parietal or occipital lobe lesions unless there was also motor involvement. It may be that these lesions affect the descending pathways to the sphincters, rather than the bladder itself, but there seems to be little documentation of possible connections between the areas concerned with bladder and with sphincteric control.

Brutkowski (1965) pointed out that the frontal cortex has a role to play in the control of a number of visceral activities in humans and animals, particularly with respect to the function of the gastrointestinal tract and the bladder. Langworthy and Hesser (1936) found that lesions of the frontal cortex cause changes in bladder function, and Gjone and Setekleiv (1963) found that electrical stimulation of the cerebral cortex of the cat could give rise to bladder contractions or bladder relaxation. Areas in front of and behind the central sulcus and the cingulate and orbital gyri could elicit these reactions (Fig. 8.15). These responses in cats fit in reasonably well with the findings in humans, except for the relationship of the orbital cortex to micturition, which, when lesioned by leucotomies or destroyed by neoplasm in humans, do not give rise to disturbances of micturition or defecation.

Lesions in pathways concerned with the voluntary control of sphincters also give rise to difficulties in voiding. Nathan (1976) describes the effects of upper motoneuron lesions on control of the sphincters. When lesions occur in the paracentral lobule of the cortex, patients experience a profound spasticity of the lower limbs, associated with an inability to relax the levatores ani, perineal muscles and external sphincters. The paracentral lobule syndrome is relatively uncommon, and upper motoneuron lesions are usually accompanied by no disturbance of micturition or defecation; conversely, the inability to initiate or stop micturition can occur without any signs of upper motoneuron lesions affecting the lower limbs.

Patients with the paracentral lobule syndrome have difficulty in initiating micturition. This is presumed to be due to a corticospinal tract lesion causing hyperactivity of the sphincteric apparatus, and it is not clear whether there is in fact any disturbance of bladder function per se. The conclusion one can draw is that the sphincteric motoneurons are influenced by descending pathways in much the same manner as are motoneurons innervating the lower limbs. In that regard, the reports of periodic interruptions of flow in patients with Parkinson's disease (Mundy and Blaivas 1984) indicate that the sphincteric motoneurons may also be under the influence of the extrapyramidal pathways responsible for parkinsonian tremor. In cats, Porter (1967) has reported that striatal units alter their activity in relation to the activity of the lower urinary tract, and often give a triphasic pattern of excitatory and inhibitory activity. It has been argued that these changes are associated with activity of the bladder, rather than the changes in somatic muscle tone that occur at high bladder pressure (see p. 214).

Sherrington (1892) found that stimulation of the motor cortex on one side caused contractions on both sides of the anal sphincter, but still greater on the contralateral side, in the monkey; this is one example of a muscle that does not have a fully crossed relationship with the motor cortex. This was attributed to the arrangement of the descending pathways, because the relationship between the pudendal nerves and the parts of the sphincter undergoing a contraction was essentially unilateral. Kuru (1965) concludes that there is a specialized area of the pons that sends tonic impulses to the external urethral sphincter; it is presumed that this area, like most pontine nuclei, receives inputs from the cortex. As in the cortical control of the skeletal muscle of the limbs, there is therefore also the possibility of direct and indirect pathways from the cortex to the sphincters.

Some of the effects of cortical stimulation may be attributable to extrapyramidal pathways that synapse in the brain stem. The final stage of a possible projection involving the reticulospinal tracts as the final stage in the pathway has been studied by Mackel (1979), who found that the ventrolateral reticulospinal pathway can inhibit the activity of the external anal sphincteric motoneurons as well as those that run in the urethral branch of the pudendal nerve. In addition, the ventromedial reticulospinal tract appears to be responsible for some excitation of sphincteric motoneurons. Both effects originate from the nucleus reticularis gigantocellularis of the medulla. McMahon et al. (1982) also demonstrated descending pathways from the raphe nuclei in the caudal

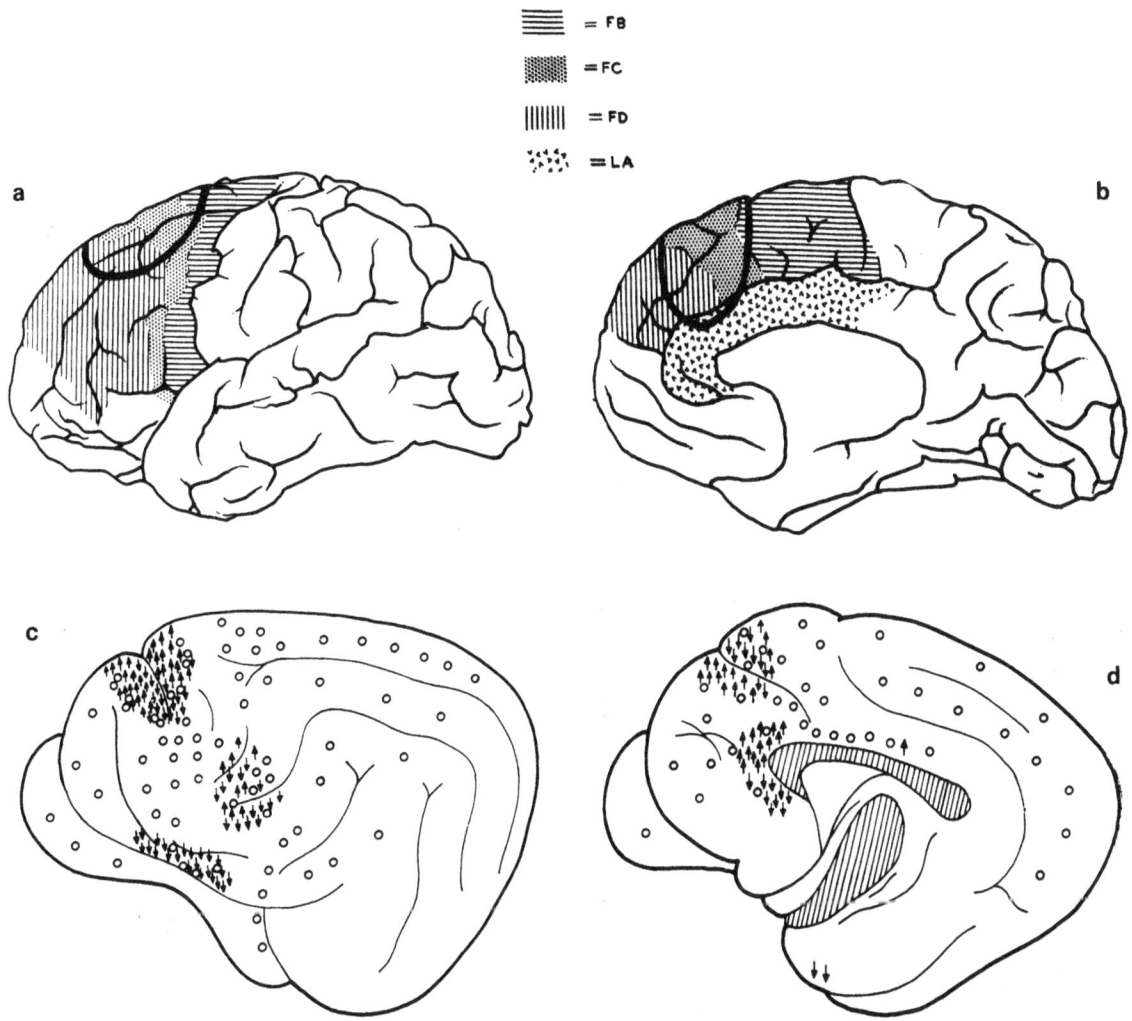

≡ = FB

▓ = FC

||||||| = FD

∴∴ = LA

Fig. 8.15.a–d. The lateral (**a, c**) and medial (**b, d**) aspects of the brain of the human (**a,b**) and the cat (**c, d**) showing the similarity in the areas involved in the control of detrusor function. The area of human cerebral cortex concerned is denoted by a *heavy black line* and includes the superior frontal gyrus and extends towards the cingulate area. In the cat the areas are subdivided into those causing bladder contraction (↑), relaxation (↓) or no response (O). In the human there appears to be no bladder representation in the orbital cortex as is apparent in the cat. However, there are areas located near the optic chiasm. (**a,b** from Andrew and Nathan 1964; **c,d** from Gjone and Setekleiv 1963)

medulla that have inhibitory influences on reflex activity in the pudendal nerve efferents. This descending pathway is probably associated with terminals that can be demonstrated to contain serotonin and substance P in Onuf's nucleus. In addition, there are descending fibres which contain noradrenaline in their terminals in Onuf's nucleus. These may originate from the locus coeruleus (Westlund and Coulter 1980), but a function for that pathway has yet to be demonstrated; it is likely to be inhibitory.

McMahon et al. (1982) also found that the inhibitory effect of the pelvic nerve on pudendal reflex activity was prolonged, which might be in keeping with looped inhibitory pathways involving the brain

stem (Fig. 8.16). Light (1985) has demonstrated that the terminals of raphe–spinal neurons are distributed in the ventral as well as the dorsal horn; there could be considerable overlap between the raphe–spinal terminals described by Light and the sphincteric interneurons and motoneurons described by McMahon and Morrison (1982b) and McMahon et al. (1982). The responses of raphe–spinal neurons to afferent stimuli have been investigated by West and Wolstencroft (1977) and Lumb and Morrison (1984). Pelvic nerve stimulation can excite raphe–spinal and reticulospinal neurons in the rat, and the latencies vary between 7.5 and 300 ms. Thus, the prolonged time course of inhibition of pudendal motoneurons may well be

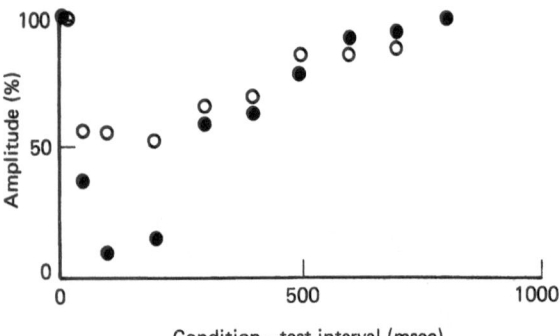

Fig. 8.16. The time course of excitability in the pudendal nerve motoneuron pool following conditioning electrical stimulation of the raphe nucleus (four × 60 μA, 0.2 ms pulses in 20 ms). The test stimulus which gave the maximal amplitude mass wave in the pudendal nerve was applied to the pelvic nerve (○) or to the contralateral pudendal nerve (●). Both of these nerves received 10 V, 1 ms pulses at 0.25/s. (McMahon et al. 1982)

related to inhibitory looped pathways involving the raphe–spinal and reticulospinal pathways.

Micturition or defecation during emotional stress may be regarded as one of the bodily manifestations of emotion, and as such is likely to be initiated within the limbic system. In some species, such as rats, mice and elephants, voiding can be induced by agitation and nervousness, while micturition is part of the normal defence reaction shown by sheep being approached by strangers. Many species have been shown to possess connections between the limbic system and the septal and preoptic regions of the hypothalamus (see Fig. 8.14). Stimulation of the hippocampus induces micturition or strong bladder contractions, whereas stimulation of the amygdala can induce either relaxation (see Fig. 8.19) or contraction (Koikegami et al. 1957; Gjone 1966). In humans, however, bilateral lesions of the limbic system in the temporal lobes, amygdala and hippocampus do not give rise to any disturbance of the voluntary control of micturition in normal circumstances (Terzian and Ore 1955). Thus, pathways from the limbic system to lower regions of the brain concerned with micturition may well be concerned not with voluntary control, but with the bodily manifestation of emotional states; this would be consistent with the general view of the limbic system.

Basal Ganglia

Current thinking suggests that the basal ganglia are concerned with the initiation of movements that are programmed by the cerebral cortex. During voluntary movement the first electrical changes that precede muscular movement can be found in the basal ganglia, then in the cortex, then in the efferents. In the older literature, attempts to elicit bladder contractions by stimulation of different parts of the basal ganglia were not successful (Kabat et al. 1936; Ranson et al. 1935). Lewin et al. (1967) found that stimulation in the globus pallidus and putamen of the basal ganglia inhibited bladder motility, and Porter (1967) showed that some striatal neurons altered their patterns of discharge when the bladder contracts or is distended. The major efferent pathway from the basal ganglia is to the cortex (Kandel and Schwartz 1981) and it may be that the sites at which Lewin et al. (1967) found detrusor inhibition may project to areas of the frontal lobe concerned with bladder control and possibly to the area of motor cortex that controls the sphincters (see p. 255). Lesions in the basal ganglia and the associated pathways through the substantia nigra give rise to spontaneous rhythmic movements, such as parkinsonian tremor, athetosis, chorea and hemiballismus (Kandel and Schwartz 1981). It is known from studies in humans that patients with Parkinson's disease can show hesitancy in micturition, and that this can be due to failure to relax the skeletal muscle of the pelvic floor and possibly the urethral sphincter (Mundy and Blaivas 1984).

Cerebellum

Barrington (1921) investigated the effects of removal of the cerebellum on micturition in three cats, and concluded that this procedure had no effect on reflex micturition. In one animal, the flow was interrupted intermittently, and these effects were attributed to damage to the brain stem caused by the bleeding during removal of the cerebellum; respiration was also spasmodic.

The influence of the cerebellum on autonomic function and on the lower urinary tract has also been examined by electrical stimulation of the cerebellar cortex, or of the intracerebellar nuclei, such as the fastigial nucleus and nucleus interpositus. These nuclei send axons to the brain stem, and generally have inhibitory effects on pontine nuclei. The latter also receive inputs from the cerebral cortex and other structures, so the effects of the cerebellar input could be modulated by one of the others. It is possibly not surprising, therefore, that the effects of stimulating the cerebellum are variable and often contradictory, depending on the experimenter and the conditions of the experiment.

Stimulation of the fastigial nucleus in the cerebellum has effects on a number of branches of the autonomic nervous system, including cardiovascular and gastrointestinal effects. Chambers (1947) found that stimulation of the fastigial nucleus gave rise to micturition, whereas Bradley and Teague (1969) and Martner (1975) found that micturition was suppressed.

The evoked potential technique has shown that neurons in the vermis of the cerebellum become active after an electrical stimulus to the pelvic nerve (Bradley and Teague 1969), or to the pudendal nerve. If the way in which this information is processed is akin to other spinal inputs, and there is no information available on this, then these cerebellar circuits will also receive information from the cortex (a copy of the cortical command to the periphery) to be used in the adjustment of muscular tone and the coordination of movements appropriate to the task being undertaken. The role of the cerebellum in coordination and in postural control might suggest that these pathways are concerned with postural control during micturition behaviour. This, however, is speculation, and, as such, is comparable with much that has been written on this subject. The role of the cerebellum in micturition is far from certain, and the basic reflex can proceed in its absence; the fact that the cerebellum has been reported to cause as well as to suppress bladder contractions is consistent with the major function of this part of the central nervous system, i.e. the coordination of motor behaviour, which will sometimes require micturition to be suppressed, and to be facilitated in other situations, such as the normal micturition posture for the species.

Development of Urinary Continence

Some observers appear to believe that the development of bladder control depends on the development of connections between the frontal and cingulate cortices and the hypothalamus (Nathan 1976). Others do not specify, but associate voluntary control with the assumption of detrusor reflex drive by the central nervous system (Bradley 1984). Others (de Groat 1975) have observed changes in the nature of pelvic nerve reflexes during the development of bladder control. In neonates of a number of species, and in kittens up to the age of 5–7 weeks, tactile stimuli to the perineum induces reflex voiding, which is mediated by a spinal pathway within the lumbosacral cord. Older cats show the normal spinobulbospinal reflex observed in adult animals, and the short latency EPSPs seen in neonates and chronic spinal animals are replaced by short latency IPSPs, followed by the longer latency response subserved by pathways through the brain stem. The perineal reflex is said to be essential to the survival of these neonates, which cannot void without tactile stimulation of the perineum; this is normally provided by the mother. It is assumed that the perineal reflex is under tonic inhibition in normal adults, as it returns if the spinal cord is sectioned.

The changes in the short latency (3–5 ms) responses in feline parasympathetic efferents between 5 and 7 weeks of age has to be mediated by spinal pathways: there is insufficient time for loops through the cortex to be involved (Fig. 8.17). Thus, there is clear evidence that the development of continence in this species is associated with changes in spinal reflex responses; similar reflexes can be seen in human paraplegics, so this mechanism cannot be dismissed as a species difference. Of course this does not rule out the possibility of changes in the connections between the cortex and the brain stem. Direct evidence on the latter, however, is lacking, and the idea that these loops are essential for voluntary control should be viewed with a certain amount of scepticism. Intuitively, the frontal cortex would seem an essential component of voluntary control, but the development of connections between the cortex and the parasympathetic efferents need not involve the specific pathways suggested by Nathan (1976) or by Bradley (1984).

Balance of Excitatory and Inhibitory Controls Over Micturition

In common with the descending control over other motor functions, the regulation of bladder motility depends on a balance between excitatory and inhibitory descending pathways, and feedback from afferents in the effector organ. It is implicit in the studies of Tang (1955) and Tang and Ruch (1956) that some balance of effects had to be achieved, and the excitability or threshold of the micturition reflex was used as a measure of how this balance had been changed by a lesion. Kuru (1965) used a similar concept, which is analogous to those current in the 1940s and 1950s and used to explain phenomena such as the role of the reticular formation in the regulation of skeletal muscle tone. This concept is still applicable in the 1980s, but with some improvement in the degree of definition of the aggregations of neurons involved, their possible neurotransmitters, and their connections. In the

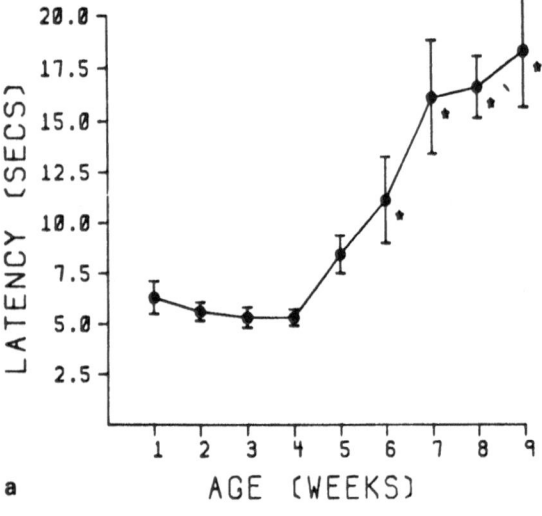

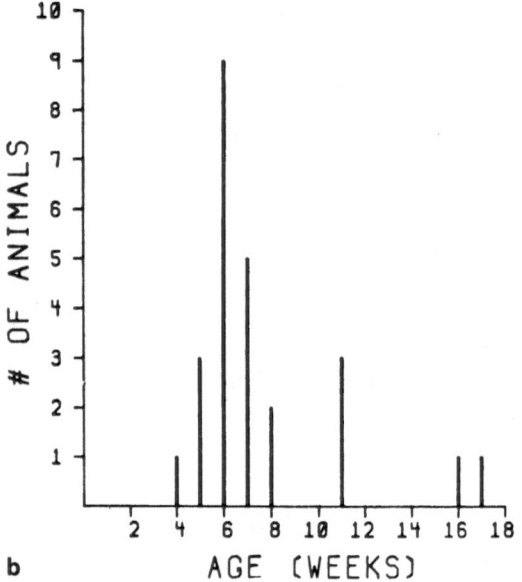

Fig. 8.17. a Latency for micturition in response to tactile stimulation of the perineal region of kittens. *Ordinate*: latency in seconds. *Abscissa*: age in weeks. Each point is the mean (± SEM) of at least 16 tests on 4 or more animals. Statistically significant difference (*P* < 0.05) between mean latency at age 2 weeks and mean latencies in older animals is indicated by *asterisk*. **b** Age at which perineal-induced micturition disappears in kittens. *Ordinate*: number of animals. *Abscissa*: age in weeks. Graph represents data on 25 kittens from 10 litters. (de Groat et al. 1975)

case of the control of tone, much more precise information is also available on the ways in which these pathways function in anaesthetized and in awake animals; the problems of working with the smaller cells and smaller total numbers of neurons in the autonomic nervous system may result in a solution to these problems still being relatively distant.

Micturition Threshold and Factors that Influence It

The evidence that the micturition reflex is subject to control is seen in its simplest form in the existence of a micturition threshold: the micturition reflex is switched off during filling, when the pressure is low, and is switched on once a threshold pressure or volume has been reached. This threshold shows that there is feedback of information regarding bladder volume or pressure which can influence whether or not stimuli (such as distension) elicit detrusor contraction. The cystometrogram can give some indication of the pressure–volume relationship when this switching occurs; more sophisticated techniques are required to show that the reflex activity is graded, depending on intravesical pressure between the threshold and a maximal level. The reflex is usually absent below 5–10 mmHg and is graded between the threshold and about 30 mmHg, at which pressure it becomes maximal; the afferent input from the bladder via the pelvic nerves is one of the important factors that determines this behaviour (Fig. 8.18).

The facilitation of parasympathetic reflex activity between 10 and 30 mmHg is accompanied by concomitant changes in the excitability of sympathetic and somatic reflexes. Evans and McPherson (1959) noted that flexor and crossed extensor reflexes and tone were enhanced below 15 mmHg and depressed maximally above 30 mmHg, and attributed this change to supraspinal inhibition. de Groat and Lalley (1972) referred to a similar range of pressures over which supraspinal inhibition of sympathetic reflexes increases to become maximal. Sato et al. (1980) were able to demonstrate reflex bladder contractions at low intravesical pressures when nociceptive afferents in the limb were stimulated; this effect was reversed, producing inhibition of bladder contractions at pressures of 30–40 cmH$_2$O. Thus, the micturition threshold, i.e. the pressure at which reflex bladder contractions are elicited, marks not only a set of conditions in which the pelvic parasympathetic reflexes elicited from the bladder are facilitated, but also one in which there occurs a powerful supraspinal inhibition of sympathetic, somatic and sphincteric reflexes elicited from the bladder. In addition, supraspinal inhibition of parasympathetic reflexes induced from somatic nociceptive afferents and of nociceptive transmission through the dorsal horn can be demonstrated as concomitant events (Sato et al. 1980; Cadden and Morrison 1984, 1987). This threshold also operates on excitatory descending pathways,

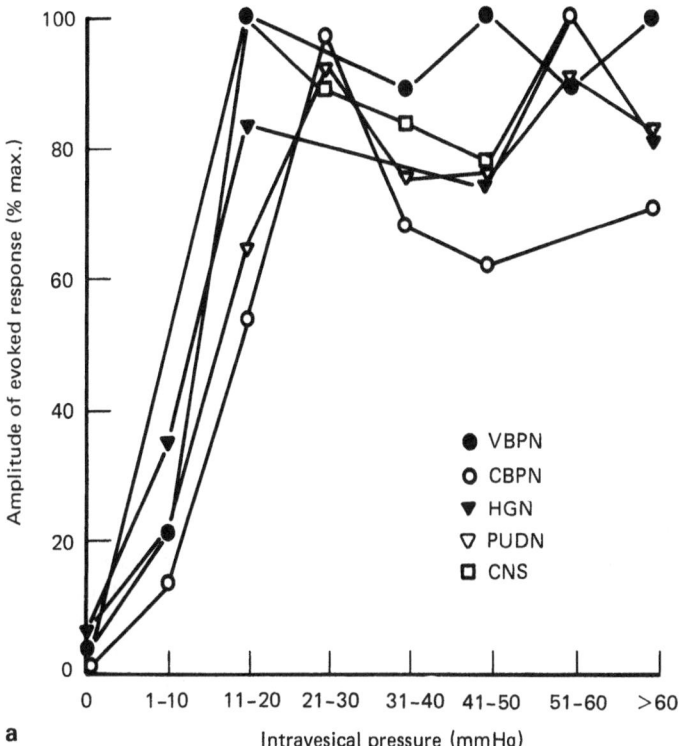

a

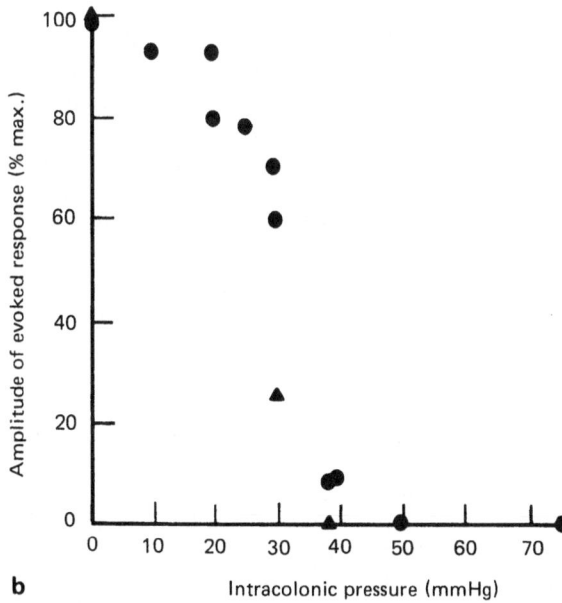

b

Fig. 8.18. a The relationship between the amplitude of 16 averaged evoked responses in the vesical parasympathetic efferents following electrical stimulation of the vesical (*VBPN*) or colonic (*CPBN*) branches of the pelvic nerve, the hypogastric nerve (*HGN*), the pudendal nerve (*PUDN*) or below a hemitransection of the spinal cord at the second cervical segment (*CNS*), at different intravesical pressures and low intracolonic pressure. The responses in all experiments have been pooled into bins of intravesical pressure as indicated on the *abscissa*. **b** The influence of intracolonic pressure on the amplitude of 16 averaged evoked responses in the vesical parasympathetic efferents at constant high bladder pressure. *Triangles* indicate responses to stimulation of the colonic branches, and *circles* responses to stimulation of the vesical branches of the pelvic nerves. (McMahon and Morrison 1982c)

such as that from the cingulate gyrus (Kremer 1947), but also on inhibitory descending pathways from the raphe nucleus (Spillane 1982). This suggests that a different set of conditions arises within the central nervous system when bladder pressure is raised, and that descending facilitation of the parasympathetic efferents and descending inhibition of sympathetic and somatic reflexes mediate

some of these changes. The conditions that elicit these effects (intravesical pressures > 30 mmHg) are similar to those that give rise to excitation of the raphe–spinal pathway (Lumb 1986; see also Chap. 4, p. 125); it is interesting that this pathway may mediate antinociceptive phenomena and inhibit sympathetic and somatic reflex activities (Lumb 1986; Gilbey et al. 1981; Wolstencroft and

Table 8.1. Effects of low (< 10 mmHg) and high (> 30 mmHg) intravesical pressure on different reflex responses

Stimulus	Reflex response	Low pressure	High pressure
Bladder distension	Bladder contractions	Absent	Present
Colonic distension	Inhibition of bladder contractions	Absent	Present
Electrical stimulation of:			
Pelvic nerves	PN evoked response	Absent/small	Large
Pelvic nerves	HGN evoked response	Large	Small
Pelvic nerves	PUDN evoked response	Present	Absent
Nociceptors	Bladder contraction	Present	Reversed
Muscle afferents	Muscle contraction	Present	Inhibited

HGN, hypogastric nerve; PN, pelvic nerve; PUDN, pudendal nerve.

West 1982). Table 8.1 summarizes the changes in neural responses seen around the micturition threshold.

A micturition threshold is seen in human paraplegics and in animals following spinal transection, an observation which suggests that one important site of interaction between the afferent input from the bladder and the efferent pathway is in the spinal cord. Admittedly, the threshold is usually higher than normal, resulting in an increased residual volume, but it is nevertheless present; its elevation is probably due to the loss of central descending influences on micturition threshold (see p. 248). Following spinal transection, however, micturition threshold can be influenced by afferent inputs from a number of somatic and visceral structures; colonic, genital and cutaneous afferent inputs to the spinal cord generally raise the threshold, and this control appears to be exerted at the final stage of the efferent pathway to the bladder within the central nervous system.

In addition to the peripheral inputs that can reset the absolute pressure at which bladder contractions occur, there are also central factors which can modify the micturition threshold. Many investigators have observed that micturition threshold is influenced by lesions of the central nervous system. The most comprehensive and systematic studies were those of Tang (1955) and Tang and Ruch (1956), who found that transhypothalamic decerebration removed an inhibitory influence, and that a further section just above the superior colliculi removed an excitatory influence attributable to the posterior hypothalamus (see Fig. 8.12). The midbrain was considered the site of a tonic inhibitory influence, and the pons facilitated the micturition reflex. More discrete lesioning showed the existence of excitatory influences from the mamillary bodies and from the dorsal tegmentum of the rostral pons,

and inhibitory influences from the midbrain tegmentum, lateral to the central grey matter at the level of the superior colliculus. Lesions in different areas of the amygdala either raise or lower the micturition threshold (Fig. 8.19; Edvardsen and Ursin 1968), and this is in keeping with the results of electrical stimulation within this complicated part of the cortex (Gjone 1966). The influence of these descending pathways may be regarded as a balance of tonic activities (as discussed already).

In humans, the intravesical volume that induces detrusor contractions is often measured during routine cystometry. Bladder contractions at low volumes are abnormal, and when this occurs in humans, the bladder is called unstable or hypersensitive, terms that may be slightly unsatisfactory because there is no definition of the cause of the abnormal responses; many factors can contribute to the hyperexcitable reflexes. Lesions of the frontal cortex lead to a hyperexcitable bladder in humans (Andrew and Nathan 1964); Fig. 8.20 shows a cystometrogram from one of their patients with a cerebral lesion, and demonstrates the presence of spontaneous bladder contractions at low bladder volumes. In cats, similar behaviour can be induced by various central nervous lesions, such as decerebration, possibly because the influence of inhibitory descending pathways has been removed. In rats, similar behaviour can be elicited by administration of chemicals that release neuropeptides from nerve endings (Maggi et al. 1986). It is probably one of the gross oversimplifications of Bradley's loop hypothesis that such behaviour is attributed to lesions of pathways between the cortex and the brain stem. The measurement of micturition threshold indicates one aspect of the functional state of the system, irrespective of whether the changes are brought about by a lesion in the cortex (see loop I in the Bradley model) or due to local

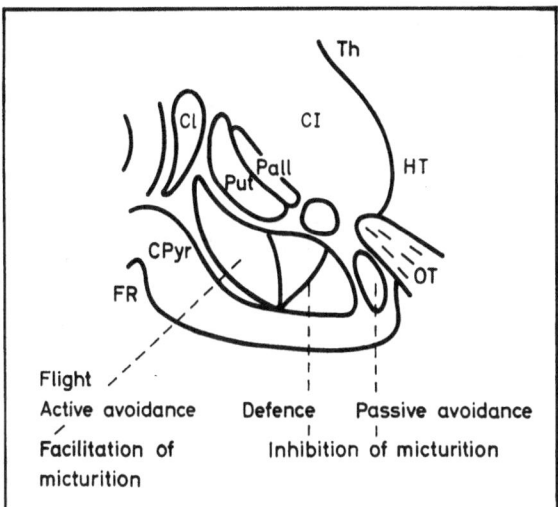

a

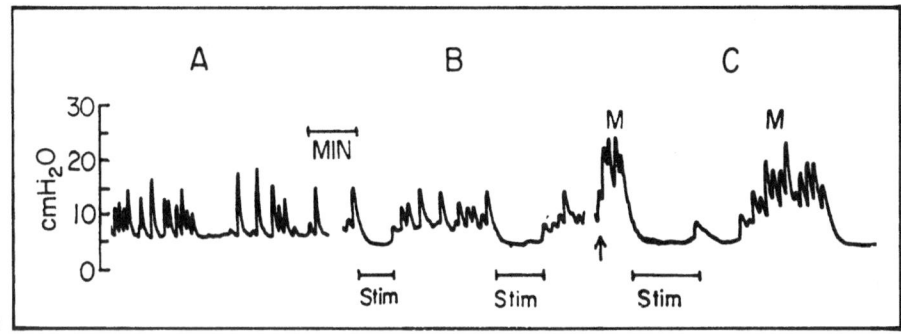

b

Fig. 8.19. **a** Review of results from amygdala stimulation and ablation experiments. *Cl*, claustrum; *CI*, internal capsule; *CPyr*, pyriform cortex; *FR* rinal fissure; *HT*, hypothalamus; *OT*, optic tract; *Pall*, pallidum; *Put*, putamen; *Th*, thalamus. **b** *A*, spontaneous changes in frequency and amplitude of the rhythmic pressure fluctuations at intravesical volume of 15 ml. *B*, stimulation of point within the amygdala. Inhibitory effects on the rhythmic activity and basic pressure level resembling the spontaneous changes in A. *C*, the inhibitory effects indicated in B are verified by interruption of micturition, following stimulation of the same point. Micturition initiated prior to stimulation by rapid filling of the bladder with 30 ml at *arrow* (**a** from Edvardsen and Ursin 1968; **b** from Gjone 1966)

spinal afferent inputs (McMahon and Morrison 1982c). The changes that may give rise to raising or lowering of the micturition threshold are summarized in Table 8.2.

Possible Mechanisms Underlying the Micturition Threshold

Another useful analogy in our thinking about how functional control over the threshold of reflex events may be exerted comes from transistor technology. Computers make use of logical AND gates; these devices allow the passage of a signal from source A if source A is active at the same time as source B. A signal from A is powerless unless it occurs at the same time as source B. Similarly, a signal from B is ineffective by itself. In essence, this sort of device can be used as a switch to enable or disable a pathway. In terms of the micturition reflex, de Groat (1975) described switching mechanisms controlling parasympathetic and sympathetic reflex activity which were dependent on the bladder pressure to operate the switches. Supraspinal inhibition of the spinal sympathetic reflex was switched on, and the activity of the spinobulbospinal micturition

reflex was enhanced. Depression of spinal reflexes and concomitant enhancement of supraspinal reflexes is not uncommon in the sympathetic branch of the autonomic nervous system, and another example, but not one initiated by raising bladder pressure, is quoted by Dembowsky et al. (1981). McMahon and Morrison (1982c) and McMahon and Spillane (1982) also made use of the idea of the logical gate in the micturition reflex (Fig. 8.21). They found that descending pathways from the brain stem could evoke activity in the parasympathetic efferents, but only if the bladder pressure was above a certain level. It was proposed that elevation of bladder pressure operated a logical gate that allowed descending impulses to excite the parasympathetic efferents, and that the pathways that controlled micturition threshold were different from those that mediated the micturition reflex. The gate behaves as if it were closed at bladder pressures below 5 mmHg and maximally open around 25–30 mmHg; between these values the transfer of activity is graded. This gate is known to modulate the transfer of activity from the descending pathways in the brain stem which cause bladder contractions, from the inhibitory raphe–spinal pathways, and in reflex pathways activated by elec-

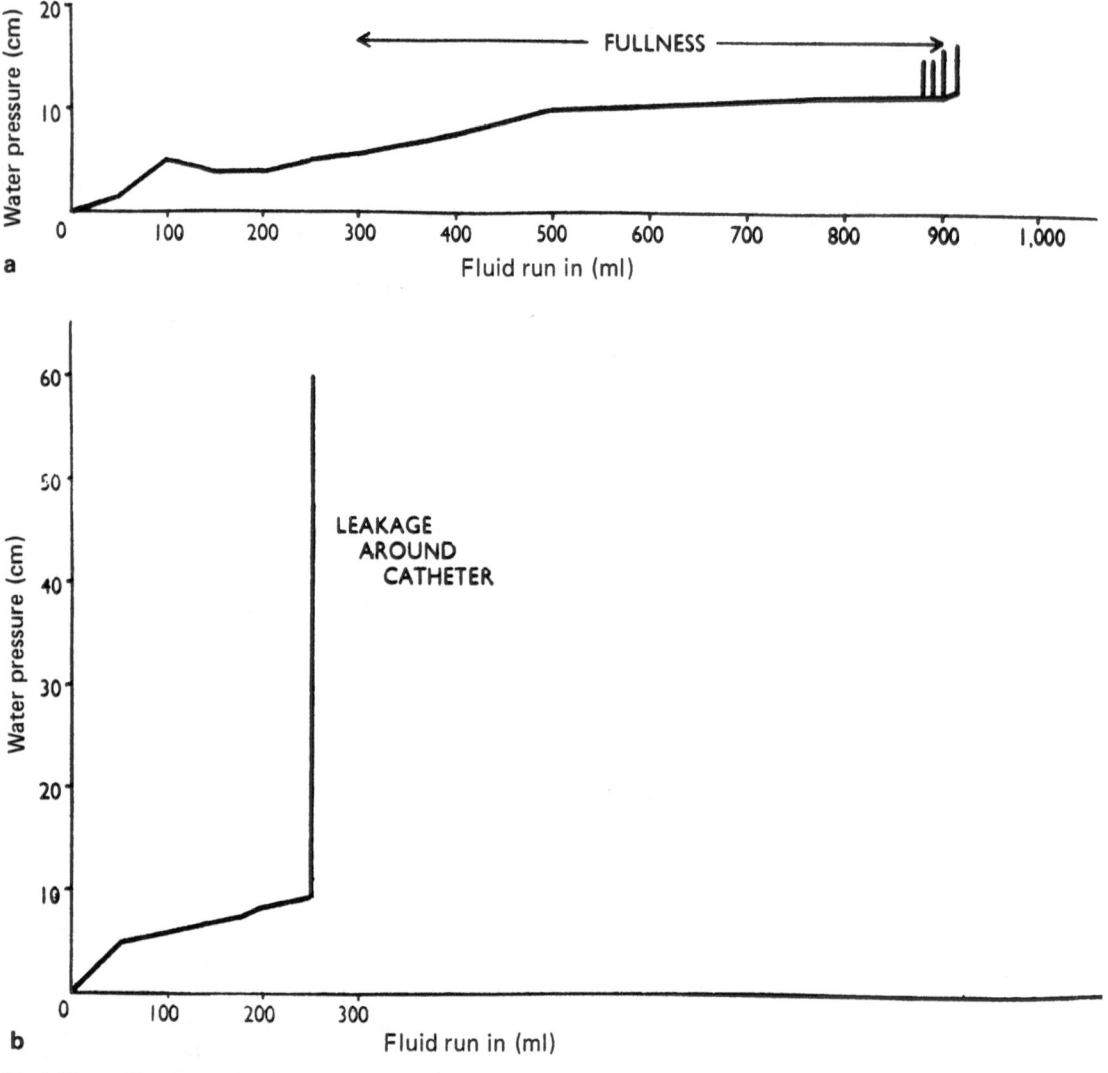

Fig. 8.20a,b. The effects of a cerebral lesion on bladder function. Cystometrograms performed before (**a**) and 5 days after (**b**) a modified Lyerly–Poppen leucotomy. (Andrew and Nathan 1964)

Table 8.2. Factors that influence the micturition threshold

	Effect on micturition threshold
Peripheral:	
PN afferents from bladder	Decrease
PN afferents from colon	Increase
HGN afferents	No effect
PUDN afferents from perineum	Increase
Central:	
Raphe–spinal pathway	Increase
Reticulospinal pathway	Increase
Pons	Decrease
Midbrain	Increase
Posterior hypothalamus	Decrease
Frontal cortex	Increase (predominantly)
Cingulate cortex	Increase (predominantly)

HGN, hypogastric nerve; PN, pelvic nerve; PUDN, pudenal nerve.

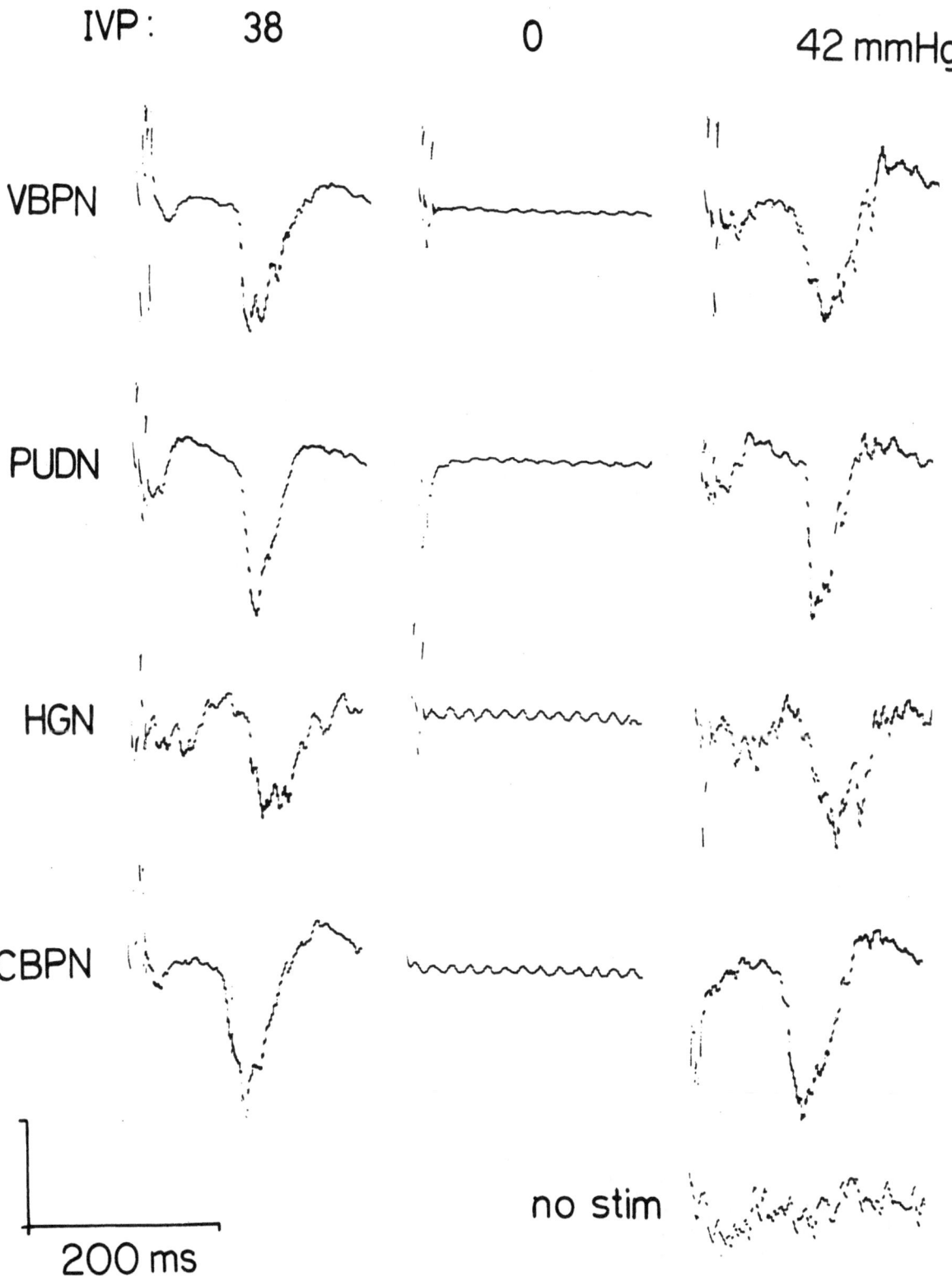

Fig. 8.21. The average of 16 responses in vesical parasympathetic efferents following electrical stimulation (at start of sweep) of vesical branches of the pelvic nerve (*VBPN, top row*), pudendal nerve (*PUDN, second row*), hypogastric nerve (*HGN, third row*) and colonic branches of the pelvic nerve (*CBPN, fourth row*). An averaged control response using no electrical stimulation is shown on the *bottom right*. Responses were recorded at 42 mmHg (*third column*). Vertical calibration: 60 μV for responses in *first row*, and 30 μV for all others. All responses obtained from one animal. (McMahon and Morrison 1982c)

trical stimulation of pelvic, hypogastic, pudendal and other somatic nerves; in each case, an increase in bladder pressure increases the gain of the system.

Gain in a Reflex Pathway

The concept of the gain of a reflex pathway is also one that is useful: in control systems applications the gain is the ratio of the output to the input, i.e. an amplification factor that can be measured, even though the structure of the system is complex, or largely unknown (Milhorn 1966). If some stimulus to the reflex is present, the reflex may either be absent (in which case the gain is zero), or weak or powerful (when the gain is low or high, respectively). Physiologically, such modulation of the gain can be achieved at synapses, by modifying the output of transmitter from presynaptic endings (the process of presynaptic inhibition), or by having pathways that either depolarize or hyperpolarize the soma of individual postsynaptic neurons. In addition, recruitment of additional motoneurons contributes to the overall gain. The effects of an increase in gain are akin to an increase in amplification in a system. In railway terms, the analogy might be the ability to alter the power generated by the locomotive, and therefore the size of its load.

The gain of a system is important, particularly if there is an element of positive feedback in the response of the system. Positive feedback appears to be present in the control of micturition: the ratio of output to input of different portions of the reflex pathway increases in circumstances that occur during micturition. For instance, the pelvic parasympathetic discharge is dependent on bladder pressure above the micturition threshold, because of the firing of pelvic nerve afferents; in addition, transmission through the pelvic ganglia increases as the rate of efferent firing increases (see Fig. 8.23); furthermore the reflex response to electrical stimulation of pelvic nerve afferents increases as the rate of stimulation increases. All of these occurrences tend to switch the bladder "on" i.e. to maximal activity for the length–tension relationship of bladder muscle at that volume. The effects of positive feedback depend on the gain of the system; if it is less than unity, then the response achieves a new, but stable level, because successive increments in input produce less output (Fig. 8.22). If the gain is greater than unity, then the output increases progressively until the system reaches its extreme limits and becomes "out of control" (Milhorn 1966).

Gain of the Micturition Reflex

In the late 1960s electrical methods became available in animals to study the activity of the micturition reflex. A given electrical stimulus to afferent

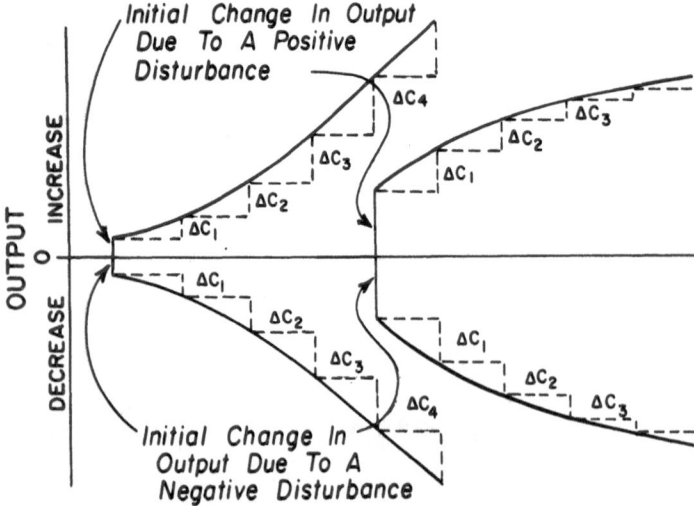

Fig. 8.22. The effects of gain on positive feedback. This figure shows the contribution of a positive feedback control system following an initial disturbance (positive or negative) to the system; in this case the control system adds to the effects of the disturbance. The effect depends on the gain of the feedback loop: on the *left* the gain of the system is greater than unity, and the output would theoretically continue to increase towards infinity; on the *right* the gain of the system is less than unity and the output tends towards a plateau level. The contribution of the control system ($\Delta C1$, $\Delta C2$, etc.) increases progressively (*left*) when the gain is greater than unity, and diminishes successively (*right*) if the gain is less than unity. (Milhorn 1966)

fibres in a pelvic nerve gave rise to a burst of activity in parasympathetic efferents to the bladder, and the response could be averaged and quantitated; changes in the size of the response to a given, constant, stimulus were due to changes in the excitability, or gain, of the reflex. Thus, the main factor which increased the gain of the reflex was the pressure (or volume) in the bladder, i.e. the response of parasympathetic efferents to a given electrical stimulus could be switched on, and made larger, by gradually increasing the pressure in the bladder (de Groat and Ryall 1969). McMahon and Morrison (1982c) servocontrolled the bladder pressure at a series of fixed levels and found that the reflex excitability or gain was increased in a graded manner between bladder pressures of about 5 mmHg and about 30 mmHg. A similar phenomenon has been found to affect parasympathetic activity evoked from the brain stem and the cervical cord (Spillane 1982). There are a number of possibilities regarding the origin of this pressure-dependent increase in the gain of the reflex. McMahon and Morrison (1982c) thought it was a property of the final stages of signal processing in the reflex, i.e. it was dependent on connections between the descending pathway from the brain stem and the parasympathetic efferents, or on interneurons that interact with them; it was suggested that the afferents from the bladder might modulate transmission at this site.

Another possibility is that the change in the gain of the reflex is a result of pre-existing activity in the efferent pathway; we have no direct evidence on this point in respect of the micturition reflex, but this mechanism has been implicated by Matthews (1986) in the automatic gain compensation of the human stretch reflex. Matthews considered the reflex EMG response that could be induced in a muscle by applying pulsed vibrations to its tendon; the size of the response was increased in proportion to the resting activity in the muscle, and he called this phenomenon automatic compensation of reflex gain. Within any individual motoneuron, the frequency of impulses will depend on the initial level of depolarization, and any given synaptic input would be expected to produce a greater change in firing frequency as the pre-existing discharge rate increases. He also believes that this property is dependent on the ways in which recruitment is induced by increasing the level of excitation. The properties of the motoneuron pool, such as the distribution of thresholds, sizes and synaptic drive, also have to be considered. Similar alterations in the gain of the lung inflation reflex induced by changing the resting activity of respiratory muscles have been described by Newsom Davis and Sears (1970). One point that needs to be made as a result of this comparison is

that very little is known about the differences between parasympathetic efferent motoneurons. It is known that they divide in their peripheral course (see Chap. 3, p. 69), but properties akin to those considered in the size hypothesis for somatic motoneurons (Henneman and Olson 1965), and the distribution of thresholds have not been adequately studied. Matthews (1986) demonstrated the existence of a similar gain control in the stretch reflex during reciprocal inhibition; McMahon and Morrison (1982c) showed that reciprocal inhibition from the colon on to the micturition reflex also affected the gain of the micturition reflex, and colonic pressure between about 15 and 45 mmHg was the effective range of bowel pressures (see Fig. 8.18). In neither instance was it considered necessary to postulate a supraspinal mechanism to explain these events.

Filters

The function of filters in a pathway carrying information can be to remove certain dynamic components from the signal; thus, a high pass filter allows high frequency (rapidly changing) signals to pass onward, and filters out information that is concerned with slowly rising or falling signals. In Chapter 4, spinothalamic tract neurons that responded to sudden changes in bladder pressure were described (see p. 116): the information about the resting, steady state pressure in the bladder had been removed between the primary afferents and the spinothalamic tract cells, and this illustrates the function of a high pass filter that selects information concerned with rapidly changing events. Other neuronal systems allow slow changes in discharge to be transmitted, while eliminating any rapidly changing signals. One site where this occurs was mentioned in Chapter 5 (see p. 135), in the pelvic ganglia, where de Groat and Saum (1972) had shown that repetitive electrical stimulation of preganglionic efferents resulted in a gradual enhancement of transmission of information to the postganglionic fibres (see Fig. 5.4, p. 138). These two types of filters have functions analogous to the treble and bass tone controls on a stereo system. de Groat (1975) argued that the filtering characteristics of the efferent parasympathetic pathway to the bladder blocked output to the bladder when the level of efferent activity was low, and facilitated transmission when the output was high. The central pathways from the bladder also show filtering characteristics: the micturition reflex pathway shows facilitation at rates of stimulation above $5 \, s^{-1}$, whereas the sympathetic reflex pathway fails to transmit at these rates. This introduces a degree of damping into the parasympathetic system, and filters this sort of information from the sympathetic output (de Groat and Ryall 1969; de Groat 1975; Morrison and Spillane 1981; Spillane 1982).

Role of Feedback in Voluntary Motor Control of the Lower Urinary Tract

Afferent input from the bladder can modify the efferent discharge in the parasympathetic and sympathetic pathways to the viscus. This has been looked at mainly as a means of reflex modulation (de Groat 1975; McMahon and Morrison 1982c), but might also be regarded as a feedback mechanism. Activation of descending pathways can produce different degrees of excitation of parasympathetic efferents, depending on bladder pressure; the afferent input that monitors the effects of this "command" signal will vary as a consequence, and the level of output is likely to depend on the gain of the reflex, if the feedback signal contributes to positive feedback. The afferent input that results from a command might therefore be expected to vary with the volume of urine in the bladder. The modulation of the sensory experience by the command signal (or the "corollary discharge") was discussed in Chapter 4 (see p. 110); however, the same afferent information could also be used by parallel pathways to modify the efferent control, such that when a descending excitatory pathway to the bladder is active, the feedback effect is greater at higher intravesical pressures or volumes. The importance of such a mechanism (i.e. the effect on the force of contraction) would depend on the overall gain of the system, and that may be determined by a number of different descending pathways, whose individual contributions could alter during micturition. The potential within this system to modulate the power of the efferent signal to the bladder (or urethra) by feedback could be of great consequence in the consideration of mechanisms underlying the occurrence of spontaneous contractions (e.g. in the unstable bladder or enuresis), frequency (e.g. in the urge syndrome) or hesitancy. Control systems analysis could provide an approach that allows analysis of the characteristics of the system without having to have details of every central mechanism that might affect the effector organs.

Andersson and Grillner (1981) describe how feedback modulation adjusts the complex responses of the somatic motor system controlling limb muscles. In their experimental studies on locomotion, signals from the periphery add to or subtract from the general excitability of motoneurons during the stepping cycle, and the step cycle duration and some of its components are under special feedback control. They suggest that the step cycle originates from some central pattern generator; the central pattern generator may be regarded as a source of rhythmic movement whose precise mechanisms may or may not be known, and may operate independently of feedback from the effector organs. In this instance (spinal kittens), the central pattern generator in the cord is susceptible to movements of the hind limbs when the animal is on a treadmill, and the peripheral input can modulate the sequence of muscular movements. For instance, the peripheral input can speed up or slow down the speed of the movements, and can change the pattern of movement from alternation to more or less simultaneous movements, as in a gallop. In the bladder situation, there are a number of examples of how feedback can modulate the spontaneous rhythmic drive that is present in anaesthetized cats. For instance, McPherson (1966) described a relation between the size of the bladder contractions and their frequency; and de Groat and Ryall (1969) and McMahon and Morrison (1982c) described how intravesical pressure can alter size and frequency of contractions and the excitability of the micturition reflex. In McPherson's study, there was a range of pressures over which the frequency of contractions was related to resting pressure; in addition there was an inverse relationship between the peak pressure in a series of bladder contractions and their frequency. A similar process could be seen in the studies of Floyd et al. (1982); if one of a series of spontaneous bladder contractions was prevented from occurring by stimulation of afferents from the colon, the subsequent contraction (after the cessation of stimulation) achieved higher pressure than had occurred previously. The mechanism of this relationship is not known; one possibility is that high pressures bring about a prolonged inhibitory state that delays any subsequent contraction. Prolonged inhibitory effects of raising bladder pressure on the transmission of sensory information have been described (Cadden and Morrison 1984), but little is known about possible parallel effects on the excitability of bladder reflexes.

Whether or not a central pattern generator that might determine the frequency of bladder contractions is present in normal, incontinent or spinal animals is not known; equally, the possibility of spontaneous bladder activity being the result of oscillation in a feedback loop has not been seriously considered. At the moment we do not know much about the role of feedback from the bladder in the process of voluntary micturition; nevertheless, the presence of feedback and the effects of gating on the parasympathetic efferent pathway apply to all reflex controls over this system that have so far been studied. It would be surprising if these feedback controls did not also contribute to descending influences that are active in the voluntary process of micturition.

We have gone full circle in our discussion of the central control of micturition; this chapter has been concerned with the function of individual anatomical paths in the central nervous system, and the function of the whole system has been regarded as a synthesis of many components, central and peripheral. Some of these descending parallel pathways may participate in reflex or control loops with different transfer functions, i.e. different thresholds, gains or filtering capacities, and act on one or more branches of the autonomic innervation of the bladder or the innervation of the urethral sphincters. In the periphery, the two efferent systems, parasympathetic and sympathetic, are themselves able to interact in pelvic ganglia, and the possibility exists that different components of each pathway may have different effects (see Chaps. 5 and 7). The extent to which coordination of urethral activities occur reflexly or by coactivation of parallel mechanisms is not fully understood, but it seems likely that both occur. Finally, the control of the bladder cannot be considered just in terms of command signals sent from on high down parallel pathways with facilitatory or inhibitory functions. Instead, afferent feedback probably plays an important role in determining how the efferent pathways respond to commands; the gain of the reflex loop is one important factor which influences the power of the bladder contraction, particularly in loops, such as those involving the dorsolateral pons, where positive feedback is involved. The occurrence of oscillation in a negative feedback control loop is similarly affected by the gain of that loop and other factors such as dead time, but whether these can explain the occurrence of spontaneous bladder contractions in anaesthetized animals, or bladder instability in humans is unknown.

Summary

A variety of hypotheses are currently in the literature that attempt to explain the neural control of the bladder. Some are based largely on anatomy, while others involve functional concepts, which seek to distinguish between a reflex pathway and pathways that mediate some type of functional control over the threshold, gain or filtering capacity of the reflex. The threshold of the micturition reflex is a functional concept, the intravesical pressure above which reflex bladder contractions are elicited. This threshold is normally 5–10 mmHg, but is not an absolute value, as it can be reset by peripheral and central events; the pelvic nerve is essential for the setting of micturition threshold, and activity in vesical afferents lowers the threshold, but activity in colonic, genital, or perineal afferents raises it. Descending pathways from the brain also influence the micturition threshold, and the tonic activities of pathways from the forebrain, particularly the frontal and cingulate cortex, the hypothalamus, midbrain and medial parts of the pons and medulla raise the threshold, while the dorsolateral pons and the mamillary bodies have the opposite effects. The balance of central influences and the peripheral inputs determines this threshold.

The gain, amplification or power of the micturition reflex is influenced by many of the factors that set the threshold.

The gain of the reflex increases above the threshold and becomes maximal around 30 mmHg; facilitation of parasympathetic reflexes occurs in this range of intravesical pressures, and these changes appear to be due to the afferent input from the bladder along the pelvic nerve. Concomitantly, inhibition of sympathetic and somatic reflexes as well as of somatosensory transmission through the dorsal horn can be demonstrated. These inhibitory phenomena are of supraspinal origin. Concomitant inhibition of sphincteric activity is partly of spinal and partly of supraspinal origin.

The filtering characteristics of the central pathways of the micturition reflex and of the pelvic ganglia normally remove rapid transients from the parasympathetic efferent signals to the bladder, and ensure that the bladder contracts maximally only during maintained reflex or descending excitation of the efferent pathway.

References

Andersson O, Grillner S (1981) On the feedback control of the cat's hindlimb during locomotion. In: Taylor A, Prochazka A (eds) Muscle receptors and movement. Macmillan, London, pp 427–431

Andrew J, Nathan PW (1964) Lesions of the anterior frontal lobes and disturbances of micturition and defaecation. Brain 87:233–262

Appenzeller O (1982) The autonomic nervous system: an introduction to basic and clinical concepts, 3rd edn. Elsevier, New York, p 524

Bagshaw EV, Evans MH (1976) Measurement of current spread

from microelectrodes when stimulating within the central nervous system. Exp Brain Res 25:391–400

Barrington FJF (1921) The relation of the hind brain to micturition. Brain 44:23–53

Barris RW, Lehuman HR (1953) Bilateral anterior cingulate gyrus lesions. Syndrome of the anterior cingulate gyri. Neurology 3:44–52

Bors E, Comarr AE (1971) Neurological urology. Karger, New York

Bradley WE (1984) Neurophysiology of the urinary bladder. In: Bradley, WE, Hald T (eds) The urinary bladder neurology and dynamics. Williams and Wilkins, Baltimore, pp 22–36

Bradley WE, Scott FB (1978) Physiology of the urinary bladder. In: Harrison JH, Gittes RF, Perlwitter AD, Stamey TA, Walsh PC (eds) Campbell's urology, 4th edn. Saunders, Philadelphia, pp 87–124

Bradley WE, Teague CT (1968) Spinal cord organisation of micturition reflex afferents. Exp Neurol 22:504–516

Bradley WE, Teague CT (1969) Cerebellar control of the urinary bladder. Exp Neurol 23:399–411

Brodal A (1981) Neurological anatomy in relation to clinical medicine, 3rd edn. Oxford University Press, Oxford

Brutkowski S (1965) Functions of the prefrontal cortex in animals. Physiol Rev 45:721–746

Cadden SW, Morrison JFB (1984) The effects of visceral distension on the activities of lumbar dorsal horn neurones in the rat. J Physiol 350:71P

Cadden SW, Morrison JFB (1987) Effects of visceral distension on the activities of neurones receiving cutaneous inputs in the rat lumbar dorsal horn: comparison with effects of remote noxious somatic stimuli. (Submitted to Brain Res)

Carstens E (1982) Inhibition of spinal dorsal horn neuronal responses to noxious skin heating by medial hypothalamic stimulation in the cat. J Neurophysiol 48:808–822

Chambers WW Jr (1947) Electrical stimulation of the interior of the cerebellum in the cat. Am J Anat 80:55–93

de Groat WC (1971) Inhibition and excitation of sacral parasympathetic neurons by visceral and cutaneous stimuli in the cat. Brain Res 33:499–503

de Groat WC (1975) Nervous control of the urinary bladder of the cat. Brain Res 87:201–211

de Groat WC (1976) Mechanisms underlying the recurrent inhibition in the sacral parasympathetic outflow to the urinary bladder. J Physiol 257:503–514

de Groat WC (1986) Spinal cord projections and neuropeptides in visceral afferent neurons. In: Cervero F, Morrison JFB (eds) Visceral sensation. Elsevier, Amsterdam, pp 165–187 (Progress in brain research, vol 67)

de Groat WC, Lalley PM (1972) Reflex firing in the lumbar sympathetic outflow to activation of vesical afferent fibres. J Physiol 226:289–309

de Groat WC, Ryall RW (1968) Recurrent inhibition in sacral parasympathetic pathways to the bladder. J Physiol 196:579–591

de Groat WC, Ryall RW (1969) Reflexes to the sacral parasympathetic neurones concerned with micturition in the cat. J Physiol 200:87–108

de Groat WC, Saum WR (1972) Sympathetic inhibition of the urinary bladder and of pelvic ganglionic transmission in the cat. J Physiol 220:297–314

de Groat WC, Saum WR (1976) Synaptic transmission in parasympathetic ganglia in the urinary bladder in the cat. J Physiol 256:137–158

de Groat WC, Theobald RJ (1976) Reflex activation of sympathetic pathways to vesical smooth muscle and parasympathetic ganglia by electrical stimulation of vesical afferents. J Physiol 259:223–238

de Groat WC, Douglas JW, Glass J, Simonds W, Weimer B,

Werner P (1975) Changes in somato-vesical reflexes during postnatal development in the kitten. Brain Res 94:150–154

de Groat WC, Nadelhaft I, Milne RJ, Booth AM, Morgan C, Thor K (1981) Organisation of the sacral parasympathetic reflex pathways to the urinary bladder and large intestine. J Auton Nerv Syst 3:135–160

de Groat WC, Booth AM, Milne RJ, Roppolo JR (1982) Parasympathetic preganglionic neurons in the sacral spinal cord. J Auton Nerv Syst 5:23–43

de Groat WC, Kawatani M, Hisamitsu T, Lowe I, Morgan C, Roppolo J, Booth AM, Nadelhaft I, Kuo D, Thor K (1983) The role of neuropeptides in the sacral autonomic reflex pathways of the cat. J Auton Nerv Syst 7:339–350

Dembowsky K, Lackner K, Czachurski J, Seller H (1981) Tonic catecholaminergic inhibition of the spinal somato-sympathetic reflexes originating in the ventrolateral medulla oblongata. J Auton Nerv Syst 3:277–290

Edvardsen P, Ursin T (1968) Micturition thresholds in cats with amygdala lesions. Exp Neurol 21:495–501

Elam W, Thoren P, Svensson TH (1986) Locus coeruleus neurones and sympathetic nerves: activation by visceral afferents. Brain Res 375: 117–125

Enoch DM, Kerr FWL (1967a) Hypothalamic vasopressor and vesicopressor pathways. I. Functional studies. Arch Neurol (Chicago) 16:290–306

Enoch DM, Kerr FWL (1967b) Hypothalamic vasopressor and vesicopressor pathways. II. Anatomic study of their course and connections. Arch Neurol (Chicago) 16:307–320

Evans MH, McPherson A (1959) The effects of distension of the bladder on somatic reflexes in the cat. J Physiol 146:438–458

Fields HL, Basbaum AI (1978) Brainstem control of spinal pain-transmission neurones. Annu Rev Physiol 40:217–248

Floyd K, McMahon SB, Morrison JFB (1982) Inhibitory interactions between colonic and vesical afferents in the micturition reflex of the cat. J Physiol 322:45–52

Giesler GJ, Liebeskind JC (1976) Inhibition of visceral pain by electrical stimulation of the periaqueductal gray matter. Pain 2:43–48

Gilbey MP, Coote JH, Macleod VH, Peterson DF (1981) Inhibition of sympathetic activity by stimulating in the raphe nuclei and the role of 5-hydroxytryptamine in this effect. Brain Res 226:131–142

Gjone R (1966) Excitatory and inhibitory bladder responses to stimulation of 'limbic', diencephalic and mesencephalic structures in the cat. Acta Physiol Scand 66:91–102

Gjone R, Setekleiv J (1963) Excitatory and inhibitory responses to stimulation of the cerebral cortex in the cat. Acta Physiol Scand 59: 337–349

Hammond DL, Proudfit HK (1980) Effects of locus coeruleus lesions on morphine-induced antinociception. Brain Res 188:79–91

Henneman E, Olson CB (1965) Relations between structure and function in design of skeletal muscles. J Neurophysiol 28:581–598

Hess WR (1947) Vegetative Funktionen und Zwischenhirn. Helv Physiol Pharmacol Acta Suppl IV:1–65

Hess WR, Brügger M (1943a) Das subkortikale Zentrum der affektiven Abwehrreaktion. Helv Physiol Pharmacol Acta 1:511–532

Hess WR, Brügger M (1943b) Der Mitkions- und der Defäktionsakt als Erfolg zentraler Reisung. Helv Physiol Pharmacol Acta 1:533–547

Jänig W (1983) The autonomic nervous system. In: Schmidt RF, Thews G (eds) Human physiology. Springer, Berlin Heidelberg New York, pp 111–144

Kabat H, Magoun HW, Ranson SW (1936) Reaction of the bladder to stimulation of points in the forebrain and midbrain. J Comp Neurol 63:211–239

Kandel ER, Schwartz JH (1981) Principles of neural science. Arnold, London

Karplus JP, Kreidl A (1909) Gehirn und Sympathicus. I. Zwischenhirnbasis und Halssympathicus. Arch Physiol (Bonn) 129:138–144

Koikegami H, Dodo T, Mochida Y, Takahashi H (1957) Stimulation experiments on the amygdaloid nuclear complex and related structures. Effects upon renal volume, urinary secretion, movements of the urinary bladder, blood pressure and respiratory movements. Folia Psychiatr Neurol Jpn 11:157–206

Kremer WF (1947) Autonomic and somatic reactions induced by stimulation of the cingulate gyrus in dogs. J Neurophysiol 10:371–379

Kuru M (1965) Nervous control of micturition. Physiol Rev 45:425–494

Kuru M, Ozaki H, Kurati T (1961) Effect of simultaneous simulations of the bulbar vesico-constrictor and vesico-relaxer centers. J Comp Neurol 116:195–208

Kuru M, Koyama Y, Ozaki H (1963) Part of the brainstem controlling the tone of the external urethral sphincter. Proc Jpn Acad 39:530–533

Langworthy OR, Kolb LC (1933) The encephalic control of tone in musculature of the urinary bladder. Brain 56:371–382

Langworthy OR, Hesser FH (1936) An experimental study of micturition released from cerebral control. Am J Physiol 115:694–700

Lewin RJ, Dillard GV, Porter RW (1967) Extrapyramidal inhibition of the urinary bladder. Brain Res 4:301–307

Lichternstein R (1912) Über die zentrale Blaseninnervation, ein Beitrag zur Physiologie des Zwischenhirnes. Wien Klin Wochenschr 25:1248–1249

Light AR (1985) The spinal terminations of single, physiologically characterised axons originating in the ponto-medullary raphe of the cat. J Comp Neurol 234:536–548

Loewy AD (1982) Descending pathways to sympathetic preganglionic neurones. In: Kuypers HGJM, Martin GF (eds) Descending pathways to the spinal cord. Elsevier, Amsterdam, pp 267–277 (Progress in brain research, vol 57)

Lumb BM (1986) Brainstem control of visceral afferent pathways in the spinal cord. In: Cervero F, Morrison JFB (eds) Visceral sensation. Elsevier, Amsterdam, pp 279–293 (Progress in brain research, vol 67)

Lumb BM, Morrison JFB (1984) Convergence of visceral and somatic information on to identified reticulo- and raphespinal neurones in the rat. J Physiol 357:33P

Mackel R (1979) Segmental and descending control of the external urethral and anal sphincters in the cat. J Physiol 294:105–122

Maggi CA, Santicioli P, Borsini F, Giuliani S, Meli A (1986) The role of capsaicin-sensitive innervation of the rat urinary bladder in the activation of micturition reflex. Naunyn-Schmiedebergs Arch Pharmacol 332:276–283

Magoun HW, Ranson SW, Hetherington A (1938) Descending connections from the hypothalamus. Arch Neurol Psychiatry 39:1127–1149

Martner J (1975) Influences on the defaecation and micturition reflexes by the cerebellar fastigial nucleus. Acta Physiol Scand 94:95–104

Matsumoto H (1957) On the autonomic control of superior colliculus and limbic area to the bladder and rectum. Wakayama Igaku 8:65–80

Matthews PBC (1986) Observations on the automatic gain compensation of reflex gain on varying the pre-existing level of motor discharge in man. J Physiol 374:73–90

McMahon SB (1986) Sensory-motor integration in urinary bladder function. In: Cervero F, Morrison JFB (eds) Visceral sensation. Elsevier, Amsterdam, pp 245–253

McMahon SB, Morrison JFB (1982a) Spinal neurones with long projections activated from the abdominal viscera of the cat. J Physiol 332:1–20

McMahon SB, Morrison JFB (1982b) Two groups of spinal interneurones that respond to stimulation of the abdominal viscera of the cat. J Physiol 332:21–34

McMahon SB, Morrison JFB (1982c) Factors that determine the excitability of parasympathetic reflexes to the bladder. J Physiol 332:35–43

McMahon SB, Spillane Kathy (1982) Brainstem influences on the parasympathetic supply to the urinary bladder of the cat. Brain Res 234:237–249

McMahon SB, Morrison JFB, Spillane Kathy (1982) An electrophysiological study of somatic and visceral convergence in the reflex control of the external sphincters. J Physiol 328:379–387

McPherson A (1966) The effects of somatic stimuli on the bladder of the cat. J Physiol 185:185–204

Milhorn HT (1966) The application of control theory to physiological systems. Saunders, Philadelphia, p 376

Morrison JFB (1982) The neural control of the bladder. In: Bloom SR, Polak JM, Lindenlaub E (eds) Systemic role of regulatory peptides. Schattauer, Stuttgart, pp 381–396

Morrison JFB, Spillane K (1981) The inhibitory influence of raphe stimulation on vesical parasympathetic reflexes in the cat. J Physiol 317:84P

Morrison JFB, Spillane Kathy (1986) Neuropharmacological studies on descending inhibitory controls over the micturition reflex. J Auton Nerv Syst (suppl):393–397

Mukai K (1959) Experimental studies on the bladder contraction point in gyrus cinguli and projection fibres from this point. Osaka Daigaku Igaku Zasshi 11:1645–1653

Mundy AR (1984) Clinical physiology of the bladder, urethra and pelvic floor. In: Mundy AR, Stephenson TP, Wein AJ (eds) Urodynamics, principles, practice and application. Churchill Livingstone, Edinburgh, pp 14–25

Mundy AR, Blaivas JG (1984) Nontraumatic neurological disorders. In: Mundy AR, Stephenson TP, Wein AJ (eds) Urodynamics, principles, practice and application. Churchill Livingstone, Edinburgh, pp 278–287

Nathan PW (1976) The central nervous connections of the bladder. In: Williams DI, Chisholm GD (eds) Scientific foundations of urology, vol II. Heinemann, London, pp 51–58

Nishiyama K (1959) Experimental studies on the bladder responses following electrical stimulation of the hypothalamus and its descending pathways. Osaka Daiguka Igaku Zasshi 11:4801–4808

Newsom Davis J, Sears T (1970) The proprioceptive control of the intercostal muscles during their voluntary activation. J Physiol 209:711–738

Pfeifer B (1919) Über corticale Blasenstörungen und deren Lokalisation bei Hirnverletzten. Z Neurol Psychiatry 46:223–284

Porter RW (1967) A pallidal response to detrusor contraction. Brain Res 4:381–383

Porter RW, Pazo JH, Dillard GV (1971) Triphasic brain stem response to detrusor contraction. Brain Res 35:119–126

Ranson SW, Kabat H, Magoun HW (1935) Autonomic responses to electrical stimulation of hypothalamus preoptic region and septum. Arch Neurol Psychiatry 33:474–477

Ruch TC (1965) The urinary bladder. In: Ruch TC, Patton HD (eds) Physiology and biophysics. Saunders, Philadelphia, pp 1010–1021

Saper CB, Loewy AD, Swanson LW, Cowan WM (1976) Direct hypothalamo-autonomic connections. Brain Res 117:305–312

Sato A, Sato Y, Schmidt RF (1980) Reflex bladder activity induced by electrical stimulation of hind limb somatic afferents in the cat. J Auton Nerv Syst 1:229–241

Satoh K, Tohyama M, Tetsuro S, Yamamoto K, Shimizu N (1978a) Descending projection of the nucleus tegmentalis laterodorsalis to the spinal cord; studied by the horseradish peroxidase method following 6-hydroxy-dopa administration. Neurosci Lett 8:9–15

Satoh K, Shimizu N, Tohyama M, Maeda T (1978b) Localisation of the micturition reflex at dorsolateral pontine tegmentum of the rat. Neurosci Lett 8:27–33

Sherrington CS (1892) Notes on the arrangement of some motor fibres in the lumbo-sacral plexus. J Physiol 13:621–772

Skyltety FM (1959) Relation of periaqueductal grey matter to stomach and bladder motility. Neurology 9: 190–198

Spillane Kathy (1982) Brainstem influences on pelvic parasympathetic reflexes. PhD Thesis, University of Leeds

Takebayashi H, Kobai Y, Ushida A, Matsumoto H, Yoshioka T, Arimoto K, Shimizu K, Matsuda A, Kawashima A, Yinya Y, Fujita I, Maeshima S (1957) Superior colliculus, playing roles in the optokinetic and neurovegetative mechanism. Wakayama Med Rep 4:1–12

Tang PC (1955) Levels of brainstem and diencephalon controlling micturition reflex. J Neurophysiol 18:583–595

Tang PC, Ruch TC (1956) Localisation of brainstem and diencephalic areas controlling the micturition reflex. J Comp Neurol 106:213–245

Terzian H, Ore GD (1955) Syndrome of Kluver and Bucy. Reproduced in man by bilateral removal of the temporal lobes. Neurology 5:373–380

Wang SC, Ranson SW (1939a) Autonomic responses to stimulation of the lower brainstem. J Comp Neurol 71:437–455

Wang SC, Ranson SW (1939b) Descending pathways from the hypothalamus to the medulla and spinal cord. Observations on blood pressure and bladder responses. J Comp Neurol 71:457–472

West DC, Wolstencroft JH (1977) Location and conduction velocity of raphe-spinal neurones in nucleus raphe magnus and raphe pallidus in the cat. Neurosci Lett 5:147–151

West DC, Wolstencroft JH (1982) Descending inhibition of the flexor withdrawal reflexes in the decerebrate cat. J Physiol 327:60P

Westlund KN, Coulter JD (1980) Descending projections of the locus coeruleus and subcoeruleus/medial parabrachial nuclei in the monkey: axonal transport studies and dopamine-β-hydroxylase immunocytochemistry. Brain Res Rev 2:235–264

Wolstencroft JH, West DC (1982) Functional characteristics of raphespinal and other projections from nucleus raphe magnus. In: Sjolund B, Bjorklund A (eds) Brain stem control of spinal mechanisms. Elsevier, Amsterdam, pp 359–380

Yamamoto S, Araki K (1962) Intra-abdominal pressure response to medullary stimulation in cats. Exp Neurol 5:110–119

Yamamoto S, Araki K, Kikuchi M (1961) Abdominal muscle reflexes of pelvic origin in cats. Exp Neurol 4:345–357

Yamamoto S, Araki K, Kikuchi M (1962) Electromyographic study of reflex activity in abdominal wall muscles and diaphragm following pelvic afferent excitation in cats. Tohoku J Exp Med 75:375–383

Section 3

CLINICAL PHYSIOLOGY

Urodynamics
Clinical Neurophysiology
Human Physiology

Introduction

In this section the function of the lower urinary tract in the intact subject is reviewed. In Chapter 9 the hydrodynamics of urine storage and of voiding are discussed and the normal values for these hydrodynamic parameters are summarized. Chapter 10 describes the extent to which the electrical activity of nerve and muscle within the relevant areas can be assessed usefully by clinical neurophysiological techniques. Finally, in Chapter 11 there is a summary of physiology, with emphasis on the human, which is intended to provide some integration of the information derived from previous chapters.

In descriptions of human lower urinary tract physiology it is usually, as here, true that the investigations concentrate on hydrodynamics and neurophysiology. If one regards the system from a phylogenetic point of view it is very obvious that the system is involved in far more than just the storage and voiding of urine. First there is the intimate relation of the urological with the genital tract, involving significant changes of function with age and sexual activity. In addition, micturition has great social significance in animals, particularly where olfaction is the dominant sense, and, though territory marking is the most obvious example of this, many other social activities are linked with micturition in some way. Because of the relation between olfaction and the limbic system it is not surprising that, even in humans, micturition is strongly related to emotion—just one of the factors that makes it a difficult subject to investigate.

Micturition is an autonomic activity largely, but only very lately, under voluntary control. It is unique in the range of coordination between somatic and autonomic nervous systems. It is perhaps because of the social significance of micturition in animals that it has been possible for humans and some domesticated animals to develop this exceptional degree of voluntary control. However, it is not so surprising that under some circumstances the control can become tenuous.

All the factors mentioned make it difficult to construct proper scientific studies of urinary function in vivo. Yates (1980) has constructively criticized the quality of research in this area. In particular, he recommends that investigators should look at the methods being used in parallel disciplines and in other, nonbiological, areas. His comments remain very pertinent and deserve reappraisal.

Reference

Yates FE (1980) Urodynamics and the study of female incontinence: the state of the art as perceived by an outsider, with a modest proposal. In: Zinner NR, Sterling AM (eds) Female incontinence. Prog Clin Biol Res 78:1–14

9 · Urodynamics

Michael Torrens

Urine Flow Studies
Cystometry: Filling Phase
Cystometry: Voiding Phase
Urethrometry

Clinical measurement of lower urinary tract function has been expanding in its influence for 30 years. This has been allowed by the increasing availability of inexpensive, reliable measuring systems. The procedure is now a standard investigation relevant to patients from urology, gynaecology, neurology and geriatric departments. Such investigations are included in the science of *urodynamics*, which involves the study of the function and dysfunction of the urinary tract by any appropriate method. It encompasses the morphological, physiological, biochemical and hydrodynamic aspects of urine transport and storage.

The techniques described in this chapter will be concerned mainly with hydrodynamic aspects and in particular with pressures generated within the lower urinary tract and with urinary flow. Other measurement techniques are occasionally relevant, as outlined later. The advent of microcomputers may change the spectrum of investigation by providing on-line analysis both of the primary data and also of derived functions, allowing a degree of automated diagnosis. This makes the definitions of normal values particularly important.

Much of the contents of this chapter will be familiar to those who are established urodynamic investigators. The purpose therefore is to familiarize other clinicians and physiologists with the activities in the clinical area, which are seldom conducted with reference to any academic department. In addition the normal findings in the human will be reviewed in detail. As these are still rather inadequate perhaps this book will stimulate further studies of the normal state wherever appropriate.

Urine Flow Studies

Uroflowmetry is the simplest and least invasive test. As urine flow reflects the final result of the micturition process it is also the most useful test, giving the maximum information for the least interference. However, whilst uroflowmetry produces a high index of suspicion as to a diagnosis in many cases, it does not permit exact localization of the suspected abnormality. In addition to direct measurement of urine flow rate the characteristics of the urinary stream may be assessed by measuring its residual energy or momentum flux (see p. 284).

Urine flow studies in normal male subjects have been performed by von Garrelts (1957b), Susset et al. (1973), Kondo et al. (1978), Drach et al. (1979a), Siroky et al. (1979) and Rollema (1981) who includes an extensive review. Studies in females are few, but include those of Backman (1965), Walter et al. (1979) and Drach et al. (1979b). Gierup (1970) performed tests on children.

Flowmeters

The techniques for measurement of urine flow depend most often on measuring urine volume with respect to time and deriving, electronically, the rate of change of volume with time which is the flow rate. Volume can be measured by weight, by pressure at the bottom of a vessel or by depth, using a capacitance registering dipstick. In the last quoted method the solutes in urine conduct electricity across the capacitor, reducing its effective area as the level of urine rises. The capacitance thus falls in proportion to the volume.

Urine flow can be measured directly by an electromagnetic flowmeter. Another common technique involves urine falling onto a disc rotating at a constant speed. The mass of the urine tends to slow the rotation, but a servomotor keeps the speed constant. The power necessary to do this is proportional to the urine flow rate. The urine volume can be derived by integration of the flow rate.

Characteristics of Uroflowmetry

Figure 9.1 summarizes the main characteristics of uroflowmetry:

1. Delay time
2. Time to maximum flow rate
3. Maximum flow rate
4. Voided volume
5. Average flow rate
6. Flow curve pattern

Delay Time

Delay time is the time from the initiation of the voluntary cerebral activity related to voiding to the recording of urinary flow. This is usually less than 10 s. The time may be prolonged in cases of infravesical obstruction or psychological inhibition.

Time to Maximum Flow Rate

The ascending limb of the flow trace rises rapidly. A very gradual rise indicates a deficiency of opening of the bladder neck due either to bladder neck obstruction or to a weak detrusor muscle. Time to maximum flow should not exceed the first one-third of the total flow time. An excessively rapid rise in flow is also abnormal. If the ascending limb of the flow trace is almost vertical it suggests that flow is initiated by voluntary relaxation of the urethral sphincter after the bladder neck has already been opened by a strong involuntary detrusor contraction.

Maximum Flow Rate

The maximum flow rate is the single figure most usually recorded from a flow trace. The value is very dependent on the voided volume and varies also with age and sex. This is discussed in more detail in the sections on nomograms. A low flow may indicate infravesical obstruction or impaired detrusor function. However, the bladder may com-

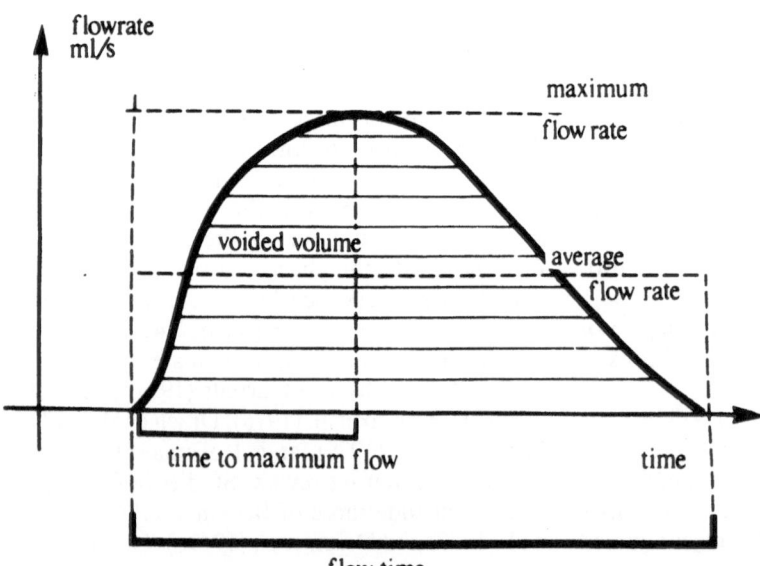

Fig. 9.1. The terminology relating to the description of urinary flow. (Abrams et al. 1983)

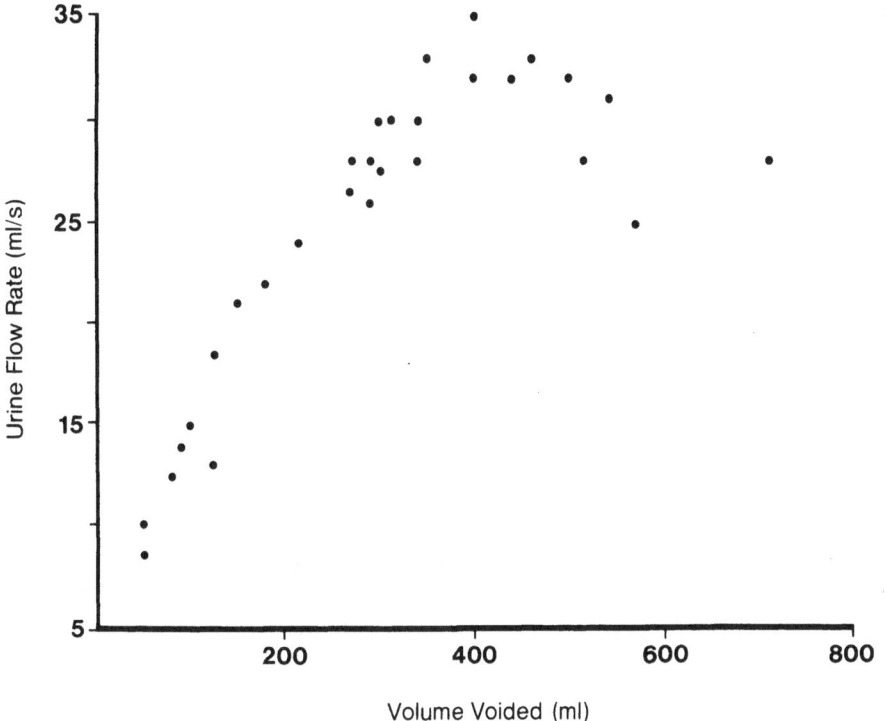

Fig. 9.2. The relationship between flow rate and voided volume in one normal subject. Above 400 ml the efficiency of the detrusor decreases. (Abrams et al. 1983)

pensate for obstruction by increasing the strength of contraction, thus producing an elevated voiding pressure. This may allow normal flow rates in spite of infravesical obstruction.

Voided Volume

The maximum flow rate rises proportional to the square root of the voided volume up to a certain limit, usually around 400–500 ml. Beyond this point the bladder muscle contracts inefficiently and the flow rate tends to decrease (Fig. 9.2).

Average Flow Rate

The average flow rate is the voided volume divided by the time over which measurable flow actually occurs. Average flow rate seems to show less scatter when graphically represented against voided volume (Siroky et al. 1979).

Flow Curve Pattern in Diagnosis

The qualitative assessment of the appearance of the flow curve may give useful diagnostic information.

Various types of flow tracing are shown in Fig. 9.3 and some explanation of the various patterns is given in the figure legend.

Flow Curve Pattern in Normal Persons

Not only is the maximum flow rate dependent upon volume voided, but so is the pattern of the flow trace. Figure 9.4 shows the superimposition of various flow rate tracings in one person at different volumes. It will be seen that the slope of the ascending limb remains relatively constant, but that the slope of the descending limb becomes considerably flatter as the voided volume increases. It is possible that some conclusion may be drawn from the slope of the descending limb as to the efficiency of the muscle when it is relatively overstretched. This may give some clues as to the liability of this muscle to decompensate in later life if subjected to infravesical obstruction.

Rollema (1981) has defined the variability of the flow curve pattern in a group of 142 normal volunteers. The slope of the ascending and descending limb respectively, can be represented in part as the time to maximum flow (tQ_{max}) and time of the descending limb (t_{desc}). Rollema has shown t_{desc} to

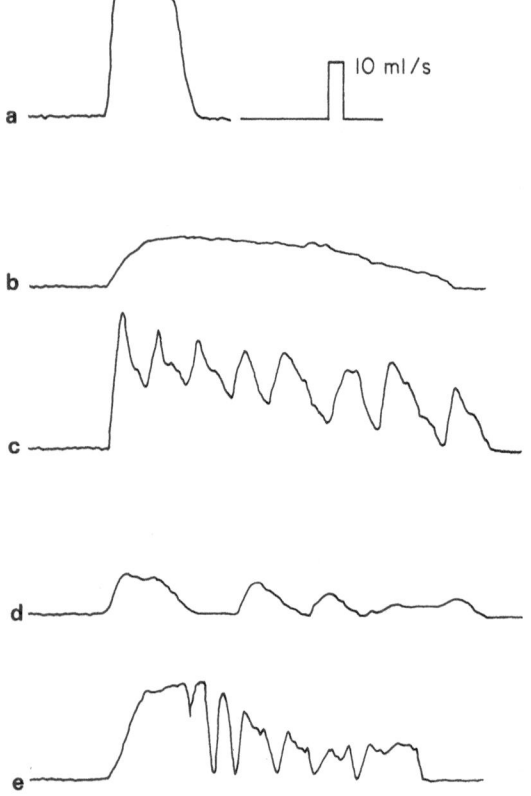

Fig. 9.3a–e. Various flow curve patterns. **a** Normal; **b** obstructed; **c** regular waves due to voluntary straining; **d** irregular low waves caused by inadequate intermittent detrusor contraction; **e** rapid fluctuations (interruptions) caused by sphincter contraction, detrusor sphincter dyssynergia or nervousness.

have the best discriminating ability of all the variables he studied in the diagnosis of obstruction due to benign prostatic hypertrophy.

Normal Values

Volume Dependence

The maximum flow rate is the only value so far submitted to an extensive quantitative investigation in normal subjects. The most definitive studies came from the Karolinska Hospital, Stockholm, between 1958 and 1970, where von Garrelts (1957b) investigated 125 normal males. He identified the correlation between the maximum flow rate and the square root of the volume voided. The spread of normal values in relation to age and volume voided is shown in Fig. 9.5. He noted that deviation from the regression line was most marked when the volume was high and appreciably less for volumes below 220 ml. He considered that analysis of average flow rate added nothing to his previous conclusions.

A similar type of analysis of urinary flow during micturition in normal women was carried out by Backman (1965) in 151 subjects. In addition to categorizing normal values (Fig. 9.6) he identified a slight difference between first and second voidings and showed that parity had no obvious effect on urinary flow.

Comparisons between males and females showed that the regression coefficients, reflecting the

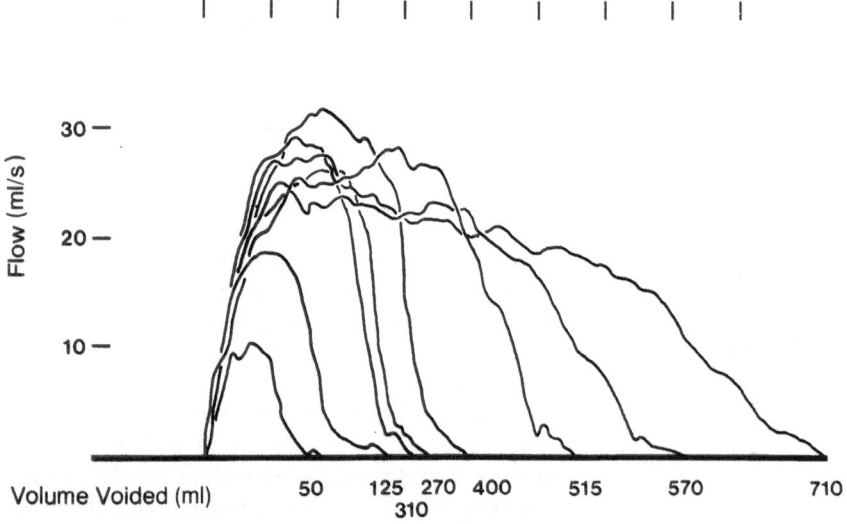

Fig. 9.4. The superimposition of various flow rate tracings. Examples taken from the same subject as Fig. 9.2. (Abrams et al. 1983)

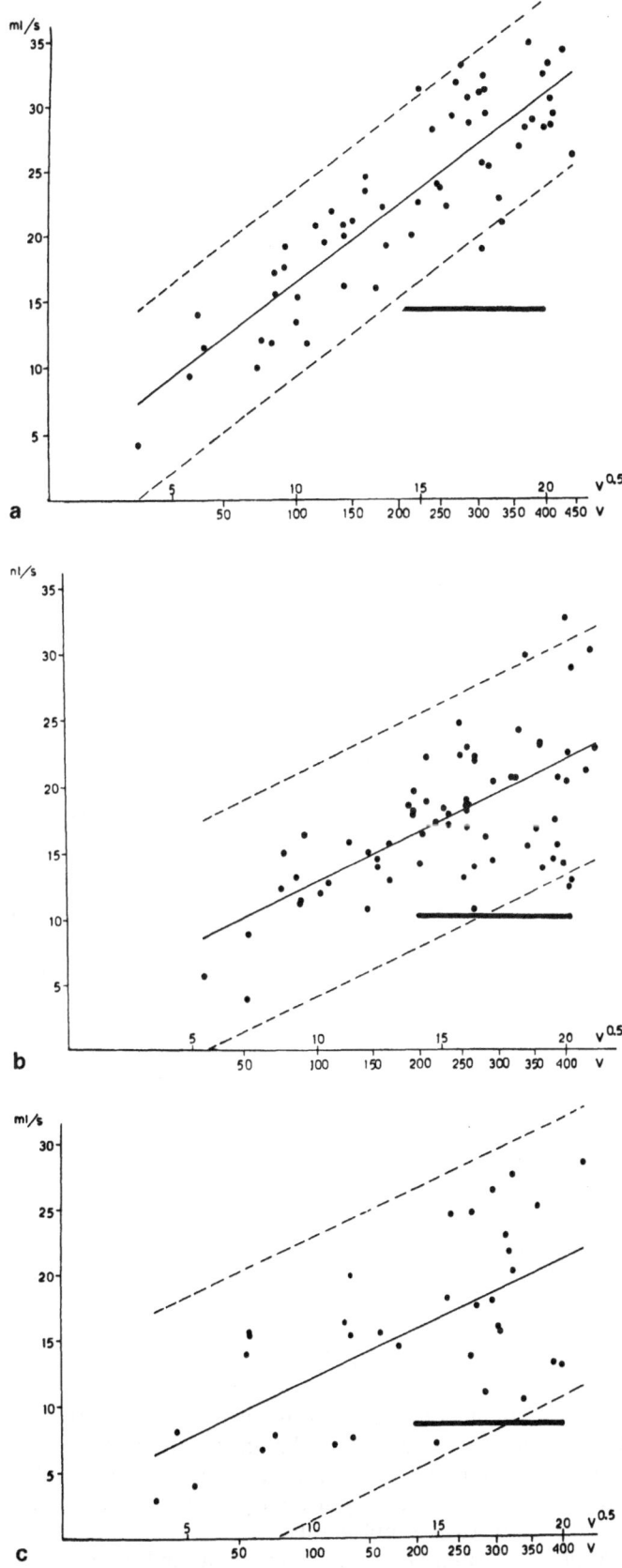

Fig. 9.5a–c. The correlation between urine flow rate (ml/s) and volume voided (*V*) in males: **a** under 25 years, **b** 25–50 years, **c** over 50 years. The values for voided volume raised to the power 0.5 are marked on the *abscissa*. The *regression line* and *dotted lines* indicating ±2 SD are also marked. The *horizontal line* indicates an acceptable lower limit of normal in the volume range 200–400 ml. (von Garrelts 1975b)

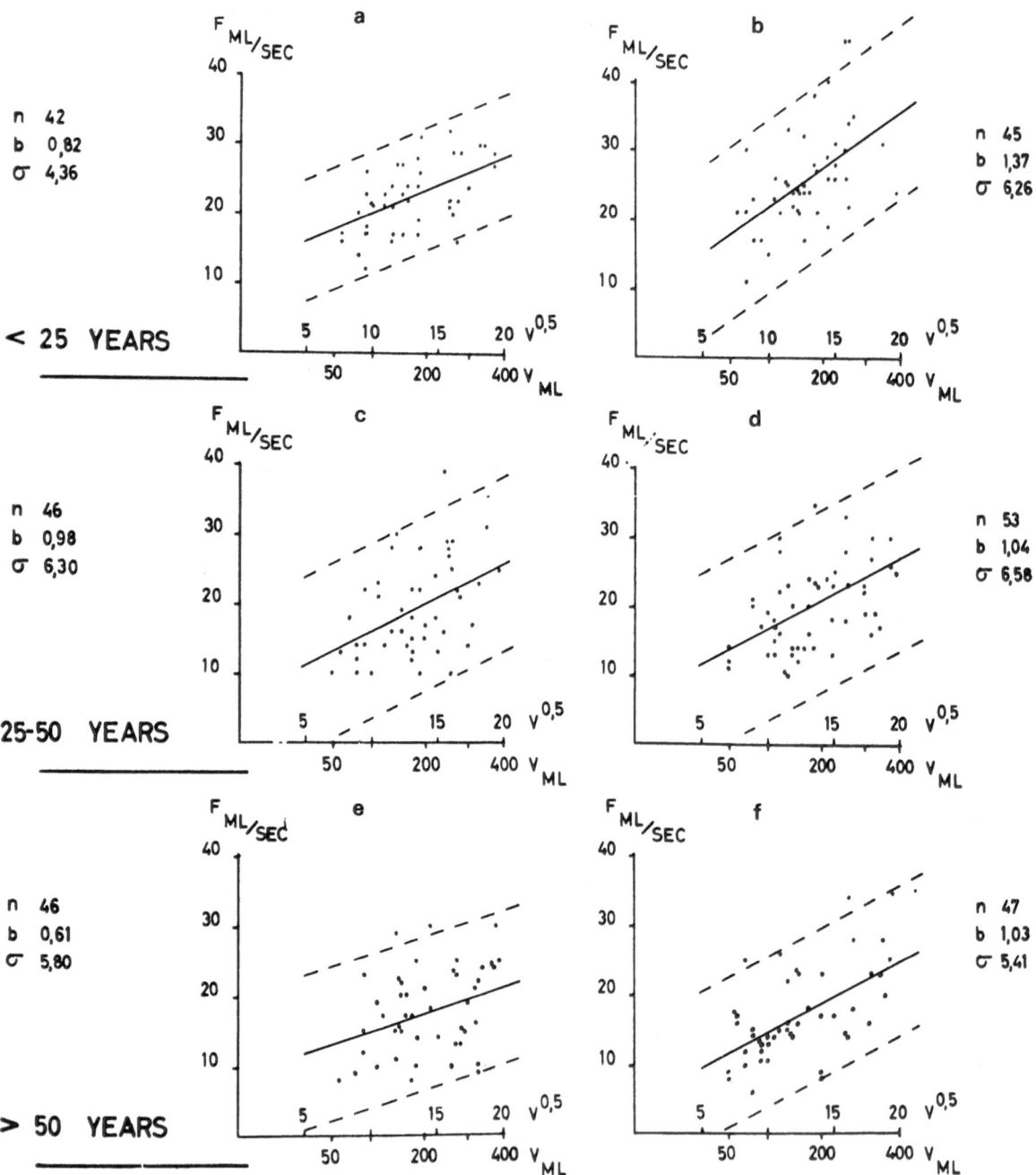

Fig. 9.6a–f. The correlation between urine flow rate (F ml/sec) and volume voided (V ml) in females. Diagrams **a, c** and **e** represent the first, and **b, d** and **f** the second micturition occasions. n, number of examinations; b, regression coefficient; σ, standard deviation. Other notation as in Fig. 9.5. (Backman 1965)

magnitude of the increase in flow with increasing volume, were roughly the same in comparable age groups in the two series (Table 9.1). Although the mean value for maximum flow at different volumes is significantly higher in women, inspection of the original figures suggests that the minimum acceptable value for normal flow rate, assuming a voided volume of at least 200 ml, is the same in either sex at the same age. Backman also made the observation that women were able to augment their urinary stream by straining whereas men were less able to accomplish this.

Urinary flow studies in infants and children were undertaken by Gierup (1970). He studied 216 boys

Table 9.1. Data on the mean flow rates ($\bar{x}$) at 200 ml volume voided, standard deviation (SD) and regression coefficient (b) in males and females of all ages (Backman 1965; Gierup 1970)

Age (years)	Males			Females		
	$\bar{x}$	SD	b	$\bar{x}$	SD	b
<7	17.4	3.50	0.80	18.5	4.33	1.08
7–10	18.9	4.47	0.98	22.9	5.95	1.56
10–13	17.4	4.82	0.71	22.0	5.15	1.38
13–16	20.8	6.10	1.03	24.1	6.38	1.58
<25	23	3.55	1.41	28	6.26	1.37
25–50	16	4.42	0.92	21	6.58	1.04
>50	15	5.39	0.90	19	3.41	1.03

and 180 girls between the ages of 3 days and 16 years. The results demonstrate that, as expected, there is a gradual increase in volume voided with age and that this overrides any other factor in its effect on maximum flow. No other growth parameter was relevant. Table 9.1 compares boys and girls. It is evident that there is little statistical difference between the sexes, but the tendency is for the flow in boys to remain much the same, whereas that in girls gradually increases more with age independent of volume voided. This may be related to increasing distensibility of the urethra in the female.

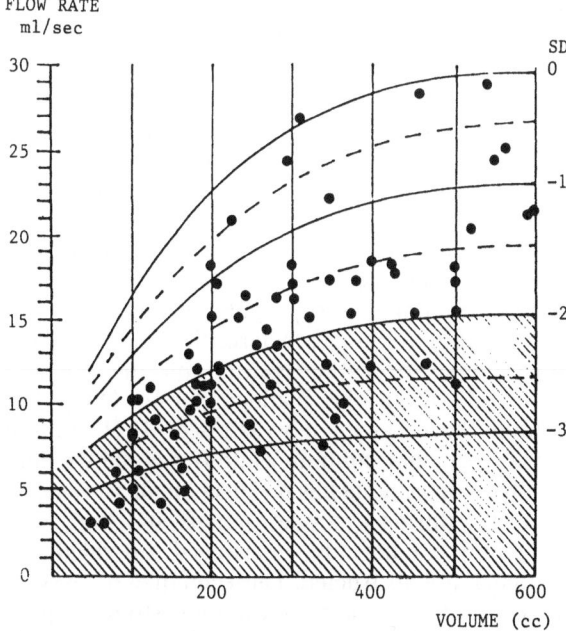

Fig. 9.7. Maximum flow rates from 56 asymptomatic men aged 55–78 years superimposed on the Siroky nomogram. The values are similar to Fig. 9.5c and illustrate the importance of constructing nomograms related to age. (Abrams 1984)

Gierup also notes that the prolonged flat-topped plateau curve, which suggests obstruction in adults, is a normal finding around puberty in both sexes.

Age Dependence

One disadvantage of the nomograms produced from the Karolinska Hospital and described in the previous section is that they do not take into account the closer grouping, and therefore reduced standard deviation, in the flow rates at lower voided volumes. This problem was addressed by Siroky et al. (1979), but unfortunately they used a rather small series and did not take into account the changes associated with age. Abrams (1984) reassessed the definition of the lower limit of normal along the lines of the Siroky nomogram and came to the conclusion that the flow rates of uncomplaining males aged over 55 years were almost all less than "normal" when judged by Siroky's criteria, half of them being more than two standard deviations below normal (Fig. 9.7).

The same problem with the definition of normality in older age groups was encountered by Rollema (1981). He produced a detailed analysis of flow parameters in volunteers and obstructed patients and found no significant age dependence for any of the variables he studied in the volunteer group (age range 17–53 years). He then assessed the discriminatory potential of these variables against a patient group with outflow tract obstruction (age range 46–78 years). Such a comparison is clearly invalid when judged in the light of the Abrams (1984) figures.

In an attempt to clarify the issue the results from the Bristol Urodynamic Unit have been analysed (M. J. Torrens and L. Kavanagh 1986, unpublished work). 161 males and 651 females, judged to be functionally normal (but not asymptomatic) and

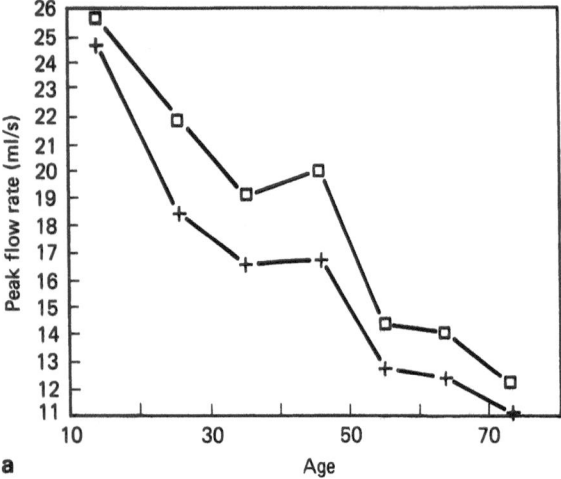

a

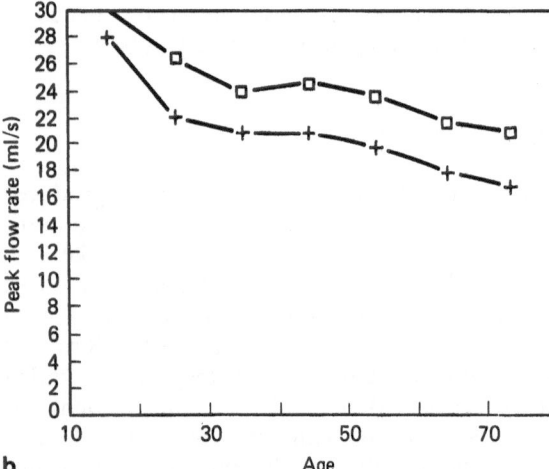

b

Fig. 9.8a, b. The correlation between maximal (peak) flow rate and age in 161 males (**a**) and 651 females (**b**) from the Bristol series, illustrating values from "free" flow rates (*squares*) and flow rates during pressure–flow studies (*crosses*).

Momentum Flux

The term "momentum flux" describes the force of the urinary stream directed perpendicular (horizontal) to a flat (vertical) surface. Its value depends on the mechanical properties of the external urethral meatus, and it is necessary for the flow rate to be high enough to distend the meatus fully. The technique has been described and evaluated by Meyhoff et al. (1980a–c) and discussed by Griffiths and Scholtmeijer (1982), who emphasize some of the limitations of its use. Its value lies only in the diagnosis of meatal stenosis.

Cystometry: Filling Phase

Evaluation of the storage capacity of the bladder requires a test which reproduces as closely as possible the circumstances of bladder filling. Cystometry is the method by which the pressure–volume relationship of the bladder is measured. Zero reference for all pressures is the level of the superior edge of the symphysis pubis. Present techniques allow the continuous recording of the pressure within the bladder during natural or artificial filling.

Intravesical pressure (P_{ves}) is the pressure measured within the bladder. *Abdominal pressure* (P_{abd}) is the pressure surrounding the bladder and this is usually measured as rectal pressure. *Detrusor pressure* (P_{det}) is that proportion of the intravesical pressure produced by active and passive bladder wall properties. Detrusor pressure is calculated by electronically subtracting the abdominal pressure from the intravesical pressure. This derivation is extremely important in order to be able to distinguish intrinsic and extrinsic contributions to intravesical pressure.

Access for pressure measurement is most commonly by transurethral catheterization. Occasionally a percutaneous suprapubic catheter is used. The effect of urethral catheterization and/or topical anaesthesia for catheterization on the pressure recorded within the bladder during filling has seldom been adequately assessed. P. H. Powell (1980, personal communication) has shown that urethral anaesthesia has no effect on measurements recorded during cystometrograms and that pressure–flow studies are unchanged provided the patient is not obstructed. The techniques of pressure registration and their use during cystometry have been reviewed by Abrams et al. (1983).

It is appropriate to describe cystometrograms in terms of *capacity, compliance, contractility* and *sensation*.

with a voided volume between 200 and 400 ml, had age correlated with free flow rates which were not part of a pressure–flow study and subsequently with those recorded during pressure–flow measurement. The results, presented in Fig. 9.8 and Table 9.2, show a clear age dependence in both sexes. It is also interesting to observe that the flow rate after bladder filling by infusion during cystometry is uniformly decreased, despite the fact that the voided volume remains in the physiological range of 200–400 ml. This suggests that rapid bladder filling (at 50 ml/min in this case) may alter bladder contractility, for the fine epidural catheter inserted per urethram for pressure recording should not interfere with flow.

Table 9.2. Mean maximum flow rate and volume voided ($\pm$SD) in males and females related to age (from the Bristol series)

Age (years)	Males			Females		
	n	Qmax (ml/s)	V (ml)	n	Qmax (ml/s)	V (ml)
A. Free flow before catheterization						
10–19	13	26 ± 8	319 ± 67	12	30 ± 7	290 ± 66
20–29	11	22 ± 6	311 ± 69	72	26 ± 7	294 ± 62
30–39	21	19 ± 7	311 ± 71	158	24 ± 7	284 ± 64
40–49	30	20 ± 7	294 ± 58	187	25 ± 7	299 ± 64
50–59	28	24 ± 4	289 ± 55	148	24 ± 7	297 ± 65
60–69	43	14 ± 4	279 ± 60	57	22 ± 8	289 ± 63
70–79	15	12 ± 4	280 ± 68	17	21 ± 7	276 ± 62
B. During urodynamic (Pressure–flow) studies						
10–19	13	25 ± 9	355 ± 134	12	28 ± 6	423 ± 129
20–29	11	18 ± 9	330 ± 178	72	22 ± 9	312 ± 135
30–39	21	17 ± 7	316 ± 142	158	21 ± 9	343 ± 126
40–49	30	17 ± 8	347 ± 143	187	21 ± 9	360 ± 153
50–59	28	13 ± 6	298 ± 137	148	20 ± 9	343 ± 131
60–69	43	13 ± 6	318 ± 138	57	18 ± 9	335 ± 147
70–79	15	11 ± 5	354 ± 95	17	17 ± 7	291 ± 104

Capacity

It is important to emphasize that measurements of capacity are capricious. The *maximum cystometric capacity* is defined as the volume at which the patient has a normal strong desire to void. This value may have little relationship to the *average functional capacity* (the mean of the volumes voided measured under normal circumstances) or to the *maximum structural capacity* (assessed by filling the bladder under general or regional anaesthetic).

The quoted values for maximum cystometric capacity vary considerably. The range of normal in men may be 350–750 ml and for women rather less, 250–550 ml. As a cystometrogram is such a poor way of measuring bladder capacity there is no great point in putting reliance on the measurement, except to show that it falls within the normal range for the unit concerned. The assessment of bladder capacity depends upon when the investigator stops the infusion. If a subsequent voiding study is to be undertaken it is better not to overdistend the bladder. N. George and P. Lewis (1978, personal communication) concluded that the ideal capacity for filling before a voiding study was between 50% and 90% of the average functional capacity. It is our practice to stop infusion at the point when patients say they would normally be obliged to void.

The interpretation of detrusor function depends largely on cystometrograms. It is fair to ask 'how physiological is a cystometrogram?'. Very little attention has been paid to this question. Indeed the demands for convenience and the concept of "provocation" have led to exceptionally rapid infusion of the alien gases and solutions. George and Lewis analysed cystometrograms performed at a continuous filling rate of 50 ml/min, using saline at 37°C and the with patient sitting. Maximum cystometric capacity was compared with average functional capacity. Of 38 patients studied, 37 had a reduced maximum cystometric capacity compared with the average functional capacity. The mean reduction was by 41%. Cystometry therefore tends to underestimate "normal" capacity.

Cystometry also affects the residual urine. The residual volume after voiding cystometry can only be compared with that after the initial pretest void prior to catheterization. Of 38 patients, 22 had some residual urine after the pretest void, and in 18 of 22 this was reduced or abolished after cystometrogram. The reduction was greater than that to be expected if capacity were to decrease. In 9 of 22 the residual urine was greater than 500 ml. In only one patient was the residual unchanged. In other patients whose pretest voided volume was much greater than the average functional capacity (and who were not included in the figures quoted above) residual urine was the rule, present in 22 of 25. In only 3 of 22 did this persist after cystometrography.

If artificial infusion alters the results of filling cystometrograms, it may also affect contractility and hence the results of voiding studies. This has already been suggested in relation to the observation that flow rates were apparently decreased after artificial bladder filling. The conclusion from this is that cystometric capacity is unreliable and

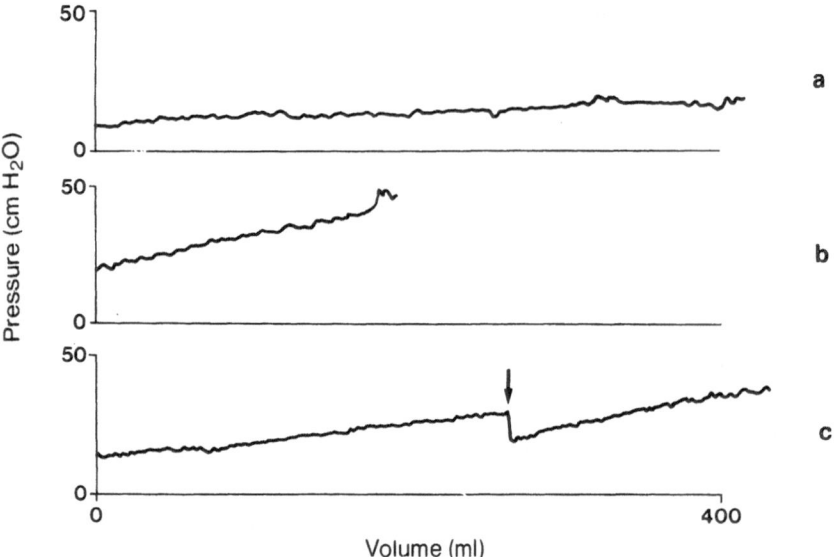

Fig. 9.9a–c. Various responses to bladder filling. **a** Normal cystometrogram; **b** constantly reduced compliance; **c** reduced compliance caused by fast filling. *Arrow* indicates when filling is stopped. The fall in pressure indicates that the low compliance is caused by the fast filling rate. (Abrams et al. 1983)

both uroflowmetry and residual urine measurements after cystometry are suspect. The bladder must not be allowed to get overstretched at any stage of the proceedings or contractility will be affected. Patients are often asked to attend for urodynamic investigation with a "full" bladder, but this may compromise the results of testing if the bladder becomes overdistended.

Compliance

Compliance indicates the change in volume for a change in pressure. It is defined as

$$C = \Delta V / \Delta P$$

where ΔV is the volume increment and ΔP is the change in pressure associated with this volume increment. Compliance during bladder filling is a more accurate observation than bladder capacity. However, compliance may not be constant throughout bladder filling (Coolsaet 1985), though it usually is over the range up to 400 ml. If the limits of structural capacity are approached the compliance will decrease.

The intravesical pressure when the bladder is empty varies according to patient position and factors such as obesity, but it is usually less than 40 cm H_2O. The detrusor pressure is independent of posture and should be less than 10 cm H_2O at zero volume. Most bladders are very compliant (accom-

modate well), even to fast-fill cystometry. Pressure increases are small; less than 10 cm H_2O at 300 ml and less than 15 cm H_2O at capacity (Fig. 9.9a). However, some bladders do appear to be less compliant (Fig. 9.9b), especially when filled fast. It must be realized that this low compliance may be artefactual (Klevmark 1974). If this situation of relatively low compliance is found during testing then the cystometrogram should be repeated at a slow fill rate or, alternatively, the infusion may be stopped and the detrusor pressure observed. If the pressure falls (Fig. 9.9c) then the low compliance is a phenomenon caused by the fast filling. Not enough work has been done on the assessment of compliance during infusion at different rates and it may be that this will prove an additional means of evaluating bladder wall function.

Compliance corresponds to what has been called, rather imprecisely, "tone" by some previous investigators. The term tone implies some degree of neuromuscular activity and it is evident that there are both active and passive influences contributing to compliance. The passive or structural characteristics of the bladder wall during distension can only be assessed after the abolition of all neurogenic activity by appropriate anaesthesia or ganglion blockade. Low compliance due to physical indistensibility is associated with intramural fibrosis typically following chronic infection, radiotherapy or bladder wall hypertrophy associated with obstruction.

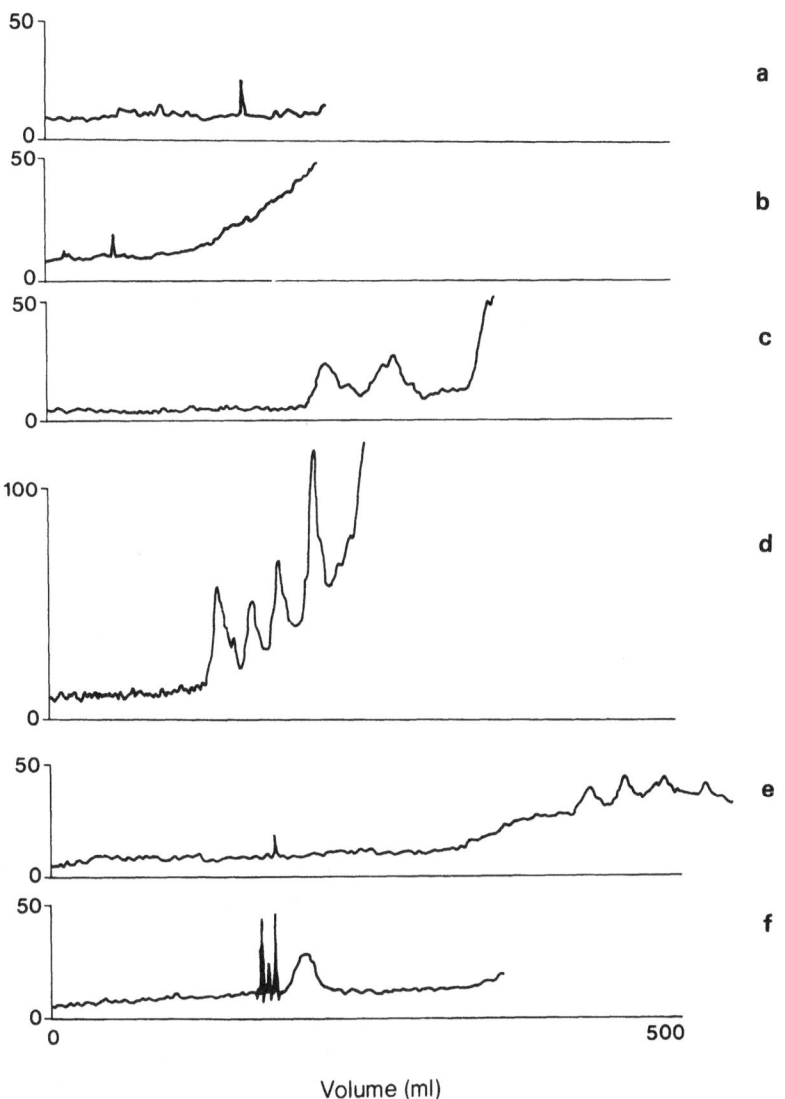

Fig. 9.10a-f. Various types of filling cystometrograms. **a** Hypersensitive cystometrogram, **b** Cystometrogram with decreased compliance in the terminal phase, described by some workers as "unstable" and by others as "hypersensitive". **c** The commonest form of unstable cystometrogram with overactivity occurring as capacity is reached. **d** Unstable cystometrogram with decreased compliance in the overactive phase usually caused by muscle hypertrophy. **e** Cystometrogram with an abnormally large capacity and with unstable activity occurring only in the terminal phase, associated often with a neurological deficit. **f** An unstable contraction occurring after stress, in this case coughing.

Low compliance may also be associated with detrusor muscle activity. If the volume increase is gradual and continuous the cystometrogram in such cases of actively decreased compliance conforms more to a curved profile as in Fig. 9.10b. By contrast the pressure increase in cases of passively decreased compliance conforms more to a straight line (see Fig. 9.9b). In many cases the change in compliance may be obscured by abnormal contractility (Fig. 9.10d). Under these circumstances it is reasonable to take the line of basal pressure (between contractions) as indicative of compliance.

There is considerable scope for further investigation of the relationships between detrusor wall compliance, contractility and morphology in relation to voiding function. Further comments are made in the section on Hydrodynamics (see p. 289).

Contractility

The use of the cystometrogram to detect inappropriate contractions of the bladder is perhaps the most significant application of the technique in the clinical environment. Nevertheless, in normal subjects the bladder should not contract during filling under any circumstances. It is sometimes argued that the unphysiological circumstances of cystometry may provoke detrusor contraction, even in normal subjects. This question has, however, been investigated only rarely. Bagley and O'Shaughnessy (1985) performed cystometry on ten volunteer subjects. They concluded that all subjects were able voluntarily to inhibit the detrusor during filling and this inhibition was not accompanied by any increase in urethral sphincter activity on EMG. Three subjects (all female) were unable to void while the catheters were in situ and it therefore appeared that urodynamic investigation was more likely to overinhibit detrusor contraction than provoke it. The tests were repeated in the supine, sitting and standing positions and it was found that position change had no effect on the parameters being measured. Despite these conclusions we have, in the course of many thousands of cystometrograms, observed patients in whom a well-marked detrusor contraction occurs at the onset of filling and subsequently the bladder goes on to accept an entirely normal volume without any further change in pressure. We are prepared to regard this bladder contraction as one caused by the circumstances of the test and to regard such bladders as normal.

A detrusor that is shown objectively to contract, spontaneously or on provocation, during the filling phase whilst the patient is attempting to inhibit micturition is usually accepted as abnormal and described as "unstable". What constitutes a contraction has been the cause of some dispute but it is generally recognized to be a relatively rapid increase in pressure followed by decrease, a phasic contraction–relaxation complex. Any increase in pressure which is not followed by such a relaxation phase should be interpreted with caution. Various forms of hyperactive "unstable" cystometrogram are shown in Fig. 9.10.

A method of assessing bladder contractility by rapid injection cystometry has been described (Susset et al. 1982). In this technique, saline is injected over 3 s. The resulting pressure rise has a rapid phase, the character of which depends on the viscoelastic properties of the bladder, and there may then be an independent contraction phase which may reflect aspects of muscle function not elaborated by slower fill cystometry.

A sensitive method for investigation of bladder contractility, and the only method for identification of bladder relaxation, is called isobaric volume registration (Fall and Erlandson 1983). In this test the bladder is connected by a wide bore catheter to a fluid reservoir placed on a weight transducer. Contraction of the bladder leads to a weight increase and relaxation to a decrease.

Sensation

The description of sensory phenomena during bladder filling obtained from the patient is not particularly useful or accurate until the point is reached where the patient is uncomfortably full, wishes to void or wishes to terminate the investigation. The volume at which this sensation occurs is significant and the characteristics, whether described as pain or as an urgent desire to void, are relevant. A normal person will tolerate at least 300 ml. A cystometric capacity around 200 ml or below implies that the bladder in question is hypersensitive.

A more quantitative evaluation of bladder sensibility was obtained by Frimodt-Møller (1972). He introduced an electrode into bladders which were filled with approximately 100–200 ml saline. A constant current stimulator delivered square-wave impulses of 1 ms duration and 2.5 Hz frequency. Normal patients exhibited an electrical perception threshold of less than 10 mA. A group of neurological problems showed distinctly abnormal thresholds greater than 20 mA. These included both peripheral problems (diabetic neuropathy, radical hysterectomy, polyneuritis) and central problems (Parkinson's disease, multiple sclerosis, myelo-

Table 9.3. Various hydrodynamic variables related to urodynamics

Strain	= extension/original length	$(L - L_0)/L_0$
Force (Newtons)	= mass × acceleration	$(N = kg/s^2)$
Tension	= force/length	$(T = N/m)$
Stress, pressure (Pascals)	= force/area	$(Pa = N/m^2)$[a]
Flow	= volume/time	$(Q = ml/s)$
Work, energy (Joules)	= force × distance	$(J = Nm)$
Power (Watts)	= energy/time	$(W = Nm/s)$

[a] The accepted unit, however, is cmH_2O to allow apparatus to be calibrated easily. 1 cmH_2O is approximately equal to 100 pascals (98.07 Pa).

dysplasia). It is interesting that the mean threshold in cases of vesical prolapse and infravesical obstruction was also elevated slightly. Frimodt-Møller noted a good correlation between electrosensibility and bladder capacity.

Hydrodynamics of Filling

Various mechanical analogies have been made to explain bladder function, some of which will be discussed below. Most concentrate on the passive properties of the bladder and are therefore to be regarded with cautious interest, bearing in mind that the active properties are likely to be of greater significance even during the filling phase. Such mechanical modelling may help in the assessment of structural pathology of the bladder wall or in categorizing functional abnormalities. The terms and variables used in hydrodynamics sometimes cause confusion. These variables are summarized in Table 9.3.

Figure 9.11 represents a bladder assumed to be a thin sphere. The Laplace relation assumes that the force within the sphere (pressure acting on the area of the "cut" surface) is equal to the force in the wall (tension acting around the circumference) which is holding the two halves together

$$P\pi R^2 = T2\pi R$$

where P is the detrusor pressure, R the radius of the spherical bladder and T the tension per unit length. Therefore

$$P = \frac{2T}{R}$$

As the bladder volume increases, both T and R increase. These increases must tend to cancel each other out and there is therefore relatively little change in pressure. As the bladder is not a thin sphere the position is much more complex.

The force per unit area in the wall (stress or pressure) can be related to the wall thickness or width. The contribution of passive properties of the bladder wall to stress and therefore to detrusor pressure during filling falls into four categories.

1. No contribution. When empty the bladder dome collapses down inwards towards the base. During the initial stage of filling the wall evaginates and there is therefore no stretch and no contribution to detrusor pressure.

2. Elastic behaviour. In certain elastic materials the stress is proportional to the size of the deformation and therefore to the change in length (length dependence). This factor is represented by a spring in Fig. 9.13.

3. Viscous behaviour. In viscous material the stress depends on the rate of change in length (rate dependence). This is represented by a dashpot in Fig. 9.13. An explanation of the behaviour of the bladder during filling is given in Fig. 9.12. A slow rate of distension causes a lesser and slower increase in stress than a rapid distension. If the bladder is then maintained in its distended state (isometric) then the stress (and pressure) will gradually decrease to a value which depends on the volume and not on the previous rate of filling. This phenomenon is caused by stress relaxation in the tissue, a property that is a consequence of *viscoelasticity*—a combination of the two properties described.

4. Plastic behaviour. Plasticity implies a deformation of material which is irreversible and which only occurs when the stress exceeds a certain threshold value. Such a deformation has been observed in tests on bladder muscle strips and to some extent is reproduced in vivo. van Duyl (1985) suggests that the observed reversion of this plastic deformation on bladder emptying may be an active process related to the small spontaneous con-

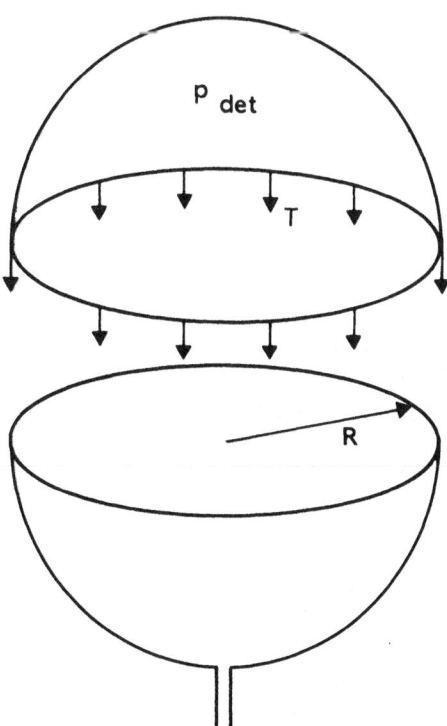

Fig. 9.11. A representation of Laplace's law. The halves of the sphere (bladder) are forced apart by the pressure (P_{det}) acting on the area (πR^2), and held together by the tension (T) acting around the circumference ($2\pi R$). (Griffiths 1984)

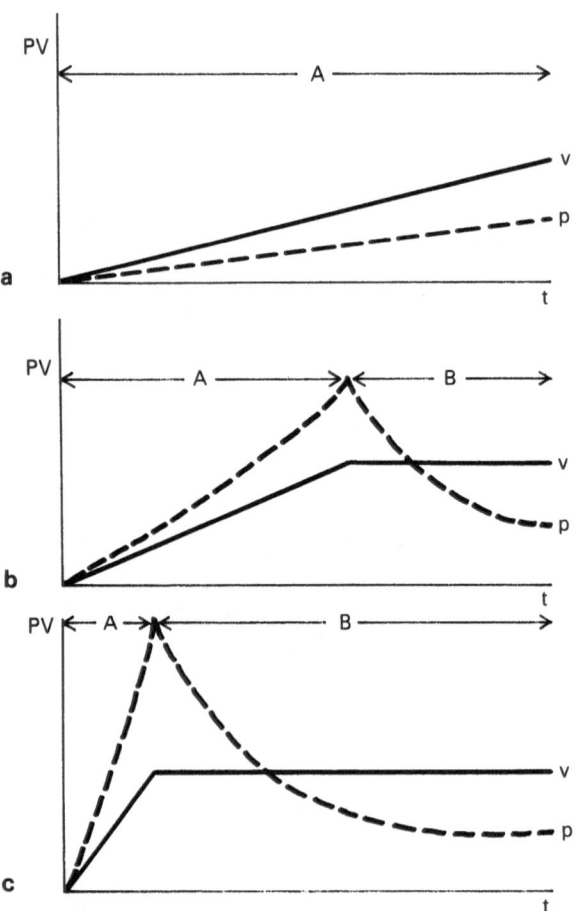

Fig. 9.12a–c. Pressure in the bladder is partly dependent on the rate of filling. Fast filling (c) results in higher pressure increase than slow filling (a, b). When filling is stopped the pressure decreases to a level appropriate to the volume infused. *V*, volume; *P*, pressure; *t*, time; *A*, volume increasing; *B*, volume constant.

tractions that can be observed in bladder muscle even when it appears, from pressure measurements, to be at rest. He argues that the properties of plasticity and contractility should be combined as a "plastocontractile" element, thus setting the basis for linking the mechanical properties of the bladder wall during filling and emptying.

Coolsaet (1984), describing experiments on bladder wall strips, reported that the behaviour could be described mathematically by three exponential functions and a constant. This can be represented as a mathematical model (Fig. 9.13) with three viscoelastic elements representing the exponential functions and with the constant represented by a parallel elastic element E_0. This complex acts in series with the element of plasticity according to van Mastrigt et al. (1978). van Duyl (1985) considers that the behaviour many be represented more accurately by placing a plastocontractile element in

series only with the elastic element E_0. Whether this latter single model representing both passive and active function within the bladder wall will be shown to be valid remains to be determined. Figure 9.13 also explains the greater development of force in the bladder wall after more rapid distension of the bladder (for details see figure legend).

Cystometry: Voiding Phase

The study of the voiding function of the bladder involves concurrent recording of pressure and flow. The bladder and urethra have independent functional properties and in combination these characteristics determine the *pressure–flow relationships of micturition*. By knowing both factors and relating them to the normal values of each it is possible to ascertain whether voiding function itself is normal. This can be done more accurately than from either measurement alone. For example, if the flow rate is low it is not possible to distinguish between outflow tract obstruction and inadequate detrusor contraction unless the detrusor pressure is measured.

The *opening time* is the elapsed time from the initial rise in detrusor pressure to the onset of flow. This is the initial isovolumetric contraction period of micturition. The *premicturition pressure* is the pressure recorded immediately before the initial isovolumetric contraction. It will be the same as the full resting pressure if the patient has not moved following the filling cystometrogram. *Opening pressure* is the pressure recorded at the onset of measured flow. Any delay in the recording of flow because of the passage time of urine to and through the flow meter must be allowed for in interpretation. *Maximum voiding pressure* is the maximum value of the measured pressure. *Pressure at maximum flow* is the pressure recorded at the time of maximum flow rate. Again, any delay in the recording of flow rate must be allowed for. *Contraction pressure at maximum flow* is the difference between the pressure at maximum flow and the premicturition pressure. An *after contraction* is represented by a pressure increase after flow ceases at the end of micturition. Any of these pressures may be recorded as intravesical (P_{ves}) or as detrusor pressure (P_{det}). The abbreviation for detrusor contraction pressure at maximum flow (for example) becomes $P_{ves}Q$max.

A typical arrangement of apparatus for recording pressure–flow studies of voiding is shown in Fig. 9.14. Pressure recording is performed as described previously during the filling phase. For voiding studies it is essential that any transurethral

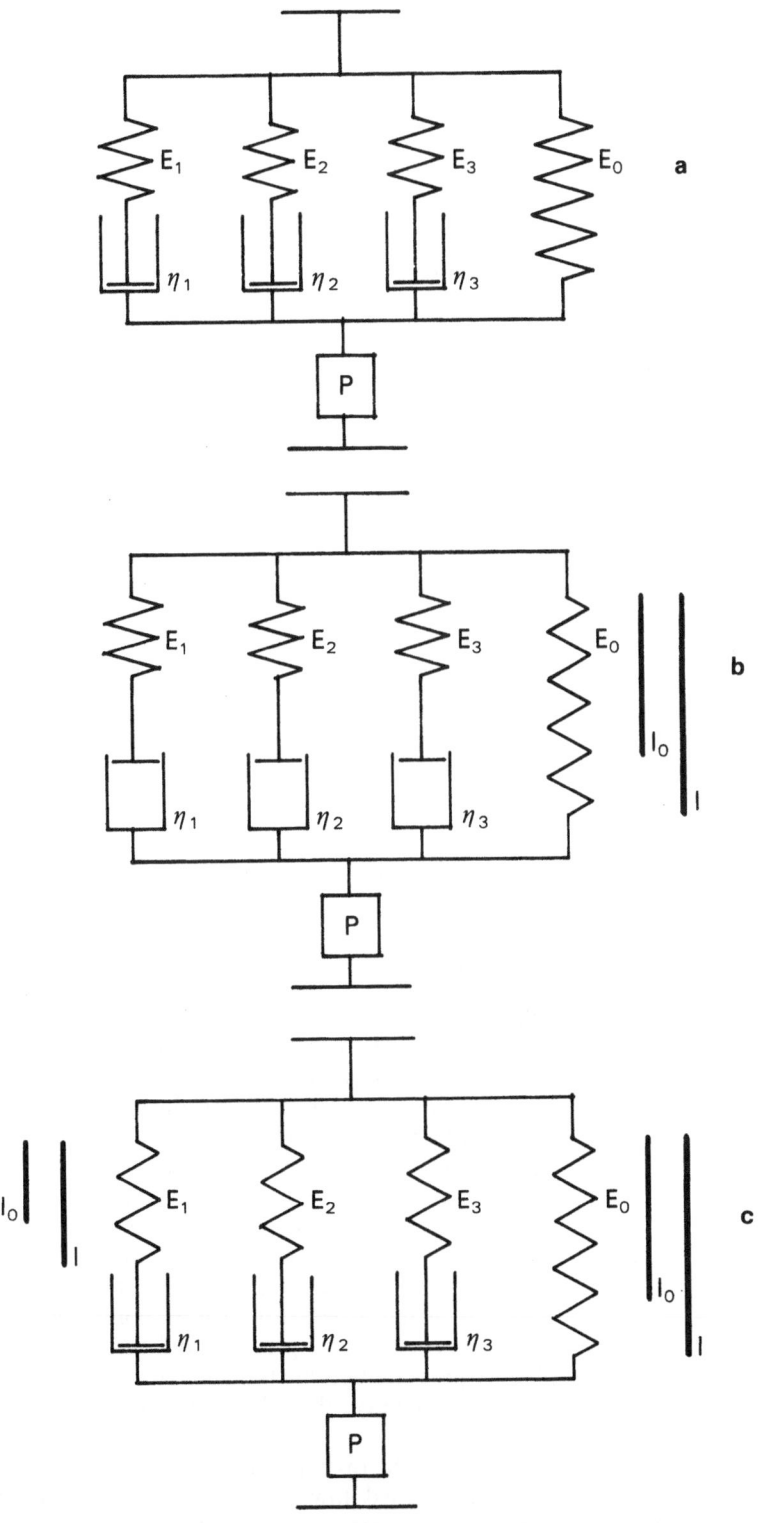

Fig. 9.13a–c. Mechanical model of the bladder wall. **a** Springs (*E*) represent elastic properties and dashpots (*η*) represent viscous properties. The force that remains when no further time-dependent phenomena are involved is represented by a parallel elastic element (E_0). A plastic element (*P*) must be added in series to denote the element of deformation on stretching. **b** When stretched slowly the dashpots can follow the elongation. The springs E_{1-3} are not required to change and so no force develops in them. The force is represented only by elongation of the spring E_0. This is equivalent to Fig. 9.12a. l_0, unstretched length; *l*, stretched length. **c** When stretched rapidly the dashpots cannot follow the elongation and so the springs elongate producing more force in the bladder wall and (temporarily) a higher pressure. This is equivalent to Fig. 9.12c. (After Coolsaet 1984)

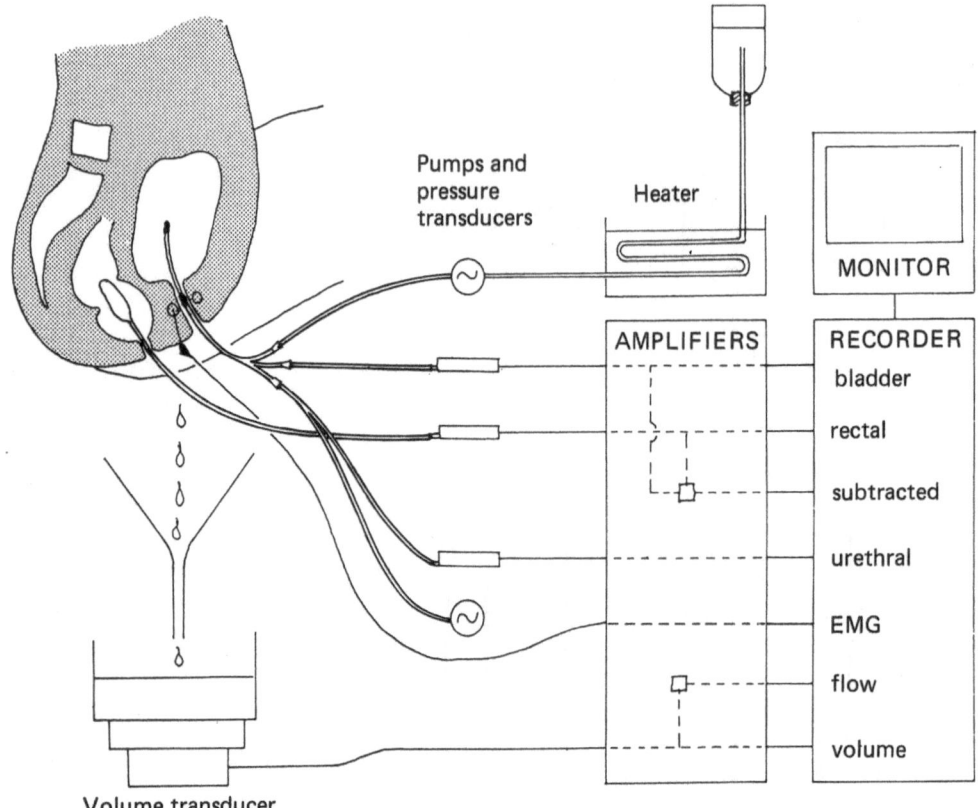

Fig. 9.14. The arrangement of apparatus for the clinical measurement of bladder and urethral function during voiding.

catheter is small enough to have no significant obstructive effect. Backman et al. (1966) have shown that tubes less than 1.55 mm external diameter are satisfactory in this respect. Gleason and Bottacini (1984) showed that a 5-Fr. catheter was .not. Uroflowmetry is performed as described earlier.

Table 9.4. Mean intravesical pressure at the onset of urinary flow. Figures in parentheses indicate either the normal range or ±2.SD

	n	Opening pressure (cmH$_2$O)
Males		
Denny-Brown and Robertson (1933)	1	(18–43)
Claridge and Shuttleworth (1964)	8	48 (15–85)
Backman et al. (1966)	10	74 (44–76)
Scott et al. (1964)	14	53 (32–74)
Females		
Zinner and Paquin (1963)	19	41 (23–85)
Backman et al.	15	59 (28–90)
Walter et al. (1979)	11	45 (39–61)
Sjoberg and Nyman (1981)	19	37 (20–54)

Opening Pressure

Although opening pressure is defined as the pressure recorded at the onset of measured flow, the delay between the actual bladder neck opening and the recording of flow by the transducer varies considerably between the sexes and under experimental circumstances. It is therefore difficult to make a uniform and accurate correction for the delay. It is perhaps better to consider the opening pressure to be measured at the point at which bladder neck opening is visualized on a video screen (if this is being concurrently performed) or the point at which the measured urethral pressure and the measured bladder pressure become the same (if simultaneous urethrocystometry, described later, is being performed).

In females the opening pressure measured at the onset of recorded flow varies from 37 to 52 cmH$_2$O according to various authors (Table 9.4). Sjoberg and Nyman (1981) have shown that opening pressure is independent of the voided volume. Backman et al. (1966) have shown that it is independent of age. Zinner and Paquin (1963) have revealed that the distribution of pressures is not normal; the skew deviation tends to produce a median pressure slightly lower than the mean pressure.

It is not surprising that there is considerable variability in opening pressure in females. Tanagho and Miller (1970) have shown that the initiation of voiding follows three patterns: (a) urethral pressure decreases and bladder pressure increases until the same value is achieved; (b) urethral pressure decreases with no change in bladder pressure; and (c) urethral pressure is maintained at a higher level until the bladder pressure increases to meet it. This again emphasizes the need for studies of opening pressure in the female, which may well be related very strongly to urethral function, to be performed with simultaneous urethral pressure measurement.

The opening pressures in men are significantly higher than those in women (Table 9.4). Claridge and Shuttleworth (1964) and Andersen (1982) have shown that opening pressure is elevated still further in cases of bladder outlet obstruction. The degree to which the pressure is elevated, however, does not correspond to the severity of the obstruction nor to the response to surgery (Jensen et al. 1983). This conclusion is not surprising in view of the variability of the bladder in its compensatory hypertrophy following obstruction.

It is instructive to reflect that one of the best studies of opening pressure in males was that performed by Denny-Brown and Robertson (1933). Their technique involved the concurrent measurement of bladder and urethral pressure. In one subject it was shown that the opening pressure varied from 18 to 43 cmH$_2$O at different micturitions and was inversely proportional to bladder volume. The latter correlation was not confirmed in other subjects. Now that more reliable methods of assessing bladder neck opening are available such as those mentioned already, the significance of opening pressure may well be ripe for reappraisal.

Sjoberg and Nyman (1981) have shown that the opening detrusor pressure is decreased as the bladder volume is increased. However, the reason for this is that the opening intravesical pressure, as noted already, is not changed with volume whereas the (hydrostatic) premicturition pressure increases with bladder volume. It follows that less of an *increment* of detrusor pressure is required as initial bladder volume increases.

Pressure During Voiding

The commonly recorded pressures during micturition are maximum intravesical pressure (P_{ves}max), maximum detrusor pressure (P_{det}max), intravesical pressure at maximum flow ($P_{ves}Q$max) and detrusor pressure at maximum flow ($P_{det}Q$max). In general, these pressures are independent of bladder volume, although maximum intravesical pressure has been reported both to increase (Sjoberg and Nyman 1981) and to decrease (Arbuckle and Paquin 1963) as bladder volume increases.

The pressure most usually recorded as the one which is most stable and relevant to voiding behaviour is the pressure at maximum urinary flow. The various values for this in normal persons are recorded in Table 9.5.

Analysis of $P_{ves}Q$max in the Bristol population (M. J. Torrens and L. Kavanagh 1986, unpublished

Table 9.5. Mean pressures during micturition in normal subjects. Figures in parentheses indicate either the normal range or ± 2SD

	n	P_{ves}max (cmH$_2$O)	$P_{ves}Q$max (cmH$_2$O)	$P_{det}Q$max (cmH$_2$O)
Males				
von Garrelts (1957a)	10	79 (65–98)		
Claridge and Shuttleworth (1964)				
Scott et al. (1964)	14	67 (41–93)		
Smith (1968)	13		77 (56–93)	
Frimodt-Møller and Hald (1972)	11		70 (50–91)	
Females				
Zinner and Paquin (1963)	19	53 (23–85)		
Arbuckle and Paquin (1963)	27	56 (29–85)	45 (23–67)	
Backman et al. (1966)[a]	15	72 (31–106)	63 (31–87)	23 (13–71)
Smith (1968)[a]	41		64 (43–84)	
Frimodt-Møller and Hald (1972)	11		54 (28–80)	
Walter et al. (1979)[a]	11	61 (44–105)		
Sjoberg and Nyman (1981)[a]	19	57 (25–89)	46 (28–64)	27 (9–37)

[a] Patients over 60 years included in series.

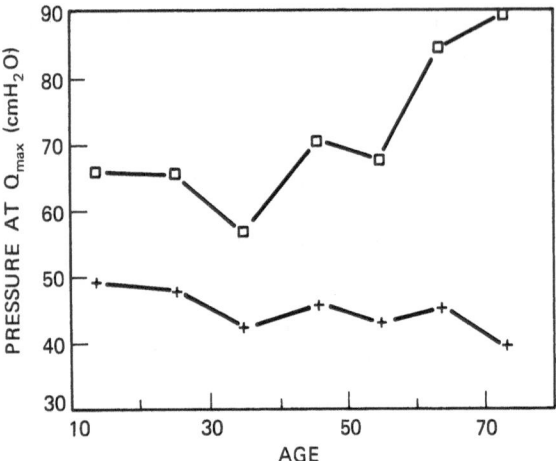

Fig. 9.15. The variation of intravesical pressure at maximum flow with age in males (*squares*) and females (*crosses*) from the Bristol series (121 males and 651 females).

Table 9.6. Mean intravesical pressure at peak flow and corresponding mean maximum flow rate (both $\pm$SD) in normal patients from the Bristol series as illustrated in Fig. 9.15

Age (years)	n	$P_{ves}Q$max (cmH$_2$O)	Qmax (ml/s)
Males			
10–19	13	66 ± 30	25 ± 9
20–29	11	66 ± 47	18 ± 9
30–39	21	57 ± 28	17 ± 7
40–49	30	70 ± 19	17 ± 8
50–59	28	68 ± 33	13 ± 6
60–69	43	84 ± 42	13 ± 6
70–79	15	89 ± 33	11 ± 5
Females			
10–19	12	49 ± 18	28 ± 6
20–29	72	48 ± 22	22 ± 9
30–39	158	42 ± 23	21 ± 9
40–49	187	46 ± 26	21 ± 9
50–59	148	43 ± 23	20 ± 9
60–69	57	45 ± 27	18 ± 9
70–79	17	40 ± 27	17 ± 7

work) shows a variation with age and differences between the sexes. Figure 9.15 and Table 9.6 show how $P_{ves}Q$max increases with age in males and decreases in females.

Bladder pressure during voiding is elevated in cases of bladder outlet obstruction so long as the muscles of the detrusor can produce compensatory hypertrophy. Voiding pressure will be reduced when the detrusor muscle is damaged in any way or when its innervation is compromised. Because the process of bladder overdistension which may follow outlet obstruction will itself compromise the contractility of the bladder it is not possible to correlate elevated voiding pressure with the degree of infravesical obstruction in all cases. Smith (1968) has found that the detrusor pressure is an accurate parameter in the assessment of urethral obstruction with P_{det} max > 100 cmH$_2$O being indicative of obstruction.

Detrusor Contractility

The contractile power of the detrusor muscle is dissipated both as pressure and flow. Assessment of contractility cannot therefore be made from pressure alone so long as flow is occurring. Muscle contraction has two characteristics—strength and speed. In the detrusor the strength of contraction can be estimated from the magnitude of the detrusor pressure generated under isovolumetric conditions, i.e. when there is no flow. The speed is less easy to assess, but the maximum velocity of shortening is related to the potential maximum flow rate which can be estimated as described in the section of hydrodynamics.

The isovolumetric detrusor pressure can be assessed by causing the flow to be suddenly interrupted during micturition, usually by contracting the urethral sphincter. This has been named the stop test (Griffiths 1977; Whiteside and Bates 1979) and the characteristic pressure trace is illustrated in Fig. 9.16. Griffiths (1980) estimates the normal range to be between 50 and 100 cmH$_2$O in adults of both sexes with a range in a group of 87 patients (Griffiths and van Mastrigt 1985) of 26–280 cmH$_2$O. This is a relatively simple test, but does depend on the efficiency with which the patient is able to interrupt the urinary stream. Flow interruptions may also involve some degree of detrusor inhibition which may influence the results.

Hydrodynamics of Voiding

Early attempts to identify the physical characteristics of urinary flow involved the calculation of the Reynolds number and the application of Bernoulli's law and assumed that the urethra was a round, straight, uniform, rigid tube. Under these

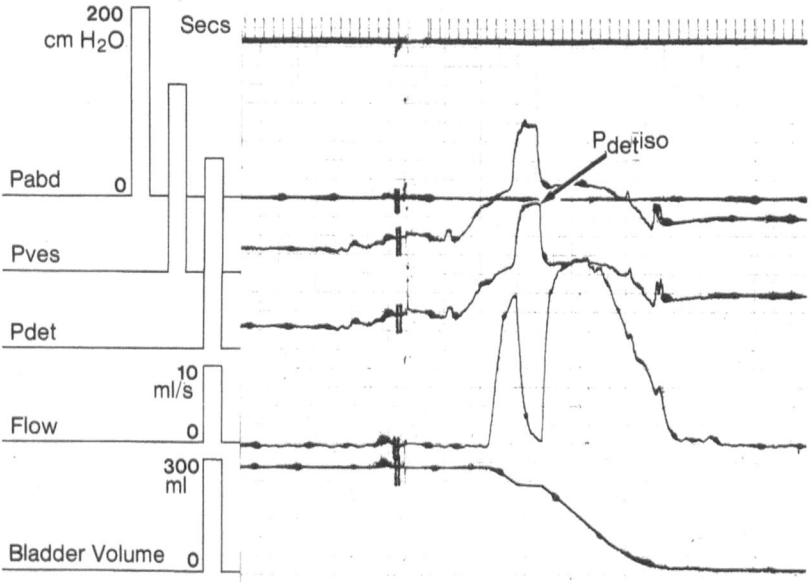

Fig. 9.16. A pressure–flow study of interrupted micturition (stop test) showing the isovolumetric pressure increase (P_{det} iso) generated by a normal bladder. (Abrams et al. 1983)

circumstances the relationship between pressure P and flow Q becomes (Backmann 1965)

$$P = \frac{Q^2 8\gamma}{\pi^2 D^4} \left(\frac{\lambda L}{D} + 1 \right)$$

including the effects of density (γ), Viscosity (λ), urethral length (L) and urethral diameter (D). This can be simplified to

$$P = Q^2 R \qquad \text{or} \qquad R = \frac{P}{Q^2}$$

where R is a composite resistance factor which varies (inversely) with urethral diameter. If the residual energy of the stream E_r is subtracted from the intravesical (pump) pressure a more appropriate formula might be

$$R = \frac{P - E_r}{Q^2}$$

However, the urethra is actually an irregular, variably distensible, curved tube. Resistance factors alone cannot quantify urethral function in a physiologically meaningful way because they depend on pressure. The invalidity of such factors is discussed by Schafer (1985). Resistance is only of secondary importance when defining the relation between pressure and flow and, because it depends on the square of the flow rate, it is heavily dependent on the voided volume. One of the reasons that resistance is not constant for the same urethra at different times is that urethral distensibility varies.

Griffiths (1980) introduced the concept of the *urethral resistance relation* by continuously plotting out pressure against flow during micturition. Under clinical circumstances a correction is needed to remove the delay inherent in recording flow, usually 1–2 s after the appropriate corresponding pressure. If this is done the results show plots as in Figs. 9.17 and 9.18. Flow does not commence until a certain urethral opening pressure P_{UO} is reached. This value characterizes the outlet elasticity and distensibility. After flow is established, *synchronous changes* in pressure and flow rate correspond to changes in detrusor power and *inverse changes* (e.g. pressure increase and flow decrease) correspond to changes in outlet function (see Fig. 9.21). Schafer (1985) notes that under certain circumstances

$$P = P_{UO} + \frac{1}{C} Q^2$$

Where C is a constant strongly related to the cross-sectional area of the urethra. This is in fact similar to $P = RQ^2$ derived from Bernoulli's equation, where R is also strongly related (inversely) to effective urethral diameter. This is perhaps because the urethral resistance relation reflects the function of a relaxed bladder outlet exhibiting constant mechanical properties during voiding (i.e. more like a uniform rigid tube).

In practice, the usually observed form of the pressure–flow plot does not correspond to the characteristic urethral resistance relation except under optimal conditions. The optimal plot,

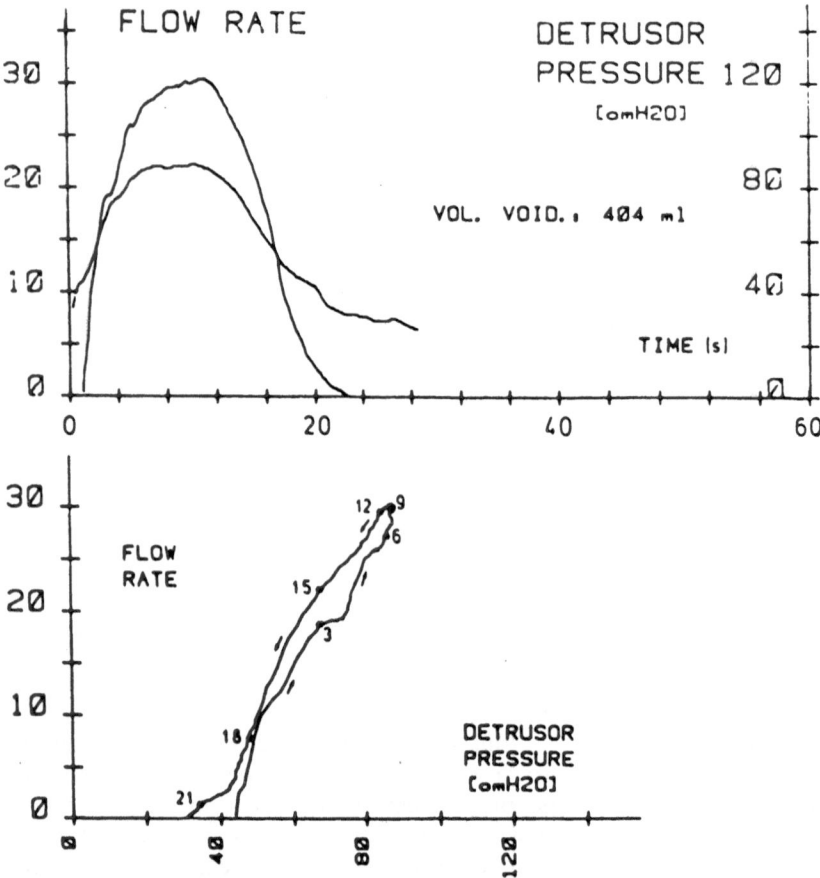

Fig. 9.17. Normal voiding dynamics presented as pressure–flow traces with respect to time (*upper trace*). The data is filtered and corrected for the time lag in measuring flow and displayed as pressure plotted against flow (*lower trace*). Time marks in seconds are annotated. The curve approximates to the urethral resistance relation (URR). (Schafer 1985)

described by Schafer as the *passive urethral resistance relation* (PURR) can be regarded as the plot of flow rate against pressure under conditions of perfect urethral relaxation (see Fig. 9.21). The difference between the observed flow rate and the predicted or optimal flow rate can be used to quantify the interference with relaxation caused by dynamic urethral factors such as sphincter contraction. This *dynamic urethral resistance relation* (DURR) is written

$$DURR = \frac{Q\,(\text{predicted}) - Q\,(\text{observed})}{Q\,(\text{observed})}$$

related to time (Fig. 9.19). It is very similar to the integrated EMG trace which confirms its dependence on active urethral factors.

The advantage of the PURR is that it is based on two factors quite independent of bladder volume (urethral opening pressure and urethral cross-sectional area). This makes its use in cases of

obstruction with small voided volumes quite reliable. Prostatic obstruction has its greatest effect by elevation of the urethral opening pressure P_{UO} with a relatively normal curve of urethral resistance relation (compressive type, Fig. 9.20). A urethral stricture alters the cross-sectional area of the urethra and therefore causes a flattened urethral resistance relation curve (constrictive type, Fig. 9.20). From severe prostatic obstruction there is a combination of compressive and constrictive features.

Pressure measurements may not reveal the contractility of the bladder while flow is occurring because the flow will influence the measured pressure. Contractility is best judged under isometric conditions before flow begins or when it has been interrupted suddenly in mid void. Figures 9.16 and 9.18 show the effect of interruption of voiding. During the time when flow is rapidly decreasing or subsequently increasing the pressure–flow plot describes a different curve from that of the urethral

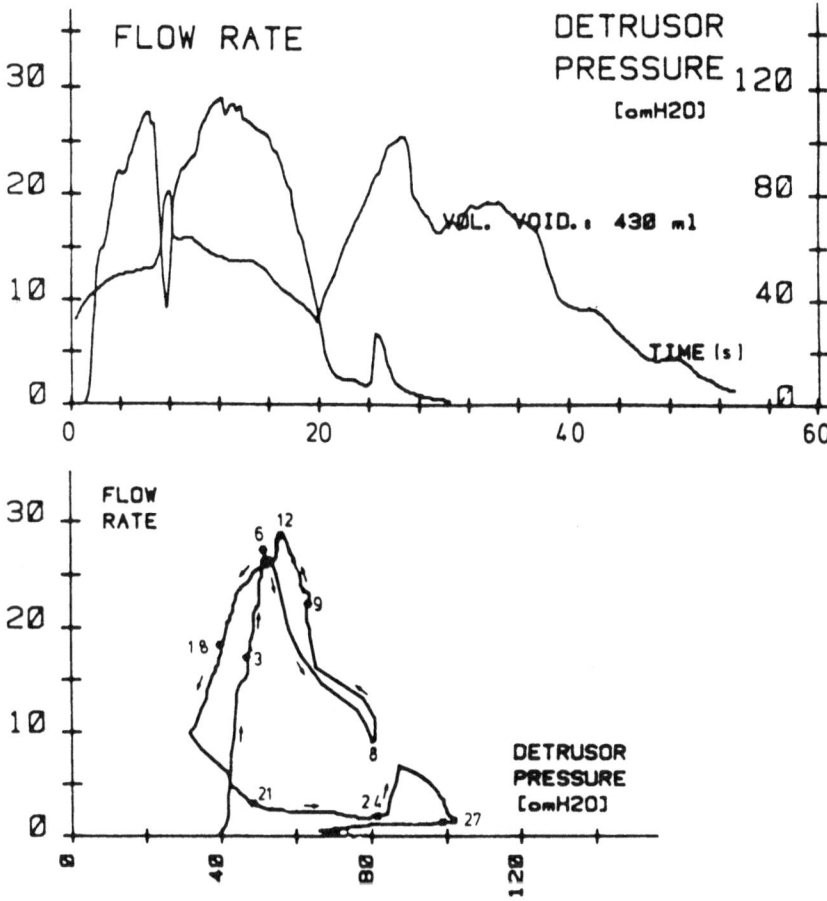

Fig. 9.18. Similar tracings to Fig. 9.17 from a patient in whom micturition is briefly interrupted by changes at the bladder outlet. Owing to this interruption the pressure–flow plot describes part of a different curve—the bladder output relation (BOR). (Schafer 1985)

resistance relation (see Figs. 9.18 and 9.22). The behaviour of the pressure–flow plot when approaching isometric conditions has been described by Griffiths (1980) as the *bladder output relation* (BOR). The theoretical maximum isometric detrusor pressure can be predicted by projecting the BOR curve to the point where flow Q is zero (Fig. 9.21). Griffiths has shown that the maximum isometric detrusor pressure is not greatly dependent on bladder volume in normal cases. However, it is well known that flow rate is so dependent. Extrapolation of the curve to the intercept on the flow axis, to measure the potential maximum flow, is therefore difficult and inaccurate.

If pressure is converted to detrusor force (strength) and flow to speed of shortening of detrusor muscle, an almost exactly similar curve to the BOR relates detrusor strength and speed. Further explanation of URR and BOR is given in Fig. 9.21 and its accompanying legend.

Urethrometry

The role of the urethra is to control and convey the flow of urine. Its principal properties are its distensibility and occlusibility. Distension occurs because of the potential relaxation, compliance and elasticity of the urethra allowing urinary flow. Urethral occlusion is provided by intrinsic and extrinsic, mainly muscular, forces acting on the soft and vascular submucosal "cushion" to close and seal the folds on urethral mucosa.

Observations, especially cine- and video-urethrography, have suggested that there are two separate continence zones, one at the bladder neck and the other at the level of the urethral sphincter. These correspond respectively to the areas which used to be called internal and external sphincters. Either one alone can maintain continence despite the incompetence of the other. Measurements have

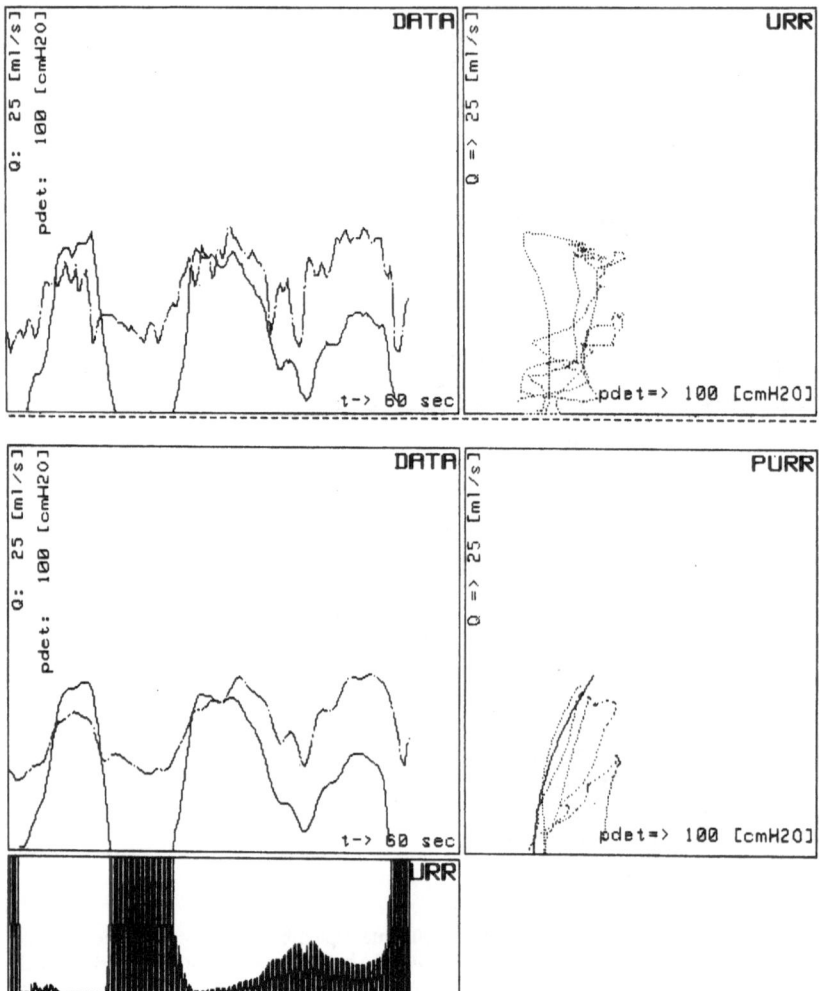

Fig. 9.19. Tracings illustrating the derivation of PURR and DURR according to Schafer (1985). The raw data (*upper traces*) are filtered and corrected for time lag (*middle traces*) and an ideal URR is fitted. In the *lowest trace* the difference between the observed flow and the flow predicted from PURR is plotted (*DURR*). It is minimal during the first flow period when outflow tract relaxation is nearest to optimal.

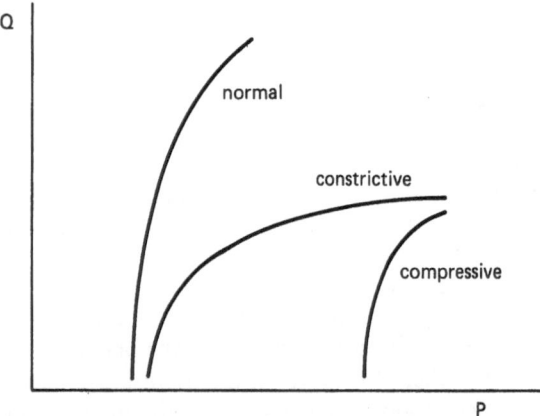

Fig. 9.20. Illustrations of the urethral resistance relation in normal, constrictive (urethral stricture) and compressive (prostatic hypertrophy) circumstances.

therefore concentrated on the function of these two zones. Most measurements have been of urethral pressure. Urethral function is also assessed by electromyography (see Chap. 10, p. 309) and indirectly by pressure–flow studies as described above.

Urethral Pressure

Urethral pressure is partly generated by the urethral wall itself (intrinsic pressure) and partly transmitted from outside the urethra (extrinsic urethral pressure). The latter is considered separately below. Early studies of urethral pressure (Bors 1948) were performed by measuring the resistance to retrograde infusion of fluid up the urethra. The technique

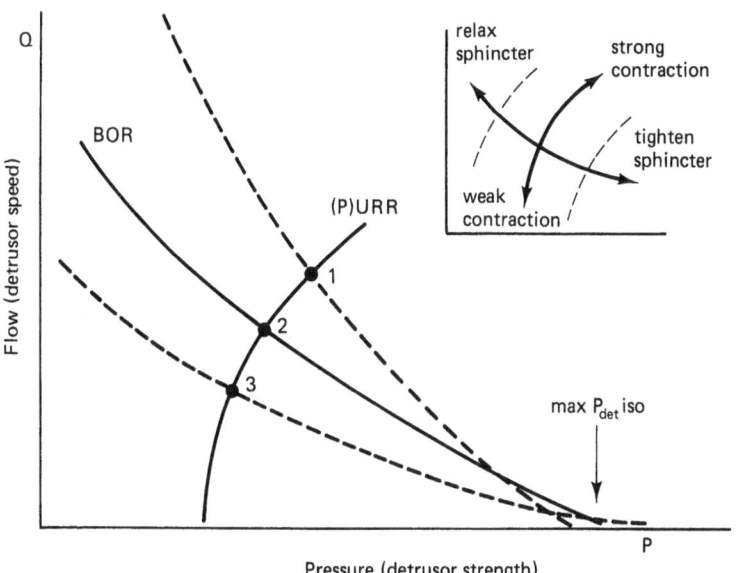

Fig. 9.21. The simultaneous solution of the urethral resistance relation (*URR*) and bladder output relation (*BOR*). As the bladder empties the detrusor contraction weakens and the flow rate and detrusor pressure usually follow the trajectory *1–2–3* along the URR. If the urethra is tightened the URR will be altered and the point will move to a higher pressure/lower flow state along the BOR (*inset*). The inverse changes will occur on urethral relaxation. Although the detrusor contraction is weaker at point 3, the isometric detrusor pressure following interruption of the stream usually increases as the bladder empties because, though the force gets smaller, the area of the bladder it acts on gets smaller faster. This explains the overlapping of BOR traces (*lower right*). (After Griffiths 1980)

is relatively inaccurate and the results very variable, depending on the reaction of the subject and the speed of infusion (Brindley et al. 1974) and the technique has been superseded by measurements recorded diectly from various points within the urethra. These measurements may be taken sequentially at all points through the urethra during a steady state period and reproduced in the form of a functional urethral profile (static assessment). Alternatively, recordings may be taken from one part of the urethra over a period of time during which events related to urine storage and micturition may occur (dynamic assessment). The measurement of urethral pressure in females has been reviewed by Hilton (1984).

Urethral Pressure Profile

The urethral pressure profile (UPP) indicates the intraluminal pressure along the length of the urethra with the bladder at rest. Figure 9.22 shows the terminology for urethral pressure measurement. The zero pressure reference point is the superior edge of the symphysis pubis. When describing the method it is necessary to specify the exact technique by which the measurements are made, particularly the type of recording catheter, the rate of catheter withdrawal, the bladder volume and the position of the patient. *Maximum urethral pressure* is the maximum pressure of the measured profile, *maximum urethral closure pressure* is the difference between the maximum urethral pressure and the intravesical pressure. *Functional profile length* is the length of the urethra along which the urethral pressure exceeds intravesical pressure.

There are three main methods for urethral pressure profile measurement:

1. The resistance to fluid or gas perfusion (Brown and Wickham 1969)
2. Small intraluminal balloons (Enhorning 1961)
3. Catheter tip transducers (Asmussen and Ulmsten 1975)

A review of these techniques has been provided by Abrams et al. (1983).

Brown and Wickham (1969) introduced the concept of a recording catheter with a lateral eye kept open by a constant flow of saline and attached to a measuring transducer during its withdrawal. They showed that the pressure registration error was less than 5 cmH$_2$O when the flow rate was greater than 2 ml/min. Harrison and Constable

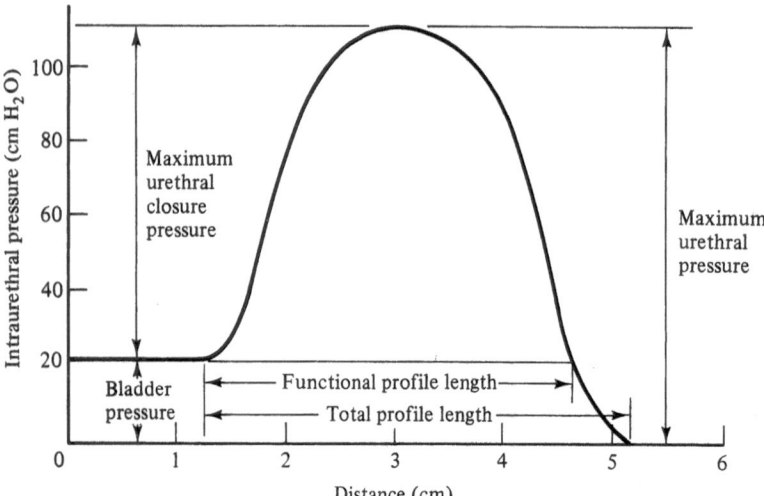

Fig. 9.22. A schematic representation of the urethral pressure profile giving the terminology recommended by the International Continence Society. (Abrams et al. 1983)

(1970) improved the method by incorporating an automatic position registering mechanism and a constant inflow by a motor-driven syringe. The profile was displayed on an X–Y plotter. Collins (1972) developed a motorized catheter withdrawal machine to move the catheter at a constant speed (Fig. 9.23). The technique of using small balloons, developed by Enhorning (1961) and Enhorning et al. (1964), was also used by Donker et al. (1962). However, balloons have disadvantages both in relationship to size, taking readings over an area of several millimetres, and also because of technical problems such as the difficulty of excluding air from the system.

Shelley and Warrell (1965) developed a urethral 'force' gauge and most recent studies have been performed with strain gauge catheter tip pressure transducers. Such transducers are sensitive not only to pressure, but also to occlusive force and, because of the rigidity of the carrying catheter may generate artefacts during recording if the urethra is not exactly straight (Plevnik et al. 1985).

Urethrocystometry

A dynamic assessment of urethral pressure changes with time and in relationship to bladder pressure

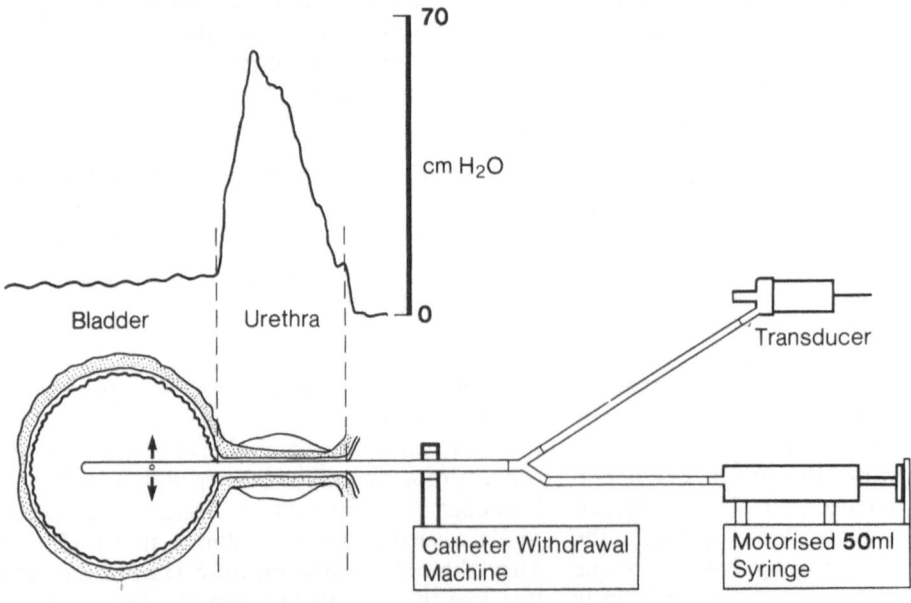

Fig. 9.23. A urethral pressure profile performed by the perfusion technique. (Abrams et al. 1983)

changes is very useful in the examination of the
initiation of micturition and of the cause for episodes of incontinence. The most accurate registration is by two catheter-mounted transducers with
one recording point in the bladder and one (or
more) in the urethra, usually located at the point of
maximum pressure. The mechanism of initiation of
micturition will be discussed in detail later in
Chapter 11 (see p. 347).

Normal Urethral Pressure Values

Studies of urethral pressure in normal female subjects have been undertaken by Enhorning (1961),
Beck and Maughan (1964), Shelley and Warrell
(1965), Edwards and Malvern (1974), Henriksson
et al. (1977), Obrink et al. (1977), Plante and Susset
(1980), Rud (1980b) and Hilton and Stanton (1982,
1983). Observations in normal males are infrequent
(Abrams and Torrens 1977; Yalla and Fraser 1982)
mainly because pathological alterations in male
urethral pressure are relatively rare.

The relation between maximum urethral pressure
and age in continent subjects (M. J. Torrens and L.
Kavanagh 1986, unpublished work) is shown in
Fig. 9.24 and Table 9.7. In a series of 161 males
there is a slight elevation of pressure in the 30–50
years age range and a slight reduction of pressure
over the age of 50 years.

In a group of 651 females there is a steady
reduction of pressure from the second decade
onwards. These results correlate closely with those
of Rud (1980b), but do not show the more profound
reduction in old age which would be evident if the
whole of a random population was analysed. The
reason for this is that the proportion of incontinent
subjects also increases with age.

Rud (1980b) showed that the functional urethral
length in females, which averaged 25–27 mm in
youth and old age, increased slightly to 31 mm in
the fourth decade. It is likely that these changes in
the urethra are related to oestrogen levels. Rud
(1980a) showed that the administration of oral
oestrogen caused a significant increase in maximum
urethral pressure and urethral length at rest. van
Geelen et al. (1981) studied the urethra at different
phases of the menstrual cycle. They could not
confirm any increase in pressure during those
phases when the serum levels of oestrogen were
shown to be high, but did show an increase in the
functional length of the urethra at these times.

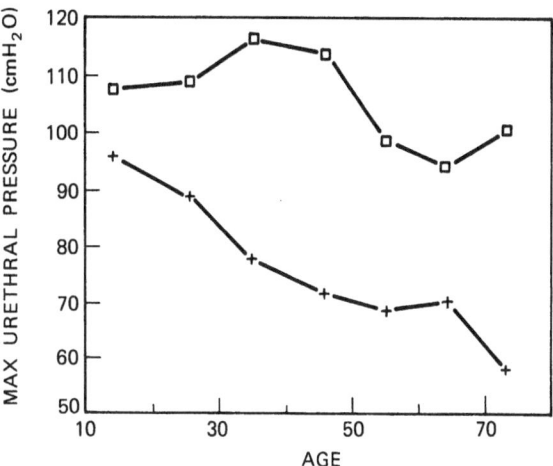

Fig. 9.24. The correlation between age and maximum urethral
pressure in 161 males (*squares*) and 651 females (*crosses*) from
the Bristol series.

Table 9.7. Mean maximum urethral pressure
($\pm$ SD) in normal patients from the Bristol
series as illustrated in Fig. 9.24

Age (years)	n	P_{ura}max (cmH$_2$O)
Males		
10–19	13	108 ± 31
20–29	11	109 ± 31
30–39	21	116 ± 27
40–49	30	114 ± 24
50–59	28	99 ± 28
60–69	43	94 ± 30
70–79	15	101 ± 25
Females		
10–19	12	96 ± 20
20–29	72	89 ± 25
30–39	158	78 ± 47
40–49	187	72 ± 53
50–59	148	69 ± 69
60–69	57	70 ± 59
70–79	17	58 ± 18

Transmission of Abdominal Pressure to the Urethra

Collins (1972) studied the transmission of pressure
produced by the Valsalva manoeuvre to the urethra
and the concept that this pressure transmission
assisted urethral closure during stress was
developed. A more detailed assessment by Heidler
et al. (1979) demonstrated that not only was 100%
of abdominal pressure transmitted to the urethra in
most subjects, but the transmission was in fact more
than 100%, indicating an active mechanism. About

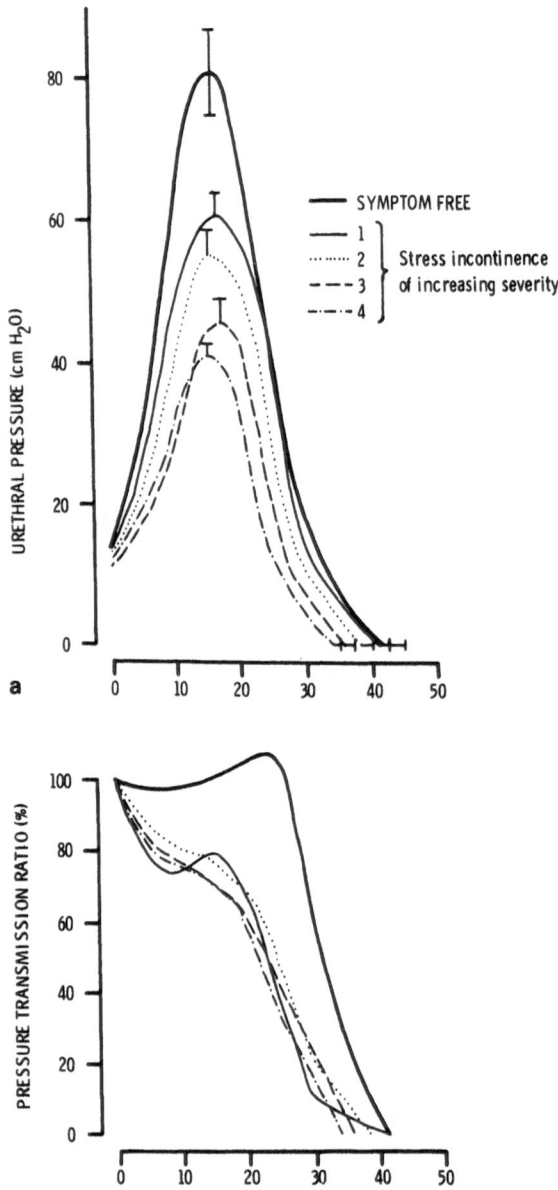

Fig. 9.25. a Averaged urethral pressure profiles in groups of stress-incontinent and symptom-free women. **b** Averaged pressure transmission ratio (on coughing) in the same groups. The ratio is greater than 100% in the more normal group. (Hilton and Stanton 1982)

or strain is responsible for the apparent high pressure transmission. It follows that lack of this reflex contraction of the pelvic floor will be just as important in the predisposition towards incontinence as a lack of transmission of abdominal pressure to the proximal urethra.

Urethral Pressure Gradient During Voiding

The term "profile" was probably introduced by Zinner et al. (1963) who studied the pressure in the urethra at various points during voiding. They identified the presence of a significant pressure gradient across the urethral sphincter area which they considered was the narrowest segment and therefore the segment controlling flow. This has been termed the "compressive zone" by Griffiths (1980). Yalla et al. (1980) confirmed that the area of the sphincteric or membranous urethra in men was the narrowest segment, but both Yalla et al. (1980) and Woodside (1982) find that the narrowest part of the female urethra is distal to the area of the greatest concentration of urethral muscle and conclude that the flow rate is controlled almost entirely by the diameter of the distal urethral segment. These observations are confirmed by studies of the urine flow rate, exit velocity of the urinary stream and effective urethral diameter conducted by Gleason and Bottacini (1982). Despite these observations the numbers of studies of pressure gradients in the urethra remain very small.

A. D. Desmond (1986, personal communication) has devised a microcomputer-based method by which static urethral pressure profiles (UPP) and voiding or micturition urethral pressure profiles (MUPP) can be compared. This allows a better assessment of the level of any obstruction by comparing the site and amount of pressure decrease along the urethra (as the profile catheter is withdrawn during micturition) to the position of the bladder neck and the distal urethral sphincter (maximum urethral pressure) on the static profile. The results of such studies are shown in Fig. 9.27 (normal) and Fig. 9.28 (bladder neck obstruction).

Bladder Neck Function

The best way to assess bladder neck opening and closure remains its visualization during video-cystourethrography (VCUG). Under normal circumstances the urethra will be invisible during the filling phase; the bladder neck will funnel open rapidly during voiding and close even more rapidly when voiding is complete or is interrupted, "milking

one-third of continent women have a pressure transmission ratio of more than 100% and the result in a group of continent and incontinent subjects is shown in Fig. 9.25 (Hilton and Stanton 1982). Constantinou and Govan (1980) have shown that in some young normal women the pressure transmission ratio can be as high as 140%. Concurrent recording of sphincter EMG (Fig. 9.26) suggests that active muscle contraction during a cough reflex

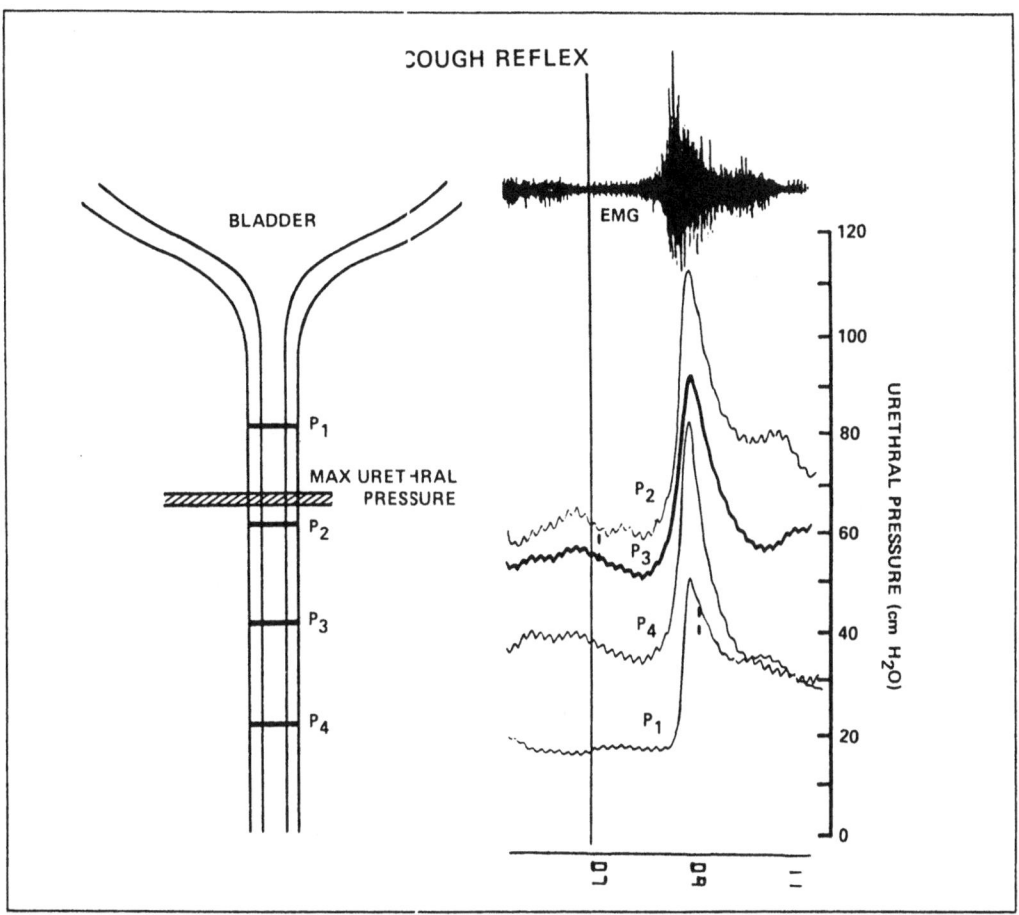

Fig. 9.26. Recording of the pressure at four points in the urethra (P_1–P_4) during a cough to show that the effect is greater at the level of the urethral sphincter (P_2) than in the proximal intra-abdominal urethra (P_1), suggesting an active component associated with muscle contraction. (Constantinou and Covan 1980)

back" any urine remaining proximal to the urethral sphincter. Such visualization allows a complete, though subjective, appreciation of the position, shape and speed of function of the bladder neck area.

Fonda et al. (1985) have managed to quantitate urethral shape from densitometry measurements performed on VCUG. This may allow new insight into bladder neck and urethral function for, with increasing use of digital radiography, it will be possible to compute the three-dimensionally reconstructed shape of the urethra as it changes during voiding.

Various attempts have been made to measure bladder neck activity, concentrating mainly on its competence during physical stress. Detection of fluid leakage has been achieved by the use of stress–pressure profiles (see Fig. 9.25; Henriksson and Ulmsten 1978; Hilton and Stanton 1982), by the fluid bridge test (Sutherst and Brown 1979) and

improved by the measurement of urethral conductance (Plevnik et al. 1983). All these methods have significant technical problems due to transducer orientation, catheter stiffness or catheter movement in relation to the urethra.

Urethral Sensation

The subjective sensory threshold within the urethra can be measured by applying the output from a constant current stimulator across a pair of platinum electrodes mounted on a urethral catheter (Kieswetter 1977; Powell and Feneley 1980). The normal threshold for a stimulus (20 Hz, 0.3 ms) is in the range 4–8 mA. The threshold is decreased in cases of infection and "hypersensitivity". It is increased in various neuropathic groups, but also in obstruction caused by benign prostatic hypertrophy and in a subgroup of enuretic subjects who may

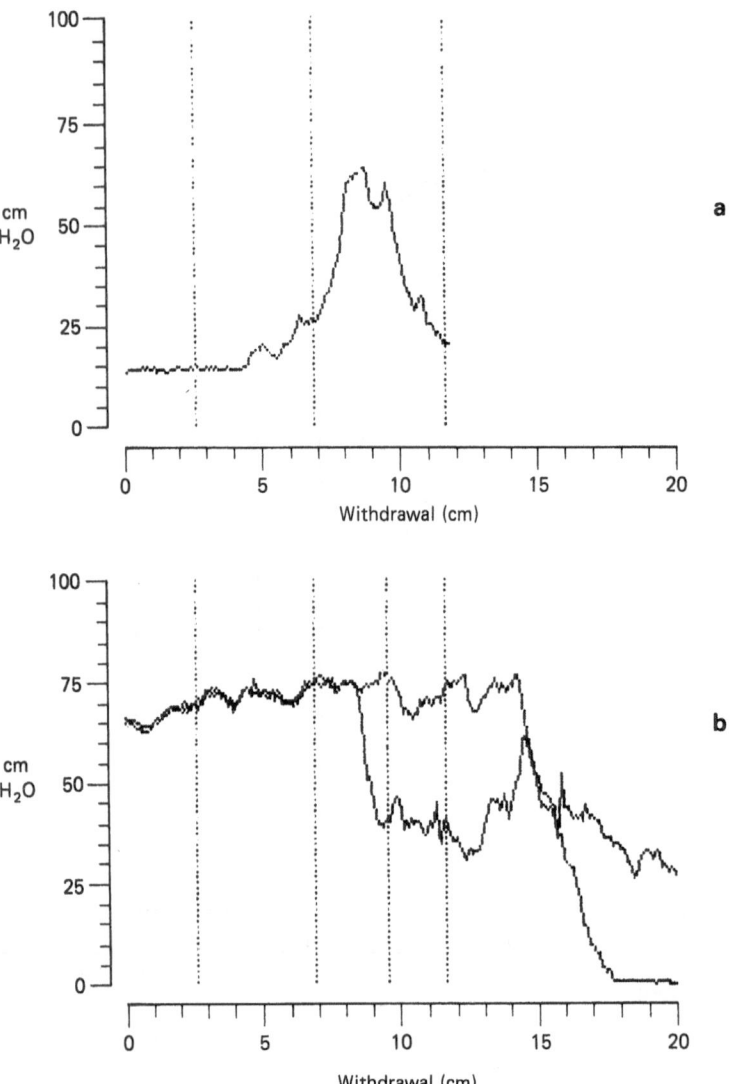

Fig. 9.27. a The UPP of a 65-year-old man with symptoms of urinary frequency and equivocal clinical evidence of urinary outflow obstruction. He was found to have unobstructed voiding on pressure–flow studies. The UPP was performed with 200 ml in the bladder and demonstrates a normal trace with a low amplitude prostatic plateau. **b** The MUPP in the same patient. The traces remain superimposed until the urethral catheter is beyond the prostatomembranous vertical cursor. This establishes that the bladder pressure and proximal urethral pressures are isobaric on voiding, indicating a normally functioning unobstructed bladder neck. (Courtesy of A.D. Desmond)

have a sensory defect as part of the reason for their dysfunction.

Murray (1982) noted that the sensory threshold declined during bladder filling, especially after the patients had experienced a first desire to void. He suggested that this change in sensitivity, absent in various cases with voiding dysfunction, played an integral part in the interaction of detrusor and urethra. Gerstenberg (1982) has found that the administration of emepronium bromide (25 mg

i.m.) causes an elevation of the urethral sensory threshold.

Urethral Compliance

Another new and interesting technique involves the measurement of the relation between urethral pressure and cross-sectional area (Colstrup et al. 1983). A highly distensible balloon is introduced into the

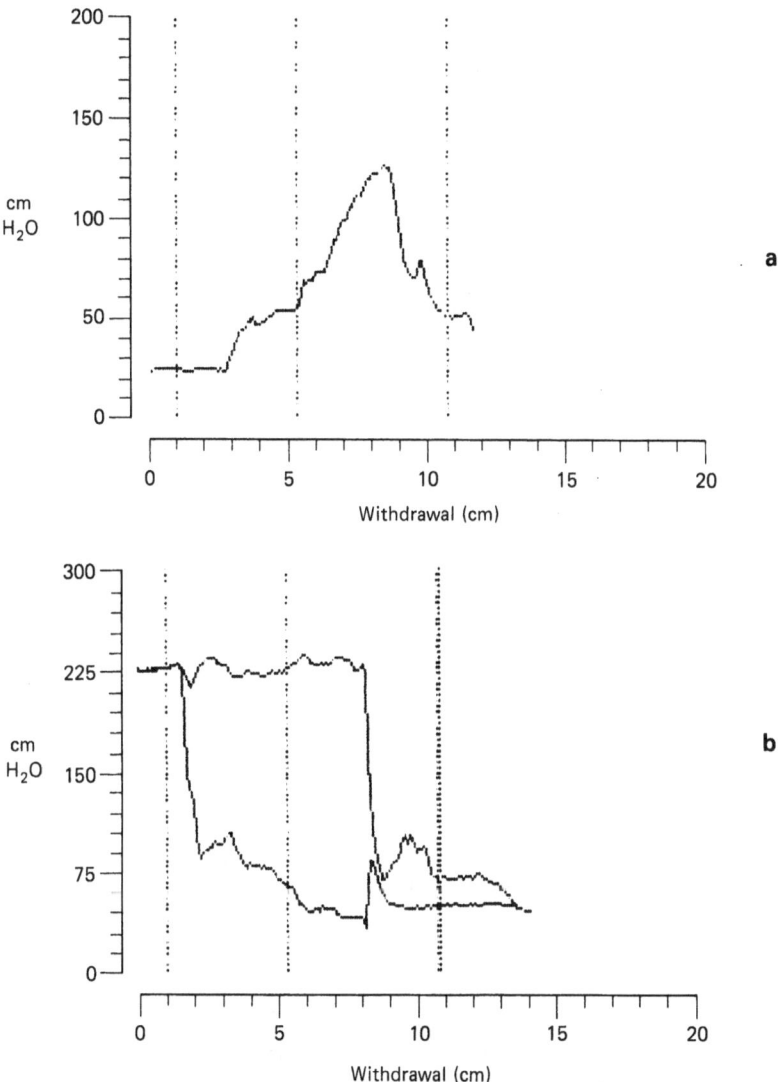

Fig. 9.28. a The UPP of a 55-year-old man with a typical history of obstructed micturition, which was performed after 200 ml had been infused into his bladder. Vertical cursors are positioned on the profile to mark sites that correspond to the bladder, the prostatomembranous junction and the urethra below the level of the distal sphincter mechanism. **b** The MUPP in the same patient. Initially, both catheter lumens are in the bladder and record the unsubtracted voiding pressure. As the urethral catheter is withdrawn, it passes through a grossly obstructed bladder neck and records a marked drop in pressure (163 cmH$_2$O). This is shown graphically by a wide separation of the traces, which occurs well proximal to the vertical cursor positioned at the level of the prostatomembranous junction. (Courtesy of A.D. Desmond)

urethra, the pressure within it is recorded directly and the cross-sectional area derived by an impedance method as the balloon is distended with fluid. The urethra exhibits hysteresis (the pressure at a given volume is higher during inflation than deflation) and Colstrup (1984a) concludes that this is due to an active, probably neuromuscular effect. The compliance decreases towards the mid and distal portions of the female urethra. Colstrup (1984b) also measured the dynamic response to

urethral distension. He found the additional pressure response to rapid change in volume (similar to that in Fig. 9.12c) to settle to the "balance" point more slowly than would be expected if the urethral muscle was solely responsible for it. He considered that elastic fibres and the vascular bed of the urethral submucosa could also contribute, as no doubt does the overall plasticity of the tissues involved. This technique deserves appraisal in stress incontinence and in recurrent urinary infections.

References

Abrams PH (1984) The practice of urodynamics. In: Mundy AR, Stephenson TP, Wein AJ (eds) Urodynamics, principles, practice and application. Churchill Livingstone, Edinburgh pp 76–92

Abrams PH, Torrens MJ (1977) Urethral closure pressure in the male. An analysis of 280 patients. Urol Int 32:137–145

Abrams PH, Feneley RCL, Torrens MJ (1983) Urodynamics. Springer, Berlin Heidelberg New York

Andersen JT (1982) Prostatism. Clinical, radiological and urodynamic aspects. Neurourol Urodyn 1:241–293

Arbuckle LP, Paquin AJ (1963) Urinary outflow tract resistance in normal human females. Invest Urol 1:216–228

Asmussen M, Ulmsten V (1975) Simultaneous urethrocystometry and urethral pressure profile measurement with a new technique. Acta Obstet Gynecol Scand 54:385–386

Backman KA (1965) Urinary flow during micturition in normal women. Acta Chir Scand 130:357–370

Backman KA, von Garrelts B, Sundblad R (1966) Micturition in normal women. Studies of pressure and flow. Acta Chir Scand 132:403–412

Bagley NA, O'Shaughnessy EJ (1985) Urodynamic evaluation of voluntary detrusor response in healthy subjects. Arch Phys Med Rehabil 66:160–163

Beck RP, Maughan GB (1964) Simultaneous intraurethral and intravesical pressure studies in normal women and in those with stress incontinence. Am J Obstet Gynecol 89:746–753

Brindley GS, Rushton DN, Craggs MD (1974) The pressure exerted by the external sphincter of the urethra when its motor nerve fibres are stimulated electrically. Br J Urol 46:453–462

Bors E (1948) A simple sphincterometer. J Urol 60:287–289

Brown M, Wickham JEA (1969) The urethral pressure profile. Br J Urol 41:211–217

Claridge M, Shuttleworth KED (1964) The dynamics of obstructed micturition. Invest Urol 2:188–199

Collins CD (1972) Electrical stimulation in the treatment of incontinence. ChM Thesis, University of Sheffield

Colstrup H (1984a) Rigidity of the resting female urethra. Part 1. Static measurements. J Urol 132:78–81

Colstrup H (1984b) Rigidity of the resting female urethra. Part 2. Dynamic measurements. J Urol 132:82–86

Colstrup H, Mortensen SO, Kristensen JK (1983) A new method for the investigation of the closure function of the resting female urethra. J Urol 130:507–511

Constantinou CE, Govan DE (1980) Urodynamic analysis of urethral, vesical and perivesical pressure distribution in the healthy female. Urol Int 35:53–72

Coolsaet BLRA (1984) Cystometry. In: Stanton SL (ed) Clinical gynecologic urology. Mosby, St Louis, pp 59–81

Coolsaet B (1985) Bladder compliance and detrusor activity during the collection phase. Neurourol Urodyn 4:263–273

Denny-Brown, D, Robertson EG (1933) On the physiology of micturition. Brain 56:149–189

Donker PJ, Ivanovici F, Noach EL (1972) Analyses of the urethral pressure profile by means of electromyography and the administration of drugs. Br J Urol 44:180–193

Drach GW, Layton TN, Binard WJ (1979a) Male peak urinary flow rate: relationships to volume voided and age. J Urol 122:210–214

Drach GW, Ignatoff J, Layton T (1979b) Peak urinary flow rate: observations in female subjects and comparison to male subjects. J Urol 122:215–219

Edwards L, Malvern J (1974) Studies of intraurethral pressures in normal and incontinent women. Urol Int 29:205–206

Enhorning G (1961) Simultaneous recording of intravesical and intraurethral pressure. Acta Chir Scand (Suppl) 276

Enhorning G, Miller ER, Hinman F (1964) Urethral closure studies with cine roentgenography and simultaneous bladder-urethra pressure recording. Surg Gynecol Obstet 118:507–516

Fall M, Erlandson BE (1983) Isobaric volume registration in the assessment of detrusor instability. Neurourol Urodyn 2:205–212

Fonda D, Hickery DS, Brockleburst JC (1985) Dynamic shape of the female urethra during micturition. Clinical measurements. J Urol 134:88–91

Frimodt-Møller C (1972) A new method for quantitative evaluation of bladder sensibility. Scand J Urol Nephrol 6(suppl 15):135–142

Frimodt-Møller, C, Hald T (1972) Clinical urodynamics. Scand J Urol Nephrol 6(suppl 15):143–155

Gerstenberg T (1982) Cited by Hald T, Bradley WE (eds) The urinary bladder, neurology and dynamics. Williams and Wilkins, Baltimore, p 116

Gierup J (1970) Micturition studies in infants and children. Normal urinary flow. Scand J Urol Nephrol 4: 191–207

Gleason DM, Bottacini MR (1982) Urodynamic norms in female voiding II. The flow modulation zone and voiding dysfunction. J Urol 127:495–500

Gleason DM, Bottacini MR (1984) The effect of a fine urethral pressure measuring catheter on urinary flow in females. Neurourol Urodyn 3: 162–171

Griffiths DJ (1977) Urodynamic assessment of bladder function. Br J Urol 49:29–36

Griffiths DJ (1980) Urodynamics, the mechanics and hydrodynamics of the lower urinary tract. Adam Hilger, Bristol

Griffiths DJ (1984) Hydrodynamics and mechanics of the bladder and urethra. In: Mundy AR, Stephenson TP, Wein AJ (eds) Urodynamics, principles, practice and application. Churchill Livingstone, Edinburgh, pp 42–49

Griffiths, DJ, van Mastrigt R (1985) The routine assessment of detrusor contraction strength. Neurourol Urodyn 4: 77–78

Griffiths DJ, Scholtmeijer RJ (1982) Momentum flux measurements in boys. A clinical evaluation. Neurourol Urodyn 1: 173–182

Harrison NW, Constable AR (1970) Urethral pressure measurement, a modified technique. Br J Urol 42:229–233

Heidler H, Wolk H, Jonas U (1979) Urethral closure mechanism under stress conditions. Eur Urol 5: 110–112

Henriksson L, Ulmsten U (1978) A urodynamic evaluation of the effects of abdominal urethrocystopexy and vaginal sling urethroplasty in women with stress incontinence. Am J Obstet Gynecol 113:78–82

Henriksson L, Ulmsten U, Andersson KE (1977) The effect of changes of posture on the urethral closure pressure in healthy women. Scand J Urol Nephrol 11:201–206

Hilton P (1984) Urethral pressure measurement. In: Stanton SL (ed) Clinical gynecologic urology. Mosby, St Louis pp 110–126

Hilton P, Stanton S (1982) Urethral pressure measurements in women without urinary symptoms. Proceedings of the 12th Annual Meeting of the International Continence Society, Leiden, pp 12–15

Hilton P, Stanton S (1983) Urethral pressure measurement by microtransducer. The results in symptom free women and in those with genuine stress incontinence. Br J Obstet Gynaecol 90:919–933

Jensen KME, Bruskewitz RC, Iversen P, Madsen PO (1983) Predictive value of voiding pressures in benign prostatic hyperplasia. Neurourol Urodyn 2: 117–125

Kieswetter H (1977) Mucosal sensory threshold of urinary bladder and urethra measured electrically. Urol Int 32:437–448

Klevmark B (1974) Motility of the urinary bladder in cats during filling at physiological rates 1. Intravesical pressure patterns studied by a new method of cystometry. Acta Physiol Scand 90: 565–577

Kondo A, Mitsuya H, Torii H (1978) Computer analysis of micturition parameters and accuracy of uroflowmeter. Urol Int 33:337–344

Meyhoff HH, Griffiths DJ, Nordling J, Hald T (1980a) Accuracy and interpretation of results from the DISA momentum flux meter. Urol Res 8:37–41

Meyhoff HH, Griffiths DJ, Nordling J, Hald T (1980b) Momentum flux values at maximum flow in normal males. Urol Res 8: 67–70

Meyhoff HH, Griffiths DJ, Nordling J, Hald T (1980c) Clinical applications of momentum flux measurements with special reference to detrusor hyperreflexia and meatal stenosis. Urol Res 8:71–75

Murray K (1982) Urethral sensitivity—an integral component of the storage phase of the micturition cycle. Neurourol Urodyn 1:193–197

Obrink A, Bunne G, Ulmsten U (1977) Intraurethral and intravesical pressure in continent women. Acta Obstet Gynecol Scand 56:525–529

Plante P, Susset J (1980) Studies of female urethral pressure profile. Part 1. Normal urethral pressure profile. J Urol 123:64–69

Plevnik S, Brown M, Sutherst JR, Vrtacnik P (1983) Tracking of fluid in the urethra by simultaneous electric impedance measurement at three sites. Urol Int 38:29–32

Plevnik S, Janez J, Vrtacnik P, Brown M (1985) Directional differences in urethral pressure recordings: contributions from the stiffness and weight of the recording catheter. Neurourol Urodyn 4: 117–128

Powell PH, Feneley RCL (1980) The role of urethral sensation in clinical urology. Br J Urol 52:539–541

Rollema HJ (1981) Uroflowmetry in males. Reference values and clinical application in benign prostatic hypertrophy. Thesis, University of Groningen

Rud T (1980a) The effects of estrogens and gestagens on the urethral pressure profile in urinary continent and stress incontinent women. Acta Obstet Gynecol Scand 59: 265–270

Rud T (1980b) Urethral pressure profile in continent women from childhood to old age. Acta Obstet Gynecol Scand 59:331–335

Schafer W (1985) Urethral resistance? Urodynamic concepts of physiological and pathological bladder outlet function during voiding. Neurourol Urodyn 4:161–201

Scott FB, Quesada EM, Cardus D (1964) Studies on the dynamics of micturition. Observations on healthy men. J Urol 92: 455–463

Shelley T, Warrell DW (1965) Measurement of intravesical and intraurethral pressure in normal women and in women suffering from incontinence of urine. J Obstet Gynaecol Br Common 72:926–929

Siroky MB, Olsson CA, Krane RJ (1979) The flow rate nomogram. I Development. J Urol 122:665–668

Sjoberg B, Nyman CR (1981) Hydrodynamics of micturition in healthy females: Pressure and flow at different micturition volumes. Urol Int 36:23–34

Smith JC (1968) Urethral resistance to micturition. Br J Urol 40: 125–156

Susset JG, Picker P, Fretz M, Jorest R (1973) Critical evaluation of uroflowmeters and analysis of normal curves. J Urol 109: 874–878

Susset JG, Ghoniem GM, Regnier Ch (1982) Rapid cystometry in males. Neurourol Urodyn 1:319–327

Sutherst J, Brown M (1979) A test for bladder neck competence: the fluid bridge test. Urol Int 34: 403–409

Tanagho EA, Miller ER (1970) Initiation of voiding. Br J Urol 42: 175–183

van Duyl WA (1985) A model for both the passive and active properties of urinary bladder tissue related to bladder function. Neurourol Urodyn 4: 275–283

van Geelen JM, Doesburg WH, Thomas CMG, Martin CB (1981) Urodynamic studies in the normal menstrual cycle: the relationship between hormonal changes during the menstrual cycle and the urethral pressure profile. Am J Obstet Gynecol 141:384–392

van Mastrigt R, Coolsaet B, van Duyl WA (1978) Passive properties of the urinary bladder in the collection phase. Med Biol Eng Comput 16: 471–482

von Garrelts B (1957a) Intravesical pressure and urinary flow during micturition in normal subjects. Acta Chir Scand 114:49–66

von Garrelts B (1957b) Micturition in the normal male. Acta Chir Scand 114: 197–210

Walter S, Olesen KP, Nordling J, Hald T (1979) Bladder function in urologically normal middle aged females. Scand J Urol Nephrol 13:249–258

Whiteside G, Bates P (1979) Synchronous video pressure-flow cystourethrography. Urol Clin North Am 6: 93–102

Woodside JR (1982) Micturitional static urethral pressure profilometry in women. Neurourol Urodyn 1:149–158

Yalla SV, Fraser L (1982) Total and static pressure measurements of the male lower urinary tract. Neurourol Urodyn 1:159–172

Yalla SV, Sharma GVRK, Barsamian EM (1980) Micturitional static urethral pressure profile. A method of recording urethral pressure profile during voiding and its implications. J Urol 124:649–651

Zinner NR, Paquin AJ (1963) Clinical urodynamics: 1 Studies of intravesical pressure in normal human female subjects. J Urol 90:719–730

Zinner NR, Ritter RC, Arbuckle LD, Paquin AJ (1963) Clinical urodynamics III. Pressure profiles of the female urethra. Surg Forum 14:481–483

10 · Clinical Neurophysiology

Clare J. Fowler and Chris Fowler

EMG of the Sphincters and Pelvic Floor
Spinal Sacral Reflexes and Nerve Conduction Studies
Distal Motor Latencies, Pudendal Motor Conduction Velocities and Spinal Stimulation
Lower Urinary Tract Sensation
Sensory Evoked Responses
EMG of the Bladder Detrusor

Neurophysiological testing is becoming increasingly important and more widely used in the assessment of urological problems. Simple electromyography (EMG) of the pelvic floor with surface electrodes is already considered by many to be necessary for a full urodynamic study. More sophisticated neurophysiological investigations, such as needle electrode EMG, reflex studies, nerve conduction studies and motor latencies, can give more detailed information about the site of a neurological lesion affecting bladder function. However, their clinical usefulness is still uncertain. These more complicated tests are mostly performed in specialized centres and there are sound practical reasons why this should be so. Such studies require expensive equipment and the help of a clinical neurophysiologist. The investigations tend to be uncomfortable for the patient and, as a means of providing information which helps in the planning of effective treatment, they are often disappointing.

There is also a more fundamental reason why these studies have yet to become obligatory in the investigation of bladder dysfunction. Most of the current clinical neurophysiological methods measure conduction in large myelinated nerve fibres and the electrical activity of striated muscle. The bladder is predominantly innervated by small unmyelinated autonomic fibres and has no striated muscle. Consequently, it is the striated muscle of the sphincters and the pelvic floor and its nerve supply which will usually be examined. The results may be important but they will have only indirect relevance where the problem is that of bladder rather than rhabdosphincter function.

EMG of the Sphincters and Pelvic Floor

Introduction

EMG is the recording of electrical potentials generated by the depolarization of muscle fibres. In striated muscle, where electrical activity is tightly coupled to mechanical activity, potential changes can be recorded with extracellular electrodes as the muscle is excited. Unlike skeletal muscle, which can be voluntarily relaxed to the point of complete electrical silence, the striated muscle of the urethral sphincter shows continous motor unit activity. Occlusive urethral pressure is thus maintained until the subject initiates voiding.

Muscle fibres can be classified into two main types by histochemical means (Dubowitz and Brooke 1973). The types have contractile properties which broadly correspond to so-called slow-twitch (type I) or fast-twitch (type II) muscle fibres. Slow-twitch fibres are specialized for maintaining posture, and fast-twitch fibres for phasic voluntary movements. Whereas human skeletal muscle is a homogeneous mixture of both fibre types, the striated muscle of the urethral sphincter is composed predominantly of type I slow-twitch fibres (Gosling et al. 1981). It is probably the slow-twitch, type I fibres which take part in the continuous motor unit activity.

Bladder Storage Phase

At rest, as urine collects in the bladder, EMG activity is found in the urethral sphincter, with motor units firing at a steady rate of 2–8 Hz (Fig. 10.1). This persistent activity continues during sleep and light anaesthesia (Chantraine 1973), and the resultant muscle contraction maintains urethral pressure above intravesical pressure, thus contributing to urinary continence. Coughing and other manoeuvres which increase intra-abdominal pressure, produce a marked increase in the number of motor units in the firing pattern (recruitment). Recruitment is from a population of large "phasic" units which do not discharge continuously at rest. It is also these larger phasic units which are activated by voluntary efforts to squeeze with the pelvic floor muscles.

A recording of EMG activity from either the urethral or anal sphincter during bladder filling in cystometry, shows an increase in the number of

activate motor units as the bladder is filled towards its maximum capacity. The fast rate of filling often used in such studies is unphysiological, but the relationship between bladder fullness and sphincter activity is frequently stressed (Blaivas 1983). Voluntary contraction accompanying a conscious effort to postpone micturition also produces an increased interference pattern (Chantraine 1973).

Voiding

If urethral sphincter EMG is being recorded, the first identifiable stage of micturition is an inhibition of motor unit activity, which results in relaxation of the sphincter and a reduction in urethral pressure. In a normally coordinated void this is followed, within seconds, by a detrusor contraction and a rise in intravesical pressure and expulsion of urine from the bladder (Tanagho and Miller 1970)

Coordination of sphincter and detrusor contraction is essential for normal voiding. The realization that pelvic floor EMG could be used to time the onset and duration of sphincter contraction during a voiding cystometrogram first brought the technique to the attention of urologists. In disease, disruption of normal coordination can occur. For instance, active and inappropriate contraction of the sphincter with simultaneous detrusor contraction is a frequent finding in multiple sclerosis, and is known as "detrusor-sphincter dyssynergia" (Andersen and Bradley 1976). Such disorganization of the normal reflex is thought to result from lesions anywhere from the pons to just above the sacral spinal cord (de Groat and Booth 1980). A similar picture can be seen in neurologically intact subjects with voiding difficulties associated with failure of

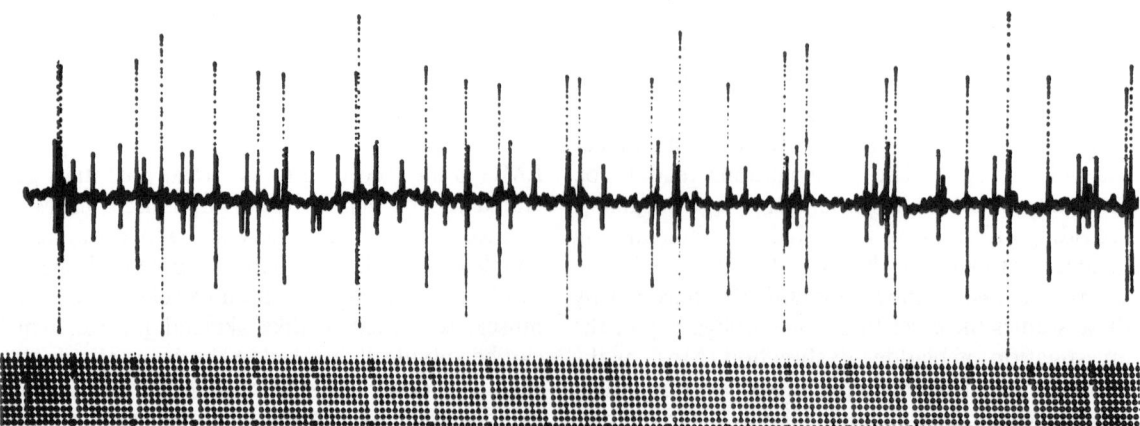

Fig. 10.1. EMG recording made with a concentric needle electrode from the striated muscle of the urethral sphincter. At least four different motor units, identified by amplitude, can be seen to be firing spontaneously.

pelvic floor relaxation (Blaivas et al. 1981). Detrusor sphincter dyssynergia is important because it may cause high intravesical pressures, ureteric reflux and progressive renal impairment.

The complex anatomy of the striated muscle of the urogenital diaphragm has already been described in this book. For the neurophysiologist, the anatomical separation of the sphincters from the main bulk of the pelvic floor muscle is of great importance. Although these anatomically distinct structures usually act in unison, by using suitable EMG electrodes, recordings can be made specifically from each sphincter and from selected pelvic floor muscles.

Methods of Recording Pelvic Floor and Sphincter EMG with Surface Electrodes

To record EMG from a chosen muscle, pick-up electrodes must be placed as close as possible to it. The most certain method is to impale the muscle on a needle or wire. However, there are serious problems with needle recordings from the urethral sphincter. The procedure is distinctly uncomfortable for the subject, and maintaining needle placement during the rigours of a cystometrogram is difficult. For many purposes, the precision of needle recording is unnecessary, and methods employing surface electrodes can be used.

Surface electrodes are arranged so that an active electrode is close to the muscle under study, and potential differences are recorded between this and a "remote" electrode. The remote electrode can be placed either upon some inactive part of the muscle, such as the tendon insertion, or at a more distant site. Alternatively, both recording electrodes can be placed over active muscle. The electrodes are connected to EMG recording apparatus and voltage changes are amplified, filtered and displayed on an oscilloscope. An audio amplifier is usually included in the apparatus to give an audible signal. The experienced electromyographer relies upon the sound produced by an EMG signal to recognize artefact, correctness of electrode placement and patterns of firing characteristic of individual motor units.

Anal Plug Electrodes

On the assumption that the sphincters and pelvic floor act synchronously, it has been convenient to record EMG activity from the anal sphincter during cystometry. Although this assumption appears correct in most physiological circumstances, careful studies show that it may not be so in partial cord lesions and multiple sclerosis (Vereecken and Verduyn 1970; Nordling and Meyhoff 1979). Despite this, recording from the anal sphincter is probably acceptable in most patients without overt neurological disease.

The anal plug electrode was used by Andersen and Bradley (1976) in their early studies of detrusor-sphincter dyssynergia. Bipolar concentric recording surfaces were mounted on a Teflon, hour-glass-shaped anal plug (a in Fig. 10.2). This was inserted into the anal canal and positioned so that the anal sphincter EMG could be recorded. Although much favoured in the past, because it is simple to insert and use, the anal plug is rather unpleasant and cumbersome for the patient. Its bulk may stimulate an urge to defecate. Children, in particular, quite rightly resent this device.

Skin Surface Electrodes Recording Perineal EMG

Before the introduction of perineal surface electrodes, urodynamic studies in children with disorders of voiding required hospital admission and a general anaesthetic for the insertion of a catheter and perineal copper wire electrodes. The need for a less invasive procedure led to the use instead of perineal skin electrodes. Monopolar electrodes are placed on the skin surface on either side of the anal orifice. Studies by Maizels and Firlit (1979), Barrett (1980), and Nielsen et al. (1985) which compared the quality of signal, ease of electrode placement and patient tolerance of various EMG recording methods all recommended the use of surface electrodes for children. Perineal skin electrodes are now increasingly preferred for recording during cystometrograms in adults as well.

A disposable surface recording electrode such as a silver chloride disc can be mounted on a self-adhesive sticker (b in Fig. 10.2). Skin electrical resistance may be reduced by washing and drying the area to which the electrode is to be applied, and recordings are further improved by the use of electrode paste. Firm adherence of the backing allows the patient to be completely mobile for testing in the sitting or standing position.

Catheter-mounted Ring Electrode

In an attempt to overcome the problem posed by the possible dissociation of urethral and anal sphincter activity in certain disease states, Nordling et al. (1978) developed a catheter-mounted ring electrode

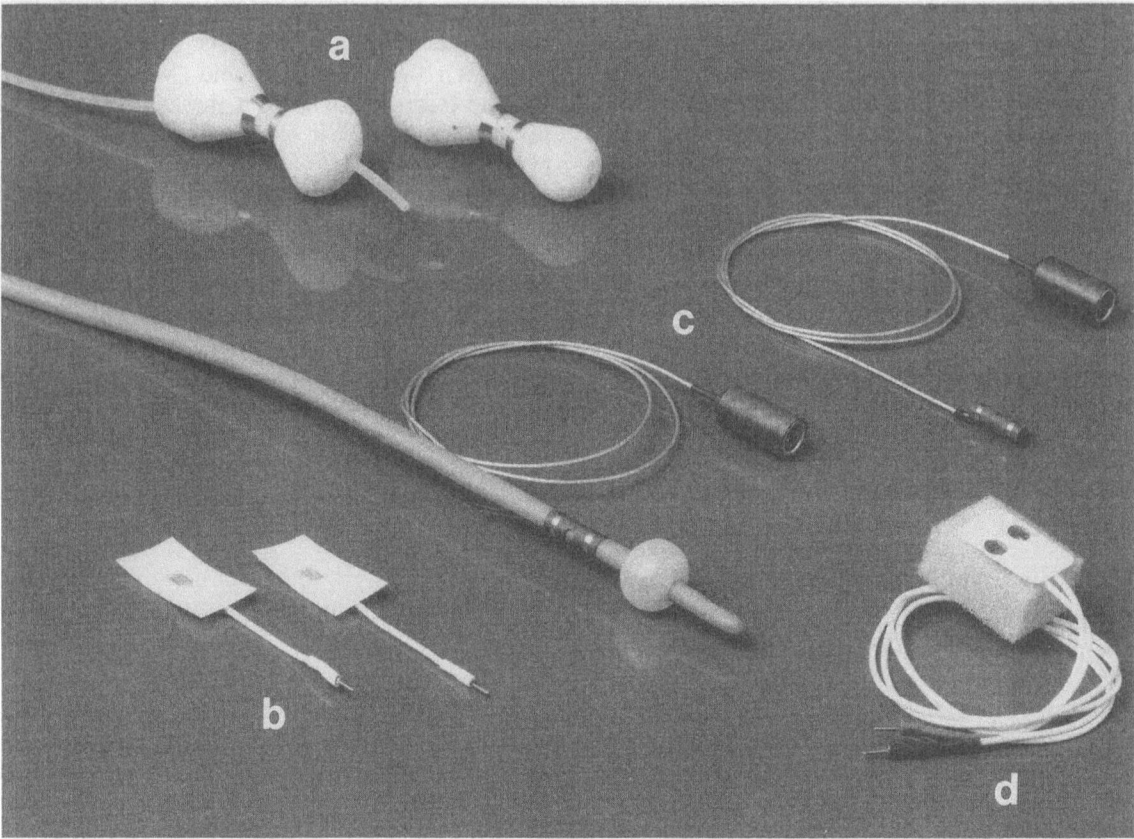

Fig. 10.2. Electrodes used to record sphincter activity: *a* Teflon anal plug; *b* disposable self-adhesive surface recording electrodes; *c* catheter-mounted ring electrode; *d* intravaginal recording electrode. (Courtesy of Dantec, Copenhagen)

with which to record urethral sphincter EMG directly. The electrode consists of a thin cylinder which fits snugly onto a Foley catheter. Wound onto the cylinder are two lengths of platinum wire separated by a distance of a centimetre or so and protected by a plastic coating (c in Fig. 10.2). The ring is positioned so that it is 2–3 cm below the balloon for women and 2–5 cm below it for men.

Although expensive, the electrode is sturdy, may be disinfected in 2% glutaraldehyde and is reusable. The disadvantages of this device are that catheterization is required for its use, and that simultaneous pressure–flow studies are impossible with it in place.

Vaginal Electrode

Lose et al. (1985) have developed a recording electrode which is applied intravaginally. This consists of a pair of silver chloride electrodes on a disposable flexible vinyl foam tampon (d in Fig. 10.2). It is inserted just behind the urethra, approximately 3 cm proximal to the external meatus. This places the recording electrodes in close proximity to the urethral striated sphincter. The device remains well in place during coughing and pelvic floor squeezing, and the electrical insulation offered by the sponge is such that only minor electrical activity from the anal sphincter is picked up.

Comparative studies using a coaxial needle electrode and the new electrode showed that the two methods compared well and that there is little movement artefact from the vaginal electrode. The method seems promising for routine urodynamics in women, because it appears to be accurate and causes minimal discomfort. There is as yet no study to show whether this recording method is able to identify when the sphincters are not acting synchronously.

Concentric Needle Electrode EMG of the Urethral Sphincter

Recent American interest in urethral sphincter concentric needle electrode (CNE) EMG (Blaivas et al.

1977; Dibenedetto and Yalla 1979; Mayo 1979) follows earlier work in Europe (Franksson and Petersen 1955; Chantraine 1966; Vereecken and Verduyn 1970). In the USA, CNE EMG has been used principally in combination with urodynamic studies to supply "kinesiological" information about sphincter activity during voiding, as the preferred method for identifying detrusor–sphincter dyssynergia (Blaivas 1983). The importance of recognizing a reduced interference pattern and the presence of polyphasic motor units has also been re-emphasized (Blaivas et al. 1977; Dibenedetto and Yalla 1979; Mayo 1979).

A combined urodynamic and electrophysiological approach has its disadvantages. Good-quality EMG requires quiet recording conditions for at least 15 min with the subject in a horizontal position. There must be a clean EMG signal, uncontaminated by electrical interference from other equipment, recording apparatus with a signal trigger and delay line, and an experienced electromyographer. These circumstances are not usually encountered during videocystometry in the average X-ray department, and a separate neurophysiological recording session is preferable.

Needle Placement

Some practice is needed to place the needle electrode for urethral sphincter EMG correctly. In women, it can best be done with the patient supine with hips flexed and abducted. The electrode (30 mm length, tip diameter 0.45 mm) is inserted at a point 1 cm lateral to the urethral meatus and guided towards the midline. Male subjects lie in the left lateral position; the needle is introduced transperineally in the midline and guided to the level of the apex of the prostate by a finger in the rectum. It is useful to listen to the audio output of the EMG machine during needle placement. As the electrode nears the striated muscle, the sound of distantly firing motor units can be heard. Once it is correctly sited, there is a burst of activity which rapidly subsides to a tonic interference pattern of 2–4 motor units (see Fig. 10.1)

The mass of intramural striated muscle is small but is anatomically separate from the pelvic floor musculature (Gosling et al. 1983). Inspection of an eviscerated cadaveric female pelvis from within the abdominal cavity, shows a distinct space between the anterior parts of pubococcygeus and the urethra, known as the "paraurethral gutter". In women, therefore, if the needle electrode recording striated muscle EMG is angled towards the midline, it must be lying in the sphincter muscle. Similarly,

in a man, the sphincter is the only striated muscle located in the midline at the apex of the prostate. Consequently, though the target is small, it can be located with some certainty.

The normal interference pattern comes from a small number of tonically firing motor units in the region of the needle tip. An increase in the number of firing units (recruitment), occurs with any manoeuvre which increases intra-abdominal pressure. Motor units are occasionally recorded which fire rhythmically with inspiration. Coughing causes a dramatic increase in the interference pattern with reflex recruitment of high amplitude phasic units. The patient can voluntarily produce a similar result when asked to "squeeze".

The steady tonic firing of motor units in the sphincter is ideal for the technique of individual motor unit analysis. By expanding the time scale of the display, the configuration of individual units becomes apparent (Fig. 10.3). Using a trigger and delay line, which are standard modules in modern EMG equipment, a motor unit can be made to appear repeatedly at the same point on the oscilloscope screen. The duration and amplitude of individual motor units can be measured from the screen with cursors, or later from film. The number of turns (phase reversals), of more than 100 μV are counted, to assess whether the motor unit is normally configured or excessively polyphasic (Fig. 10.4). Normal units have up to five such turns.

Analysis of motor units from the urethral sphincter shows that they have a characteristic appearance with a duration of less than 6 ms and an amplitude of between 0.15–0.5 mV and usually less than 1.5 mV (Franksson and Petersen 1955; Chantraine 1966; Jesel et al. 1973; Blaivas 1983; Vodusek and Light 1983). The three-dimensional histogram (Fig. 10.5) shows control data from one such study (Fowler et al. 1984).

Motor units from the anal sphincter have been shown to have very similar characteristics. Chantraine (1966) found the mean duration of motor units to be 5.5 ms, and in a similar study Bartolo et al. (1983) obtained a figure of 6.9 ± 0.3 ms.

EMG Changes of a Lower Motor Neuron Lesion

Denervation

Complete Lower Motor Neuron Lesion Following a complete lower motor neuron lesion, motor unit activity ceases in the affected striated

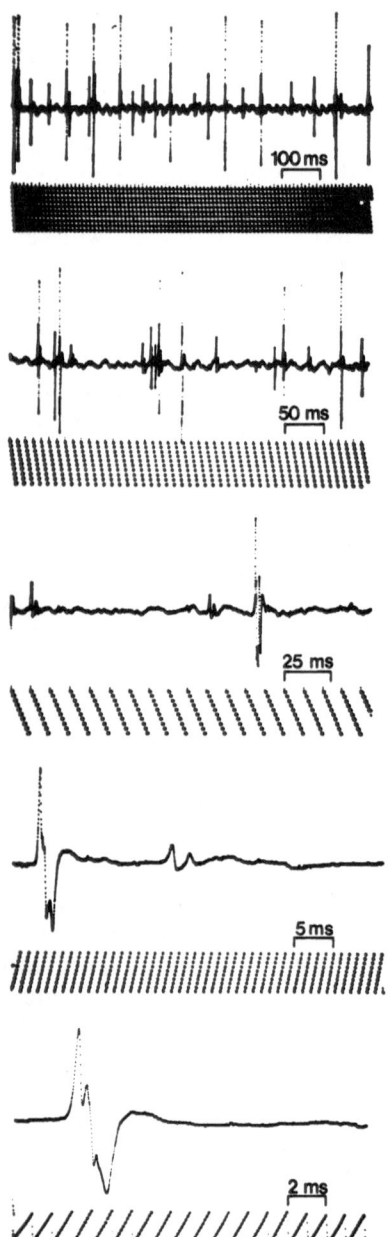

Fig. 10.3. CNE traces, with a progressively expanded time scale.

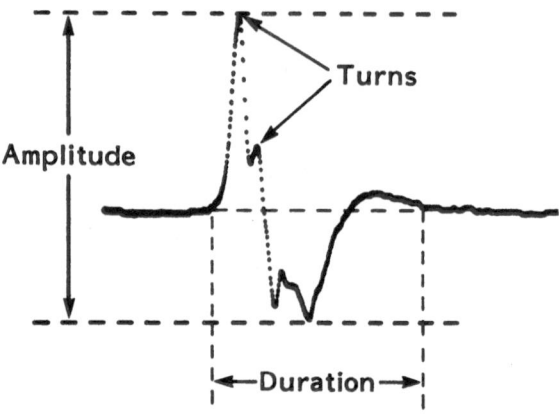

Fig 10.4. Measured features of an individual motor unit, on an expanded time scale, captured using a trigger and delay line.

the quiet, "popping" noise they produce on the audio amplifier. These potentials can only be picked up against a background of electrical silence, thus it is essential that all voluntary motor unit activity be suppressed. Some EMG experience is necessary to avoid confusing small, distant motor units with fibrillations. This is a particular problem for sphincter EMG where tonic activity of small motor units cannot be suppressed. Consequently, it is virtually impossible to identify fibrillations with certainty. In the experience of one author (C. J. F.), unequivocal fibrillations were found in one case in a series of more than 300 CNE recordings. Insertional fibrillations and spontaneous, persisting fibrillations were recordable in a man with a complete cauda equina lesion in whom no motor units were active. Because conditions for recognizing fibrillations in the urethral sphincter are so unfavourable, they should be reported with circumspection. Positive sharp waves, having the same significance as fibrillations, are also thought to arise from denervated muscle fibres, and the same reservations will apply when reporting this type of activity from the sphincter.

Partial Lower Motor Neuron Lesion A partial lower motor neuron lesion results in a diminished number of the motor units which can be deployed to maintain the strength of a muscle contraction. Thus, reduced numbers of motor units firing at abnormally high frequencies are from partially denervated muscle. On close inspection, such individual motor units show features characteristic of reinnervation.

muscle and none can be induced by attempting voluntary activation. Within 4–7 days of nerve section, electrical stimulation of the distal nerve stump also produces no response.

Fibrillations appear 12–21 days after denervation. These are short duration (0.5–2 ms), low amplitude (50–100μV), diphasic or triphasic potentials which are thought to arise from denervated muscle fibres. Fibrillations are best recognized by

CONTROLS

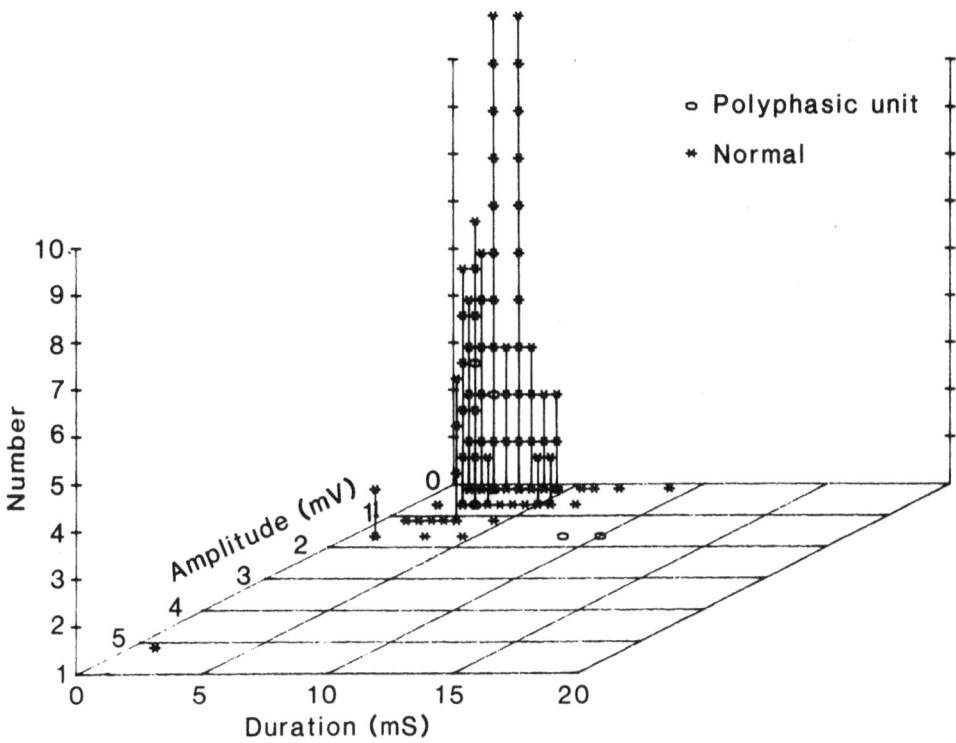

Fig. 10.5. Three-dimensional histogram showing the features of individual motor units found in 10 control subjects. (Fowler et al. 1984)

Reinnervation

Muscle becomes reinnervated either by regrowth of the axon from the site of the lesion, or by collateral reinnervation. With collateral reinnervation, motor axons which have survived a partial lesion sprout within the muscle and send out branches to denervated muscle fibres. Consequently, a single axon comes to supply a larger number of fibres and so the motor units develop a more complex waveform (Fig. 10.6). Initially, the immature axon sprouts conduct impulses slowly, which increases the duration of motor unit potentials. With time, the conduction velocity of these branches increases so that the waveform becomes more compact and of a larger amplitude than formerly.

Following complete peripheral nerve section, small low amplitude units (nascent units) are seen first as newly grown axons reach the muscle. These may later increase in amplitude and complexity.

The characteristic EMG pattern seen in established reinnervation is of a reduced number of high amplitude, polyphasic motor units firing at a rapid rate.

Changes in Motor Units of the Urethral Sphincter in Disease States

Cauda Equina Lesion

The cauda equina extends from a level of L-2 in the spinal canal to the sacral root exit foramina, a distance of approximately 15–20 cm in an adult. The considerable length of this structure renders it vulnerable to trauma, compression from lumbar disc protrusion or expanding tumours such as meningiomas or neurofibromas. A cauda equina lesion may result in perineal sensory loss, a loss of bladder function and of sphincter control.

Characteristically, urodynamic studies show an areflexic bladder and an incompetent sphincter (Pavlakis et al 1983). As would be expected in a lower motor neuron lesion, EMG with a CNE shows a reduced number of tonically firing motor units. As reinnervation is usual, the motor units tend to be polyphasic and, depending on the duration of the lesion, may be of prolonged duration or of abnormally high amplitude (Franksson and

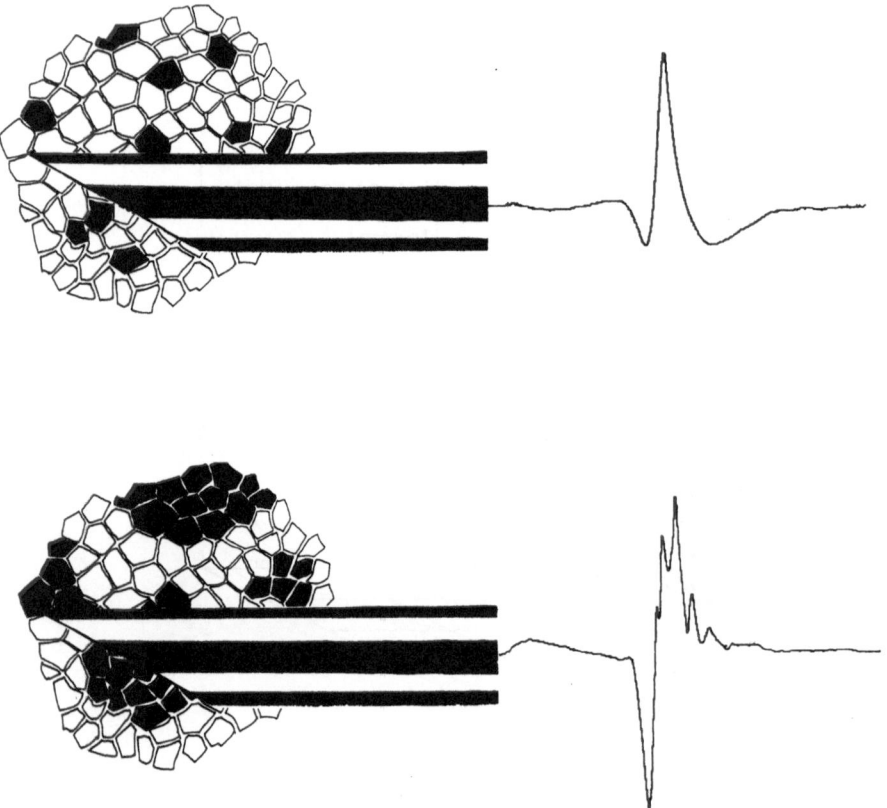

Fig 10.6. Changes in motor unit structure, recorded by CNE, following denervation and subsequent reinnervation. Muscle fibres which are part of the same motor unit are *shaded darkly*. With reinnervation, surviving motor axons sprout to innervate denervated fibres, thus producing groups of fibres all belonging to the same motor unit.

Petersen 1955; Pavlakis et al. 1983; Fowler et al. 1984).

Pelvic Nerve Injury

Disturbances of continence and micturition are common complications of radical pelvic surgery and are now thought to be due to intraoperative damage to the pelvic nerves (Mundy 1982; Kirby et al. 1986a). Analysis of individual motor units from the sphincter is particularly helpful in this condition as EMG abnormalities may be readily recognized. In addition to a depleted interference pattern, abnormal motor units can be found (Figs. 10.7, 10.8). The latter abnormality is easier to recognize with certainty than the former (Kirby et al. 1986a). Such a finding is particularly helpful in the management of patients who are not able to pass urine after radical pelvic surgery. Unlike those with a more proximal cause for the lower motor neuron lesion involving bladder innervation (e.g. a cauda equina lesion), patients with pelvic nerve injury do not have noticeable perineal sensory loss.

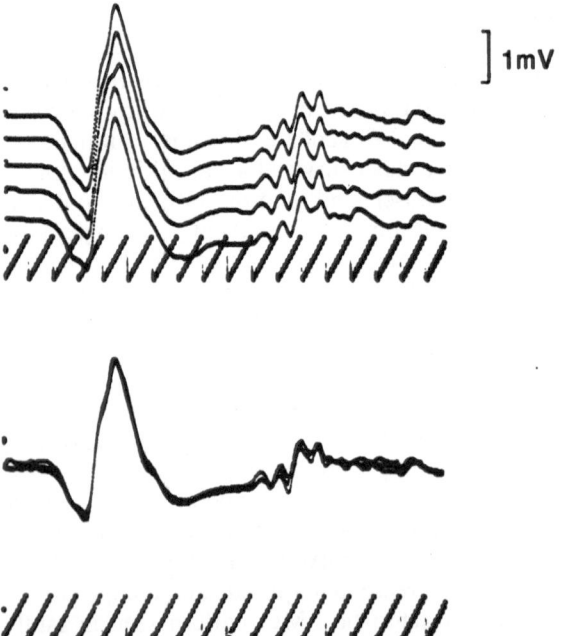

$]$ 1mV

Fig. 10.7. An individual motor unit from a patient with pelvic nerve injury after a total colectomy.

PELVIC NERVE INJURY

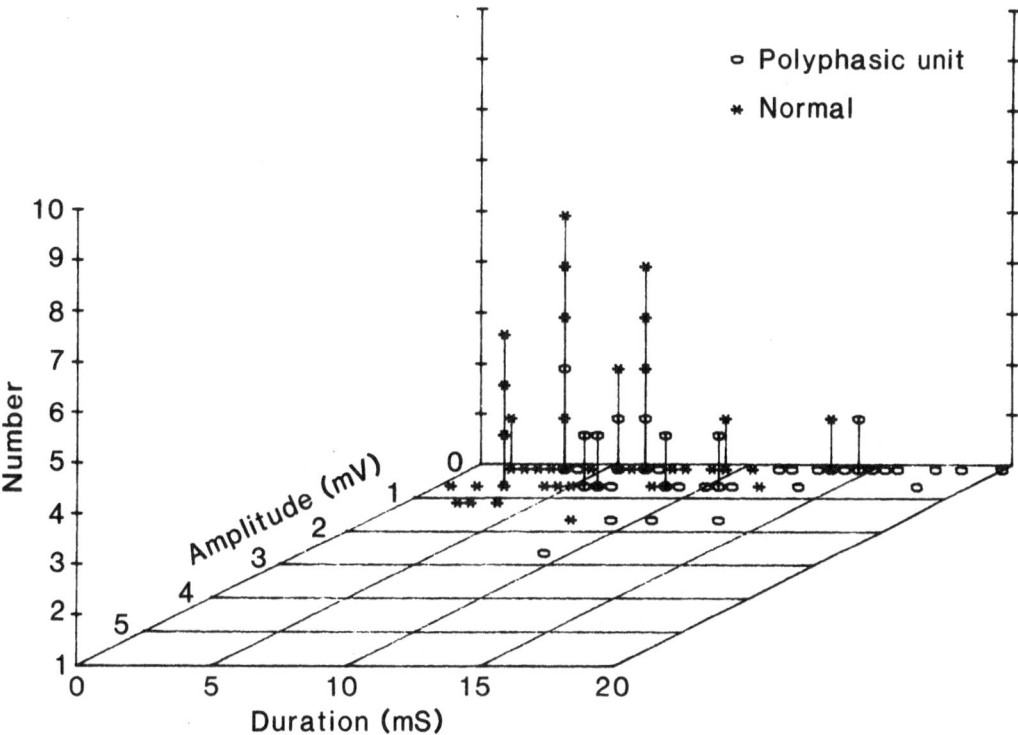

Fig. 10.8. Three-dimensional histogram showing the features of individual motor units found in 11 patients with pelvic nerve injury. (Kirby et al. 1986a)

Progressive Autonomic Failure and Multiple System Atrophy

Progressive autonomic failure and multiple system atrophy (PAF + MSA), formerly known as Shy–Drager syndrome, is an uncommon condition, in which there is a progressive loss of central autonomic neurons. The disease is characterized by incontinence as an early feature and severe orthostatic hypotension (Shy and Drager 1960; Bannister 1983). Investigation of the bladder symptoms has shown that there is a combination of detrusor hyperreflexia and profound urethral dysfunction (Kirby et al. 1986b). Neuropathological studies have demonstrated a selective loss of motor neurons in the sacral cord, in the region where the anterior horn cells of the motor neurons innervating the sphincter lie in Onuf's nucleus. Individual motor unit analysis of units from the anal sphincter (Sung et al. 1978) and the urethral sphincter (Kirby et al. 1986b) showed highly abnormal motor units indicating denervation and reinnervation. In the urethral sphincter the motor units are of very prolonged duration and polyphasic, and do not seem

to undergo compaction of the waveform to produce a high amplitude unit, as is seen in a cauda equina lesion (Figs. 10.9, 10.10).

Interestingly, the striated muscles of the urethral and anal sphincters are not affected by the denervating and reinnervating processes which affect skeletal striated muscle in motor neuron disease (Sakuta et al. 1978). This suggests that there must be some fundamental difference between the anterior horn cells innervating skeletal and sphincter striated muscle.

Myelodysplasia

Depending on the level of the abnormality, both upper and lower motor neuron lesions affecting sphincter innervation may occur with myelodysplasia.

Chantraine et al. (1966), studying mainly the anal sphincter in neonates with spina bifida, found either a normal interference pattern with heightened sacral reflexes or a pattern indicative of a lower motor neuron lesion. In the most severe cases, the

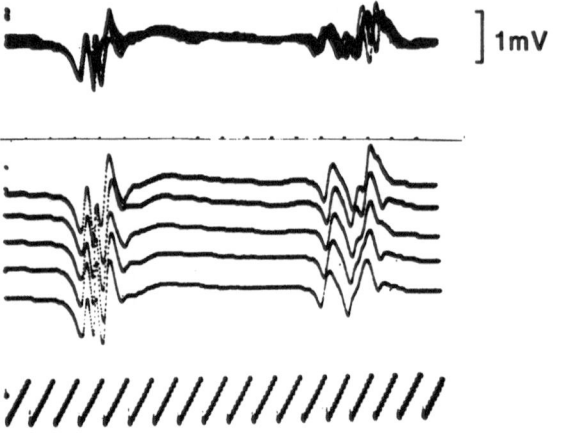

Fig. 10.9. Individual motor unit from a patient with PAF + MSA. *Upper trace* shows five traces superimposed. *Below* is a falling leaf display of the same recording. The motor unit is pathologically prolonged and polyphasic.

sphincter showed complete electrical silence, but in others polyphasic motor units were present. These authors were of the view that early operation, within 48 h of birth, could lead to an improvement in the EMG and that EMG deterioration could be considered as an indication for active surgical intervention.

More recently, Bauer et al. (1986) have followed newborn infants with myelodysplasia and shown that in half of those with a changing neurological picture, there is progressive denervation of the external urethral sphincter. This is possibly due to traction on a tethered cord as the child grows. The other half of the group showed spontaneous improvement of clinical symptoms with reinnervation of the sphincters.

Several authors have emphasized that activity known variously as "myotonia", "pseudomyotonia", complex repetitive or high frequency bizarre discharges (see below) is a common finding in CNE EMG in this condition.

Clinical Use of CNE EMG of the Sphincter

CNE EMG can be used to identify with confidence those patients who have EMG evidence of denervation and reinnervation in a lower motor neuron lesion. The uroneurophysiologist is most commonly faced with the question, "Does this patient have a

PAF + MSA

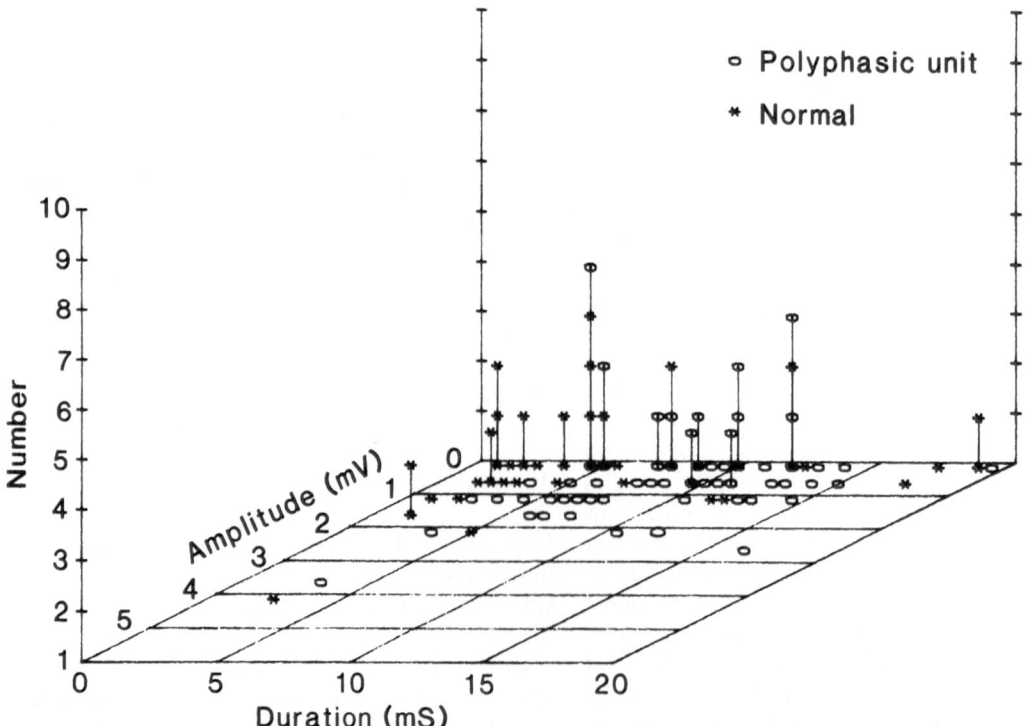

Fig. 10.10. Three-dimensional histogram showing the features of individual motor units found in 14 patients with PAF + MSA. (Kirby et al. 1986b)

neurological lesion to account for their urinary symptoms?". Finding a normal interference pattern, with normally configured motor units, from both sides of the urethral sphincter, is evidence which weighs heavily against a lower motor neuron lesion affecting the S-2 to S-4 roots.

Single Fibre Needle EMG of the Pelvic Floor

A single fibre (SF) needle electrode has similar outer dimensions to a concentric needle electrode but has only a small recording surface, 25 μm in diameter, mounted in an aperture on the side of the needle shaft. The diameter of the sphere of muscle from which an SF needle electrode records is very much smaller (300 μm) compared with that recorded by a CNE (1 mm); thus an SF electrode picks up activity from only a few muscle fibres at a time (Fig. 10.11). Muscle fibres belonging to a single motor unit are widely distributed throughout the ana-

tomical muscle structure, so that it is unlikely, in a normally innervated muscle, for an electrode with a recording volume as small as that of the SF electrode to record from more than two muscle fibres of the same motor unit. This means that potentials recorded with an SF electrode are usually singles or doubles (Fig. 10.11a). With the alterations in motor unit structure that follow denervation and subsequent reinnervation (see Fig. 10.6), the number of muscle fibres within the vicinity of the recording electrode all innervated by the same motor axon increases (Fig. 10.11b). By measuring the number of phases of each potential recorded with an SF electrode at each of 20 sampling sites, the fibre density for that muscle may be obtained.

SF EMG of the anal sphincter has shown an increased fibre density of the anal sphincter in patients with faecal incontinence (Neill and Swash 1980) and in women with stress incontinence (Anderson 1984). In both conditions it is proposed that incontinence has arisen as a result of injury to the innervation of the anal sphincter and pelvic floor and resulting muscle weakness.

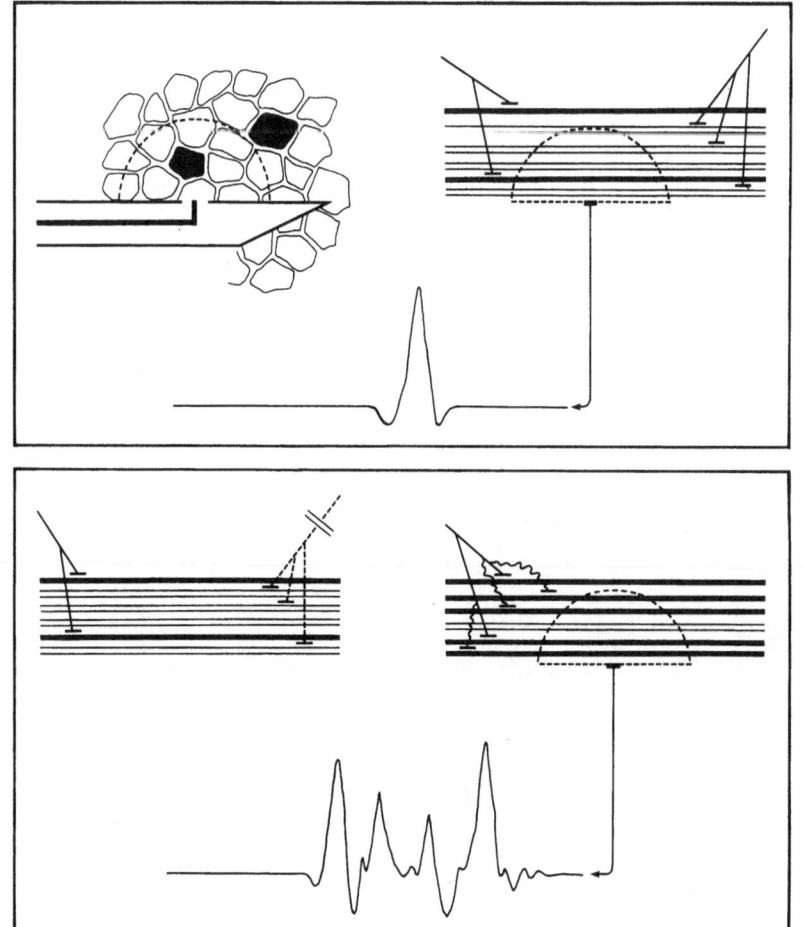

Fig. 10.11a,b. SF needle electrode recordings showing the effect of reinnervation on the number of components of a potential. **a** Normal recording; **b** recording after denervation and reinnervation by axon sprouting. (Torrens 1985)

SF EMG techniques have seldom been used to study the urethral sphincter because of the extreme difficulty of placing the needle. Locating the urethral sphincter with a concentric needle electrode is difficult, and the problem is greatly exacerbated when using an SF needle on account of the much smaller area of the recording surface. Those patients who have had SF EMG studies of the urethral sphincter also show an increase in fibre density associated with stress incontinence (Anderson 1984).

Pseudomyotonia or Decelerating Bursts and Complex Repetitive Discharges

It is not uncommon, when recording from the urethral sphincter with a needle electrode, to encounter some striking EMG activity which bears a superficial resemblance to myotonia. Detailed analysis, however, shows that it is due to repetitive firing of complexes which may fire at either a steady or decelerating rate (Fig. 10.12; Fowler et al. 1985). There is at present little agreement as to either the nomenclature or the significance of the activity.

DiBenedetto and Yalla (1979) described "high frequency bizarre discharges" and "myotonic type discharges with the typical dive-bomber sound". Because there was a similarity with myotonia recorded from skeletal muscle, they called the activity "pseudomyotonia". Butler (1979), Potenzoni et al. (1983) and Blaivas (1983) have all since referred to "pseudomyotonia", though there is no agreement in the general neurophysiological literature on the definition of the term.

In general, the essential electrophysiological difference between various types of spontaneous repetitive discharges of striated muscle, is that the activity may originate from the distal portions of the motor axon or from the direct spread of excitatory impulses between muscle fibres, ("ephaptic" transmission). The former, where repetitive discharges arise from the distal motor axon, is generally called "neuromyotonia". The best known example of this type of disorder is Isaac's syndrome—a very rare condition characterized by generalized skeletal muscle stiffness. Activity which spreads by ephaptic transmission between muscle fibres is known as "bizarre repetitive discharges" or "complex repetitive discharges" (Stalberg and Trontelj 1982). Difficulties and confusion arise because the term "pseudomyotonia" has been applied to both types of pathophysiology (Fowler et al. 1985).

In recognizing the electrophysiological properties of the repetitive sphincter activity, Dyro et al. (1983) adopted the term "complex repetitive discharges" (CRDs). Although entirely correct, the term does not encompass the most striking component of urethral activity, which is the bursts of activity which have a decelerating rate and sound so like myotonia. To resolve these difficulties, Fowler et al. (1985) introduced the term "decelerating burst and complex repetitive discharges" (DB + CRDs).

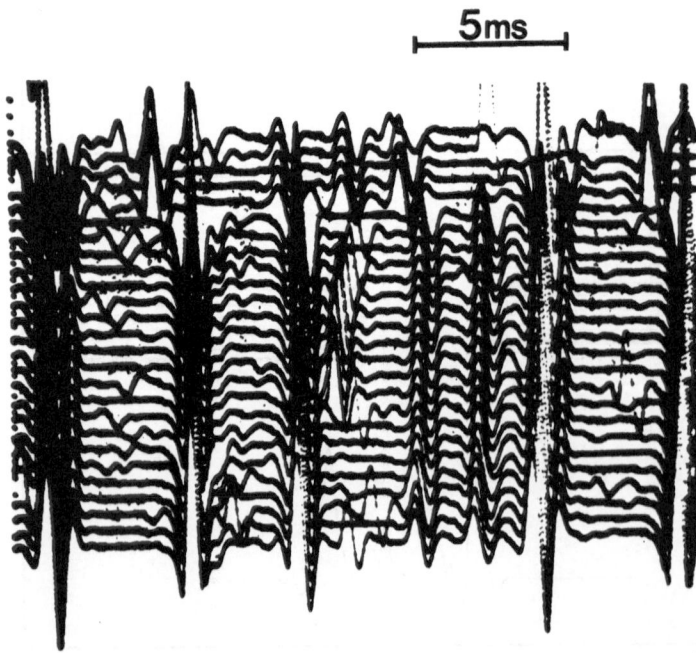

5ms

Fig. 10.12. Part of a decelerating burst discharge showing a progressive increase in the intervals between components of the complex.

Possible Significance of DB+CRDs

The significance of DB+CRDs in the urethral sphincter is uncertain. Small amounts of such activity are commonplace when recording with a needle electrode, and may occur following needle movement, similar to the "insertional" fibrillations which can be recorded from skeletal muscle. Under such circumstances and in small amounts it is probably of no importance. Butler (1979), however, described profuse amounts of such activity in three women who presented with urinary incontinence. The nine cases described by Potenzoni et al. (1983) had dissimilar anatomical-clinical conditions and included multiple sclerosis and spina bifida. Dyro et al. (1983) reported the clinical characteristics of 26 children in whom they had found such activity: 11 had myelodysplasia and 3 had spinal cord tumours; the remaining 12, all girls, had no unifying clinical picture.

Fowler and Kirby (1985) were impressed by profuse DB+CRDs in women with unexplained urinary retention. They proposed that, as with myotonia or bizarre repetitive discharges in skeletal muscle, this spontaneous activity might be associated with impairment of relaxation of the muscle of the urethral sphincter. The finding of profuse DB+CRDs in women with retention has yet to be confirmed by other laboratories. Meanwhile, the implication of this hypothesis is that CNE EMG of the urethral sphincter must be the investigation of first choice in women with unexplained urinary retention in women (Fowler and Kirby 1986).

Spinal Sacral Reflexes and Nerve Conduction Studies

There is a misleading tendency for urologists to refer to sacral reflexes as "evoked responses". To a neurophysiologist the phrase implies a potential recorded from central nervous tissue (spinal cord or cortex) which has been evoked by a peripheral stimulus. Although sacral reflexes are "evoked" by electrical stimulation, they are recorded not from the central nervous system but from muscle. They are therefore reflex responses. To avoid needless confusion, authors should stick to the well-established nomenclature of the neurophysiological literature when discussing sacral reflexes. This point has been stressed by the International Continence Society Committee for the Standardization of Terminology (Abrams et al. 1986).

Sacral Reflex Arcs

Reflex reactions in the anal and urethral sphincters and the bulbocavernosus muscle can be provoked by stimulation of the glans penis or clitoris, the mucosa of rectum, urethra, bladder, the perianal skin and even the skin of the sole of the foot. Of these reflexes, the bulbocavernosus reflex and various anal reflexes have been studied most extensively.

Reflexes are usually assessed electrophysiologically by measurement of latency. This is the time from the onset of the stimulus to the reflex muscular contraction which it produces. Latency depends upon the nerve conduction velocity in the afferent and efferent limbs of the arc and the integrity of interneuronal and synaptic connections within the central nervous system. A prolonged latency may result from a lesion anywhere within the spinal segment or peripheral conducting pathways. In addition, the amplitude of the muscle response and the intensity of the stimulus required to produce that response may also be recorded, but these values depend very much on details of technique and are much less useful.

Bulbocavernosus Reflex

Clinically, the bulbocavernosus reflex is the reflex contraction of the bulbocavernosus muscle elicited by squeezing the glans penis. Rushworth (1967) introduced a neurophysiological method for studying this reflex by recording bulbocavernosus EMG following electrical stimulation of the glans.

In the method used by Ertekin and Reel (1976), the subject lies in the lithotomy position and a concentric needle electrode is inserted into the bulbocavernosus muscle in the midline just behind the scrotum. In the resting state, unlike the sphincters, the bulbocavernosus is electrically silent. On coughing, attempting to erect the penis slightly or squeezing the glans, motor units are recruited and the muscle may be identified with certainty. A simultaneous response can be recorded from the anal or urethral sphincter on stimulation of the glans or clitoris. Its neurophysiological characteristics are almost identical with those of the reflex recorded from the bulbocavernosus muscle itself.

In control subjects the bulbocavernosus reflex has a latency which is relatively constant with repetitive stimuli, has a simple stable waveform and does not habituate (Ertekin and Reel 1976). It appears with stimulation intensities which are not painful; therefore, it is not simply a nociceptive reflex.

Anal Reflex

The anal reflex is the dimpling of perianal skin caused by contraction of the anal sphincter in response to pricking the anal mucosa or perianal skin. Pedersen et al. (1978) took recordings from the external anal sphincter using a bipolar needle electrode and a train of intense shocks applied to the perianal skin. Because the stimulating and recording electrodes were so close to each other, there was a very large stimulus artefact which saturated the amplifiers. To overcome this, a delay of 10–20 ms was incorporated into the recording circuit. These authors showed that the latency of the response which they recorded depended on the stimulus intensity, being approximately 200 ms at threshold, and reducing to 50 ms with increasing stimulus intensity. The anal reflex was present in normal subjects, although it was occasionally weak, particularly in the elderly.

Using a similar recording and stimulating arrangement to that described by Pedersen et al. (1978), Vodusek et al. (1983) found a late reflex anal response in all control subjects but only by using high (often painful) perianal stimulation. The response tended to habituate, and this, taken in conjunction with the long latency of approximately 55 ms has been taken to indicate a polysynaptic pathway (Vodusek et al. 1983).

Henry and Swash (1978), using a method similar to that of Pedersen et al. (1978) but without a recording delay, reported a much earlier reflex with a mean latency of 8.3 ms in 13 normal subjects. The presence of this early response was confirmed subsequently by Pedersen et al. (1982). They found a reaction at 2–8 ms in some subjects.

These early anal responses have a latency which is too short for a reflex and too long for stimulation of direct efferents (Pedersen et al. 1982). Such responses are probably the result of direct nerve stimulation causing an antidromic volley in the efferent nerve to the point of branching, then travelling antegradely in the collateral division of the pudendal nerve. This would explain their high threshold, constant latency, strictly ipsilateral appearance and persistence with epidural anaesthesia (Pedersen 1985).

Other Methods for Sacral Arc Reflex Recordings

Bradley et al. (1974) recorded a reflex response from the anal sphincter with an anal plug electrode following an electrical stimulus to the urethra or bladder mucosa via a stimulating electrode mounted on an indwelling catheter. The authors referred to this technique as "electromyelography". The response, recorded by averaging EMG responses in the anal sphincter, had a latency of 50–80 ms. The latency was similar whether the bladder mucosa or the urethra was stimulated. Interestingly, patients with normal supraspinal innervation could abolish the response by relaxation of the perineum, indicating intact function of inhibitory descending tracts.

Galloway et al. (1985) stimulated the dorsal nerve of the penis and recorded from the urethral or anal sphincter with surface electrodes. The arrangement of stimulating and recording electrodes was then reversed and stimuli were given to the urethra with recordings from the anal sphincter. Dorsal nerve stimulation with measurement of response in the anal sphincter in this way gives a mean latency of 40.2 ms, urethral recording 39.8 ms, and urethral stimulation with anal recording 68.9 ms.

Practical Points and Limitations

Unlike muscle responses resulting from direct nerve stimulation when, with a supramaximal stimulus, the response has a fixed minimal latency, a reflex muscle response has a variable latency. This variability arises in the synaptic component of the reflex arc. An oligosynaptic reflex such as the bulbocavernosus reflex, has a more stable latency than does a polysynaptic reflex such as the anal reflex. As with other muscle recordings, either surface or needle electrodes may be used. If surface electrodes are used and the response is not clearly discernible against background activity, discrimination of the response can be enhanced by averaging multiple sweeps. If a needle electrode is used, superimposition or a "raster" display will demonstrate repeated responses to a stimulus, so that the minimum latency can be measured.

To be of clinical use a reflex must be easily elicited in control subjects so that a failure to obtain the reflex can be ascribed to an interruption of the reflex arc rather than a technical failure or a variation of normal. Some uncertainty remains as to what proportion of neurologically intact subjects have a reflex anal or urethral contraction following stimulation of the glans or clitoris. Erterkin and Reel (1976) were able to record a response to stimulation of the glans penis from the anal sphincter in only 3 (21%) of 14 potent volunteers, although a bulbocavernosus response was present in all of them. By contrast, other authors (Vodusek et al. 1982; Bilkey et al. 1983) recorded a response in the urethral or anal sphincters to stimulation of the glans or clitoris in all their "control" subjects. This may be because their control groups were comprised of patients referred with complaints warranting neurophysiological investigations; although such a group can quite legitimately provide figures for control data, an absent reflex might have excluded a subject from the "control" group.

A final difficulty arises because of the multiplicity of reflex responses which can be recorded from the sphincters. It is quite possible for delay to be obscured when one trace merges with another. Latency measurements are highly dependent upon the precise details of recording. Thus, each laboratory must establish its own set of control data.

Clinical Use of Sacral Reflexes

The sacral reflexes may be used to test the integrity of the afferent and efferent pathways and central connections of the lower sacral segments. The finding of normal sacral reflexes and normal EMG of the sphincters or pelvic floor effectively excludes a lower motor neuron lesion of S-2 to S-4.

Bulbocavernosus Reflex in the Investigation of Impotence

The relevance of the bulbocavernosus reflex to normal male sexual function is dubious. Erection and seminal emission depend largely on autonomic nerve function, whereas the neurophysiologically recorded bulbocavernosus reflex depends on electrical stimulation of large myelinated axons. Nevertheless, the reflex has often been used in the investigation of impotence.

In the group studied by Ertekin and Reel (1976), there were 16 men with a polyneuropathy and impotence, caused either by diabetes or alcohol. The psychosensory threshold was not increased and the reflex threshold was not higher than in controls. The reflex latency was within the control range in the diabetics but was significantly delayed in alcoholics. A response in the external anal sphincter was obtained in 33% of cases. Vodusek et al. (1982) examined 12 patients with neurogenic erectile impotence who had a polyneuropathy, pelvic trauma, or lumbosacral disc disease. The bulbocavernosus response could be elicited in all patients but was of a prolonged latency in eight.

Ertekin and Reel (1976) found that in a group of 19 patients with impotence resulting from suprasacral cord lesions the psychosensory threshold was high but the reflex threshold was the same as in controls, as was the mean latency. This is in contrast to the findings of Bilkey et al. (1983), who found a significantly shorter reflex latency of the bulbocavernosus response of 108 control subjects compared with 44 patients with suprasacral spinal cord injuries. This rather unexpected result has not been explained. Ertekin and Reel (1976) found that in two-thirds of their group with spinal cord involvement the EMG reflex response was dense and prolonged. Also they were able to obtain a response in the anal sphincter in 62.5% of the patients, compared with 21% in the control group, indicating a facilitation of the reflex response, as would be expected from an upper motor neuron lesion.

Distal Motor Latencies, Pudendal Motor Conduction Velocities and Spinal Stimulation

Motor latency is the time taken from the instant of stimulation of a motor nerve to the first measurable muscle response. To measure motor latency, the stimulus is set to an intensity which is just greater than that necessary to stimulate all the fibres in the nerve (supramaximal stimulus). Because motor latency is measured as the time to the first onset of response, such studies inevitably examine only the fastest conducting motor fibres in a nerve. If even a small proportion of the fastest conducting fibres are intact, latency measurements will be unaltered. A prolonged latency is only seen either when there has been such extensive nerve fibre loss that the fastest conducting have been completely eliminated or when conduction in these same fibres has been slowed by a demyelinating process.

When studying peripheral nerve in a limb, stimuli may be given at two separate sites, and the conduction time (the time taken by the impulse to travel between stimulation sites) calculated. This is done by subtracting the distal motor latency from the proximal motor latency. The conduction distance between stimulating sites is measured and this, divided by the calculated conduction time, gives a conduction velocity for the nerve. By calculating the nerve conduction time, only those factors which depend upon distance and the conduction velocity in the main nerve truck are considered.

$$\text{Nerve conduction velocity (m/s)} = \frac{\text{Conduction Distance (mm)}}{\text{Conduction time (ms)}}$$

where

Conduction time = Proximal motor latency − Distal motor latency (ms)

The latency figures from which the conduction velocity is derived depend upon an inconveniently large number of variables. Amongst these are the time taken by the current to depolarize the nerve locally, conduction in the relatively slow conducting thin terminal intramuscular branches, the time taken for neuromuscular transmission of the depolarizing impulse, as well as the conduction time down the main nerve trunk.

Latency =
Current + Conduction + Intramuscular + Neuromuscular
utilization   in main   conduction   transmission
time   nerve trunk

Conduction velocities are therefore more valuable than motor latencies but, in practical terms, require at least two well-separated stimulation sites. This is not possible in studies of the pudendal and pelvic nerves, which lie deep within the pelvis. Distal or "terminal" motor latencies must suffice. Even so, there are still considerable problems of stimulus access. All techniques which require direct nerve stimulation within the pelvis are inevitably somewhat invasive and may be uncomfortable for the subject.

Chantraine et al. (1973) described a method for pudendal nerve stimulation that required two 70 mm stainless steel needle electrodes to be inserted vertically down to the lesser sciatic foramen below the ischial spine from the sacral skin surface, in close approximation to the pudendal nerve. The authors state quite frankly that "it is a painful examination and the latency from the point of stimulation is difficult to measure". By using two stimulation sites the technique was able to provide a motor conduction velocity for the pudendal nerve of 56 m/s. This figure is similar to that obtained for conduction velocities in limb nerves.

Jelasic et al. (1975) described their method for stimulating the sacral roots by introducing stimulating electrodes through the sacral hiatus to a level of the third to fourth sacral foramen. However, in 25% of the subjects they could not get a response, and they wrote that the next stage of development of this method would be to introduce stimulating electrodes through lumbar puncture needles.

Swash and his colleagues at St Mark's Hospital, London, have developed several methods for measuring distal motor latencies within the pelvis (Snooks and Swash 1985).

Pudendal Nerve Terminal Motor Latency to Anal Sphincter

Kiff and Swash (1984) devised a method for stimulating the pudendal nerve transrectally, using a bipolar stimulating electrode mounted on a finger-stall. Also, mounted towards the base of the finger-stall, are two circular recording electrodes which lodge against the anal sphincter. The stimulator is inserted into the rectum and the examiner localizes the pudendal nerve at the level of the ischial spine. Stimuli are given so that a maximum amplitude of response from the anal sphincter is produced.

Perineal Terminal Motor Latency to Urethral Sphincter

Using the same techniques of transrectal stimulation but recording from the urethral sphincter with a catheter-mounted ring electrode, the distal motor latency for the perineal branch of the pudendal nerve can be measured (Snooks and Swash 1984).

Transcutaneous Lumbar Spinal Stimulation

Merton and Morton (1980) reported a method for direct stimulation of the cerebral cortex using very brief, high voltage stimuli (500–1500 V, time constant decay 50 μs). These remarkable innovators then showed that the same stimulator could be used for direct spinal cord stimulation (Merton et al. 1984). Adopting this technique, Snooks and Swash (1984) developed methods for recording the latency of response in the anal and urethral sphincters and puborectalis muscle, following transcutaneous lumbar stimulation. Responses from the sphincters were measured with an anal plug electrode and a urethral catheter-mounted ring electrode respectively. The puborectalis response was measured using another fingerstall device which has two recording electrodes mounted on the tip. This was inserted per rectum and held in close approximation to the muscle bar on the posterior wall of the anorectal angulation, whilst transcutaneously 1000 V lumbar stimulation is applied.

By stimulating over the lumbar spine at two separate levels, L-1 and L-4 (Fig. 10.13), "spinal latency ratio" (SLR) may be derived.

$$SLR = \frac{\text{Latency to puborectalis after spinal stimulation at L-1}}{\text{Latency to puborectalis after spinal stimulation at L-4}}$$

From this it is possible to recognize delayed conduction within the cauda equina from an abnormal spinal latency ratio value. If there is a prolonged distal motor latency, latencies from both sites will be abnormal, whereas a lesion of the cauda equina

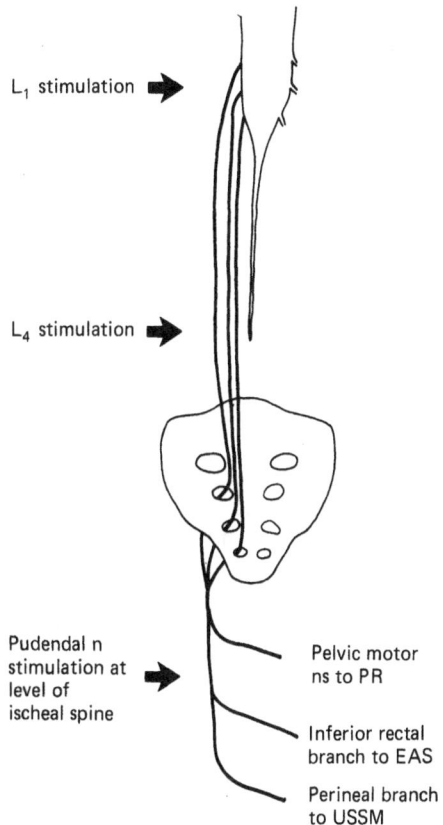

L₁ stimulation

L₄ stimulation

Pudendal n
stimulation at
level of
ischeal spine

Pelvic motor
ns to PR

Inferior rectal
branch to EAS

Perineal branch
to USSM

Fig. 10.13. Diagram showing the sacrum and some sacral nerve roots with the stimulation sites used by Swash and co-workers indicated by *arrows. ns*, nerves; *PR*, puborectalis; *EAS*, external anal sphincter; *USSM*, urethral striated sphincter musculature.

between the two stimulation sites results in a L-1 response delay only.

Clinical Applications

Swash and his colleagues have examined several pelvic floor disorders using the methods described above. They have shown slowing of pudendal motor latency in idiopathic anorectal incontinence (Snooks et al 1984b), and slowing of the terminal latencies of both the pudendal and perineal branch in double incontinence (Snooks et al. 1984a). Their methods of spinal conduction have been used to demonstrate cauda equina lesions and other lower sacral innervation abnormalities. Another study demonstrated a slowing of distal motor latency in the pudendal nerve in 20% of women following childbirth with vaginal delivery (Snooks et al 1984c). The authors suggest that childbirth is a major cause of anorectal and urinary incontinence in women.

Although these techniques have been so highly fruitful, some question remains as to the pathophysiological explanation for the finding of prolonged distal motor latencies. For the reasons already outlined, a prolonged distal motor latency cannot be interpreted as evidence of denervation.

Future Developments

A magnetic stimulator has recently been developed which can excite the motor cortex by inducing electrical current in the brain (Barker et al. 1985). Stimulation over the scalp with this device is as effective as electrical stimulation but has the overwhelming advantage of being completely painless. Magnetic stimulation of the spinal cord is proving more difficult, but doubtless this will eventually prove possible and then methods like those pioneered by Swash and his colleagues are likely to become standard for all uroneurological laboratories. Direct cauda equina stimulation provides a more precise means of assessing pelvic floor innervation than reflex arc latencies.

Lower Urinary Tract Sensation

Continence and micturition depend upon a complex of interrelated reflexes acting at a subcortical level, coupled with the conscious awareness of bladder fullness which is necessary to time the void appropriately. Thus, normal function depends heavily upon afferent pathways, and abnormalities of lower urinary tract sensation might be expected to produce a range of clinically important consequences. Unfortunately, the standard cystometrogram yields little information about sensory function. The routine bladder measurements for "first sensation", "urge to void" and post-micturition residue are influenced by bladder capacity and bladder wall compliance as well as the rate of filling. They are imprecise indicators of proprioceptive function. There have therefore been numerous attempts to measure sensation more directly.

The quantitation of sensation in the bladder is not easy. In the ideal test, the stimulus operates upon a defined sensory modality. In the bladder, there is a special difficulty because the physiology of bladder sensation is so poorly understood and the stimulator must also be small enough to be applied transurethrally. Most workers have therefore resorted to electrical stimulation for studies in the urinary tract.

Though an electric current is an unphysiological stimulus, it can be applied with little difficulty in the urethra and bladder using simple electrodes. These may be inserted blindly on a catheter, when the results may be open to criticism because the exact site of stimulation is in doubt. Application under vision has the disadvantage that conventional endoscopes are relatively uncomfortable to pass without a general anaesthetic. The continued presence of the endoscope in the urethra during testing may itself inflict a sensory stimulus comparable with that generated by the electrodes.

Frimodt-Møller (1972) passed wire electrodes into the bladder either suprapubically or trans-urethrally. The metal stimulating surface of the electrode was not itself in contact with the vesical wall. Instead, the bladder was filled with saline to provide a conductive path from the stimulating electrode to the bladder mucosa. A series of threshold measurements were made as a pulsed constant current was applied between this monopolar electrode and a reference electrode on the leg. Using a square-wave stimulus of 1 ms at 2.5 Hz, patients reported a "tingling", "tickling" or burning sensation as the threshold to perception was exceeded. This was less than 10 mA in control subjects.

By using a double-barrelled catheter bearing two wire electrodes, Kieswetter (1977) was able to measure mucosal sensitivity to bipolar stimulation, both in the bladder and in the posterior urethra, which is the site of the most potent facultative receptors for the micturition reflex. Using this method, normal bladder perception ranged from 3 to 10 mA with a mean value of 5 mA. The threshold in the posterior urethra was considerably lower at 1–2 mA with a mean of 1.5 mA. This technique has been applied by a number of workers (Powell and Feneley 1980; Murray 1982; Galloway et al. 1985). The results in normal subjects are consistent throughout these studies despite minor variations in the waveform of the stimulus applied. Moreover, Murray (1982) was able to discern a variation in posterior urethral electrosensitivity which decreased as the bladder was filled and he related this to the physiology of continence.

Abnormalities of Electrosensitivity

Workers from Frimodt-Møller onwards have been able to find abnormalities of electrosensitivity which they related to clinical disease. Hyposensitivity was found after cord lesions, pelvic nerve injury (Kieswetter 1977) and in some patients with nocturnal enuresis (Galloway et al. 1985). More surprising, perhaps, is the discovery of a state of hypersensitivity in patients with so-called sensory frequency. Such a state of reduced threshold to perception has not been reported from studies of sensation elsewhere in the body.

Difficulties with Sensory Testing

The measurement of sensory threshold depends upon the patient's state of awareness and cooperation. To some extent this can be monitored by preliminary measurements of the threshold to electrostimulation on the skin.

In conditions of reduced bladder sensory perception, the electrical stimulus to the bladder may have to be so high as to endanger the patient, through induced cardiac dysrhythmias. Kieswetter (1977) recommends a maximum stimulus of 20 mA. For the same reason, the heart must not be included in the current path.

Sensory Evoked Responses

Sensory evoked responses are potential changes that can be recorded from cortex or spinal cord, in response to stimuli at a distant site. "Averaging" of multiple responses is necessary so that the evoked potential, usually a few microvolts in amplitude, can be identified against background activity of equal amplitude. Averaging is a very powerful technique, and much used in clinical neurophysiology for recording both from peripheral nerve and the central nervous system. Recordings are made following each of a series of stimuli. The individual responses are added to each other in an accumulating sum so that activity which is time-locked to the stimulus increases in clarity. At the same time, background activity, which occurs randomly in relation to the stimulus progressively lessens in prominence (Fig. 10.14).

Despite the name, "somatosensory" evoked responses have little relevance to the process of perception of physiological sensations. This important point seems to have escaped many authors who claim that sensory evoked responses from stimulation of the lower urinary tract offer an objective means of assessing bladder sensory function. It is true that evoked potentials test afferent neuronal conducting pathways. However, to examine sensory function, measurement of thresholds for sensation should be made, not evoked potentials. When

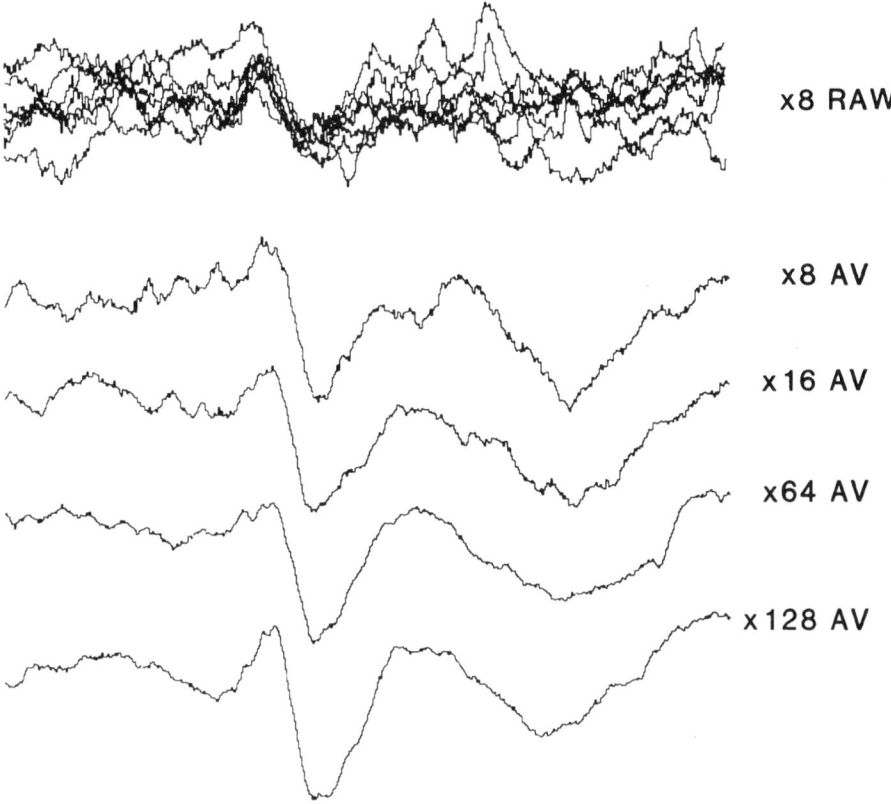

Fig. 10.14. Signal averaging: Responses time-locked to the stimulus are summated so that they stand out against random background signals.

recording somatosensory evoked potentials, stimuli are given directly to a nerve at three to four times the level of intensity needed for first feeling. Little or no attention is given to the subject's perception of such a stimulus. Only the potential change over the cortex is being examined. The clinical use of visual evoked responses illustrates the point well. Visual evoked responses are not used to measure how well a subject can see: a reading chart is used for that. The major use of visual evoked potentials is to detect slowing of conduction resulting from central nervous system abnormalities, such as the demyelination of multiple sclerosis.

Averaging of cortical responses is a technique that has been much used since the introduction of the principle by the late Prof. George Dawson in 1954. In clinical neurophysiological practice, the response evoked by stimulation of a peripheral nerve such as the median or tibial is frequently recorded. No major technological advance was needed, therefore, to record cortical responses evoked by pudendal nerve stimulation. Surprisingly, however, such a method was not attempted until quite recently (Gerstenberg et al. 1981; Haldeman et al. 1981; Kaplan et al. 1981).

Although different methods and sites for genito-urinary stimulation have been adopted by various groups of workers, basically similar recording methods have been used throughout.

Method for Obtaining Cortical Evoked Potentials

Scalp Electrode Placement

To record cortical evoked responses, scalp electrodes are placed over the sensory homunculus location for sensory afferents from midline urogenital structures. The exact placement of the active recording electrodes is important to maximize the amplitude of a response. The surface site which optimally records activity from deep, medial interhemisphere fissure structures is an electrode placed at the vertex (Badr et al. 1982) or 2 cm posterior to it (Haldeman et al. 1982). The electrode placement corresponding to the vertex is Cz, on the 10–20 international recording system.

As with all evoked potential recordings, a "reference" electrode must be used. The evoked potential is the potential difference recorded between an active and a reference electrode. If a scalp reference electrode is used, the reference electrode will inevitably contribute to the potential itself. However, provided the same reference site is always used, this is not a significant problem. A midline frontal reference electrode (Fz) has been commonly employed.

Stimulation

Stimulation methods have varied between laboratories engaged in this type of work. Kaplan (1981) used a concentric needle electrode in the external urethral sphincter as a means of stimulating the pudendal nerve. Haldeman et al. (1982) and Opsomer et al. (1985) used transcutaneous pudendal nerve stimulation with either bipolar stimulating electrodes over the dorsal nerve of the penis/clitoris or a pair of ring electrodes around the base of the penis. Badr et al. (1984) used monopolar stimulation of the bladder with a specially designed suction cathode electrode attached to the bladder wall, with an anode on the thigh. Most recently, Sarica and Karacan (1986) have used the device introduced by Andersen et al. (1976) for "electromyelography".

A constant current or a constant voltage stimulator may be used and a train of stimuli, at a frequency of 1–5/s, given. As with stimulation of somatic peripheral nerves, a stimulus greater than that of threshold is necessary and an intensity of between 2.5–4.0 times threshold is commonly used. Because the response is recorded by averaging, approximately 1000 responses are necessary for potentials to be sufficiently well defined.

Waveform of Response

Cortically recorded evoked potentials have fundamental similarities irrespective of the stimulating site, and conventions have been adopted for naming components of their waveforms. Upgoing waveforms are negative and downgoing, positive. The first positive component of the waveform is referred to as P1, and the second P2 etc. The latencies of all the various components can then be measured.

The number of components which may be reliably identified in an evoked response depends very much on the quality and details of the recording technique. It is therefore very difficult to compare genitourinary evoked responses recorded

from different laboratories using a variety of stimulating and recording techniques.

Clinical Uses

It has been proposed that somatosensory evoked potentials from stimulating the pudendal nerve will prove useful in differentiating between a peripheral or central cause of bladder dysfunction (Kaplan 1981; Haldeman et al. 1982; Opsomer et al. 1985). Employed in this manner, potentials recorded from the scalp following pudendal nerve stimulation have no particular advantage over responses evoked by tibial nerve stimulation. Cortical evoked responses to both stimulation sites will be absent in cord or other central nervous system lesions. Of much greater interest is the information that can be obtained from examining features of the response that are specific to the chosen stimulation site. For example, Sarica and Karacan (1986) have shown a marked difference in latency of response between stimulation of the distal and proximal part of the urethra and suggest that the more prolonged response from vesicourethral junction stimulation is due to more slowly conducting visceral afferents. Conduction in this important group of fibres has hitherto been unmeasurable. Also there is an intriguing report by Badr et al. (1982) that the later positive wave, P2, which may be recorded from monopolar bladder stimulation, is significantly prolonged in patients with uninhibited bladders.

EMG of the Bladder Detrusor

It is notoriously difficult to record EMG from the detrusor muscle. Several research groups have wrestled with the problem, some of them expending much time and effort merely to record an artefact.

Boyce (1952) used ball electrodes placed transurethrally via a plastic electrode introducer to record a high amplitude, low frequency (0.1–1 Hz) activity. This activity, which varied as a "bladder wave" during voiding, was dismissed as movement artefact by Brunsting (1958). Similar low frequency potentials were obtained using wire hook electrodes inserted transvaginally by Stanton and his associates (1973). Though they accepted that movement artefact was impossible to exclude during voiding, they claimed that detrusor EMG recorded in this way provided useful information in difficult cases of disturbed micturition.

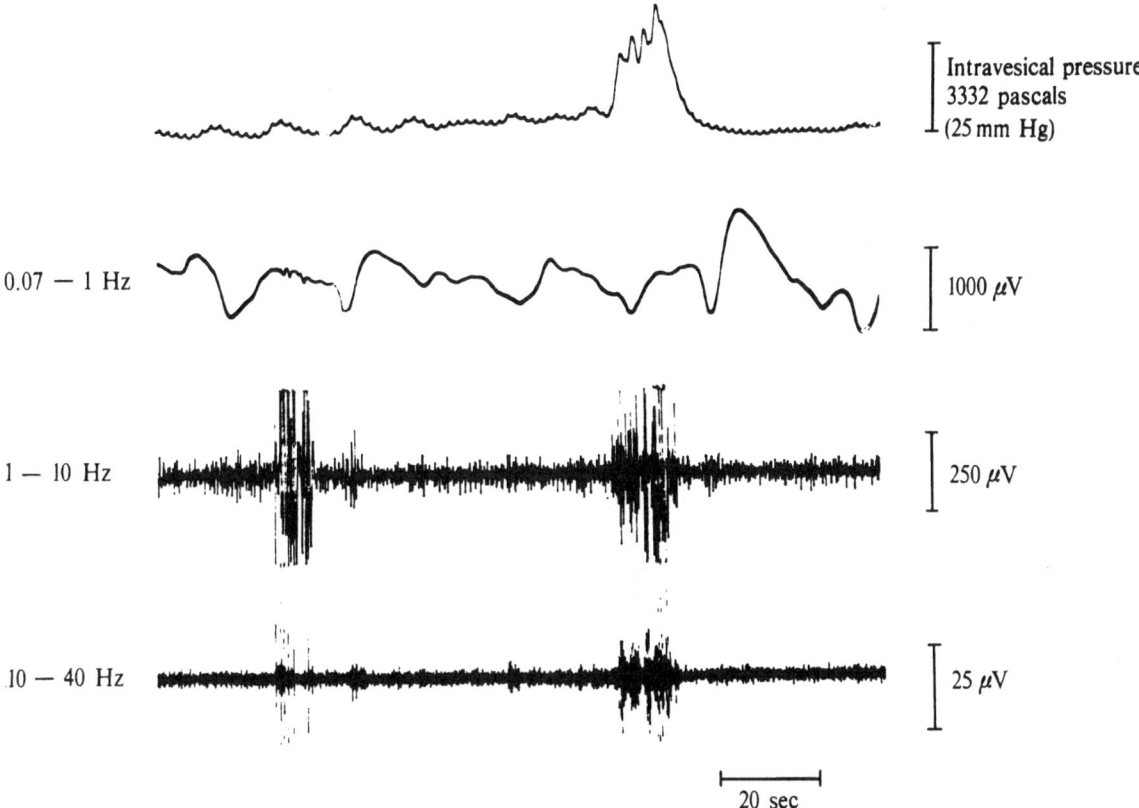

Fig. 10.15. Intravesical pressure changes and bladder electrical activity during reflex voiding in an anaesthetized cat. 1–10 Hz band activity is dissociated from reflex voiding. There is a small increase in 10–40 Hz activity in the absence of voiding because of carry-over of adjacent 1–10 Hz activity. (Craggs and Stevenson 1976)

A high frequency (70–200 Hz), low amplitude activity which increased during voiding was described by Frankson and Peterson (1953). Jones and his colleagues (1974) recorded similar potentials from needle electrodes placed endoscopically in the bladder. This activity was diminished in patients with lower motor neuron lesions affecting the bladder and it was also abnormal in patients with upper motor neuron lesions. A similar method and findings were reported by Nanninga and Kaplan (1978), who believed that the technique offered "greater clinical application in areas of bladder dysfunction and evaluation of drugs on the bladder".

Craggs and Stephenson (1976) stimulated the sacral ventral roots of the cat and the baboon to record "the real bladder electromyogram" (Fig. 10.15). Using platinum wire electrodes introduced via the exposed serosal surface of the bladder with filters set to record activity between 0.07 and 1 Hz, profuse random high amplitude (1000 μV) activity was found. They concluded that this was an artefact, probably resulting partly from fluid movement around the tip of the electrodes, since it was present in isolated dead bladder which had been kept in saline for 24 h. Other low frequency activity was attributed to movement artefact. Some high frequency activity (40–200 Hz) was shown to be due to pick up of skeletal muscle EMG, since it was particularly noticeable during sacral ventral root stimulation but abolished by succinylcholine. Ultimately, only activity in the 10–40 Hz range was seen to correlate well with contractions of the bladder sufficient to effect voiding (Fig. 10.15). In the face of these difficulties, the results of the few studies performed more recently (Nanninga and Kaplan 1978; Takaiwa et al. 1983; Takaiwa and Shiraiwa 1984) must be treated with some caution.

Once true electromyographic recordings have been made from the detrusor muscle, the problem of interpretation remains. Whereas extensive study over many years has elucidated the tightly coupled relationship between extracellularly recorded electrical activity and contractile mechanisms in striated muscle, very much less is known about the relationship of these two events in smooth muscle. Clinical striated muscle EMG is an advanced science, and from examination of recorded electrical activity it is possible to make deductions concerning the anatomical structure of the muscle, to distinguish between myopathic and neuropathic disease and to infer the anatomical structure of -motor units. There is no such knowledge upon which to base the interpretation of EMG recordings from smooth muscle.

The potential use of detrusor EMG is not to identify when the muscle is active, since that kinesiological information can undoubtedly be obtained more simply by monitoring intravesical pressure. Instead, detrusor EMG is likely to offer a means of identifying primary disorders of the detrusor muscle or abnormalities of peripheral innervation. A further promising area of research is into the direct effect of drugs on detrusor EMG, as therapies which modulate bladder activity while preserving controlling mechanisms would be of great potential value. However, until much more is known about the correlation of EMG and detrusor activity in humans, such studies remain a distant prospect.

References

Abrams P, Blaivas JG, Stanton SL, Andersen JT, Fowler Clare J, Gerstenberg T, Murray K (1986) Sixth report on the standardisation of terminology of the lower urinary tract function. World J Urol 4:2–5

Andersen JT, Bradley WE, (1976) The syndrome of detrusor-sphincter dyssynergia. J Urol 116(4):493–495

Andersen JT, Bradley WE, Timm GW (1976) Electrophysiological techniques for the study of urethral and vesical innervation. Scand J Urol Nephrol 10:189–194

Anderson RS (1984) Neuromuscular dysfunction; a significant factor in female genuine stress incontinence. Thesis, University of Bristol

Badr GG, Carlsson C-A, Fall M, Friberg S, Linstrom L, Ohlsson B (1982) Cortical evoked potentials following the stimulation of the urinary bladder in man. Electroencephalogr Clin Neurophysiol 54:494–498

Badr GG, Fall M, Carlsson C-A, Lindstrom L, Friberg S, Ohlsson B (1984) Cortical evoked potentials obtained after stimulation of the lower urinary tract. J Urol 131:306–309

Bannister R (ed) (1983) Autonomic failure. A textbook of clinical disorders of the autonomic nervous system. Oxford University Press, Oxford

Baker AT, Jalinous R, Freeston IL (1985) Non-invasive magnetic stimulation of human motor cortex. Lancet I:1106–1107

Barrett DM (1980) Disposable (infant) surface electromyogram electrode in urodynamics: a simultaneous comparative study of electrodes. J Urol 124:663–665

Bartolo DCC, Jarratt JA, Read NW (1983) The use of conventional electromyography to assess external sphincter neuropathy in man. J Neurol Neurosurg Psychiatry 46:1115–1118

Bauer SB, Dyro FM, Krarup C (1986) The urodynamic assessment of newborns with myelodysplasia. Proceedings of 16th Annual Meeting of the International Continence Society, Boston, p 331–332

Bilkey WJ, Awad EA, Smith AD (1983) Clinical application of sacral reflex latency. J Urol 129:1187–1189

Blaivas JG (1983) Sphincter electromyography. Neurourol Urodyn 2:269–288

Blaivas JG, Labib KL, Bauer SB, Retik AB (1977) A new approach to electromyography of the external urethral sphincter. J Urol 117:773–777

Blaivas JG, Sinha HP, Zayed AAH, Labib KB (1981) Detrusor-external sphincter dyssynergia. J Urol 125:545–548

Boyce WH (1952) Bladder electromyography: a new approach to the diagnosis of urinary bladder dysfunction. J Urol 67:650–668

Bradley WE, Brantley Scott F, Timm GW (1974) Sphincter electromyography. Urol Clin North Am 1(1):69–80

Brunsting CD (1958) An interpretation of the urinary bladder "electrocystogram" as artefact. J Urol 79:165–170

Butler WJ (1979) Pseudomyotonia of the periurethral sphincter in women with urinary incontinence. J Urol 122:838–840

Chantraine A (1966) Electromygraphie des sphincters stries uretal et anal humains. Renal Neurol (Paris) 115:396–403

Chantraine A (1973) EMG examination of the anal and urethral sphincters. In: Desmedt JE (ed) New developments in electromyography and clinical neurophysiology. Karger, Basel, pp 421–432

Chantraine A, Lloyd K, Swinyard CA (1966) The sphincter ani externus in spina bifida and myelomeningocele. J Urol 95:250–256

Chantraine A, Level J, Onkelinx A (1973) Motor conduction velocity in the internal pudendal nerves. In: Desmedt JE (ed) New developments in electromyography and clinical neurophysiology. Karger, Basel, pp 433–438

Craggs MD, Stephenson JD (1976) The real bladder electromyogram. Br J Urol 48:443–451

de Groat WC, Booth AM (1980) Physiology of the urinary bladder and urethra. Ann Intern Med 92(2):312–315

Dibenedetto M, Yalla SV (1979) Electrodiagnosis of striated urethral sphincter dysfunction. J Urol 122:361–365

Dubowitz V, Brooke MH (eds) (1973) Muscle biopsy—a modern approach. Saunders, London

Dyro FM, Bauer SB, Hallett M, Khoshbin S (1983) Complex repetitive discharges in the external urethral sphincter in a pediatric population. Neurourol Urodyn 2:39–44

Ertekin C, Reel F (1976) Bulbocavernosus reflex in normal men and patients with neurogenic bladder and/or impotence. J Neurol Sci 28:1–15

Fowler Clare J, Kirby RS (1985) Abnormal electromyographic activity (decelerating burst and complex repetitive discharges) in the striated muscle of the urethral sphincter in 5 women with persisting urinary retention. Br J Urol 57:67–70

Fowler Clare J, Kirby RS (1986) Electromyography of the urethral sphincter in women with urinary retention. Lancet I:1455–1457

Fowler Clare J, Kirby RS, Harrison MJG, Milroy EJG, Turner-Warwick R (1984) Individual motor unit analysis of the diagnosis of disorders of urethral sphincter innervation. J Neurol Neurosurg Psychiatry 47:637–641

Fowler Clare J, Kirby RS, Harrison MJG (1985) Decelerating burst and complex repetitive discharges in the striated muscle of the urethral sphincter, associated with urinary retention in women. J Neurol Neurosurg Psychiatry 48:1004–1009

Franksson C, Petersen I (1953) Electromyographic recording from the normal human urinary bladder, internal urethral sphincter and ureter. Acta Physiol Scand 106:15–156

Franksson C, Petersen I (1955) Electromyographic investigation of disturbances in the striated muscle of the urethral sphincter. Br J Urol 27:154–161

Frimødt-Moller C (1972) A new method for quantitative evaluation of bladder sensitivity. Scand J Urol Nephrol 6:135–142

Galloway NTM, Chisholm GD, McInnes A (1985) Patterns and significance of the sacral evoked response (the urologist's knee jerk). Br J Urol 57:145–147

Gerstenberg T, Hald T, Meyhoff HH (1981) Urinary cerebral evoked potential mediated through urethral sensory nerves. Prog Clin Biol Res 78:141–142

Gosling JA, Dixon JS, Critchley HOD, Thompson S-A (1981) A comparative study of the human external sphincter and periurethral levator ani muscles. Br J Urol 53:35–41

Gosling JA, Dixon JS, Humpherson JR (1983) Functional anatomy of the urinary tract. Churchill Livingstone, Edinburgh

Haldeman S, Bradley WE, Johnson BK (1981) Pudendal somatosensory evoked potential (PER) Neurology 31:152

Haldeman S, Bradley WE, Bhatia N (1982) Evoked responses from the pudendal nerve. J Urol 128:974–980

Henry M, Swash M (1978) Assessment of pelvic floor disorders and incontinence by electrophysiological recording of the anal reflex. Lancet I:1290–1291

Jelasic F, Fischer D, Allert ML (1975) Diagnosis of neurologic disorders of micturition without evidence of the pathological process by measuring conduction velocity of sacral roots. Urol Int 30:100–102

Jesel M, Isch-Treussard C, Isch F (1973) Electromyography of striated muscle of anal and urethral sphincters. In: Desdedt JE (ed) New developments in electromyography and clinical neurophysiology. Karger, Basel, pp 406–420

Jones WG, La Joie WJ, Cosgrove MD (1974) Electromyography in pathologic bladder. Urology 2:186–189

Kaplan PE (1981) A somatosensory evoked response obtained after stimulation of the contralateral nerve. Electromyogr Clin Neurophysiol 21:585–587

Kieswetter H (1977) Mucosal sensory threshold of urinary bladder and urethra measured electrically. Urol Int 32:437–448

Kiff ES, Swash M (1984) Normal proximal and delayed distal conduction in the pudendal nerves of patients with idiopathic (neurogenic) faecal incontinence. J Neurol Neurosurg Psychiatry 47:820–823

Kirby RS, Fowler CJ, Gilpin SA, Gosling JA, Milnoy RJG, Turner-Warwick (1986a) Bladder muscle biopsy and urethral sphincter EMG in patients with bladder dysfunction after pelvic surgery. J R Soc Med 79:270–273

Kirby R, Fowler Clare, Gosling J, Bannister R (1986b) Urethrovesical dysfunction in progressive autonomic failure with multiple system atrophy. J Neurol Neurosurg Psychiatry 49:554–562

Lose G, Tanko A, Colstrup H, and Andersen JT (1985) Urethral sphincter electromyography with vaginal surface electrodes: a comparison with sphincter electromography recorded via periurethral coaxial, anal sphincter needle and perianal surface electrodes. J Urol 133:815–818

Maizels M, Firlit CF (1979) Paediatric urodynamics: a clinical comparison of surface versus needle pelvic floor/external sphincter electromyography. J Urol 122:518–522

Mayo ME (1979) The value of sphincter electromyography in urodynamics. J Urol 122:357–360

Merton PA, Morton HB (1980) Stimulation of the cerebral cortex in the intact human subject. Nature 285:227

Merton PA, Hill DK, Morton HB, Marsden CD (1984) Scope of a technique for electrical stimulation of human brain, spinal cord and muscle. Lancet II:597–600

Mundy AR (1982) An anatomical explanation for bladder dysfunction following rectal and uterine surgery. Br J Urol 54:501–504

Murray K (1982) Urethral sensitivity—an integral component of the storage phase of the micturition cycle. Neurourol Urodyn 1:193–197

Nanninga JB, Kaplan R (1978) Experience with measurement of bladder electrical activity. J Urol 120:82–85

Neill ME, Swash M (1980) Increased motor unit fibre density in the external anal sphincter muscle in ano-rectal incontinence: a single fibre EMG study. J Neurol Neurosurg Psychiatry 43:343–347

Nielsen KK, Kristensen ES, Qvist N, Jensen KM-E, Dalsgard J, Krarup T, Pedersen D (1985) A comparative study of various electrodes in electromyography of the striated urethral and anal sphincter in children. Br J Urol 57:557–559

Nordling J, Meyhoff HH (1979) Dissociation of urethral and anal sphincter activity in neurogenic bladder dysfunction. J Urol 122:352–355

Nordling J, Meyhoff HH, Walter S, Andersen JT (1978) Urethral electromyography using a new ring electrode. J Urol 120:571–573

Opsomer RJ, Top M, Wese FX, Van Cangh PJ (1985) Cerebral evoked potentials from the pudendal nerve. Preliminary results. In: Proceedings of the 15th Annual Meeting of the International Continence Society, London, p 38

Pavlakis AF, Siroky MB, Goldstein I, Krane RJ (1983) Neurourologic findings in conus medullaris and cauda equina injury. Arch Neurol 40:570–573

Pedersen E (1985) The anal reflex. In: Henry MM, Swash M (eds) Coloproctology and the pelvic floor. Butterworths, London, pp 104–111

Pedersen E, Harving H, Klema B, Torring B (1978) Human anal reflexes. J Neurol Neurosurg Psychiatry 41:813–818

Pedersen E, Klemar B, Schroder HD, Torring J (1982) Anal sphincter responses after perianal electrical stimulation. J Neurol Neurosurg Psychiatry 45:770–773

Potenzoni D, Juvarra G, Bettoni L, Stagni G (1983) Pseudomyotonia of the striated urethral sphincter. J Urol 130:512–513

Powell PH, Feneley RCL (1980) The role of urethral sensation in clinical urology. Br J Urol 52:539–541

Rushworth G (1967) Diagnostic value of the electromyographic study of reflex activity in man. Electroencephalogr Clin Neurophysiol [Suppl] 25:65–73

Sakuta M, Nakanishi T, Toyokura Y (1978) Anal muscle electromyograms differ in amyotrophic lateral sclerosis and Shy–Drager syndrome. Neurology 28:1289–1293

Sarica Y, Karacan I (1986) Cerebral responses evoked by stimulation of the vesico-urethral junction in normal subjects. Electroencephalogr Clin Neurophysiol 65:440–446

Shy GM, Drager GA (1960) A neurological syndrome associated with orthostatic hypotension. Arch Neurol 2:511–527

Snooks SJ, Swash M (1984) Perineal nerve and transcutaneous spinal stimulation: new methods for investigation of the urethral striated sphincter musculature. Br J Urol 56:406–409

Snooks SJ, Swash M (1985) Pudendal nerve terminal motor latency and spinal stimulation. In: Henry MM, Swash M (eds) Coloproctology and the pelvic floor. Butterworths, London, pp 112–124

Snooks SJ, Barnes PRH, Swash M (1984a) Damage to the innervation of the voluntary anal and periurethral musculature in incontinence. J Neurol Neurosurg Psychiatry 47:406–409

Snooks SJ, Henry MM, Swash M (1984b) Anorectal incontinence and rectal prolapse: differential assessment of the innervation to puborectalis and external anal sphincter muscles. Gut 26:470–476

Snooks SJ, Setchell M, Swash M, Henry MM (1984c) Injury to innervation of pelvic floor sphincter musculature in childbirth. Lancet II: 546–550

Stalberg E, Trontelj JV (1982) Abnormal discharges generated within the motor unit as observed with single-fibre electromyography. In: Culp WJ, Ochoa J (eds) Abnormal nerves and muscles as impulse generators. Oxford University Press, Oxford, pp 443–474

Stanton SL, Hill DW, Williams JP (1973) Electromyography of the detrusor muscle. Br J Urol 45:289–298

Sung JH, Mastri AR, Segal E (1978) Pathology of the Shy–Drager syndrome. J Neuropathol Exp Neurol 38:149–190

Takaiwa M, Shiraiwa Y (1984) A new technique of vesical electromyogram with cystometrogram and urethral electromyogram. Urol Int 39:217–221

Takaiwa M, Shiraiwa Y, Katahira K, Tsukahara S (1983) A new urinary bladder electromyogram technique. Urol Int 38:1–4

Tanagho EA, Miller ER (1970) Initiation of voiding. Br J Urol 42:175–183

Torrens MJ (1985) Neurophysiology of bladder function. In:

Whitfield HN, Hendry WF (eds) Textbook of genitourinary surgery. Churchill Livingstone, Edinburgh

Vereecken RL, Verduyn H (1970) The electrical activity of the paraurethral and perineal muscles in normal and pathological conditions. Br J Urol 42:457–463

Vodusek DB, Light KL (1983) The motor nerve supply of the external urethral sphincter muscles: an electrophysiologic study. Neurourol Urodyn 2:193–200

Vodusek DB, Janko M, Lokar J (1982) EMG, single fibre EMG and sacral reflexes in assessment of sacral nervous system lesions. J Neurol Neurosurg Psychiatry 45:1064–1066

Vodusek DB, Janko M, Lokar J (1983) Direct and reflex responses in perineal muscles on electrical stimulation. J Neurol Neurosurg Psychiatry 46:67–71

11 · Human Physiology

Michael Torrens

Introduction

The purpose of this chapter is to summarize and integrate the more important information about human bladder physiology in order to provide an overall view of present concepts. Many of the more controversial issues have been identified and discussed in the previous chapters. This presentation will be more didactic and consequently might soon be outdated. The following account, therefore, concentrates on those aspects of present knowledge which are the most practically useful.

Neural Control Mechanisms

Control systems were reviewed at length in Section 2. A number of cerebral areas have influences on the process of micturition, as summarized in Fig. 11.1. A simple starting concept is that these areas produce varying levels of facilitation and inhibition which control micturition through some final common pathway. This concept has been further developed by ideas derived from systems control theory, those of gating and switching, gain and filtration. Also, as will become evident, the process of controlling the activities of the lower urinary tract takes place at all levels of the central and peripheral nervous system.

There is common ground between the various viewpoints discussed in Chapter 8. Certain general areas of the nervous system seem to be more significant in relation to urinary function than others, and lesions between these areas have different effects. These areas are:

1. Superior frontal gyrus/septal area
2. Cingulate/limbic area
3. Paracentral lobule
4. Pontine tegmentum
5. Lumbosacral spinal cord
6. Autonomic ganglia in hypogastric and pelvic plexus

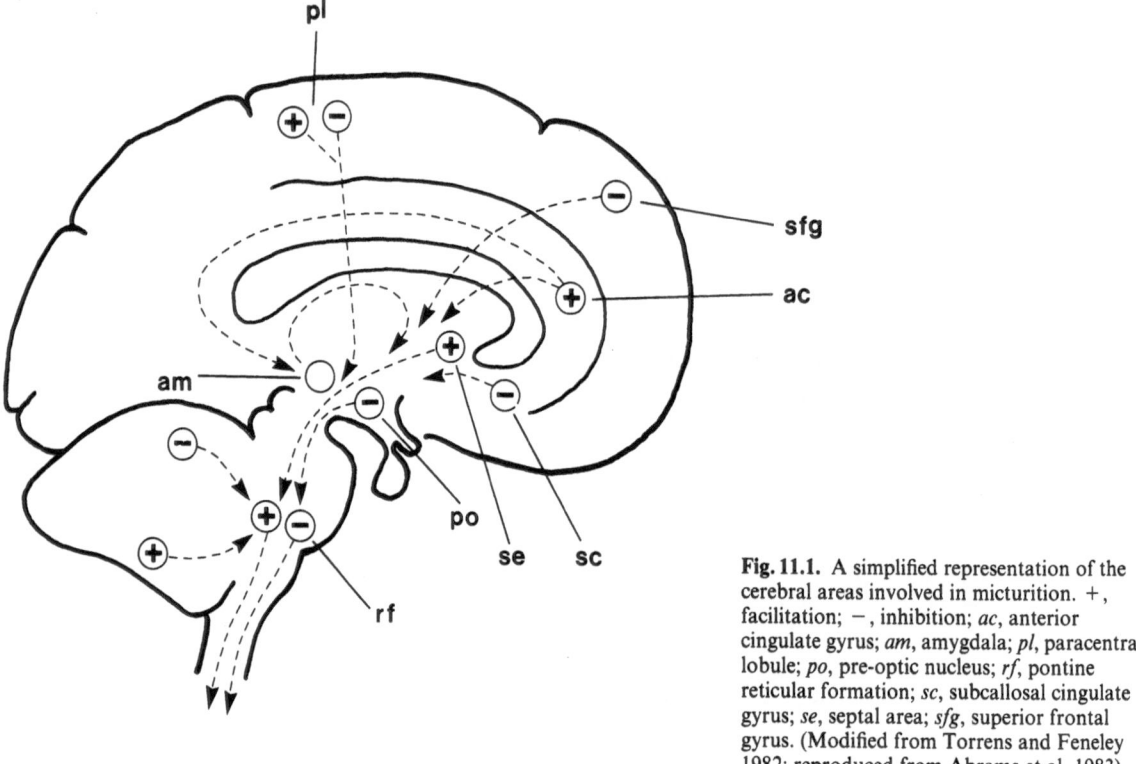

Fig. 11.1. A simplified representation of the cerebral areas involved in micturition. +, facilitation; −, inhibition; *ac*, anterior cingulate gyrus; *am*, amygdala; *pl*, paracentral lobule; *po*, pre-optic nucleus; *rf*, pontine reticular formation; *sc*, subcallosal cingulate gyrus; *se*, septal area; *sfg*, superior frontal gyrus. (Modified from Torrens and Feneley 1982; reproduced from Abrams et al. 1983)

For descriptive purposes these can be divided into an upper voluntary system (1–3) and a lower involuntary or autonomous system (4–6). The inter-relationship between these areas is shown in Fig. 11.2 and the effect of lesions is summarized in the caption.

Upper Voluntary Control System

Superior Frontal Gyrus/Septal Area

The superior frontal gyrus/septal area appears to be the most significant area for detrusor control in the human and is the only region of the cortex consistently associated with detrusor dysfunction when damaged (Andrew and Nathan 1964; Maurice-Williams 1974). The effect varies from urge incontinence to socially inappropriate or unconscious incontinence—but always with coordinated bladder and sphincter activity. Figure 11.3 illustrates the similarities between the cortical representation of the detrusor in man and cat.

The importance of the septal area is inferred from the experiments by Hess (1947), using implanted electrodes. In conscious animals stimulation of this area produces not only micturition but the carrying out prior to micturition of the coordinated social

behaviour associated with it. Similar experiments have seldom been conducted in humans, but in one series Sem-Jacobsen (1968) discovered responses (urgency) in only 2 out of 47 patients, despite a wide range of stimulation sites (Fig. 11.4). It is possible that these sites represent projection fibres to the superior frontal gyrus.

Cingulate and Limbic Area

It is assumed that the involvement of the cingulate area and associated parts of the limbic system is more concerned with the behavioural aspects of micturition, which are perhaps less prominent in humans than in certain lower animals who use urination for territorial marking. Certainly many hundreds of individual stimulations in the anterior cingulate region in conscious patients at the Burden Hospital, Bristol (R. Cooper 1987, personal communication) have produced no effects on the urinary system.

Paracentral Lobule

The area of the paracentral lobule corresponds with the cortical motor representation of the pelvic floor,

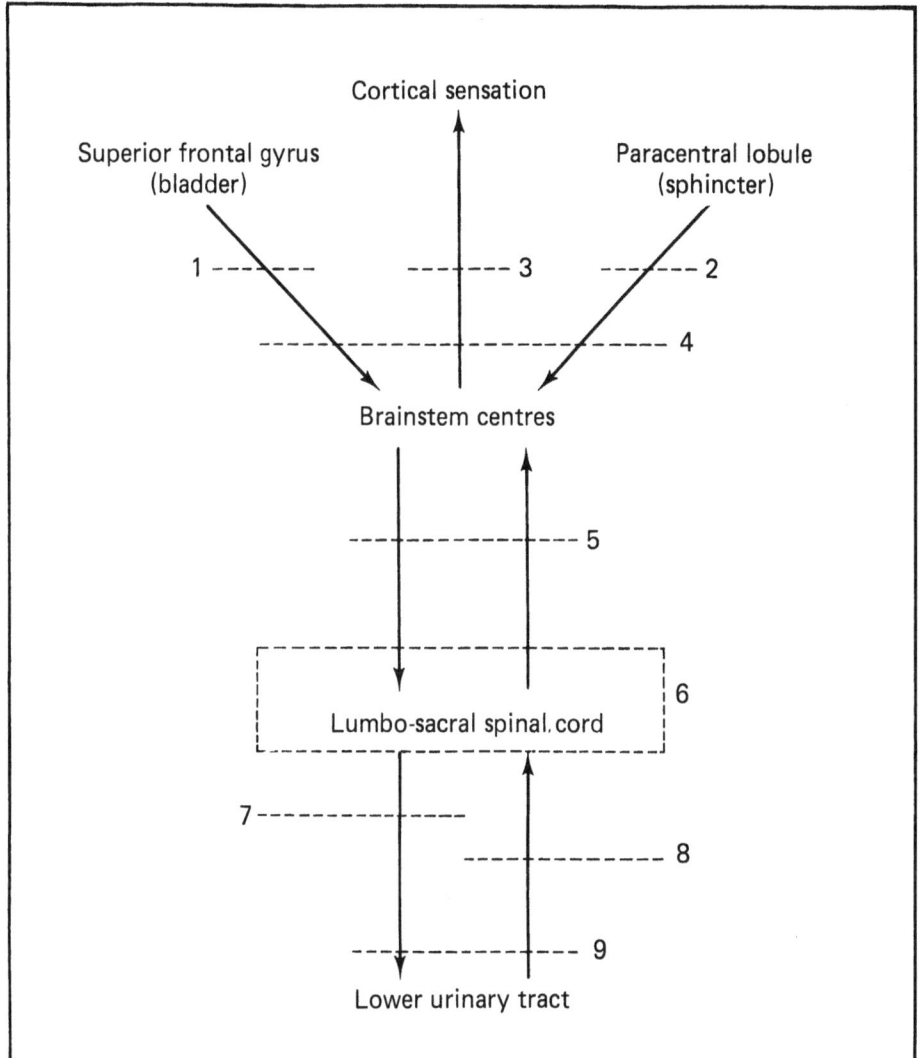

Fig. 11.2. Simplified scheme of interaction of various levels of the nervous system in micturition. The locations of certain possible nervous lesions are denoted by numbers and explained as follows:

1. Lesions isolating the superior frontal gyrus prevent voluntary postponement of voiding. If sensation is intact this produces urge incontinence. If the lesion is larger there will be additional loss of social concern about incontinence.

2. Lesions isolating the paracentral lobule, sometimes associated with a hemiparesis, will cause spasticity of the urethral sphincter and retention. This will be painless if sensation is abolished. Minor degrees of this syndrome may cause difficulty in the initiation of micturition.

3. Pathways of sensation are not known accurately. In theory, an isolated lesion of sensation above the brain stem would lead to unconscious incontinence. Defective central conduction of sensory information would explain noctural enuresis.

4. Lesions above the brain stem centres lead to involuntary voiding that is coordinated with sphincter relaxation.

5. Lesions below brain stem centres but above the lumbosacral spinal cord lead, after a period of bladder paralysis associated with spinal shock, to involuntary reflex voiding that is not coordinated with sphincter relaxation (detrusor/sphincter dyssynergia).

6. Lesions destroying the lumbosacral cord or the complete nervous connections between the central and peripheral nervous system result in a paralysed bladder which contracts only weakly in an autonomous fashion because of its remaining ganglionic innervation. However, if the lumbar sympathetic outflow is preserved in the presence of conus and/or cauda equina destruction then there may be some residual sympathetic tone in the bladder neck and urethra which may be sufficient to be obstructive.

7. A lesion of the efferent fibres alone leads to a bladder of decreased capacity and decreased compliance associated experimentally with an increased number of adrenergic nerves.

8. A lesion confined to the afferent fibres produces a bladder which is areflexic with increased compliance and capacity.

9. As there are ganglion cells in the bladder wall it is technically impossible to decentralize the bladder completely, but congenital absence of bladder ganglia may exist producing megacystis.

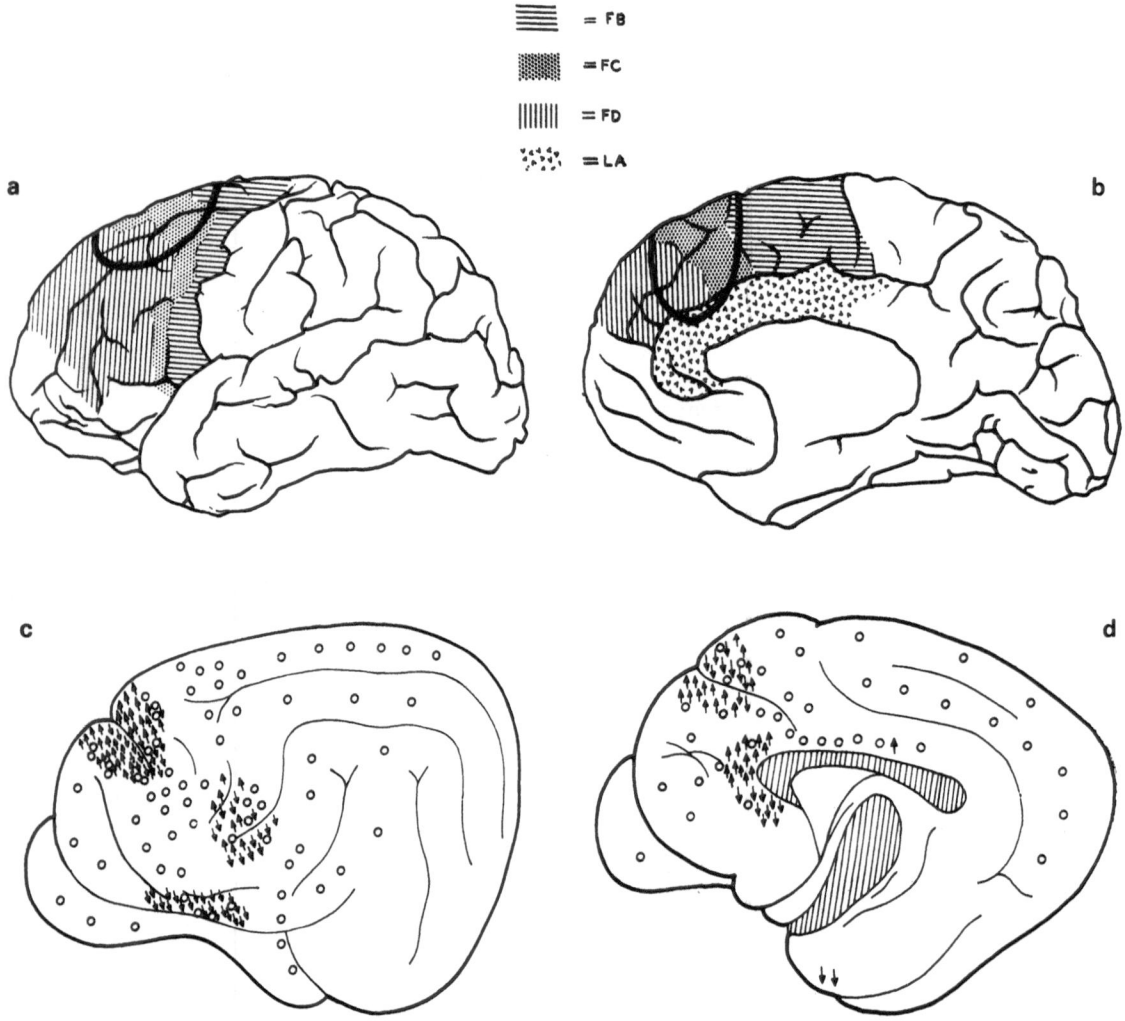

Fig. 11.3a–d. The lateral (**a, c**) and medial (**b, d**) aspects of the brain of the human (**a,b**) and the cat (**c, d**) showing the similarity in the areas involved in the control of detrusor function. The area of human cerebral cortex concerned is denoted by a *heavy black line* and includes the superior frontal gyrus and extends towards the cingulate area. In the cat the areas are subdivided into those causing bladder contraction (↑), relaxation (↓) or no response (O). In the human there appears to be no bladder representation in the orbital cortex as is apparent in the cat. However, there are areas located near the optic chiasm. (**a,b** from Andrew and Nathan 1964; **c,d** from Gjone and Setekleiv 1963)

Lower Involuntary System

The lower involuntary system consists of the peripheral nerves and their intra-axial cord and brain stem control mechanisms, mainly a spinobulbospinal (SBS) reflex. The present extent of understanding, already discussed in Section 2, will be summarized mainly by diagrams which present the hypotheses derived from animal work, together with comments on the extent to which these correlate with available evidence in humans.

Pontine Tegmentum

It is generally agreed that the level of the pons and adjacent medulla is that at which the coordination and integration of bladder and sphincter activity occurs in the adult. Lesions above this level lead to coordinated incontinence, and lesions below this level to incoordination and detrusor sphincter dyssynergia. The evidence for the localization of this centre has been considered earlier (see Chap. 8, p. 250).

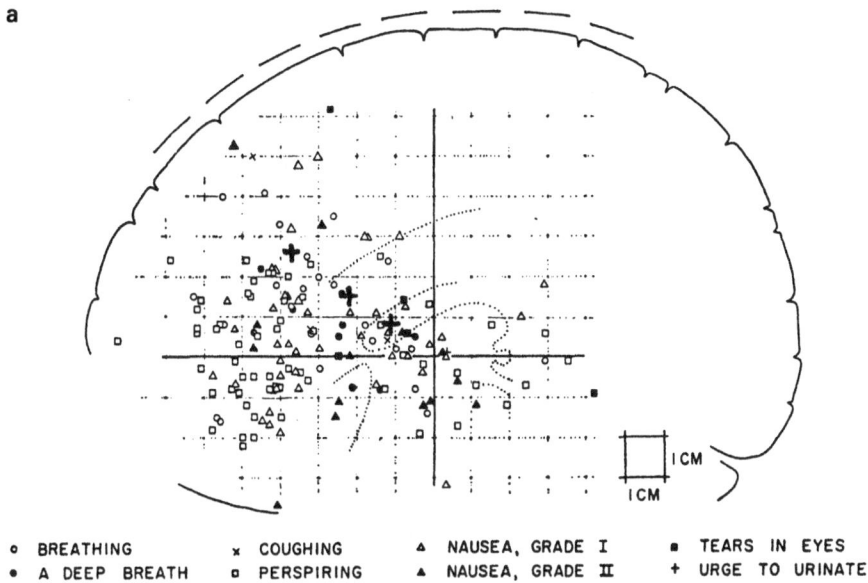

a

| ○ BREATHING | × COUGHING | △ NAUSEA, GRADE I | ▪ TEARS IN EYES |
| ● A DEEP BREATH | ▫ PERSPIRING | ▲ NAUSEA, GRADE II | + URGE TO URINATE |

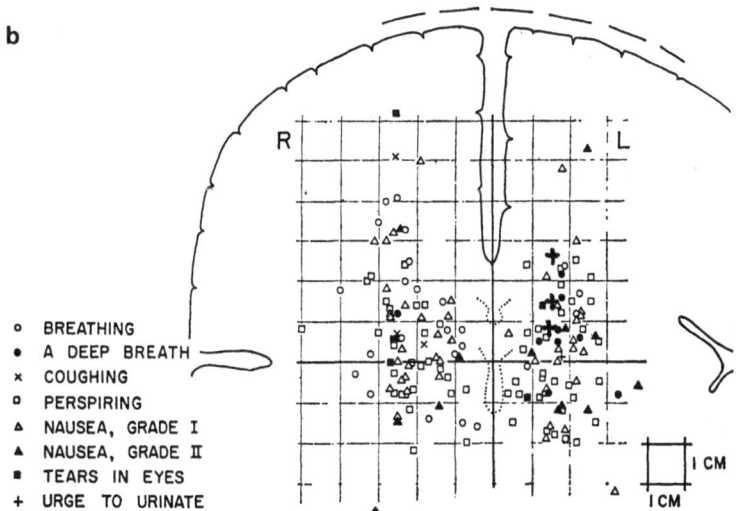

b

○ BREATHING
● A DEEP BREATH
× COUGHING
▫ PERSPIRING
△ NAUSEA, GRADE I
▲ NAUSEA, GRADE II
▪ TEARS IN EYES
+ URGE TO URINATE

Fig. 11.4a,b. The plotting of the responses to stimulation of conscious human subjects with implanted electrodes plotted in the sagittal plane (**a**) and coronal plane (**b**). Various vegetative responses are shown; those for urgency (+) are deliberately made prominent. (Sem-Jacobsen 1968)

In the cat, destruction of this area, sometimes termed the pontine detrusor nucleus, produces sustained urinary retention. For obvious reasons, studies in humans on the effects of lesions in the pons are difficult to interpret. The advent of magnetic resonance scanning capable of identifying small areas of demyelination during life may help us to understand the interesting functional disturbances of micturition that occur in multiple sclerosis.

In the most simplistic terms the pontine centre can be regarded as a switch which is normally under higher cerebral control for initiating coordinated micturition.

Spinal and Peripheral Systems

The activities of the spinal cord, peripheral nerves and autonomic ganglia are best considered together in relation to afferent and efferent activity. The various functions are summarized in Figs. 11.5 and 11.6.

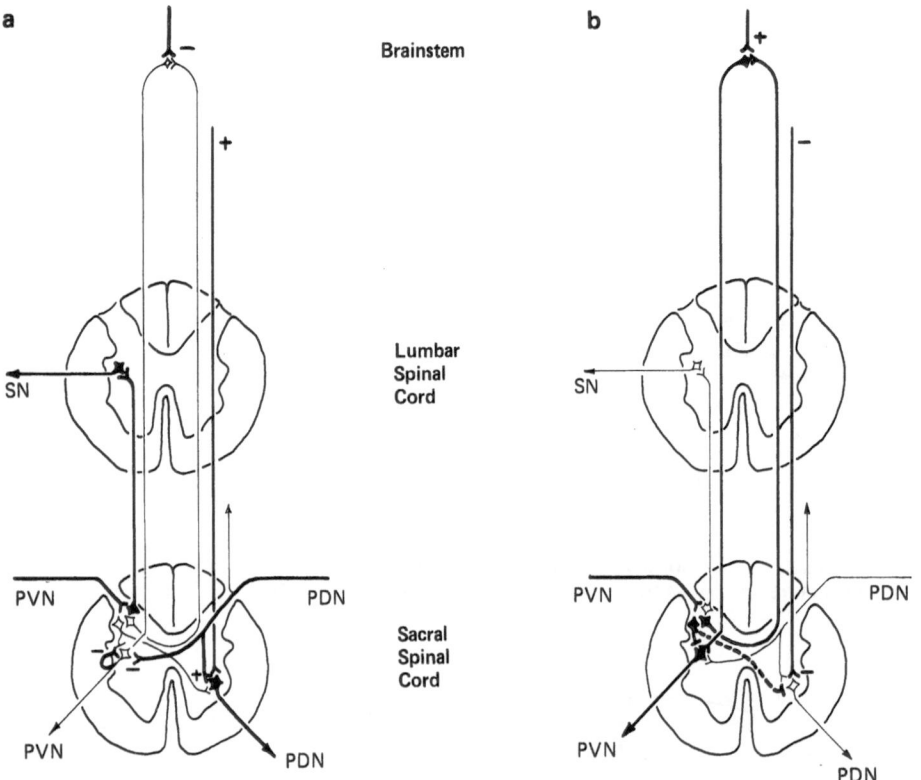

Fig. 11.5a,b. Some neural influences acting on the lower urinary tract during the bladder filling phase (**a**) and during voiding (**b**). Active pathways denoted by *heavy lines*. *Dotted lines* indicate neurons which may have a gating activity but which do not normally constitute a reflex pathway. *SN*, sympathetic nerve; *PVN*, pelvic nerve; *PDN*, pudendal nerve.

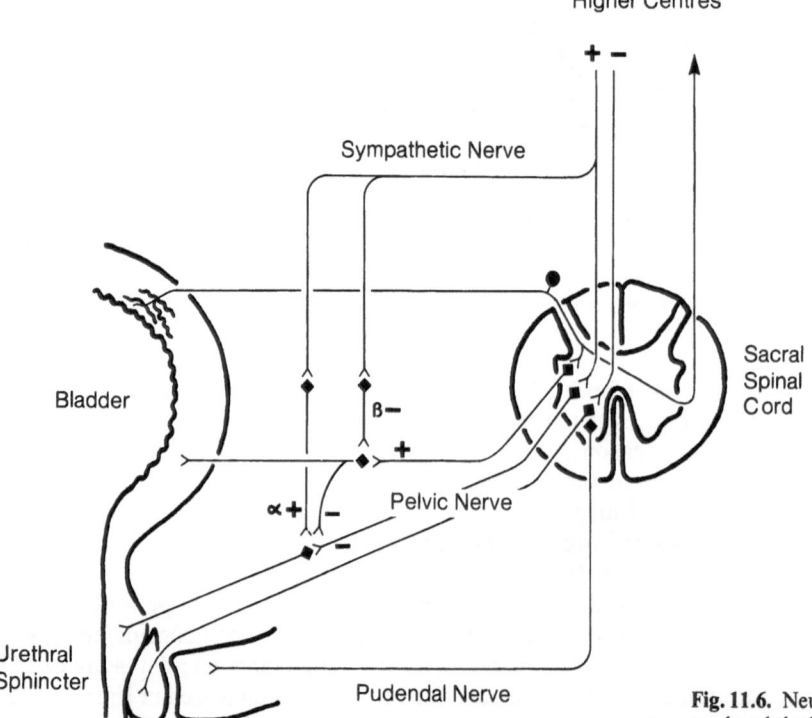

Fig. 11.6. Neural organization between the spinal cord and the lower urinary tract. (Torrens 1984)

Afferent Pathways The relevant visceral afferents arise from mechanoreceptors and nocioreceptors in the bladder, urethra, colon and rectum. These run in the pelvic and hypogastric nerves via A-delta and C-fibres. The corresponding somatic afferents from the urethra and pelvic floor run with the pudendal nerve. In addition to transmitting pain, temperature and proprioceptive information these nerves are said to convey information about the presence or absence of urinary flow.

Spinal Cord Activity The afferent impulses on arrival at the spinal cord are involved, directly or indirectly, in spinal or supraspinal reflex activity. The neurons in the spinal cord that receive afferent projections from the bladder and colon are of three main types (McMahon 1979):

1. Local sacral interneurons mainly concerned with the vesicosphincteric reflexes, except after spinal cord injury, when they may become the main reflex arc and then seem to be associated with the C-fibre afferents
2. Lumbosacral neurons with short ascending axons involved in viscerosympathetic reflexes
3. Neurons with long ascending axons that reach the brain stem and probably mediate SBS reflexes as well as sensation

A certain amount of sensory processing takes place in these interneurons. Some respond in a way that is directly analagous to primary afferent activity but others exhibit convergence of afferents from different sources. For example, A-delta and C-fibres may project onto the same cell, afferents from different visceral systems (e.g. bladder and colon) may converge on the same cell or, alternatively, visceral and somatic afferents may converge together. Following this convergence the interneuron responds either by summating the afferent activity or discriminating between the afferent activity, reacting by stimulation from one source (e.g. bladder) and inhibition from another (e.g. colon). In addition, descending pathways from the brain stem may exert an inhibitory influence on the cells, thus controlling the transmission of sensory information through the dorsal horn and exerting an effect on the micturition reflex. This information has been derived from experiments in the cat and it is likely that the complexity and multiplicity of interactions is even greater in the human.

It is likely that the pathways for sensation centripetally in the spinal cord are variable from one individual to another. Most observations have been made on patients who have undergone an anterolateral cordotomy. Nathan (1976) considers that specific sensation from the bladder and urethra ascends in the spinothalamic tracts. He states that if these are sectioned the only remaining sensation is that of imminence of micturition, which probably arises from the pelvic floor and ascends in the posterior column. Hitchcock et al. (1974) have demonstrated that bladder sensation may sometimes be preserved after spinothalamic tractotomy which has produced bilateral sacral analgesia. Kuru (1965) combines information from animal experiments and observations in humans and emphasizes the contribution of spinobulbar fibres in the paramedian superficial layer of the dorsal column. There seems no particular reason why the afferent projection in humans should be more localized than it is in animals. McMahon and Morrison (1982) have shown that the localization of long ascending projections from pelvic viscera (obtained by microelectrode recording from the spinal cord of cats) are not grouped closely in the cord and change their position between various spinal levels (see Chap. 4, p. 114).

Descending pathways from the brain stem to the bladder are assumed, from animal experiments, to run with the reticulospinal tracts. Kuru (1965) suggests that the vesicoconstrictor and vesicorelaxor fibres are separately located in the lateral and ventral reticulospinal tracts respectively. The facilitatory vesicoconstrictor pathway is concentrated in an area posterior to the dentate ligament insertion in the lower spinal cord but becomes progressively more anterior in the cervical region (Kerr and Alexander 1964). This seems to support the observation that the tract is more likely to be injured in a high cordotomy operation. The usual response to cordotomy, whether unilateral or bilateral, is an increase in the involuntary contractility or instability of the bladder. In one case reported by Hitchcock et al. (1974) unilateral cordotomy was followed by a cystographic appearance suggesting increased tone in the bladder wall only on the same side as the cord section. Since most cordotomies are restricted to the anterior 4 mm of the spinal cord surface this tends to confirm the fact that inhibitory (vesicorelaxor) fibres are located more anteriorly than any facilitatory pathways (see also Chap. 7, p. 195).

Sacral Micturition Centre It has already been emphasized that the micturition "centre" in the sacral spinal cord does not transmit a spinal detrusor reflex. Except perhaps in infants it cannot fully coordinate the parasympathetic, sympathetic and somatic activities related to filling and voiding.

However, while the absolute control of inhibition and facilitation is by higher centres, it is likely that the type of bladder response and therefore the completeness of emptying is largely controlled by neural interaction at the level of the sacral cord and peripheral ganglia. The various local facilitatory and inhibitory influences are described below.

The usual response to spinal cord injury is complete bladder paralysis for a time. Presumably the synaptic sites on the dendrites of the detrusor motor nucleus which are occupied by the descending fibres from the spinal cord are vacated and it takes some time for these to be reoccupied by local axonal sprouting. The rearrangement of synaptic contacts is one aspect of *neuroplasticity*. There is some evidence to suggest that the way synaptic contacts are reoccupied is in response to demand, and it is possible that the beneficial effects of electrical stimulation are due to the artificial manipulation of this neuroplasticity. Currently such stimulation is conducted on a somewhat empirical basis, but undoubtedly in the future the techniques will be refined to produce the most appropriate response. In this area of research lies perhaps the greatest hope for bladder rehabilitation.

Parasympathetic Efferent Pathways The parasympathetic fibres in the anterior sacral roots 2–4 are the principal motor supply to the bladder. The third sacral root is dominant in most cases in humans. Section of posterior roots profoundly depresses bladder function though compliance remains normal. The effects are usually permanent. Anterior root section alone tends to cause a decrease in bladder wall compliance and in animals this is accompanied by an increase in the proportion of adrenergic nerves in the bladder (Sundin and Dahlstrom 1973).

These parasympathetic fibres are cholinergic in humans, and the atropine resistance that has been described in a large number of species does not seem to occur significantly in baboons or humans (Brindley and Craggs 1976).

Somatic Efferents The motor neurons innervating the rhabdosphincter arise in the lateral part of the anterior horn close to the parasympathetic efferent motor neurons in the area called Onuf's nucleus (see Chap. 3, p. 73). de Groat and Kawatani (1985) note that sacral autonomic and somatic sphincter motor neurons are closely linked physiologically and anatomically. In addition to their location they receive a unique distribution of neuropeptide terminals and have an unusual and characteristic dendritic pattern. Also they do not degenerate in motor neuron disease, unlike other α-motor neurons.

Sympathetic Efferent Pathways Sympathetic efferents arising from the lower part of the thoracolumbar sympathetic outflow may synapse in the hypogastric and pelvic ganglia. Some also have their synapses in the ganglia of the sympathetic chain, and postganglionic fibres from these synapses run with the pelvic nerves, a fact which complicates the interpretation of the effect of pelvic nerve stimulation on more peripheral function.

Neurotransmitters and Receptors

The process of neurotransmission was discussed in Chapters 5 and 6. The activity in parasympathetic and sympathetic nerves is summarized in Fig. 11.7. The neurotransmitters are contained in vesicles which tend to congregate in areas of neuronal varicosity. When the neuronal membrane is depolarized the transmitter is liberated by the process known as exocytosis (Fig. 11.8). What happens on the liberation of transmitter into the synaptic cleft depends upon a number of factors:

1. Which excitatory receptors are adjacent (e.g. α 1, β)
2. Speed of breakdown or re-uptake
3. Presence of inhibitory or excitatory cotransmitters
4. Presence of prejunctional inhibitory receptors (e.g. α 2)

The increasing complexity of the concept of chemical transmission was reviewed in Fig. 5.3 (see p. 137). It is evident that the function of the lower urinary tract depends not only on the distribution of nerves but also on the distribution of receptors. The organization of the important receptor subgroups is shown in Fig. 11.9, which emphasizes the importance of the topographical representation of receptors (which is different from the distribution of the related nerves).

An attempt to summarize the full complexity of the situation is made in Fig. 11.10. Here some of the transmitters, their sites of action and their receptor types, together with the resulting effect (facilitation or inhibition), are superimposed on a diagram of the innervation of the lower urinary tract similar to that in Fig. 11.6. This summary is necessarily incomplete and represents only the beginning of our knowledge of the chemical interactions controlling the bladder and urethra.

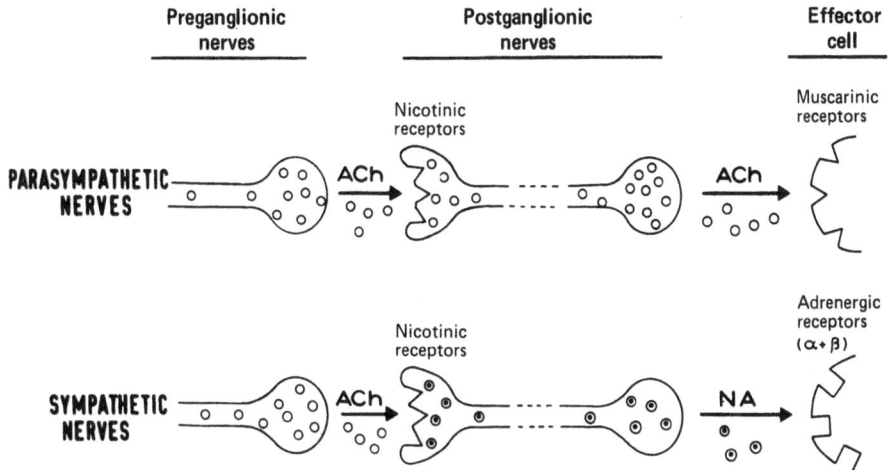

Fig. 11.7. Neurotransmission in sympathetic and parasympathetic nerves. Acetylcholine (*ACh*) and noradrenaline (*NA*) are the autonomic transmitters at the sites shown. (Caine 1984)

Filling Phase of the Micturition Cycle

Bladder Factors

Physical Distensibility

If the micturition cycle is represented as a pressure–volume loop the filling phase occurs without much significant rise in the intravesical pressure (stages 1–3 Fig. 11.11). This requires that the bladder is distensible and that there is no active detrusor contraction. At the same time the bladder neck and urethra must remain closed. The factors contributing to this state are outlined below.

The mechanism of compliance and the hydrodynamics of bladder filling were discussed in Chapter 9. The ability of smooth muscle to increase greatly in length without an exactly proportional increase in force depends to some extent on the cross-linkages between the individual muscle cells. As outlined in Chapter 6, (see p. 163), the collagen fibres between the cells are arranged in such a way

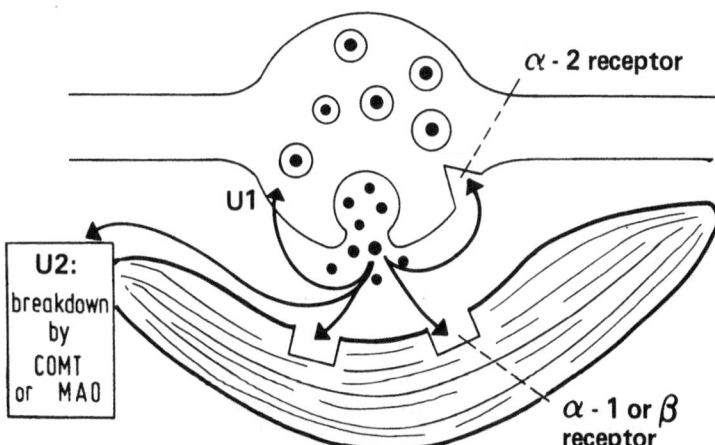

Fig. 11.8. The release of noradrenaline by exocytosis and its inactivation. *U1* (uptake 1) reabsorption into nerve terminal; *U2* (uptake 2 of little importance in neural noradrenaline), uptake into tissues and deactivation by enzymes monoaminoxidase (*MAO*) and catechol-O-methyl transphorase (*COMT*); α-2, presynaptic α receptors; α-1, post-synaptic adrenergic receptors. (Caine 1984)

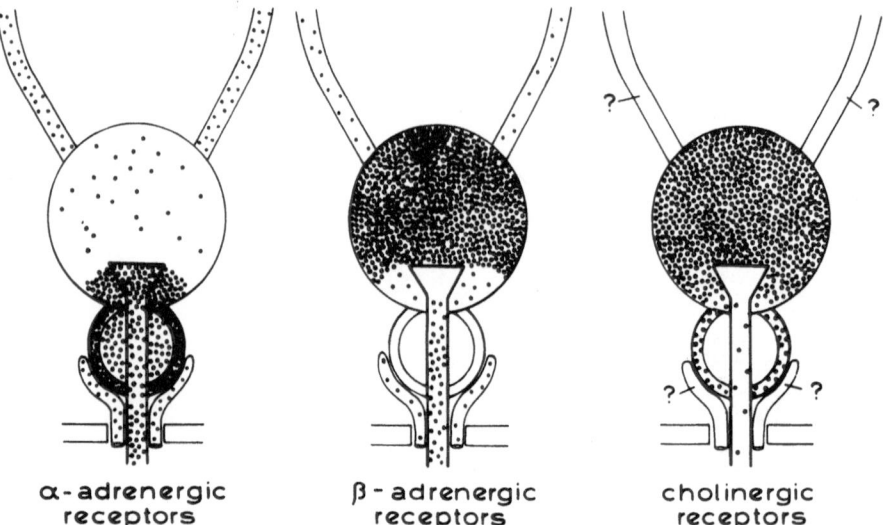

Fig. 11.9. Distribution of autonomic receptors in the human urinary tract. The relative densities are roughly indicated. α receptors are α 1 in type; β receptors are thought to be β 2. (Caine 1984)

that they can become unwound and/or more oblique as the muscle cells slide apart from one another. It is only beyond the range of normal capacity that this type of potential distensibility becomes exhausted. Conditions which increase the collagen in bladder muscle such as infection, radiotherapy and the hypertrophy associated with instability all interfere with distensibility.

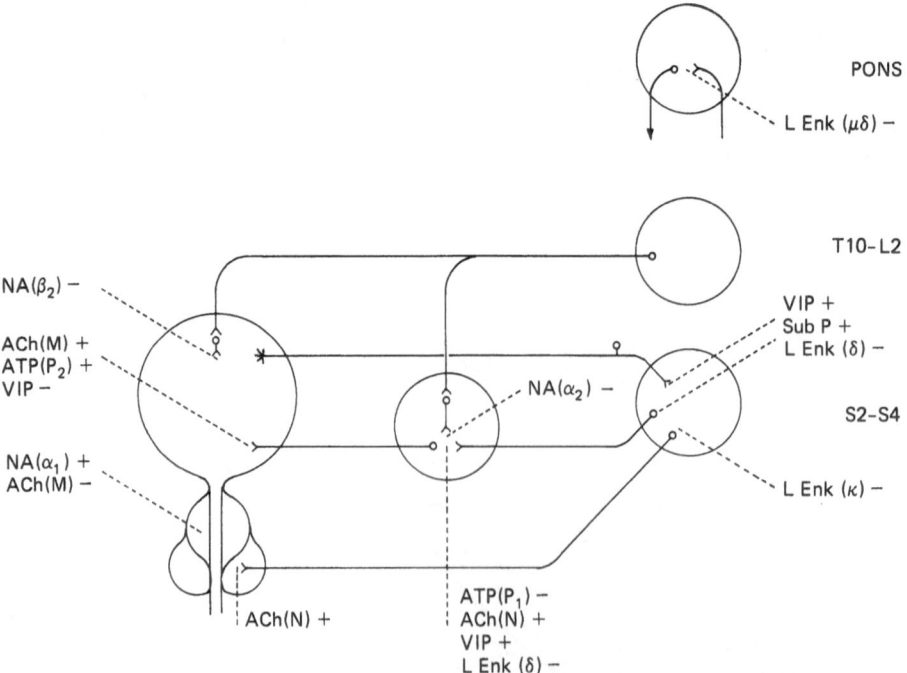

Fig. 11.10. The sites of action of various neurotransmitters and the receptors relevant to them in the lower urinary tract based largely on information from de Groat and Kawatani (1985), together with the inhibitory or facilitatory action produced. *NA*, noradrenaline; *ACh*, acetylcholine; *ATP*, adenosine triphosphate; *VIP*, vasoactive intestinal peptide; *LEnk*, leucine enkephalin; *SubP*, substance P. Receptor types *in parentheses*. *N*, nicotinic; *M*, muscarinic; μ, δ, κ, opioid receptor subtypes; *P*, purinergic; α, β, adrenoceptors; +, facilitation; −, inhibition. This figure is not intended to be exhaustive and omits (for the sake of simplicity) some of the influences on the urethral smooth muscle, particularly at a ganglionic level.

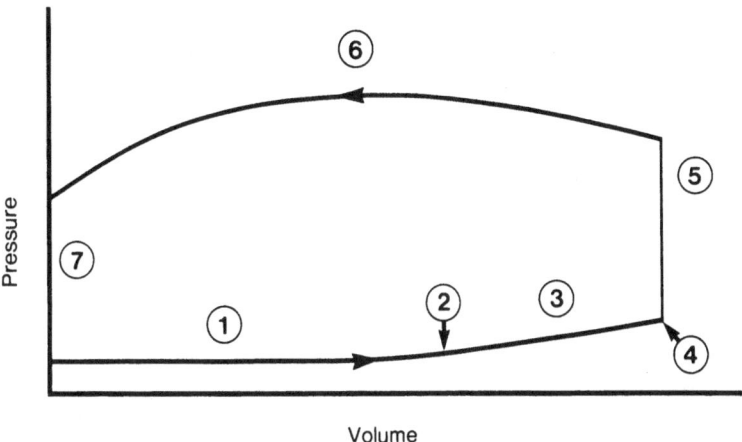

Fig. 11.11. Micturition cycle represented as a pressure–volume loop. Various phases, indicated by numbers are as follows:

1. Accommodation—partly because of intrinsic properties of bladder wall and partly because of unconscious nerve-mediated inhibition.
2. First conscious sensation—mediated by tension receptors, since no rise in detrusor pressure is necessary.
3. Postponement—cortical function lost in certain cerebral lesions.
4. Initiation of voiding—coordinated by pontine micturition centre.
5. Isometric detrusor contraction—phase before flow starts.
6. Sustained detrusor contraction—phase in which detrusor pressure remains relatively constant throughout voiding.
7. Relaxation phase, majority of which occurs after voiding has ended—in some cases detrusor pressure may rise as the bladder neck closes (isometric after contraction). (Torrens 1984)

Slow Filling Rate

The normal rate of filling of the bladder is around 2 ml per minute. At this rate of strain the dynamic characteristics of the bladder wall allow distension with the minimum of pressure increase (see Chap. 9, p. 289). In addition, the mechano-receptors for tension within the bladder wall respond differently at different rates of filling as well as at different volumes (see Chap. 4, p. 102). The phenomenon of urinary frequency during a rapid diuresis is familiar to everyone.

Inhibition of Detrusor Contraction

From a certain point, which is usually assumed to be the first desire to void, some control of detrusor "reflex" contraction is required. This takes place at several levels within the nervous system. Cerebral areas, both conscious and unconscious, control the pontine micturition centre (see Fig. 11.1). Local afferent activity influences interneurons in the spinal cord (see Fig. 11.5). The firing of the detrusor motor neurons is controlled by local recurrent inhibition. Transmission of impulses to the postganglionic parasympathetic nerves is controlled in the autonomic ganglia by several mechanisms. The

phenomenon of temporal facilitation (de Groat and Booth 1980) means that until a large number of impulses arrive at the ganglia the characteristic amplification effect does not occur and the resulting output from the ganglia is not enough to cause detrusor contraction. In addition, there is ganglionic inhibition by adrenergic and purinergic transmitters (see Fig. 11.10). Sympathetic inhibition is provoked both by afferent activity in the pelvic nerves (Edvardsen 1967) and also in pudendal nerves (Kock and Pompeius 1963) (see Fig. 11.5). The resulting preganglionic sympathetic activity is thought to stimulate the small intensely fluorescent (SIF) cells within the vesical ganglia. These release noradrenaline, which activates presynaptic α 2 adrenoceptors thus inhibiting the release of acetylcholine. In addition, direct inhibition by hyperpolarization of the postganglionic parasympathetic neurons may occur.

Urethral Factors

The normal urethral closure mechanism maintains a positive urethral closure pressure during bladder filling, even in the presence of increased abdominal pressure. Videocystography in normal persons shows that continence is maintained at the bladder neck. This can be regarded as the *proximal urethral closure mechanism*. If this vesicourethral junction is incompetent, continence may still be maintained at the high pressure zone in the mid-urethra about 2–3 cm distally. This zone corresponds to the maximum condensation of muscle, both smooth and striated, and may be regarded as the *distal urethral closure mechanism*. Whether it is really valid to separate two parts of the urethra in this way from a physiological point of view is debatable; the normal urethra probably works as one unit. However, from a practical standpoint it is useful because the two urethral areas may not be abnormal simultaneously, and the distinction helps in the classification of dysfunction.

Many factors have been thought to contribute to urethral closure. These are listed below and then considered separately.

Bladder neck/proximal urethral closure mechanism
    Muscular occlusion
    Elastic tissue
    Muscular and ligamentous suspension
    Abdominal pressure transmission
Distal/urethral closure mechanism
    Muscular occlusion
    Submucosal properties
    Hormonal factors
    Length/tension
    Elasticity

Proximal Urethral Closure Mechanism

Muscular Occlusion—Anatomical Configuration of the Bladder Neck The original concept of sphincter organization in both the bladder and the bowel was of an internal (autonomic) sphincter and an external (somatic) sphincter. This idea was supported by observations in man that stimulation of the presacral nerve caused visible contraction and pallor of the region of the bladder neck and internal urethral meatus when viewed directly (Learmonth 1931). However, it was soon realized that radical sympathectomy for conditions such as hypertension had no effect on micturition, though it did cause retrograde ejaculation. Alternative explanations were therefore sought. A number of gross anatomical studies (Woodburne 1960; Tanagho and Smith 1966; Hutch 1972) have emphasized the complex arrangement of loops and slings of detrusor muscle around the bladder neck, suggesting that these have a part to play both in the occlusion and the opening of this region. Hutch in particular produced a theoretical interpretation of bladder neck function which depended upon bladder base rigidity. A flat bladder base remained competent, whereas a "broken" base allowed bladder neck funelling and voiding. Despite the fact that this theory was based on videoscopic investigation it does not really conform to the way the bladder neck behaves at the initiation of micturition (see p. 347).

The distribution of adrenergic fibres has been discussed in Chapter 1. In summary, Gosling et al. (1977) found adrenergic fibres in the region of the bladder neck only in the male and deduced that the area served as a genital sphincter. This is supported by evidence that there is usually no significant occlusive pressure at the bladder neck in the male (or female) and that the pressure rises only in association with penile erection (Fig. 11.12). On the other hand, Ek et al. (1977b) showed adrenergic fibres in both male and female urethras and demonstrated

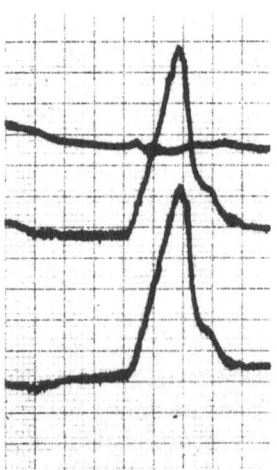

Resting Profile

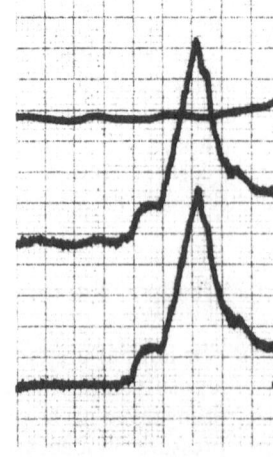

Profile on Erection

Fig. 11.12. Urethral pressure profile demonstrating elevation of pressure in the region of the bladder neck/preprostatic sphincter during penile erection. (Abrams et al. 1983)

that the isolated urethral muscle of both sexes responded in the same way to α adrenergic stimulation (Ek et al. 1977a). Certainly, the bladder neck in males and females seems to work in a similar way on videoscopy despite the structural and histochemical differences.

Apart from its action as a genital sphincter in males, the understanding of the neural control of the bladder neck is very confused. Nordling (1983) concluded from a study of neuropathic bladders that *parasympathetic* (cauda equina) lesions were more likely to be associated with bladder neck incompetence that *sympathetic* lesions, though the *addition* of urethral obstruction or a sympathetic lesion made such incompetence much more likely. This observation contradicts the findings of McGuire and Wagner (1977) and Blaivas and Barbalios (1983). Brindley (1986) reports that a combination of sacral anaesthetic blockade and phenoxybenzamine rendered him incontinent, though neither were effective alone. He also described how sacral anterior root stimulation causes relaxation of the bladder neck even after a large dose of atropine, suggesting that there is a mechanism for bladder neck opening independent of detrusor contraction. A great deal remains to be learned.

Elastic Tissue Undoubtedly, the considerable amount of collagen and elastic tissue present around the bladder neck (Gosling et al. 1983) contributes to the passive closure of the proximal urethra. However, if this elastic tissue was the major factor producing occlusion, then it would almost certainly interfere also with voiding.

Muscular and Ligamentous Suspension of the Bladder Neck The bladder neck and urethra are held in position both by the muscles in the pelvic floor and by the various ligaments containing smooth muscle that run from the back of the pubis to the urethra (and vagina and prostate) (DeLancey 1986a, b). Videoscopic examination shows that bladder neck elevation is a very important part of urethral closure. In fact, the ascent can be forceful and rapid and has been described as a "kick". At the same time any residual fluid in the proximal urethra is emptied back into the bladder. Despite the force of this movement, pressure registration at or just below the bladder neck does not show any corresponding rise. This leads to the inescapable conclusion that, despite the fact that active muscular activity must be involved, its primary role at the bladder neck is to change the configuration of the area and not to produce occlusive pressure. It

is therefore necessary to assume that, provided the configuration of the bladder neck is in the closed state, the maintenance of continence is by passive factors such as transmission of pressure, discussed below.

Abdominal Pressure Transmission The measurement of pressure transmission to the urethra from the abdominal cavity was discussed in Chapter 9 (see p. 301). This pressure transmission can only occur if the proximal urethra is maintained within the area above the pelvic floor which is effectively part of the abdominal cavity. The adequacy of transmission therefore depends upon adequate bladder neck suspension. The major importance of this concept is confirmed by the fact that successful management of bladder neck incompetence depends almost entirely on procedures to improve bladder neck suspension.

Distal/Urethral Closure Mechanism

The distal urethral closure mechanism is composed of a condensation of smooth muscle and also of striated muscle (rhabdosphincter) together with the nonmuscular components of the urethra at this point. The anatomy and innervation of the area was discussed in Chapters 1 and 2. One of the main problems requiring an explanation is that the innervation of the area in humans is mainly cholinergic, with either modest (*Ek et al. 1977a, b*) or very sparse (Gosling et al. 1977) adrenergic innervation. In contrast, the receptor activity in the urethra is mainly α adrenergic. This will be discussed below and again in relation to urethral relaxation (p. 348). Mattiasson (1984) presents a good discussion of the neuropharmacology of urethral function.

In contrast to the bladder neck the distal/urethral closure mechanism is identified by a measurable occlusive pressure in mid-urethra. This pressure does not remain constant. Sørensen et al. (1986) have identified the fluctuations in pressure that occur with time in healthy female volunteers. They identified normal pressure variations between 3 and 66 cm of water occurring in a rhythmic way at three frequency ranges. They consider that these pressure variations are a normal occurrence; however, the observation that these pressure changes seem more prominent in incontinent patients has prompted other investigators to suggest that there is a pathological variation as well (Kulseng-Hansen 1983). Variations in pressure seem to be more obvious when the bladder is relatively full.

The urethral pressure also varies with the extent of bladder filling. Rud et al. (1978) showed that when the bladder filled from 200 ml to maximum cystometric capacity (mean 400 ml) there was an increase in urethral pressure from 90 cm of water to 105 cm of water. Elevations of urethral pressure have also been shown in association with change in posture (George and Feneley 1978).

Muscular Occlusion Both the smooth muscle and the slow twitch component of the striated muscle of the rhabdosphincter are adapted to maintain occlusive pressure over a long period of time. Short-time augmentation of this pressure, e.g. in response to stress, is performed by the fast twitch component of the rhabdosphincter and the pelvic floor.

The relative contributions of smooth and striated muscle to the urethral pressure are disputed. Donker et al. (1972) showed only 17% decrease in urethral pressure after the administration of suxamethonium to three normal females and suggested that the sympathetic component was dominant. Doyle and Briscoe (1976) showed no significant effect of either suxamethonium or of competitive neuromuscular blocking agents on the urethra. Rud et al. (1980), in contrast, showed that there was a reduction of pressure of about 33% after alcuronium. Murray (1986) assessed a number of competitive neuromuscular blockers (alcuronium, pancuronium, vecuronium) and showed a reduction in urethral pressure of 15%–39% additional to that caused by the induction agent. When interpreting the results of studies using neuromuscular blockers it should be remembered that they are not entirely specific for striated muscle but do have some minor autonomic effects. These include ganglion blocking effects, muscarinic receptor blockade, inhibition of noradrenaline re-uptake and possibly stimulation of noradrenaline release (Marshall et al. 1980). Also the innervation of the rhabdosphincter is not typical of striated muscle in general.

Rather more than one-third of the urethral pressure is dependent on alpha adrenergic activity. Donker et al. (1972) observed a reduction of 65% in the urethral pressure after the administration of phentolamine 10 mg i.v. to female volunteers. There was no concurrent reduction in blood pressure. This observation, confirmed by others (e.g. Nordling 1978), has always been surprising in view of the fact that the smooth muscle in the female urethra is mainly longitudinal and has few adrenergic nerves. However, the α adrenergic blockade may influence the local vasculature, the pelvic ganglia or the striated muscle, as well as blocking the effects of circulating noradrenaline. This problem was investigated further by Nordling (1983), who concluded that circulating noradrenaline was not important and that urethral pressure was controlled by neural activity. Clonidine decreased both urethral pressure and the urethral sphincter activity on electromyography (EMG), suggesting that α adrenergic receptors in the central nervous system are involved in the regulation of striated urethral muscle tone. He also concluded that the α adrenoceptors on urethral vessels were important in their contribution to urethral pressure, thus confirming the work of Raz et al. (1972). The α adrenergic component seems to be more important in women than in men. Mattiasson et al. (1984) found that after phentolamine the pressure in male volunteers was reduced by only 10%, compared with 36% in females.

Submucosal Properties The urethral submucosa consists of connective tissue supporting a plexus of veins. Berkow (1953) likened this area in the female to the corpus spongiosum of the male and considered it might have some role in the maintenance of continence. At the very least this soft submucosal cushion allows the contour of the urethral mucosa to change when closed and assume its typical stellate or crenated crescentic shape. Various authors, however, have suggested a more active contribution to urethral pressure than this passive role.

Pulsatile variations of urethral pressure have been noted in women, particularly pre-menopausally, and occur to a lesser extent in some male urethras. Experiments in the dog by Raz et al. (1972) suggested that complete obstruction of the arterial supply to the urethra reduced the urethral pressure by 30%. Experiments of a similar nature during surgery in the human by Rud et al. (1980) suggested that 28% of the urethral pressure was due to an effect exerted by the urethral vascular bed. Rud also comments on the fact that there is a disappearance of urethral pulsation at the initiation of micturition, which suggests that part of urethral "relaxation" (see p. 348) is due to reduction of pressure within the vascular bed. This concept would require some active control over the pressure of blood in the submucosal veins. Huisman and Salome (1977) suggested that there were arteriovenous shunts in this area which might well be under neurological control. This finding, however, has not been confirmed by the extensive investigations of Gosling et al. (1983), who consider that blood vessels show no unusual or specialized features and that it is unlikely that a vascular component contributes significantly to urethral resistance. This opinion is supported by the work of Downie and

Lautt (1986), who showed that stimulation of the hypogastric nerve produced no change in the vascular resistance of the (perfused) urethra, though it did produce urethral constriction.

Hormonal Factors The variations in urethral function in normal subjects dependent on the hormonal changes in the menstrual cycle were discussed in Chapter 9 (see p. 301). In summary, high oestrogen levels have a greater effect on urethral length than on urethral pressure. Even though Rud (1980) found little change in urethral function measurements after treatment with oestrogens there is general agreement that the decline in oestrogen levels post-menopausally contributes to the gradual deterioration of urethral function that has been measured with increasing age (see Fig. 9.24, p. 301). Oestrogen receptors have been identified in the urethra. Low oestrogen levels result in changes in the mucosa, the submucosal blood vessels and probably the muscle function as well.

Despite this conviction about the importance of oestrogens, however, it must be reiterated that no comprehensive objective studies have shown any really useful improvement in urethral function after oestrogen administration.

Urethral Length, Mural Tension and Elasticity Many studies have shown the correlation between urethral length and impaired urethral closure in females, though length alone is not a diagnostic parameter for stress incontinence. It is unlikely that length is of primary importance; more likely the maintenance of a correct urethral length allows the muscle and other tissue in the urethral wall to act more efficiently to occlude the lumen. It is easy to imagine that a shortened "baggy" urethra cannot be occluded so adequately.

Bissado and Finkbeiner (1980) review the relationship between length/tension and the response to stimulation in detrusor and urethral smooth muscle and confirm that contraction is more efficient when the muscle is elongated. It follows that it is less efficient when it is shortened. The restoration of urethral length during operations to treat stress incontinence may well be significant for this reason.

Voiding Phase of the Micturition Cycle

Normal voiding is a voluntary act which results in sustained contraction of the bladder and relaxation of the urethra until the bladder is empty. This is represented by stages 4–7 in Fig. 11.11. The typical sequence of urodynamic events is summarized in Fig. 11.13. Voiding can be divided into the pre-micturition phase or initiation of voiding, the micturition phase or continuation of voiding and the post-micturition phase or cessation of voiding. The caption to Fig. 11.13 gives further details.

Initiation of Voiding

The sequence of events at the initiation of voiding usually follows the sequence summarized in Fig. 11.13. Tanagho and Miller (1970) described the progressive drop in urethral closure pressure that occurred slightly before a corresponding increase in detrusor pressure with urinary flow when the detrusor pressure exceeded the urethral pressure. They also noted other patterns. The urethral pressure could remain constant, with detrusor contraction eventually exceeding the urethral pressure. Alternatively the urethral pressure could drop, *without* a corresponding rise in detrusor pressure. Whatever the pressure changes, the process is accompanied, in normal subjects, by the bladder outlet assuming a funnelled shape. This funnelling normally begins before there is any change in detrusor pressure, making it relatively unlikely that the detrusor contraction pulls open the bladder neck. Rud et al. (1978) have documented carefully the changes occurring at the initiation of micturition in females. The intraurethral pressure decreases for approximately 3 s before the bladder begins to contract and the bladder contraction continues for 3 s before urine begins to flow. Alterations from this pattern of activity are usually associated with some urinary dysfunction such as stress incontinence, where initiation of micturition may be by urethral pressure decrease only or by straining.

How the bladder neck relaxes is still not entirely clear. Tanagho and Miller (1970) were inclined to believe that the pressure drop was due to striated muscle relaxation as they assumed the smooth muscle in the uretha would only contract or relax at the same time as that in the bladder. Certainly there is a diminution in EMG activity as soon as the pre-micturition phase starts. Tanagho noted that the administration of curare abolished the pre-voiding drop in urethral pressure and allowed funnelling of the bladder outlet during bladder filling. The conclusion from these studies was that the bladder neck opened passively as a result of pelvic floor relaxation. More recent studies have shown that the intrinsic urethral musculature, both smooth and striated, can certainly relax. Torrens (1978)

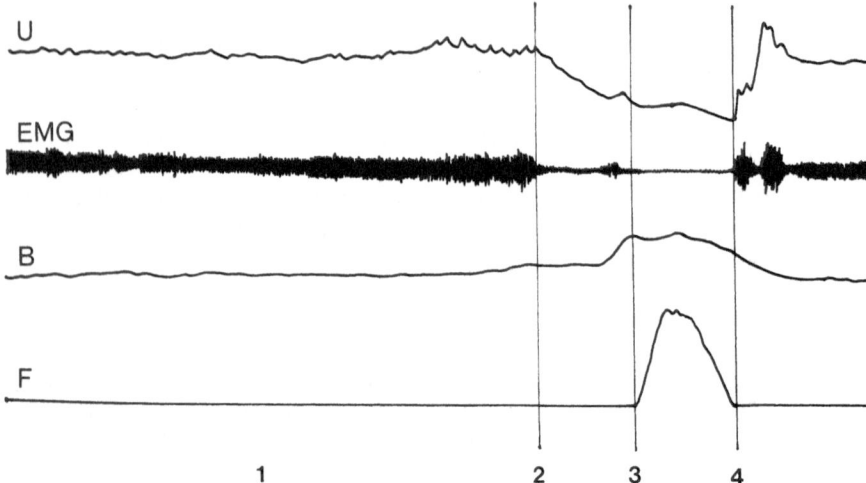

Fig. 11.13. The sequence of events observed during a normal void. *U*, maximum urethral pressure; *EMG*, urethral sphincter electromyogram; *B*, intravesical pressure; *F*, urine flow rate. Various phases, indicated by numbers, are as follows:

1. Urethral pressure, EMG discharge and bladder pressure all increase as bladder capacity is approached.

2. At the initiation of micturition EMG silence occurs, the urethral pressure progressively falls and, after a delay of a few seconds, the bladder pressure begins to increase. Descent of the bladder base is initiated by relaxation of the pelvic floor muscles and continued by relaxation of the urethra. The bladder neck sometimes opens before intravesical pressure rises.

3. Intravesical pressure is maintained at a constant level throughout voiding and the EMG should remain silent.

4. At the termination of voiding the urethra is emptied by voluntary muscle activity. The urethral pressure increases usually above its previous level and then settles to a normal resting zone. The bladder continues to contract for a short while. Sometimes at this point there is a short elevation of bladder pressure, the "after contraction".

showed that stimulation of the sacral nerves produced a reduction in urethral pressure which was independent of any afferent activity (because the sacral nerve had been cut) and was also independent of bladder contraction (see Fig. 5.11, p. 146). It has been noted that this response can be blocked by β 2 antagonists and the same drugs can cause an elevation of resting urethral pressure (Laval et al. 1977; Murray 1986). An alternative explanation is provided by Mattiasson (1984), who found that in human urethral muscle preparations prejunctional muscarinic cholinoceptors on the adrenergic nerve endings were shown to inhibit noradrenaline release. This means that increased parasympathetic activity during micturition could inhibit sympathetically mediated urethral tone. As noted earlier, Rud et al. (1980) consider that the urethral pressure contribution from the vascular element may also drop. Observation of bladder neck and urethral opening and closure on video-scopy seems to support the idea that the urethra has an intrinsic muscular action which cannot be explained entirely by the effects from the rather distant pelvic floor. Ghoneim et al. (1975) showed that bladder distension produced an additional drop in urethral resistance to flow after striated

muscle function had been abolished by succinylcholine.

The significance of the trigone in relation to bladder neck opening is not clear. Certainly the structural difference observed in this area might be expected to have some relevance. Recently, Speakman et al. (1986) have identified differences in behaviour between the detrusor, superficial trigone and deep trigone. In particular, they identified a nonadrenergic, noncholinergic influence causing both contraction and relaxation. This reinforces the idea that local variations in the chemical balance of innervation in different parts of the urinary tract may well be responsible for allowing the coordination of its behaviour.

Continuation of Voiding

Whilst the initiation of voiding is a voluntary act, its continuation depends on a more automatic activity, since the individual is only vaguely aware of the capacity of the bladder. Micturition can be initiated almost regardless of the capacity of the bladder. Once the SBS reflex has been switched on by the higher centres it would seem that it can continue to be regulated and controlled by the local afferent activity with a very fine degree of precision. The various aspects of this reflex control were discussed in Chapters 7 and 8. It seems likely that it is the afferent input from tension receptors in the bladder which is most responsible for the control of the voiding reflex. An input due to the presence of urine in the posterior urethra (Barrington's 2nd reflex) seems to be rather more dubious.

Termination of Micturition

Under normal circumstances the bladder is very accurate in its ability to void exactly the capacity within itself with a voiding pressure which is maintained at a very constant level. Occasionally, however, the bladder seems to over- or under-estimate the capacity it has to deal with. Under-estimation results in the bladder continuing to contract after it is empty, resulting in an isometric increase in contraction pressure known as "after contraction"; this seems to have no identified pathological significance. On the other hand, the pattern of voiding in some individuals seems to indicate a degree of underactivity of the detrusor which means that the contraction fails before the whole capacity has been voided, leading to a habitual residual urine. This state of affairs is represented by a relatively low pressure/low flow urodynamic pattern and may become important in later life when obstruction supervenes and produces effects more rapidly than is usual.

References

Abrams PA, Feneley RCL, Torrens MJ (1983) Urodynamics. Springer, Berlin Heidelberg New York

Andrew J, Nathan PW (1964) Lesions of the anterior frontal lobe and disturbances of micturition and defaecation. Brain 87:233–261

Berkow SG (1953) The corpus spongiosum of the urethra: its possible role in urinary control and stress incontinence in women. Am J Obstet Gynecol 65:346–351

Bissado NK, Finkbeiner AE (1980) Smooth muscle physiology and effect of bladder and urethra length/tension on response to stimulation. Urology 16:323–330

Blaivas JG, Barbalios GA (1983) Neurological implications of the pathologically open bladder neck. J. Urol 129:780–782

Brindley GS (1986) Sacral root and hypogastric plexus stimulators and what these models tell us about autonomic actions on the bladder and urethra. Clin Sci 70:415–445, 555

Brindley GS, Craggs MD (1976) The effects of atropine on the urinary bladder of the baboon and of man. J Physiol 256:55P

Caine M (1984) The pharmacology of the urinary tract. Springer, Berlin Heidelberg New York

de Groat WC, Booth AN (1980) Inhibition and facilitation in parasympathetic ganglia. Fed Proc 39:2990–2996

de Groat WC, Kawatani M (1985) Neural control of the urinary bladder: possible relationship between peptidergic inhibitory mechanisms and detrusor instability. Neurourol Urodyn 4:285–300

DeLancey JOL (1986a) Anatomy of the extrinsic continence mechanism. Proceedings of the 16th Annual Meeting of the International Continence Society, Boston, pp 22–24

DeLancey JOL (1986b) A correlative study of paraurethral anatomy. Obstet Gynecol (in press)

Donker PJ, Invanocivi F, Noach EL (1972) Analyses of the urethral pressure profile by means of electromyography and the administration of drugs. Br J Urol 44:180–193

Downie JW, Lautt WW (1986) Is sympathetic control of the urethra mediated through vasomotor action? Neurourol Urodyn 5:219–225

Doyle PT, Briscoe CE (1976) The effect of drugs and anaesthetic agents on the urinary bladder and sphincter. Br J Urol 48:329–334

Edvardsen P (1967) Nervous control of urinary bladder in cats. Acta Neurol Scand 43:543–563

Ek A, Alm P, Andersson K-E, Persson CGA (1977a) Adrenoceptor and cholinoceptor mediated responses of the isolated human urethra. Scand J Urol Nephrol 11:97–102

Ek A, Alm P, Andersson K-E, Persson CGA (1977b) Adrenergic and cholinergic nerves of the human urethra and urinary bladder. A histochemical study. Acta Physiol Scand 99:345–352

George NJR, Feneley RCL (1978) The importance of postural influences on urethral musculature. Proceedings of the 8th Annual Meeting of the International Continence Society, Manchester, p 117

Ghoneim MA, Freitin JA, Gagnon DJ, Susset JG (1975) The influence of vesical distension on urethral resistance to flow: the expulsion phase. Br J Urol 47:663–667

Gjone R, Setekleiv J (1963) Excitatory and inhibitory responses to stimulation of the cerebral cortex in the cat. Acta Physiol Scand 95:337–348

Gosling JA, Dixon JS, Lendon RJ (1977) The autonomic innervation of the male and female bladder neck and proximal urethra. J Urol 118:302–305

Gosling JA, Dixon JS, Humpherson JR (1983) Functional anatomy of the urinary tract. Gower, London, p 5.17

Hess WR (1947) Vegetative Functioen und Zwischenhirn. Helv Physiol Pharmacol Acta Suppl IV:1–65

Hitchcock E, Newsome D, Salama M (1974) The somatotopic representation of the micturition pathways in the spinal cord of man. Br J Surg 61:395–401

Huisman AB, Salome AJ (1977) Morphology of the female urethra in relation to age. Proceedings of the 7th Annual Meeting of the International Continence Society, Portoroz, pp 123–124

Hutch JA (1972) Anatomy and physiology of the bladder trigone and urethra. Meredith, New York

Kerr FW, Alexander S (1964) Descending autonomic pathways in the spinal cord. Arch Neurol 10:249–261

Kock NG, Pompeius R (1963) Inhibition of vesical motor activity induced by anal stimulation. Acta Chir Scand 126:244–250

Kulseng-Hansen S (1983) Prevalence and pattern of unstable urethral pressure in 174 gynaecological patients referred for urodynamic investigation. Am J Obstet Gynecol 146:895–900

Kuru M (1965) Nervous control of micturition. Physiol Rev 45:424–494

Laval KU, Hannappel J, Lutzeyer W (1977) Effects of β adrenergic stimulating and blocking agents on the dynamics of the bladder outlet. Proceedings of the 7th Annual Meeting of the International Continence Society, Portoroz, p 129

Learmonth JR (1931) A contribution to the neurophysiology of the urinary bladder in man. Brain 54: 147–176

Marshall IG, Agoston S, Booij LHDJ, Durant NN, Foldes FF (1980) Pharmacology of org NC45 compared with other non-depolarising neuromuscular blocking drugs. Br J Anaesth 52 (Suppl 1):11S–19S

Mattiasson A (1984) On the peripheral nervous control of the lower urinary tract. Thesis, University of Lund, p 125

Maurice-Williams RS (1974) Micturition symptoms in frontal tumours. J Neurol Neurosurg Psychiatry 37:431–436

McGuire EJ, Wagner FC (1977) The effects of sacral denervation on bladder and urethral function. Surg Gynecol Obstet 144:343–346

McMahon SB (1979) An electrophysiological study of spinal neurones activated by stimulation of the abdominal viscera. PhD Thesis, University of Leeds

McMahon SB, Morrison JFB (1982) Spinal neurones with long projections activated from the abdominal viscera of the cat. J Physiol 322:1–20

Murray A (1986) Predicting the outcome of surgery for urethral incompetence in women by an intraoperative fluid bridge test. A feasibility study. MD Thesis, University of Liverpool

Nathan PW (1976) The central nervous connections of the bladder. In: Williams DI, Chisholm G (eds) The scientific foundations of urology. Heinemann, London

Nordling J (1978) Alpha blockers and urethral pressure in neurological patients. Urol Int 33:304–309

Nordling J (1983) Influence of the sympathetic nervous system on lower urinary tract in man. Neurourol Urodyn 2:3–26

Raz S, Caine M, Zeigler M (1972) The vascular component in the production of intraurethral pressure. J Urol 108:93–96

Rud T (1980) The effects of estrogens and gestagens on the urethral pressure profile in urinary continent and stress incontinent women. Acta Obstet Gynecol Scand 59:265–270

Rud TF, Ulmsten U, Andersson K-E (1978) Initiation of voiding in healthy women and those with stress incontinence. Acta Obstet Gynecol Scand 57:457–462

Rud TE, Andersson K-E, Asmussen M, Hunting A, Ulmsten U (1980) Factors maintaining intraurethral pressure in women. Invest Urol 17:343–347

Sem-Jacobsen CW (1968) Depth electrographic stimulation of the human brain and behaviour. Thomas, Springfield, Ill

Sørensen S, Kirkeby HJ, Stødkilde-Jørgensen H, Djurhuus JC (1986) Continuous recording of urethral activity in healthy female volunteers. Neurourol Urodyn 5:5–16

Speakman MJ, Walmsley D, Brading AF (1986) The importance of non-adrenergic non-cholinergic transmission in the human bladder trigone. Proceedings of the 16th Annual Meeting of the International Continence Society, Boston, pp 2–3

Sundin T, Dahlstrom A (1973) The sympathetic innervation of the urinary bladder and urethra in the normal state and after parasympathetic denervation at the spinal root level. Scand J Urol Nephrol 7:131–149

Tanagho FA, Miller ER (1970) The initiation of voiding. Br J Urol 42:175–183

Tanagho FA, Smith DR (1966) The anatomy and function of the bladder neck. Br J Urol 38:54–71

Torrens MJ (1978) Urethral sphincteric responses to stimulation of the sacral nerves in the human female. Urol Int 33:22–26

Torrens MJ (1984) Neurophysiology. In: Stanton SL (ed) Clinical gynecologic urology. Mosby, St Louis, pp 13–21

Torrens MJ, Feneley RCL (1982) Rehabilitation and management of the neuropathic bladder. In: Illis LS, Sedgewick EM, Glanville HJ (eds) Rehabilitation of the neurological patient. Oxford, Blackwell Scientific

Woodburne RT (1960) Structure and function of the urinary bladder. J Urol 84:79–85

Subject Index